Mathematical Foundations of Quantum Field Theory

Other Related Titles from World Scientific

Lectures of Sidney Coleman on Quantum Field Theory:
Foreword by David Kaiser
edited by Bryan Gin-ge Chen, David Derbes, David Griffiths,
Brian Hill, Richard Sohn and Yuan-Sen Ting
ISBN: 978-981-4632-53-9
ISBN: 978-981-4635-50-9 (pbk)

Quantum Field Theory
by Harald Fritzsch
ISBN: 978-981-3141-72-8

Homotopical Quantum Field Theory
by Donald Yau
ISBN: 978-981-121-285-7

From Classical Mechanics to Quantum Field Theory, a Tutorial
by Manuel Asorey, Elisa Ercolessi and Valter Moretti
ISBN: 978-981-121-048-8

Mathematical Foundations of Quantum Field Theory

Albert Schwarz

University of California at Davis, USA

World Scientific

NEW JERSEY · LONDON · SINGAPORE · BEIJING · SHANGHAI · HONG KONG · TAIPEI · CHENNAI · TOKYO

Published by

World Scientific Publishing Co. Pte. Ltd.

5 Toh Tuck Link, Singapore 596224

USA office: 27 Warren Street, Suite 401-402, Hackensack, NJ 07601

UK office: 57 Shelton Street, Covent Garden, London WC2H 9HE

Library of Congress Cataloging-in-Publication Data
Names: Shvart͡s, A. S. (Al′bert Solomonovich), author.
Title: Mathematical foundations of quantum field theory /
    Albert Schwarz, University of California at Davis.
Description: New Jersey : World Scientific, [2020] |
    Includes bibliographical references.
Identifiers: LCCN 2019034999 | ISBN 9789813278639 (hardcover)
Subjects: LCSH: Quantum field theory.
Classification: LCC QC174.45 .S3295 2020 | DDC 530.14/3--dc23
LC record available at https://lccn.loc.gov/2019034999

British Library Cataloguing-in-Publication Data
A catalogue record for this book is available from the British Library.

First published 2020 (hardcover)
Reprinted 2025 (in paperback edition)
ISBN 9789819814060 (pbk)

For any available supplementary material, please visit
https://www.worldscientific.com/worldscibooks/10.1142/11222#t=suppl

Desk Editor: Ng Kah Fee

Typeset by Stallion Press
Email: enquiries@stallionpress.com

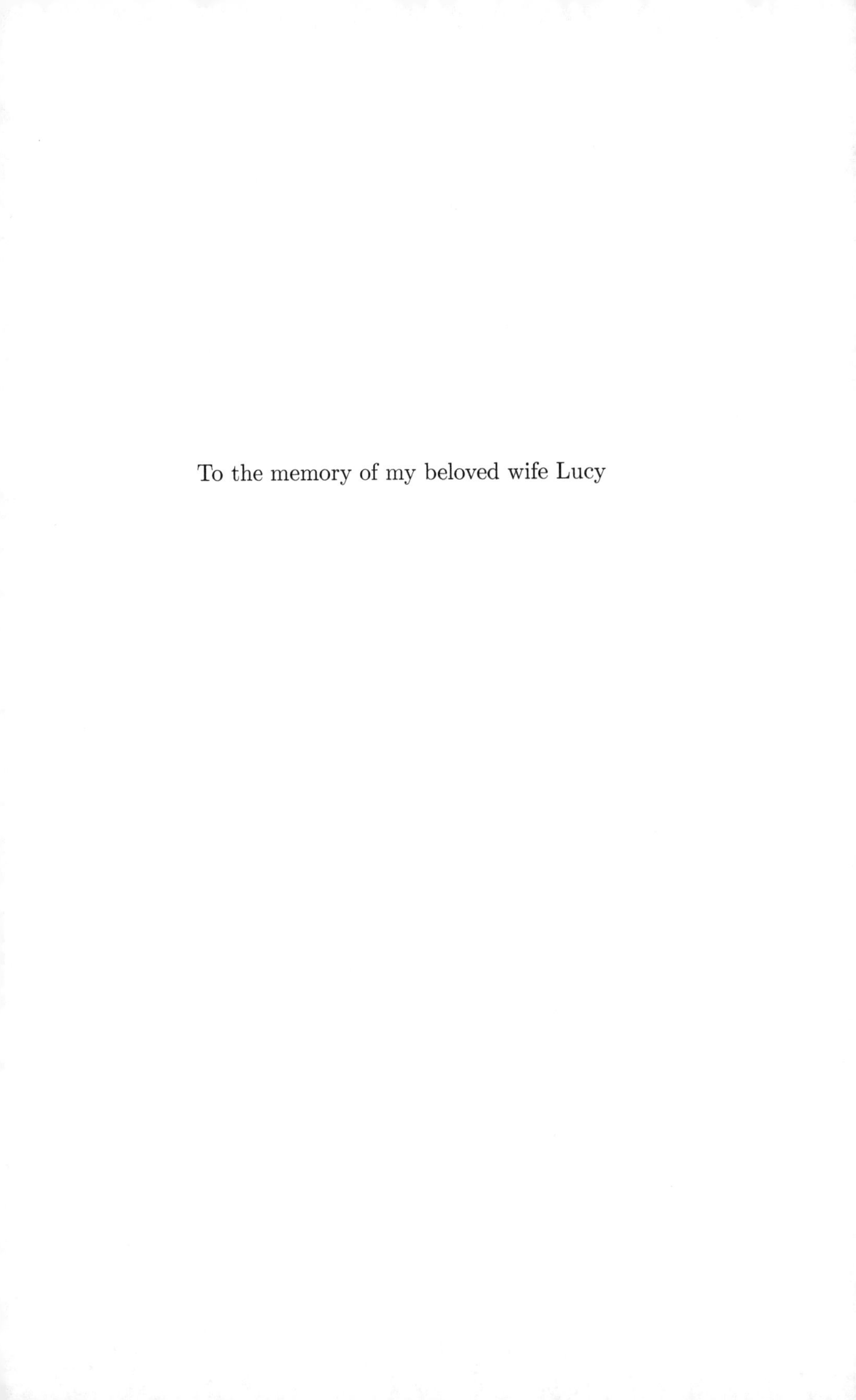

To the memory of my beloved wife Lucy

Preface

This book is addressed to mathematicians and physicists, who are interested in clear exposition in the foundations of quantum field theory. I have tried very hard to satisfy both categories of readers. I wanted the book to be accessible to a mathematician who does not know quantum mechanics and interesting to a physicist who specializes in quantum field theory. I aimed to have the rigor of the proofs to be sufficient for a mathematician, but not so much that it disturbed the reading for a physicist. I hope that this attempt to satisfy these criteria is successful at least partly.

In this book, we talk almost exclusively about the results of quantum field theory that do not depend on the assumption that the theory is Lorentz-invariant (Lorentz-invariant theories are analyzed only at the end of the book). This is the most essential difference that sets this book apart from other books. Another important feature of the book is the consideration of both the Hamiltonian and axiomatic approaches to quantum field theory; we also establish the relation between them.

In some existing books on quantum field theory, one can easily find examples where the rules of the game (the main definitions) change in the process of calculation. We can also see formal manipulations with meaningless expressions, but in the result, we somehow obtain a meaningful answer. This makes the study of quantum field theory much more difficult for a mathematically inclined reader. Of course, the reader understands that in changing the rules of the game, physicists do not imitate the characters in the books of Lewis Caroll;

instead, they are guided by physical intuition, broad use of analogies, and experiment. However, even if a mathematician recognizes that physicists are doing the right things, this does not solve his problems.

I have tried to give an exposition of the main notions of quantum field theory in such a way that without aiming for full mathematical rigor, we instead obtain maximal clarity. However, in most cases, a qualified mathematician should be able to fill in the details following the proof outlines sketched in this book.

The first two chapters of the book and Section 3.1 contain a short introduction to quantum mechanics, intended for mathematicians. In Sections 3.1, 6.1, 6.2 and 6.3, we present some basic facts about Fock space and operators on it; the rest of the book builds on these facts.

Chapter 4 is dedicated to the study of the operator of evolution in the interaction representation and its adiabatic analog. Chapter 5 presents the theory of potential scattering in quantum mechanics.

Section 8.1 of Chapter 8, is devoted to translation-invariant Hamiltonians and their operator realizations. The quantization of classical translation-invariant system with an infinite number of degrees of freedom is studied in Section 8.3 of Chapter 8.

Chapter 9 contains descriptions of different constructions of the scattering matrix of a translation-invariant Hamiltonian. The proof of the equivalence of these constructions is given in Chapter 11.

Chapters 10 and 12 introduce axiomatic scattering theory (in Chapter 12, we consider Lorentz-invariant theories).

In Chapter 11, we study translation-invariant Hamiltonians in the framework of perturbation theory; this chapter uses the results of Chapter 10 on axiomatic scattering theory and the definition of the canonical Faddeev transformation in Section 9.4 of Chapter 9. A mathematician should begin to read these sections with Section 11.5 of Chapter 11.

Chapter 13, added to the English edition, contains applications of the methods of the preceding chapters to statistical physics. An advanced reader can start with this chapter, returning in the case of necessity to Chapter 10 and to Introduction.

The mathematically inclined reader, after the first five chapters and Sections 6.1 and 6.2 of Chapter 6, can go straight to axiomatic scattering theory (Chapter 10 and Sections 12.1 and 12.2 of Chapter 12). After these, one may read Sections 8.1 and 8.3 of Chapter 8, Section 12.3 of Chapter 12, Section 11.5 of Chapter 11, and Chapter 13.

A physicist who wants to read the book on a rigorous level can find the necessary mathematical definitions and theorems in the appendix. If he is satisfied with a lower level of rigor, he should begin his reading with the fourth chapter, since the material in the first three chapters should be familiar. He can neglect the difference between pre-Hilbert and Hilbert spaces, and the difference between Hermitian operators and self-adjoint operators. A measurable function for a physicist is an arbitrary function and a measure space can be understood as the n-dimensional Euclidean space (more precisely, if functions on a set X can be integrated, then the set X can be considered as a measure space). If the physicist encounters unfamiliar mathematical notion, he can usually keep going without much harm.

The book generally does not contain references to the original papers. (I have placed references to papers only when results are formulated but not proved.)

I have received generous help in the production of this book.

I am grateful to Yu. Berezansky, F. Berezin, L. Faddeev, V. Fateev, E. Fradkin, V. Galitsky, A. Povzner, M. Polivanov, A. Rosly, V. Sushko, I. Todorov, Yu. Tyupkin, A. Vainshtein, O. Zavyalov, and other mathematicians and physicists who kindly devoted their attention to this book. I am also grateful to the translator of the book, Dmitry Shemetov, for his diligent work and for his patience.

I am also deeply indebted to my family for their support.

Contents

Introduction

The discovery of quantum field theory began with a rudimentary form of quantum electrodynamics, the basic equations of which were written at the end of the 1920s (Dirac, Heisenberg, Pauli). These equations contain a small parameter $\frac{e^2}{\hbar c} \approx \frac{1}{137}$, so it was natural to try to find a solution in the framework of perturbation theory with respect to this parameter. It turned out, however, that only in the lowest orders did the perturbation theory provide finite solutions consistent with experimental results. In higher orders, the perturbation theory resulted in divergent integrals.

Significant progress was made only two decades later, beginning with the work of Bethe (1947), Schwinger (1958) and Tomonaga (1946) and finally culminating with Feynman's (2005) method, which made it possible to extract finite results from the divergent integrals of higher orders of the perturbation theory with striking agreement to experimental results (covariant theory of renormalization). Feynman's method turned out to be applicable to whole classes of theories (to the so-called renormalizable theories); it was developed in many works, among which it is necessary to mention the article by Dyson (1949). The method was put on a solid mathematical foundation in the article by Bogolyubov and Parasyuk (1955). These groundbreaking papers started a stormy period of development of QFT in many directions. These developments did not come without delays and disappointments. Moreover, in these years, none of the fundamental problems of QFT have been solved.

Important steps forward were the discovery of non-Abelian gauge fields, quantum chromodynamics, and the Higgs mechanism. These developments led to the creation of the standard model that describes electromagnetic, weak and strong interactions in very good agreement with experiment. (The standard model assumes that neutrinos are massless; hence one should modify it to take into account the neutrino mass. This is the only deviation of standard model from the experiment that has been discovered by now.) However, physicists believe that the standard model comes from more fundamental theory. They think that it comes from the theory where electromagnetic, weak and strong interactions are on equal footing (grand unification) or from theory that also describes gravity (like string theory). Furthermore, we now significantly better understand the mathematical structure of QFT. I would like to first mention the algebraic approach to quantum field theory that is closely related to the axiomatic quantum field theory (see, for example, Bogolyubov *et al.* (1956), Lehmann *et al.* (1955), Wightman (1956), Haag (1958), Araki and Haag (1967), Haag and Kastler (1964), Ruelle (1962), Hepp (1965)). In particular, one should mention the construction of a scattering theory in the axiomatic framework. In the case of renormalizable theories, one can construct objects obeying the axioms of relativistic QFT in the framework of perturbation theory, but it is difficult to give a rigorous construction outside of this framework. Intensive development in this direction received the name of constructive QFT (see, for example, Jaffe (2000) and Velo and Wightman (2012)); it was successful in dimensions < 4, but realistic four-dimensional theories are still out of reach. Remarkable progress in the study of conformal field theories and supersymmetric theories (theories having symmetries mixing bosons and fermions) has led to a much better understanding of QFT. Very important information about QFT comes from string theory (one can obtain QFT from string theory in some limit). Today, QFT presents an extensive field of activity not only in physics but also in mathematics, being a source of numerous clearly defined mathematical problems.

This book gives an introduction to QFT. It is in many ways different from other books. In particular, we completely separate the treatment of renormalization and divergences. It became clear long ago that even in the absence of ultraviolet divergences, it is impossible to use the standard quantum mechanical definition of scattering matrix. The reason is that in QFT, there exists no natural representation of full Hamiltonian in the form $H_0 + gV$, where the first term is a free (quadratic) Hamiltonian and the second term is the interaction that can be considered as perturbation. This means that the particles described by H_0 (bare particles) are not the same as the particles described by the full Hamiltonian (dressed particles). This necessitates the change of the definition of the scattering matrix and the renormalization of Feynman diagrams. This renormalization is not related to ultraviolet divergences, a fact that is not emphasized in many textbooks.

In this book, we do not assume Lorentz invariance in the study of QFT; we discuss relativistic local theories only at the end. Therefore, one can apply many of the statements to statistical physics with minimal changes. This is the reason why I have added a new chapter devoted to the applications of methods of QFT to statistical physics. In particular, the new chapter contains the description of the formalism based on the consideration of states as positive functionals on Weyl or Clifford algebra (L-functionals) (Schwarz, 1967; Tyupkin, 1973; Schwarz, 2019b). This formalism allows us to derive the diagram techniques of thermo-field dynamics (TFD) that coincide in the case at hand with Keldysh diagrams (see Chu and Umezawa (1994) for the review of TFD and Keldysh formalism). Some of notions and results of the last chapter are based on recent papers (Schwarz, 2019b,c). I would like to mention the notion of inclusive scattering matrix and its expression in terms of generalized Green functions. (Inclusive scattering matrix is closely related to the inclusive cross-section that is necessary in the consideration of scattering in the case when the theory does not have a particle interpretation, in particular, for the consideration of collisions of thermal quasiparticles.)

The exposition in the book was influenced by the algebraic approach to quantum theory. However, the starting point is the standard Hilbert space formulation. I am including in this edition a short review of some questions of quantum theory based on algebraic approach. For some readers, it would be reasonable to start with this review, while others could read it after finishing the book or in parallel with reading the book. An advanced reader can jump to the last chapter after reading the review.

Review of algebraic approach to quantum theory

First of all some terminological conventions.

Talking about algebra, we always have in mind unital (having unit element) associative algebra over complex numbers with involution denoted by *. (An involution is an antilinear map $A \to A^*$ obeying $A^{**} = A, (AB)^* = B^* A^*$.) We assume that the algebra is a topological space and all operations are continuous, but for many of our statements, these requirements are not sufficient (one should assume that we have a Banach algebra or C^*-algebra or impose some other conditions). In our terminology, an automorphism preserves not only operations in the algebra but also the involution.

We say that h is a derivation of the algebra $\mathcal{A}$ if $h(AB) = h(A)B + Ah(B)$; we also assume that the derivation is compatible with involution.

We say that a derivation is an infinitesimal automorphism if the equation $i\frac{dA}{dt} = h(A(t))$ has a solution for all initial data $A(0)$; then a map $A(0) \to A(t)$ specifies an automorphism α_t. It is easy to check that $\alpha_{t+s} = \alpha_t \alpha_s$ (an infinitesimal automorphism generates a one-parameter group of automorphisms). One can consider infinitesimal automorphisms as elements of the Lie algebra of the group of automorphisms. (This Lie algebra should be defined on the vector space of tangent vectors to one-parameter families of automorphisms at the unit element of the group. In the infinite-dimensional case, it is not clear whether such a vector can also be considered as a tangent vector to a one-parameter group of automorphisms. In what follows, we disregard these subtleties.)

Hamiltonian formalism

The equations of motion of a three-dimensional non-relativistic particle in a potential field $U(\vec{x})$ have the form

$$\frac{d\vec{p}}{dt} = -\nabla U,$$

where $\vec{p} = m\frac{d\vec{x}}{dt}$ stands for the momentum of the particle. To solve these equations (i.e. to find the trajectory of the particle), we should know the initial data: the coordinates and the momenta at some moment of time. One says that the coordinates and the momenta specify the state of our particle at the given moment and that the equations of motion allow us to find the state of the particle at any moment if we know it at one of the moments. The equations of motion can be written in the form

$$\frac{d\vec{p}}{dt} = -\frac{\partial H}{\partial \vec{x}},$$

$$\frac{d\vec{x}}{dt} = \frac{\partial H}{\partial \vec{p}},$$

where $H = \frac{p^2}{m} + U(\vec{x})$ is called Hamiltonian function (one can say the Hamiltonian function is the energy expressed in terms of momenta and coordinates). Similar equations are valid for any mechanical system, but the number of degrees of freedom (the number of coordinates and momenta) and the Hamiltonian function can be arbitrary. This gives the so-called Hamiltonian formalism of mechanics. (In Lagrangian formalism, the state is specified by coordinates and velocities.)

In Hamiltonian formalism, the (pure) state of a classical mechanical system (at the time t) is characterized by $2n$ numbers: $p = (p_1, \ldots, p_n)$ (generalized momenta) and $q = (q^1, \ldots, q^n)$ (generalized coordinates). (Together, these numbers specify a point of $2n$ dimensional space called the phase space of the system).

More generally, we can define a state as a probability distribution on the phase space. The set $\mathcal{D}$ of probability distributions is a convex set, the pure states can be identified with extreme points of this set. Every state can be considered as a mixture of pure states. (The mixture of states $\omega_1, \ldots, \omega_n$ with probabilities $p_1, \ldots, p_n$ is the state $p_1\omega_1 + \cdots + p_n\omega_n$. If states are labeled by continuous parameter $\lambda \in \Lambda$, one defines the mixture of the states as an integral $\int \omega(\lambda)\rho(\lambda)d\lambda$ where $\rho(\lambda)$ stands for the density of the probability distribution on Λ. Note that the definition of mixture can be used for any convex set.)

The evolution of a pure state is governed by Hamiltonian equations

$$\frac{dq}{dt} = \frac{\partial H}{\partial p}, \frac{dp}{dt} = -\frac{\partial H}{\partial q}, \tag{I.1}$$

where the function $H(p, q, t)$ is called the Hamiltonian.

To write down the equation of motion for general state, we introduce the notion of Poisson bracket of two functions on the phase space by the formula

$$\{f, g\} = -\frac{\partial f}{\partial p}\frac{\partial g}{\partial q} + \frac{\partial f}{\partial q}\frac{\partial g}{\partial p}. \tag{I.2}$$

It is easy to check that the Poisson bracket is antisymmetric and satisfies Jacobi identity, hence it specifies a structure of Lie algebra on functions on phase space.[1]

Denoting by $\rho(p, q, t)$ the density of the probability distribution on the phase space at the moment t, we obtain the equation

$$\frac{d}{dt}\rho(p, q, t) = \{H, \rho(p, q, t)\}, \tag{I.3}$$

governing the evolution of state (Liouville equation). If $U(t)$ denotes the evolution operator (the operator transforming the state at the moment 0 into the state at the moment t) we can write (I.3) in the form

$$\frac{d}{dt}U(t) = LU(t),$$

where $L\rho = \{H, \rho\}$. Note that ρ is in general a generalized function on phase space. To verify (I.3), it is sufficient to check that for pure states (represented by δ-functions) it is equivalent to (I.1).

---

[1]The multiplication and the Poisson bracket specify the structure of Poisson algebra on the space of functions on the phase space (see the definition of Poisson algebra in Section 1.3). This means that the phase space is a Poisson manifold. Moreover, it is a symplectic manifold, i.e. the Poisson structure is non-degenerate. We have considered the Hamiltonian formalism on a flat symplectic manifold, but it can be considered on any symplectic manifold.

A physical quantity (an observable) can be considered as a real function $f(p, q)$ on the phase space. It follows from the chain rule that

$$\frac{d}{dt} f(p(t), q(t)) = -\frac{\partial f}{\partial p}\frac{\partial H}{\partial q} + \frac{\partial f}{\partial q}\frac{\partial H}{\partial p}. \tag{I.4}$$

One can rewrite (I.4) in the form

$$\frac{d}{dt} f(p(t), q(t)) = \{f, H\}. \tag{I.5}$$

It follows that in the case $\{f, H\} = 0$, the expression $f(p(t), q(t))$ does not depend on time (in other words, f is an integral of motion). In particular, if H does not depend on time, the function $H(p, q)$ is an integral of motion. It can be identified with the energy of the system.

Let us denote by $\mathcal{A}$ the set of all complex continuous functions on the phase space considered as an algebra with respect to the conventional addition and multiplication of functions. Formula (I.5) gives an equation for the evolution in the algebra $\mathcal{A}$.

Note that every state ω (considered as a measure on the phase space) specifies a linear functional on $\mathcal{A}$ by the formula $\omega(f) = \int f\omega$. This functional obeys the positivity condition: $\omega(f) \geq 0$ if $f \geq 0$. The evolution of states and the evolution in $\mathcal{A}$ are related by the formula $(\omega(t))(f) = \omega(f(t))$.

Quantum mechanics: An algebraic approach

The picture of the preceding section can be modified to describe quantum mechanics. The main idea is to allow non-commuting physical quantities.

The starting point is a unital associative algebra $\mathcal{A}$ over complex numbers equipped with an antilinear involution $A \to A^*$ (generalizing complex conjugation in the algebra $\mathcal{A}$ of the preceding section).

States are identified with positive linear functionals on $\mathcal{A}$ (linear functional ω is positive if $\omega(A^*A) \geq 0$). We assume that states are normalized, i.e. $\omega(1) = 1$. The set D of normalized states is convex.

The extreme points of this set are called pure states. Every state is a mixture of pure states.

Let us denote by Aut the group of automorphisms of the algebra $\mathcal{A}$ commuting with involution. This group naturally acts on states. In quantum system, the state depends on time and this dependence can be described by the evolution operator $U(t)$ transforming $\omega(0)$ into $\omega(t)$. This is the so-called Schrödinger picture; it is equivalent to Heisenberg picture where the elements of $\mathcal{A}$ depend on time, but the states do not: $\omega(t)(A) = \omega(A(t))$.

The evolution operator satisfies the equation

$$i\frac{dU}{dt} = H(t)U(t), \tag{I.6}$$

which is equivalent to the equation of motion $i\frac{d\omega}{dt} = H(t)\omega(t)$. Here, $H(t)$ stands for an element of Lie algebra of the group Aut (for infinitesimal automorphism). It plays the role of the Hamiltonian of the quantum system. If H does not depend on t, the evolution operators obey $U(t + \tau) = U(t)U(\tau)$ (constitute a one-parameter subgroup).

To specify a quantum system, we should fix an algebra with involution $\mathcal{A}$ and an infinitesimal automorphism H (or a family of infinitesimal automorphisms $H(t)$). Then the evolution is governed by (I.6).

In what follows, we assume that H does not depend on t unless the dependence on t is explicitly mentioned.

Note that an infinitesimal automorphism can be considered as a derivation of the algebra $\mathcal{A}$. However, a derivation specifies a quantum system only if it can be integrated to a one-parameter group of automorphisms.

The textbook form of quantum mechanics corresponds to the case when $\mathcal{A}$ is the algebra of bounded linear operators in Hilbert space $\mathcal{E}$ with involution defined as Hermitian conjugation. The states are specified by density matrices (positive operators with unit trace) by the formula $\omega_K(A) = \operatorname{Tr} KA$. Pure states correspond to vectors $\Psi \in \mathcal{E}$; proportional vectors specify the same state. If the vector Ψ is normalized, the corresponding state is the functional $\langle A\Psi, \Psi\rangle$.

Every unitary operator W determines an automorphism of the algebra with involution $\mathcal{A}$ by the formula $A \to W^{-1}AW$. This correspondence allows us to identify the group Aut with the group of unitary operators and infinitesimal automorphisms with self-adjoint operators. The equation of motion for the density matrix has the form $i\frac{dK}{dt} = HK$, where H is an operator on the space $\mathcal{L}$ of trace class operators in $\mathcal{E}$ defined by the formula $HK = -\hat{H}K + K\hat{H}$ and the equation of motion for the vector $\Psi \in \mathcal{E}$ representing the pure state is $i\frac{d\Psi}{dt} = \hat{H}\Psi$ where $\hat{H}$ is a self-adjoint operator (Hamiltonian). The evolution operator in $\mathcal{E}$ will be denoted by $\hat{U}(t)$; it obeys $i\frac{d\hat{U}}{dt} = \hat{H}\hat{U}$. The evolution operator in $\mathcal{L}$ will be denoted by $U(t)$; it is easy to check that $U(t)K = \hat{U}^{-1}(t)K\hat{U}(t)$.

Note that in the case at hand pure states have very simple description, therefore, very often it is convenient to work with pure states and to consider other states as mixtures of pure states. In principle, we can work only with pure states for any algebra $\mathcal{A}$, but in general, this is not convenient because the description of pure states is complicated.

For any algebra $\mathcal{A}$ and any state ω, we can construct a pre-Hilbert space $\mathcal{E}$ and a representation of $\mathcal{A}$ by operators in this space such that for some cyclic vector $\Phi \in \mathcal{E}$, we have $\omega(A) = \langle \hat{A}\Phi, \Phi \rangle$. (An element $A \in \mathcal{A}$ is represented by operator $\hat{A}$; we assume that the map $A \to \hat{A}$ is an algebra homomorphism and is compatible with involution: $\hat{A}^* = (\hat{A})^*$. The vector Φ is cyclic in the following sense: every other vector can be represented in the form $\hat{A}\Phi$ where $A \in \mathcal{A}$.) This construction (Gelfand–Naimark–Segal (GNS) construction) is essentially unique (up to equivalence).

Let us sketch the proof of this theorem. Assume that the representation we need is constructed. Let us introduce in $\mathcal{A}$ an inner product by the formula $\langle A, B \rangle = \omega(B^*A)$. It is easy to see that the map $\nu : A \to \mathcal{E}$ sending A to $\hat{A}\Phi$ preserves this inner product. It follows from cyclicity of Φ that this map is surjective; this allows us to identify $\mathcal{E}$ with the quotient of $\mathcal{A}$ with respect to zero vectors. (Recall that a zero vector is a vector that is orthogonal to all other vectors.) The obvious relation $\nu(BA) = \hat{B}\nu(A)$ allows us to describe our representation in terms of the algebra $\mathcal{A}$ and state ω. Namely, we construct $\mathcal{E}$ as $\mathcal{A}$ factorized with respect to zero vectors of the inner product $\langle A, B \rangle = \omega(B^* A)$, the operation of the multiplication from the left by A descents to the operator $\hat{A}$, the unit element of $\mathcal{A}$ corresponds to the vector Φ.

We worked with pre-Hilbert spaces, but we can take a completion of $\mathcal{E}$ to obtain a representation of $\mathcal{A}$ by operators acting in Hilbert space.

We see that every state of $\mathcal{A}$ can be represented by a vector in Hilbert space. However, we cannot consider all states as elements of the same Hilbert space.

If ω is a stationary state (a state invariant with respect to time evolution), then the group $U(t)$ descends to a group $\hat{U}(t)$ of unitary transformations of corresponding space $\mathcal{E}$. The generator $\hat{H}$ of $\hat{U}(t)$ plays the role of Hamiltonian.

We say that the stationary state ω is a ground state if the spectrum of $\hat{H}$ is non-negative.

This definition agrees with the definition of the ground state in Hilbert space formulation of quantum mechanics. Let us apply the GNS construction to the algebra of bounded operators in Hilbert space $\mathcal{E}$ and to a state represented by a vector $\Phi \in \mathcal{E}$ where Φ is an eigenvector of the Hamiltonian: $\hat{H}\Phi = E\Phi$. Then the space given by GNS construction can be identified with $\mathcal{E}$ and the generator of $\hat{U}(t)$ is equal to $\hat{H} - E$. The condition that Φ is the eigenstate with minimal eigenvalue is equivalent to the positivity of $\hat{H} - E$.

The representation containing the ground state will be called ground state representation. This representation is especially important in quantum field theory: we will consider particles as elementary excitations of ground state.

Let us define the correlation functions in a stationary state ω as functions

$$w_n(t_1, \ldots, t_n) = \omega(A_1(t_1) \ldots A_n(t_n)),$$

where $A_1, \ldots, A_n \in \mathcal{A}$.

The Green functions in the state ω are defined by the formula

$$G_n(t_1, \ldots, t_n) = \omega(T(A_1(t_1) \ldots A_n(t_n))),$$

where T stands for time ordering. It is easy to express them in terms of correlation functions.

The correlation functions in the ground state are called Wightman functions. The properties of particles in quantum field theory can be expressed in terms of these functions and/or corresponding Green functions; the same is true for scattering matrix.

Classical and quantum

To relate quantum and classical mechanics, we consider a family of algebras $\mathcal{A}_\hbar$ depending smoothly on the parameter $\hbar$ assuming that for $\hbar = 0$, we have a commutative algebra with the product that will be denoted $A \cdot B$. More precisely, we assume that all these algebras are defined on the same vector space, in other words, the addition and multiplication by a number do not depend on $\hbar$, but the multiplication of elements A, B of the algebra (denoted by $A \cdot_\hbar B$) smoothly depends on $\hbar$. The commutator $[A, B]_\hbar = A \cdot_\hbar B - B \cdot_\hbar A$ vanishes for $\hbar = 0$, therefore, we can introduce a new operation $\{A, B\}$ (Poisson bracket) using the formula

$$[A, B]_\hbar = i\{A, B\}\hbar + O(\hbar^2).$$

It is easy to verify that $\mathcal{A}_0$ (the algebra with commutative multiplication $A \cdot B$ that we have for $\hbar = 0$) with the new operation is Poisson algebra, i.e. the new operation satisfies the axioms of Lie algebra and

$$\{A \cdot B, C\} = \{A, C\} \cdot B + A \cdot \{B, C\}.$$

If there exists an involution $A \to A^*$ compatible with multiplication in all algebras $\mathcal{A}_\hbar$, then it is also compatible with the Poisson bracket (i.e. the Poisson bracket of two self-adjoint elements is again a self-adjoint element). This statement is one of the reasons why a factor i is included in the definition of the Poisson bracket.

Quantization; Weyl algebra

We have found that the classical mechanics can be obtained from quantum mechanics in the limit $\hbar \to 0$. Conversely, quantum mechanics can be obtained as a deformation of classical mechanics. We can start with Poisson algebra and deform it (this means we would like to construct a family $\mathcal{A}_\hbar$ that gives this Poisson algebra in the limit $\hbar \to 0$).

If a Poisson algebra $\mathcal{A}$ is an algebra of polynomial functions on a vector space with coordinates $(u^1, \dots, u^n)$, then the Poisson bracket can be written in the form

$$\{A, B\} = \frac{1}{2}\sigma^{kl}(u)\frac{\partial A}{\partial u^k}\frac{\partial B}{\partial u^l}. \tag{I.7}$$

It specifies a structure of Poisson manifold on the vector space. (Here, $\{u^k, u^l\} = \sigma^{kl}(u)$.)

Let us consider the case when the Poisson bracket on polynomial functions on vector space is defined by the formula (I.7) with constant coefficients σ^{kl}. Moreover, we assume that the matrix σ^{kl} is non-degenerate, then the dimension of the vector space is necessarily even. (This is the situation in Section 1.1.) Then we can define the algebra $\mathcal{A}_\hbar$ as a unital associative algebra with generators $\hat{u}^k$ and relations $\hat{u}^k\hat{u}^l - \hat{u}^l\hat{u}^k = i\sigma^{kl}$. This algebra is called Weyl algebra. If the coordinates u^k are considered as real numbers, we introduce an involution in Weyl algebra requiring that the generators $\hat{u}^k$ are self-adjoint. If the Poisson bracket is written in the form (I.2) (we can always write it in this form changing coordinates), then the Weyl algebra is generated by self-adjoint elements $(\hat{p}_1, \dots, \hat{p}_n, \hat{q}^1, \dots, \hat{q}^n)$ with relations $\hat{p}_k\hat{p}_l = \hat{p}_l\hat{p}_k$, $\hat{q}^k\hat{q}^l = \hat{q}^l\hat{q}^k$, $\hat{p}_k\hat{q}^l - \hat{q}^l\hat{p}_k = \frac{\hbar}{i}\delta_k^l$. These relations are called canonical commutation relations (CCR). Instead of self-adjoint generators $(\hat{p}_1, \dots, \hat{p}_n, \hat{q}^1, \dots, \hat{q}^n)$, one can consider generators a_k, a_k^* where $a_k = \frac{1}{\sqrt{2}}(\hat{q}^k + i\hat{p}_k)$, $a_k^* = \frac{1}{\sqrt{2}}(\hat{q}^k - i\hat{p}_k)$. These generators satisfy relations $a_k a_l = a_l a_k$, $a_k^* a_l^* = a_l^* a_k^*$, $a_k a_l^* - a_l^* a_k = \hbar\delta_{kl}$. These relations are also called CCR. To say that the family of algebras $\mathcal{A}_\hbar$ can be considered as a deformation of the commutative polynomial algebra, we should realize their elements as polynomials. This can be done in many different ways. For example, we can note that using CCR we are able to move all operators $\hat{q}^k$ to the left and all operators $\hat{p}_k$ to the right; "removing hats" in the expression, we obtain a polynomial called $(q - p)$ symbol of the element of

Weyl algebra. This representation of elements of Weyl algebra by polynomials does not agree with involution ($(q-p)$-symbol of self-adjoint element is not necessarily real), however, it is easy to modify the construction to avoid this drawback. For example, one can write down an element of Weyl algebra in terms of generators a_k, a_k^* and use CCR to move a_k^* to the left and a_k to the right. The expression we get is called normal form of the element of Weyl algebra. Considering a_k, a_k^* in the normal form as complex variables, we obtain a polynomial called Wick symbol.

We can go in opposite direction and obtain an element of Weyl algebra from a polynomial. This operation is called quantization. Quantization allows us, for example, to obtain quantum Hamiltonian from classical Hamiltonian. It is important to note that the quantization depends on the choice of symbol (we have "ordering ambiguity"). However, for some classical Hamiltonians, there exists a natural choice of corresponding quantum Hamiltonians. In particular, this is true for Hamiltonians represented as a sum of kinetic energy expressed as a function of momenta and potential energy depending on coordinates (no ordering ambiguity).

Stationary states

A state that does not depend on time is called stationary state. In what follows, we work in the formalism where the states are described by density matrices in Hilbert space $\mathcal{E}$. The state represented by a density matrix K is stationary if K obeys $HK = 0$, i.e. if K commutes with the Hamiltonian $\hat{H}$. (Recall that H acts as a commutator with $\hat{H}$.) An important particular case of a stationary state is the Gibbs state $K = Z^{-1}e^{-\beta\hat{H}}$ where $Z = Tre^{-\beta\hat{H}}$. This state corresponds to the equilibrium state with the temperature $T = \beta^{-1}$. It tends to the ground state as $T \to 0$.

Let us assume that the operator $\hat{H}$ has discrete spectrum with orthonormal basis ϕ_n of eigenvectors with eigenvalues E_n. Then it is convenient to work in representation where $\hat{H}$ is represented by a diagonal matrix with entries E_n (it is called $\hat{H}$-representation). In this representation, the eigenvectors of the operator H in the space $\mathcal{L}$ are matrices ψ_{mn} having only one non-zero entry equal to 1 in the position (m, n). Alternatively, one can define ψ_{mn} as an operator acting by the formula $\psi_{mn} x = \langle x, \phi_n \rangle \phi_m$ where $x \in \mathcal{E}$. Corresponding eigenvalues are $E_m - E_n$. We see that the vectors ϕ_n and the corresponding density matrices ψ_{nn} are stationary states. It follows that all diagonal density matrices in $\hat{H}$-representation are stationary states. Moreover, if the spectrum is simple, all stationary states are represented by diagonal matrices.

A density matrix K_0 represents a ground state if $\operatorname{Tr} \hat{H} K_0 \leq \operatorname{Tr} \hat{H} K$ for any density matrix K. If the ground state is unique, it is necessarily pure and corresponds to the eigenfunction of $\hat{H}$ with lowest eigenvalue.

Adiabatic approximation

Let us consider the case of slowly varying Hamiltonian $\hat{H}(t)$. We will assume that the energy levels $E_n(t)$ are distinct and vary continuously with t, corresponding eigenvectors will be denoted by $\phi_n(t)$. We assume that these eigenvectors constitute an orthonormal system. Then it is easy to prove that in the first approximation the evolution operator $\hat{U}(t)$ transforms eigenvector to eigenvector

$$\hat{U}(t)\phi_n(0) = e^{-i\alpha_n(t)}\phi_n(t), \quad \frac{d\alpha_n(t)}{dt} = E_n(t). \qquad (\text{I.8})$$

To verify (I.8), we check that the RHS satisfies the equation of motion up to terms that are small for slow varying $\hat{H}(t)$ (see Section 4.3 for more details).

Let us introduce the operators $\psi_{mn}(t)$ by the formula $\psi_{mn}(t)x = \langle x, \phi_n(t) \rangle \phi_m(t)$. (These operators are eigenvectors of the operator $H(t)$ in the space $\mathcal{L}$.) Applying (I.8) or analyzing directly the evolution operator $U(t)$ in $\mathcal{L}$, we obtain that $U(t)$ transforms eigenvector into eigenvector

$$U(t)\psi_{mn}(0) = e^{-i\beta_{mn}(t)}\psi_{mn}(t), \quad \frac{d\beta_{mn}(t)}{dt} = E_m(t) - E_n(t) \qquad (\text{I.9})$$

(this equation is true up to terms that can be neglected for slowly varying Hamiltonian). Note that β_{mm} does not depend on t.

Decoherence

Let us consider a quantum system (atom, molecule, etc.) described by a Hamiltonian $\hat{H}$ with simple discrete spectrum. We assume that the system is "almost closed" in the following sense: the interaction with the environment can be described as an adiabatic change of the Hamiltonian $\hat{H}$. Let us consider the evolution operator $U(T)$ assuming that $\hat{H}(0) = \hat{H}(T)$. If the Hamiltonian is time-independent, then for the eigenvector ϕ_n with the eigenvalue E_n, we can say that $\hat{U}(T)\phi_n = C_n(T)\phi_n$ where $C_n(T) = e^{-iE_nT}$. If the Hamiltonian is slowly changing, we have the same formula with $C_n(T)$ calculated from (I.8). Hence, if we started with pure stationary state, we remain in the same state. Similarly, $U(T)\psi_{mn} = C_{mn}(T)\psi_{mn}$ and the phase factor is constant for $m = n$.

It is natural to assume that the environment is random (the time-dependent Hamiltonian depends on some parameters $\lambda \in \Lambda$ with some probability distribution on Λ), then for $m \neq n$, we have a random phase factor $C_{mn}(\lambda, T)$. If we start with density matrix $K = \sum k_{mn}\psi_{mn}$ (with matrix entries k_{mn} in $\hat{H}$-representation), then the density matrix $K_\lambda(T)$ is equal to $\sum C_{mn}(\lambda, T)k_{mn}\psi_{mn}$, i.e. the matrix entries acquire phase factor $C_{mn}(\lambda, T)$. Now, we should take the mixture $\bar{K}(T)$ of states $K_\lambda(T)$ (this means that we should take the average of phase factors). It is obvious that non-diagonal entries of $\bar{K}(T)$ are smaller by absolute value than corresponding entries of K. Imposing some mild conditions on the probability distribution on Λ, one can prove that the non-diagonal entries of $\bar{K}(T)$ tend to zero as $T \to \infty$. In other words, the matrix $\bar{K}(T)$ tends to a diagonal matrix $\bar{K}$ having the same diagonal entries as K. (See Schwarz and Tyupkin (1987) and Schwarz (2019a) for more details.)

The matrix $\bar{K}$ can be considered as a mixture of pure states corresponding to the vectors ϕ_n with probabilities k_{nn}.

This phenomenon is known as decoherence.

Observables and probabilities

Until now, we did not relate the formalism of quantum mechanics to experiment. We know that (at least in some cases) quantum mechanics can be considered as a deformation of classical mechanics, therefore, one can conjecture that classical observables correspond to quantum observables. In particular, remembering that the energy, which is represented by Hamiltonian function, is an integral of motion if the system is invariant with respect to time shift should have quantum analog with similar properties. This is not quite true: a quantum system does not have a definite energy and measuring its energy by means of macroscopic device we obtain different values of energy with some probabilities. (A similar thing happens when we are trying to use a thermometer to measure the temperature of a system that is out of equilibrium. Such a system does not have a temperature, but still we can measure it; thermometer readings will be different, but we will obtain some probability distribution of these readings.) The interaction with macroscopic device leads to decoherence (the non-diagonal matrix entries of the density matrix K in $\hat{H}$-representation die) and we obtain a mixture of pure states with probabilities k_{nn}. If the density matrix K corresponds to a pure state described by a vector ϕ in $\mathcal{E}$, then in general, decoherence leads to mixture of pure states corresponding to the vectors ϕ_n with probabilities $p_n = |\langle \phi, \phi_n \rangle|^2$. However, if ϕ coincides with one of eigenfunction ϕ_k, all probabilities p_n vanish except $p_k = 1$. Therefore, we say that an eigenfunction of $\hat{H}$ with eigenvalue E has definite value of energy equal to E. For any other normalized vector ϕ, we can speak only about a probability to get some value E measuring the energy. This probability is non-zero only for eigenvalues E_n of $\hat{H}$; it is equal to $p_n = |\langle \phi, \phi_n \rangle|^2$. For a state represented by density matrix K, the probability is given by the formula $p_n = \langle K\phi_n, \phi_n \rangle$.

Other observables are represented by self-adjoint operators; the formulas we have written for energy remain valid for any observable. Again, if the physical quantity A is represented by self-adjoint operator $\hat{A}$ in $\mathcal{E}$, the quantity A has definite value a if a state is represented by an eigenvector of $\hat{A}$ with eigenvalue a. Otherwise, we can talk only about probabilities. It is convenient to write the density matrix K representing a state in $\hat{A}$-representation; if the operator $\hat{A}$ has simple discrete spectrum, the probabilities are equal to the diagonal entries of K in $\hat{A}$-representation.

In the general case, we can specify the probability distribution of the observable A by the formula

$$\overline{f(A)} = \operatorname{Tr} K f(\hat{A}), \tag{I.10}$$

where f stands for any piecewise continuous function and $\overline{f(A)} = \int f(a)\rho(a)da$ denotes the mean value of $f(A)$ with respect to the probability distribution $\rho(a)da$. (It is also called the expectation value of $f(A)$.) If the operator $\hat{A}$ has simple discrete spectrum, (I.10) is equivalent to the formulas for probabilities we gave in this case.

If $\hat{A}_1, \dots, \hat{A}_n$ are commuting self-adjoint operators, one can define the joint probability distribution of corresponding physical quantities $A_1, \dots, A_n$ using the formula

$$\overline{f(A_1, \dots, A_n)} = \operatorname{Tr} K f(\hat{A}_1, \dots, \hat{A}_n). \tag{I.11}$$

More generally, if quantum mechanical system is specified by C^*-algebra $\mathcal{A}$ and a state by a positive functional ω, then the joint probability distribution of physical quantities $A_1, \dots, A_n$ is given by the formula

$$\overline{f(A_1, \dots, A_n)} = \omega(f(\hat{A}_1, \dots, \hat{A}_n)). \tag{I.12}$$

(Physical quantities A_i correspond to self-adjoint elements $\hat{A}_i$ of $\mathcal{A}$. The assumption that $\mathcal{A}$ is a C^*-algebra guarantees that the notion of a function of a family of commuting self-adjoint elements makes sense.)

Integrals of motion

One can work either in the Schrödinger picture where states are time-dependent, but observables do not depend on time, or in the Heisenberg picture, where states do not depend on time, but the observables do. These pictures are equivalent:

$$\operatorname{Tr} K(t) f(\hat{A}) = \operatorname{Tr} K f(\hat{A}(t))$$

if $\hat{A}(t)$ obeys the Heisenberg equation

$$i\frac{d\hat{A}}{dt} = [\hat{H}, \hat{A}].$$

This implies that the observable $\hat{A}$ commuting with the Hamiltonian $\hat{H}$ is an integral of motion (corresponding probabilities do not depend on time).

Let us suppose that we have a one-parameter family $U(t)$ of symmetries of the system. (A symmetry is an automorphism preserving the equations of motion. In our case, we consider it as a unitary operator, commuting with the Hamiltonian.) The operator $i\frac{dU}{dt}$ is an integral of motion — a self-adjoint operator commuting with the Hamiltonian.

The Hamiltonian itself is an integral of motion corresponding to the time translation. We have already noted that the corresponding observable is energy.

The integral of motion corresponding to invariance with respect to spatial translations is a component of momentum.

The integral of motion corresponding to the invariance with respect to rotation around some axis is a component of angular momentum.

Weyl and Clifford algebras

In quantum theory in algebraic approach, we are starting with a unital associative algebra with involution. We have seen already that a natural candidate for this algebra is Weyl algebra. In this section, we will study this algebra and its cousin, Clifford algebra. Recall that the Weyl algebra is generated by self-adjoint elements $\hat{p}_i$, $\hat{q}^i$ with relations

$$\hat{p}_k\hat{p}_l = \hat{p}_l\hat{p}_k, \quad \hat{q}^k\hat{q}^l = \hat{q}^l\hat{q}^k, \quad \hat{p}_k\hat{q}^l - \hat{q}^l\hat{p}_k = \frac{1}{i}\delta_k^l. \tag{I.13}$$

These relations are called canonical commutation relations (CCR). Instead of self-adjoint generators $\hat{p}_i, \ldots, \hat{q}^i$, one can consider generators a_k, a_k^* where $a_k = \frac{1}{\sqrt{2}}(\hat{q}^k + i\hat{p}_k), a_k^* = \frac{1}{\sqrt{2}}(\hat{q}^k - i\hat{p}_k)$. These generators satisfy relations

$$a_k a_l = a_l a_k, a_k^* a_l^* = a_l^* a_k^*, a_k a_l^* - a_l^* a_k = \delta_{kl}. \tag{I.14}$$

Both (I.13) and (I.14) are called CCR.

For finite number of degrees of freedom, there exists only one irreducible representation of CCR. The relations (I.13) can be represented in the space $L^2(\mathbb{R}^n)$ by operators of multiplication and differentiation: $\hat{q}^k \psi(q^1, \ldots, q^n) = q^k \psi(q^1, \ldots, q^n)$, $\hat{p}_k \psi(q^1, \ldots, q^n) = \frac{1}{i} \frac{\partial}{\partial q^k} \psi(q^1, \ldots, q^n)$. In non-relativistic quantum mechanics, this representation is very convenient. If the classical Hamiltonian has the form $T(p) + V(q)$ where the kinetic energy $T(p)$ is a quadratic function of momenta and the potential energy $V(q)$ has a non-degenerate minimum, then in a neighborhood of this minimum, the classical system can be approximated by multidimensional harmonic oscillator. Similarly, for the quantized Hamiltonian $\hat{H} = T(\hat{p}) + V(\hat{q})$, we can approximate low energy levels as energy levels of quantum oscillator. The Hamiltonian of one-dimensional quantum oscillator can be written in the form

$$\hat{H} = \frac{\hat{p}^2}{2} + \frac{\hat{q}^2}{2} = a^* a + \frac{1}{2}.$$

Noting that $\hat{H}a^* = a^*(\hat{H} + 1)$, $\hat{H}a = a(\hat{H} - 1)$, we obtain that the operator a^* transforms an eigenfunction ϕ with eigenvalue E in an eigenfunction with eigenvalue $E + 1$. Similarly, the operator a either sends ϕ to zero (if ϕ is the ground state) or to an eigenfunction with eigenvalue $E - 1$. Using this fact, we can check that the ground state θ has energy $\frac{1}{2}$ and the states $\frac{1}{\sqrt{n!}} (a^*)^n \theta$ constitute an orthonormal basis consisting of eigenfunctions of $\hat{H}$ with eigenvalues $n + \frac{1}{2}$.

In appropriate coordinates, the Hamiltonian of multidimensional quantum oscillator can be considered as a sum of non-interacting one-dimensional oscillators:

$$\hat{H} = \sum \left(\omega_k a_k^* a_k + \frac{\omega_k}{2} \right);$$

again applying many times operators a_k^* to the ground state θ, we obtain a basis of eigenfunctions.

We see that it is convenient to use the operators a_k^*, a_k in the analysis of excitations of the ground state. In quantum field theory, we are interested first of all in the excitations of ground state; this is one of the many reasons why these operators are so useful.

Let us consider now the Weyl algebra with infinite number of generators. This notion can be made precise in different ways. We can consider simply an algebra with infinite number of generators obeying (I.14), but it is more convenient to start with some (pre)-Hilbert space $\mathcal{B}$ and consider an algebra with generators $a(f)$, $a^*(f)$ and relations

$$a(\lambda f + \mu g) = \lambda a(f) + \mu a(g), \ \ a^*(f) = (a(f))^*,$$
$$a(f)a(g) = a(g)a(f), \ \ a^*(f)a^*(g) = a^*(g)a^*(f), \qquad (\text{I}.15)$$
$$a(f)a^*(g) - a^*(g)a(f) = \langle f, g \rangle.$$

A representation of Weyl algebra (= representation of CCR) is a family of operators in a (pre)-Hilbert space obeying (I.15). (Note that we use the same notation for elements of Weil algebra and for operators.)

If $\mathcal{B} = l^2$, then $a(f) = \sum f_k a_k$, $a^*(f) = \sum \bar{f}_k a_k^*$ where a_k, a_k^* obey (I.14). If $\mathcal{B}$ is some space of test functions on $\mathbb{R}^n$ (for example, the Schwartz space), we can say that the formal expression $a(f) = \int f(x)a(x)dx$ specifies $a(x)$ as a generalized operator function on $\mathbb{R}^n$. We can also use this notation and terminology in cases when $\mathcal{B}$ is a space of functions on some set.

The theory of Weyl algebra is very similar to the theory of Clifford algebra (an algebra that is defined by canonical anti-commutation relations (CAR) where commutators are replaced by anticommutators). The representations of Clifford algebra are also called representations of CAR.

In the simplest form, Clifford algebra can be defined as an algebra with generators obeying

$$a_k a_l = -a_l a_k, \ \ a_k^* a_l^* = -a_l^* a_k^*, \ \ a_k a_l^* + a_l^* a_k = \delta_{kl}. \qquad (\text{I}.16)$$

More generally, to define the Clifford algebra, we start with some (pre)-Hilbert space $\mathcal{B}$ and consider an algebra with generators $a(f), a^*(f)$ and relations

$$a(\lambda f + \mu g) = \lambda a(f) + \mu a(g), \ \ a^*(f) = (a(f))^*,$$
$$a(f)a(g) = -a(g)a(f), \ \ a^*(f)a^*(g) = -a^*(g)a^*(f), \qquad (\text{I}.17)$$
$$a(f)a^*(g) + a^*(g)a(f) = \langle f, g \rangle.$$

Note that both Weyl and Clifford algebras can be extended in various ways (for example, one can introduce some norm and consider a completion with respect to this norm).

Weyl algebra or Clifford algebra (or tensor product of these algebras) can play the role of the algebra with involution $\mathcal{A}$ in the algebraic description of quantum system.

In this description, we also need an infinitesimal automorphism H of the algebra. The simplest way to construct it is to take a self-adjoint element h of the algebra and to define a derivation by the formula $HA = [h, A]$. (This construction works for any algebra; derivations obtained this way are called inner derivations. In C^*-algebra, one can prove that an inner derivation is an infinitesimal automorphism considering e^{ith}.) If the algebra is generated by a_k, a_k^* obeying (I.14) or (I.16), the element h can be represented in normal form (where a_k are from the right) by means of finite sum:

$$h = \sum_{mn} \sum_{k_1,\ldots,k_m,l_1,\ldots,l_n} \Gamma_{k_1,\ldots,k_m,l_1,\ldots,l_n} a_{k_1}^* \ldots a_{k_m}^* a_{l_1} \ldots a_{l_n}. \qquad \text{(I.18)}$$

Here, $\Gamma_{k_1,\ldots,k_m,l_1,\ldots,l_n} = \bar{\Gamma}_{l_n,\ldots,l_1,k_m,\ldots,k_1}$ (this condition guarantees that h is self-adjoint).

In the case of infinite number of generators, one can modify this construction to obtain other infinitesimal automorphisms. Namely, we can consider (I.18) as a formal expression; if the expressions $[a_k, h]$, $[a_k^*, h]$ can be regarded as elements of $\mathcal{A}$, these formulas specify a derivation of algebra. This happens if for every k, there exists only a finite number of summands in (I.18) where one of the indices is equal to k. (Then only a finite number of terms survives in the commutator.) To check that the derivation specifies a one-parameter family of automorphism, we should verify that the equations of motion $i\frac{da_k}{dt} = [a_k, h]$, $i\frac{a_k^*}{dt} = [a_k^*, h]$ have a solution. If h is a quadratic hamiltonian, the equations of motion are linear and the solution is the same as in classical theory. If h is a sum of quadratic Hamiltonian and a summand multiplied by a parameter g, then it is easy to prove that the equations of motion can be solved in the framework of perturbation theory with respect to g.

For example, we can take $h = \sum \epsilon_k a_k^* a_k$. Then we obtain an infinitesimal automorphism leading to well-defined equations of motion:

$$i\frac{da_k}{dt} = -\epsilon_k a_k, \quad i\frac{da_k^*}{dt} = \epsilon_k a_k^*.$$

If $\epsilon_k \geq 0$, the ground state representation is in this case the Fock representation (the representation containing a cyclic vector θ obeying $a_k\theta = 0$). To prove this fact, we note that in the space of Fock representation (Fock space), we have an orthogonal basis of eigenvectors $(a_1^*)^{n_1} \ldots (a_k^*)^{n_k} \ldots \theta$ with eigenvalues $\sum \epsilon_k n_k$. Here, $n_k = 0, 1, 2, \ldots$ in the case of CCR and $n_k = 0, 1$ in the case of CAR, only finite number of n_k does not vanish.

If $\epsilon_k < 0$, for some k, then in the case of CCR, the ground state does not exist. Let us consider the case of CAR. Let us denote by N the set of indices k where $\epsilon_k < 0$. If N is finite, one can find ground state Φ in the same Fock space: it obeys the conditions $a_k\Phi = 0$ for $k \notin N$, and $a_k^*\Phi = 0$ for $k \in N$. If N is infinite, we should introduce a new system of generators satisfying the same relations (make canonical transformation): $b_k = a_k$, $b_k^* = a_k^*$ for $k \notin N$, $b_k = a_k^*$, $b_k^* = a_k$ for $k \in N$. Then the ground state θ' lies in the Fock space constructed by means of b_k, b_k^* and obeys $b_k\theta' = 0$.

The above statements can be easily generalized in various directions. In particular, one can work with algebras defined by relations (I.15) or (I.17). Then the Fock representation can be defined as a representation in a (pre)-Hilbert space $\mathcal{F}$ that contains a cyclic vector θ obeying $a(f)\theta = 0$. If $\mathcal{B}$ consists of functions on $\mathcal{M}$, we can write formally $a(f) = \int f(k)a(k)dk$ (integration over $\mathcal{M}$) and consider $a(k)$, $a^*(k)$ as generalized functions on $\mathcal{M}$.

The equations of motion

$$i\frac{da(k)}{dt} = -\epsilon(k)a(k), \quad i\frac{da^*(k)}{dt} = \epsilon(k)a^*(k) \tag{I.19}$$

are coming from the formal Hamiltonian $\int \epsilon(k)a^*(k)a(k)dk$.

Again in the case $\epsilon(k) \geq 0$, the Fock representation is the ground state representation and the vector θ is the ground state. The formal Hamiltonian becomes a self-adjoint operator in Fock space.

More generally, any quadratic Hamiltonian leads to linear classical equations of motion that can also be considered as quantum equations of motion. This is true, for example, for Klein–Gordon equation and for Dirac equation. To construct the ground state representation, one should change variables to write the equations in the form (I.19) where a, a^* obey CCR or CAR.

If a formal Hamiltonian is represented as $H_0 + V$ where H_0 is a quadratic Hamiltonian, then in good situations, not only H_0 but also the full Hamiltonian is a well-defined self-adjoint operator in the ground state representation of H_0. If we started with a translation-invariant Hamiltonian without ultraviolet divergences, then usually we obtain a "good" Hamiltonian after the volume cutoff (infrared cutoff). We can calculate Wightman functions of "good" Hamiltonian (at least in the framework of perturbation theory). Then taking the limit of Wightman functions as the volume tends to infinity, we obtain Wightman functions of the original Hamiltonian (these construction can be considered as a definition of Wightman functions). Finally, using Wightman functions and an analog of GNS construction, we can construct the ground state representation of the original formal Hamiltonian and of Heisenberg operators in this space. The Heisenberg operators satisfy the equations of motion coming from the formal Hamiltonian. We say that we have obtained an operator realization of formal Hamiltonian (see Section 11.3 for more details).

Quantum field theory via the algebraic approach

Quantum field theory can be considered as a particular case of quantum mechanics.

To define quantum field theory on d-dimensional space (on $(d+1)$-dimensional space-time), we assume that the group of space-time translations acts on the algebra of observables. We can say that quantum field theory is quantum mechanics with the action of commutative Lie group on the algebra of observables $\mathcal{A}$. In other words, we assume that operators $\alpha(\mathbf{x}, t)$ where $\mathbf{x} \in \mathbb{R}^d$, $t \in \mathbb{R}$ are automorphisms of $\mathcal{A}$ preserving the involution and that $\alpha(\mathbf{x}, t)\alpha(\mathbf{x}', t') = \alpha(\mathbf{x} + \mathbf{x}', t + t')$. We will use the notation $A(\mathbf{x}, t)$ for $\alpha(\mathbf{x}, t)A$ where $A \in \mathcal{A}$.

The action of translation group on $\mathcal{A}$ induces the action of this group on the space of states.

Let us now consider a state ω that is invariant with respect to translation group. We will define (quasi-)particles as "elementary excitations" of ω. (If ω is the ground state, one uses the word "particles"; if ω is an equilibrium state, we are talking about thermal quasiparticles.) To consider collisions of (quasi-)particles, we require that ω satisfies the cluster property in some sense. Instead of cluster property, one can impose a condition of asymptotic commutativity of the algebra $\hat{\mathcal{A}}(\omega)$ that consists of operators $\hat{A}$ corresponding to the elements $A \in \mathcal{A}$ in GNS construction. In other words, one can require that the commutator $[\hat{A}(\mathbf{x}, t), \hat{B}]$ where $A, B \in \mathcal{A}$ is small for $\mathbf{x} \to \infty$.

The weakest form of cluster property is the following condition:

$$\omega(A(\mathbf{x}, t)B) = \omega(A)\omega(B) + \rho(\mathbf{x}, t), \tag{I.20}$$

where $A, B \in \mathcal{A}$ and ρ in some sense is small for $\mathbf{x} \to \infty$. For example, we can impose the condition that $\int |\rho(\mathbf{x}, t)|d\mathbf{x} < c(t)$ where $c(t)$ has at most polynomial growth. Note that (I.20) implies asymptotic commutativity in some sense: $\omega([A(\mathbf{x}, t), B])$ is small for $\mathbf{x} \to \infty$.

To formulate more general cluster property, we introduce the notion of correlation functions in the state ω :

$$w_n(\mathbf{x}_1, t_1, \ldots \mathbf{x}_n, t_n) = \omega(A_1(\mathbf{x}_1, t_1) \cdots A_n(\mathbf{x}_n, t_n)),$$

where $A_i \in \mathcal{A}$. They generalize Wightman functions of relativistic quantum field theory. We consider corresponding truncated correlation functions $w_n^T(\mathbf{x}_1, t_1, \ldots \mathbf{x}_n, t_n)$ (see Section 10.1 for definition).

We have assumed that the state ω is translation-invariant; it follows that both correlation functions and truncated correlation functions depend on differences $\mathbf{x}_i - \mathbf{x}_j, t_i - t_j$. We say that the state ω has cluster property if the truncated correlation functions are small for $\mathbf{x}_i - \mathbf{x}_j \to \infty$. A strong version of cluster property is the assumption that the truncated correlation function tends to zero faster than any power of $||\mathbf{x}_i - \mathbf{x}_j||$. Then its Fourier transform with respect to variables $\mathbf{x}_i$ has the form $\nu_n(\mathbf{p}_2, \ldots, \mathbf{p}_n, t_1, \ldots, t_n)\delta(\mathbf{p}_1 + \cdots + \mathbf{p}_n)$ where the function ν_n is smooth.

Let us show how one can define one-particle excitations of the state ω and the scattering of (quasi-)particles.

The action of translation group on $\mathcal{A}$ generates unitary representation of this group on the (pre-)Hilbert space $\mathcal{H}$ constructed from ω. Generators of this representation $\mathbf{P}$ and $-H$ are identified with momentum operator and Hamiltonian. The vector in the space $\mathcal{H}$ that corresponds to ω will be denoted by Φ. If Φ is a ground state, we say that it is the physical vacuum. If ω obeys KMS-condition

$$\omega(A(t)B) = \omega(BA(t+i\beta)), \tag{I.21}$$

we say that ω is an equilibrium state with the temperature $T = \frac{1}{\beta}$. (It is assumed that $A(t)$ can be analytically continued to the strip $0 \le \operatorname{Im} t \le \beta$.)

We say that a state σ is an excitation of ω if it coincides with ω at infinity. More precisely, we should require that $\sigma(A(\mathbf{x},t)) \to \omega(A)$ as $\mathbf{x} \to \infty$ for every $A \in \mathcal{A}$. Note that the state corresponding to any vector $A\Phi$ where $A \in \mathcal{A}$ is an excitation of ω; this follows from cluster property.

One can define *one-particle state* (one-particle excitation of the state ω) as a generalized $\mathcal{H}$-valued function $\Phi(\mathbf{p})$ obeying $\mathbf{P}\Phi(\mathbf{p}) = \mathbf{p}\Phi(\mathbf{p}), H\Phi(\mathbf{p}) = \varepsilon(\mathbf{p})\Phi(\mathbf{p})$. (More precisely, for some class of test functions $f(\mathbf{p})$, we should have a linear map $f \to \Phi(f)$ of this class into $\mathcal{H}$ obeying $\mathbf{P}\Phi(f) = \Phi(\mathbf{p}f), H\Phi(f) = \Phi(\varepsilon(\mathbf{p})f)$ where $\varepsilon(\mathbf{p})$ is a real-valued function. For definiteness, we can assume that test functions belong to the Schwartz space $\mathcal{S}(\mathbb{R}^d)$.) *We require that there exist an element* $B \in \mathcal{A}$ *such that* $\hat{B}\Phi = \Phi(\phi)$ *where* ϕ *is a* non-vanishing function. (The symbol $\hat{B}$ denotes the operator in $\mathcal{H}$ corresponding to the element $B \in \mathcal{A}$.) We also assume that there exists an element $A \in \mathcal{A}$ and a function $g(\mathbf{x},t) \in \mathcal{S}(\mathbb{R}^{d+1})$ such that $\hat{B} = \int g(\mathbf{x},t)\hat{A}(\mathbf{x},t)d\mathbf{x}dt$ (it follows from this assumption that $\hat{B}(\mathbf{x},t)$ is a smooth function of $\mathbf{x},t$).

We assume that $\Phi(f)$ is normalized (i.e. $\langle \Phi(f), \Phi(f') \rangle = \langle f, f' \rangle$).

Of course, it is possible that there are several one-particle states; to simplify the notations, we assume that there exists only one kind of particles.

Note that in relativistic theory, one-particle states can be identified with irreducible representations of the Poincaré group; our assumption means that we consider a scalar particle.

Let us fix a function $f(\mathbf{p})$. Define a function $\tilde{f}(\mathbf{x}, t)$ as the Fourier transform of the function $f(\mathbf{p})e^{-i\varepsilon(\mathbf{p})t}$ with respect to $\mathbf{p}$. Let us introduce the notation $B(f, t) = \int \tilde{f}(\mathbf{x}, t)B(\mathbf{x}, t)d\mathbf{x}$. (We assume that this expression specifies an element of $\mathcal{A}$ such that $\hat{B}(f, t) = \int \tilde{f}(\mathbf{x}, t)\hat{B}(\mathbf{x}, t)d\mathbf{x}$.) It is easy to check that

$$\hat{B}(f, t)\Phi = \int f(\mathbf{p})\phi(\mathbf{p})\Phi(\mathbf{p})d\mathbf{p} \qquad (\text{I.22})$$

does not depend on t.

Let us consider the vectors

$$\Psi(f_1, \ldots, f_n|t) = B(f_1, t) \cdots B(f_n, t)\Phi, \qquad (\text{I.23})$$

where $f_1, \ldots, f_n$ satisfy the following condition: if $f_i(\mathbf{p}) \neq 0$, $f_j(\mathbf{p}') \neq 0$, $i \neq j$, then $\nabla\varepsilon(\mathbf{p}) \neq \nabla\varepsilon(\mathbf{p}')$. (More precisely, we should assume that the distance between $\nabla\varepsilon(\mathbf{p})$ and $\nabla\varepsilon(\mathbf{p}')$ is bounded from below by a positive number.) We impose the additional requirement that $\varepsilon(\mathbf{p})$ is a smooth strictly convex function. Then one can derive from cluster property or from asymptotic commutativity of $\hat{\mathcal{A}}(\omega)$ that these vectors have limits in $\bar{\mathcal{H}}$ as $t \to \pm\infty$; these limits will be denoted by $\Psi(f_1, \ldots, f_n|\pm\infty)$.

Let us assume the vectors $\Psi(f_1, \ldots, f_n|-\infty)$ span a dense subset of $\mathcal{H}$. (One can hope that this condition is satisfied when ω is the ground state. In other cases, one should consider inclusive cross-sections and inclusive scattering matrix; see Chapter 13 and Schwarz (2019c) for more detail.) Then we can define in-operators a_{in}^+, a_{in} on $\mathcal{H}$. In particular, to define $a_{in}^+(f) = \int (f(\mathbf{p})a_{in}^+(\mathbf{p})d\mathbf{p}$, we can use the formula

$$a_{in}^+(\bar{f}\bar{\phi})\Psi(f_1, \ldots, f_n|-\infty) = \Psi(f, f_1, \ldots, f_n|-\infty),$$

The definition of out-operators is similar. Using these notions, we can define the scattering amplitudes by the formula

$$S_{mn}(\mathbf{p}_1, \ldots, \mathbf{p}_m|\mathbf{q}_1, \ldots, \mathbf{q}_n)$$

$$= \langle a_{in}^+(\mathbf{q}_1) \ldots a_{in}^+(\mathbf{q}_n)\Phi, a_{out}^+(\mathbf{p}_1) \ldots a_{out}^+(\mathbf{p}_m)\Phi\rangle.$$

Concluding remarks

In conclusion, a couple of general words about quantum theory.

A quantum mechanical system is specified by a Hamiltonian (infinitesimal generator of time evolution) and a QFT is specified by a Hamiltonian and a momentum operator (generator of spatial translations). Knowing these data, we can define a notion of particle in QFT; a particle can be interpreted as an elementary excitation of ground state. Under certain conditions, we can define the scattering of particles.

Note that string field theory can be regarded as a quantum field theory in the sense of this definition.

Quantum mechanics is a deterministic theory. This means that knowing the state at some moment and the Hamiltonian governing the evolution of the state, we can in principle predict the state at any moment — precisely as in classical mechanics. The probabilities in quantum mechanics can be explained in the same way as in classical statistical physics — they come from a random environment (see Sections "Decoherence" and "Observables and probabilities" on page xxxi for an explanation of how probabilities can be obtained from decoherence).[2]

The notion of particle in quantum field theory is a secondary notion. There exists no physical difference between elementary and composite particles. (There are theories that can be represented in two different ways: elementary particles of one approach or composite particles of another approach.) Probably, the most revealing illustration of properties of quantum particles is given by analogy with nonlinear scattering in classical field theory. Such a theory can have particle-like solutions, for example, solitons. (A soliton is a finite energy solution of the form $s(\mathbf{x} - \mathbf{v}t)$. It can be visualized as a bump moving with constant speed that does not change its form. If we have a bump moving with constant average speed, but the form of the bump changes with time, we can talk about generalized soliton or, in different terminology, about particle-like solution.) It seems that for

---

[2]We do not consider subtle questions of measurement theory. The above statement means only that the standard formulas for probabilities can be obtained from decoherence.

large class of theories, the asymptotic behavior of any finite energy solution as $t \to \pm\infty$ can be described as superposition of particle-like solutions and almost linear "tail" (see, for example, Tiupkin *et al.*, 1975; Soffer, 2006; Tao, 2009). (For integrable theories with one spatial dimension, this is a rigorous theorem; in higher dimensions, we only have experimental confirmation.) This means, in particular, that we can talk about scattering of solitons. Two solitons collide and we see a field configuration that does not resemble any particle-like solution. After that, solitons miraculously reappear (with the same velocities in integrable theories, with different velocities in non-integrable ones.) This is precisely the picture we can see considering a collision of quantum particles. It seems that for a given asymptotic behavior, one can rigorously construct a solution of equations of motion having this behavior. (In quantum field theory, this was done in Haag–Ruelle–Araki scattering theory in the relativistic case. Non-relativistic generalization of this theory was given in Fateev and Shvarts (1973); see also Chapters 10 and 13.) Both in classical and quantum integrable theories in two-dimensional space-time, one can analyze asymptotic behavior of any solution of equations of motion and rigorously justify the above picture. This can be done also in non-relativistic quantum mechanics: one can prove that asymptotically every solution behaves as a superposition of elementary and composite particles (bound states); see, for example, Hunziker and Sigal (2000). It is not known whether this statement is correct for more general theories (and it is definitely wrong for many relativistic conformal field theories).

We see that scattering of solitons eerily resembles quantum scattering. This similarity becomes even more complete if the classical theory has topological integrals of motion (i.e. the space of finite energy solutions is disconnected). Magnetic charge is an example of topological integral of motion, the solitons with minimal (by absolute value) magnetic charge are magnetic monopoles (if the charge is positive) or antimonopoles (if the charge is negative). Monopoles can annihilate with antimonopoles.

Quantum mechanics is a consistent theory that does not need an interpretation in the framework of classical field theory. I do not think that quantum scattering can be interpreted as classical scattering of solitons in random environment (although this possibility cannot be excluded). My point is that nonlinear classical scattering is very similar to quantum scattering. This idea is supported by remark that in the limit $\hbar \to 0$, the scattering of solitons can be obtained from the scattering of quantum particles.

Notation

$A \cup B$	the union of sets A and B;
$A \cap B$	the intersection of sets A and B;
$A \times B$	the cross product of sets A and B (the set of all (a, b) with $a \in A$ and $b \in B$);
E^n	the n-dimensional Euclidean space;
$L^2(X)$	the Hilbert space of square-integrable complex functions on the measure space X;
$\langle x, y \rangle$	the scalar multiplication of elements of (pre-)Hilbert space (for Euclidean space elements $\mathbf{x}, \mathbf{y} \in E^n$, this quantity will sometimes be denoted $\mathbf{xy}$);
$\mathcal{S}(E^n)$	the space of smooth functions of n variables that rapidly decay; by smooth, we mean infinitely differentiable and by rapidly decaying, we mean faster than any power;
Ω	the cube with length L edges in E^3, satisfying the equations $0 \leq x \leq L,\ 0 \leq y \leq L,\ 0 \leq z \leq L$;
T_Ω	a set of vectors with $\mathbf{k} = \frac{2\pi \mathbf{n}}{L}$, where $\mathbf{n}$ is an integer vector;
$\phi_{\mathbf{k}}$	the orthonormal basis in the space $L^2(\Omega)$, formed by the functions $\phi_{\mathbf{k}}(x) = e^{ix\mathbf{k}}$;
$H_1 + H_2$	a direct sum of Hilbert spaces;
$H_1 + \cdots +$ $H_n + \cdots$	a direct sum of an infinite sequence of Hilbert spaces;
$F(H)$	the Fock space built with the Hilbert space H (if $H = L^2(X)$, then $F(L^2(X)) = F_0 + F_1 + \cdots + F_n + \cdots$, where F_n consists of square-integrable symmetric functions in n variables in the case of bosons; in the case of fermions, the functions are antisymmetric);

$[A, B] =$ the commutator of A and B;
$AB - BA$

$[A, B]_+ =$ the anticommutator of A and B;
$AB + BA$

Two operators A and B on a Hilbert space are (Hermitian)
   conjugate to each other if $\langle Ax, y \rangle = \langle x, By \rangle$ for all x, y;

A^* (the adjoint operator) the conjugate operator to A
         with the maximal domain of definition;

A^+ the conjugate operator to A where the domain of
         definition coincides with A (for operators defined
         on the whole Hilbert space $A^* = A^+$; in this case,
         we will use A^*);

$\mathrm{slim}\, A_n$ the strong limit of a sequence of operators A_n (it is
         required that all the operators have the same
         domain). One says that $A = \mathrm{slim}\, A_n$ if
         $Ax = \lim A_n x$ for all x in the domain D;

$\mathrm{wlim}\, A_n$ the weak limit of a sequence of operators A_n. One
         says that $A = \mathrm{wlim}\, A_n$ if $\langle Ax, y \rangle = \lim \langle A_n x, y \rangle$ for
         all x, y in the domain D;

$\delta_{k,l} = \delta_l^k$ the Kronecker symbol ($\delta_{k,l} = 0$ if $k \neq l$, else it is 1);

$\delta(\xi, \eta)$ delta function (here, ξ and η belong to a measure
         space); if $\xi, \eta \in E^n$, then $\delta(\xi, \eta) = \delta(\xi - \eta)$;

$\theta(t) = 0$ the Heaviside step function; $\theta(t) = 0$ if $t < 0$, and
         $\theta(t) = 1$ with $t \geq 0$;

$\dfrac{1}{w+i0}$ the formula $\lim_{\epsilon \to 0+} \frac{1}{w+i\epsilon} = \frac{1}{i} \int \theta(t) \exp(iwt) dt$;

$\Box$ the d'Alembert operator
$$\frac{\partial^2}{\partial t^2} - \frac{\partial^2}{\partial x^2} - \frac{\partial^2}{\partial y^2} - \frac{\partial^2}{\partial z^2} = \frac{\partial^2}{\partial t^2} - \Delta;$$

$\mathrm{supp}\, f$ the support of the function f defined as the closure
         of the set of arguments where f is non-zero.

For the momentum variables, we use the notations $\mathbf{k}, \mathbf{p}, \mathbf{l}, \mathbf{q}$; for
the energy variable, we usually use ω. For coordinates, we use $x, \xi,$
y, η, and time is denoted by t or τ.

We use the system of units where the Planck constant $\hbar$ and the
speed of light c are equal to 1.

Chapter 1

Principles of Quantum Theory

1.1 Principles of quantum mechanics

Almost all modern physics theories begin with the following scheme. First, one specifies a mathematical object ψ_t that describes the state of an observed physical system at time t. Second, one obtains equations that govern the object's change with time (usually taking the form $\frac{\partial \psi_t}{\partial t} = A(\psi_t)$). Finally, one defines ways to calculate experimentally observed quantities from the object ψ_t.

Accordingly, we will formulate the postulates of quantum mechanics in this manner.

The state of a quantum system at a fixed moment in time is specified by a non-zero vector ψ taken from a complex Hilbert space R; two vectors ψ and $\psi' = C\psi$, differing only by a non-zero scalar factor C, represent the same state. By choosing $C = \|\psi\|^{-1}$, we can normalize the vector ψ' and hence a state can be represented as a normalized vector (a vector with norm equal to 1). If we don't say otherwise, we consider quantum states to be normalized vectors in the rest of the book.

1.2 Evolution of state vectors

Knowing the state of a quantum system at time t_0 (denoted by ψ_{t_0}), one can calculate the state ψ_t for any moment in time t, that is, there exists a linear operator $U(t, t_0)$, *the evolution operator*, that transforms the state vector ψ_{t_0} into the vector ψ_t.

The evolution operator $U(t, t_0)$ satisfies the group relations $U(t_2, t_1) \cdot U(t_1, t_0) = U(t_2, t_0)$, $U(t_0, t_0) = 1$. It obeys the Schrödinger

1

equation:

$$i\frac{\partial U(t, t_0)}{\partial t} = H(t)U(t, t_0), \tag{1.1}$$

with the initial condition $U(t_0, t_0) = 1$. Here, $H(t)$ is a self-adjoint operator called the *Hamiltonian operator* of the quantum system. The Schrödinger equation can also be written in the form

$$i\frac{d\psi_t}{dt} = H(t)\psi_t, \tag{1.2}$$

though less precisely, since the operator $H(t)$ may not be defined on the full Hilbert space R, while $U(t, t_0)$ is defined everywhere.

It is easy to check that $U^*(t, t_0) = U(t_0, t)$; using the group property, we can see that the operator is unitary (to be rigorous, we should assume that the evolution operator is unitary and prove that the Hamiltonian is self-adjoint using the unitarity of the evolution operator).

If the Hamiltonian does not depend on time explicitly, we can write $U(t, t_0)$ in the form $U(t, t_0) = \exp(-iH(t - t_0))$.

For a non-relativistic one-dimensional particle, the state space R can be taken to be $L^2(E^1)$, the space of square-integrable functions $\psi(x)$ of one variable x, where $-\infty < x < \infty$. The Hamiltonian operator in this case can be written as $H(t) = -\frac{1}{2m}\frac{d^2}{dx^2} + \hat{\mathcal{V}}(x, t)$ where $\hat{\mathcal{V}}(x, t)$ is the multiplication operator by the function $\mathcal{V}(x, t)$ (t plays the role of a parameter here).[1] This Hamiltonian describes the non-relativistic particle with mass m, moving in the field with time-dependent potential $\mathcal{V}(x, t)$.

1.3   Calculating the probabilities

An observable in quantum mechanics corresponds to a self-adjoint operator A on the space R. We will consider the correspondence between classical and quantum observables later in the book.

---

[1]It is easy to check that $\hat{\mathcal{V}}(x, t) = \mathcal{V}(\hat{x}, t)$, where $\hat{x}$ is the operator of multiplication by x.

An important general remark: the energy corresponds to the Hamiltonian operator H.

If we know the operator A corresponding to the observable a and a state of the quantum system, written as a normalized vector $\psi \in R$, we can obtain the probability distribution of the values obtained by measuring the observable a in the state ψ. Let us denote by $\mu(\alpha)$ the probability that after measurement we obtain the value a to be less than or equal to α. In quantum mechanics, we postulate that $\mu(\alpha) = \langle e_\alpha(A)\psi, \psi \rangle$, where $e_\alpha(\lambda)$ is a function equal to 1 when $\lambda \leq \alpha$ and 0 otherwise.

This postulate can be formulated another way: for any function f, the mean value $\overline{f(a)} = \int f(\alpha)d\mu(\alpha)$ of $f(a)$ in the state ψ is given by the formula $\overline{f(a)} = \langle f(A)\psi, \psi \rangle$. (The probability $\mu(a)$ is equal to the mean value of $e_\alpha(a)$, hence the first form of the postulate follows from the second form. To derive the second form from the first one, we note that any function f can be represented as a limit of linear combinations of the functions e_α.)

If the observables $a_1, \ldots, a_n$ correspond to commuting operators $A_1, \ldots, A_n$, then $a_1, \ldots, a_n$ are simultaneously measurable. This means that for any state ψ we can find the joint probability distribution for the values obtained when measuring $a_1, \ldots, a_n$ simultaneously. Namely, let us denote by $\mu(\alpha_1, \ldots, \alpha_n)$ the probability that in the state ψ we obtain $a_1 \leq \alpha_1, \ldots, a_n \leq \alpha_n$. We postulate this probability to be $\langle e_{\alpha_1,\ldots,\alpha_n}(A_1, \ldots, A_n)\psi, \psi \rangle$ where $e_{\alpha_1,\ldots,\alpha_n}(\lambda_1, \ldots, \lambda_n)$ is a function equal to 1 when $\lambda_1 \leq \alpha_1, \ldots, \lambda_n \leq \alpha_n$ and 0 otherwise.

As in the case of a single observable, the above postulate can be reformulated by demanding that for any function $f(a_1, \ldots, a_n)$ the mean value $\overline{f(a_1, \ldots, a_n)}$ of $f(a_1, \ldots, a_n)$ in the state ψ can be obtained by the formula $\overline{f(a_1, \ldots, a_n)} = \langle f(A_1, \ldots, A_n)\psi, \psi \rangle$.

For a single one-dimensional particle, the position observable x (the x-coordinate of the particle) corresponds to the multiplication operator $\hat{x}\psi = x\psi$. The momentum of the particle is represented by the operator $\hat{p} = \frac{1}{i}\frac{d}{dx}$, with the operator $\frac{\hat{p}^2}{2m} = -\frac{1}{2m}\frac{d^2}{dx^2}$ corresponding to the kinetic energy. The operator corresponding to the potential energy is the multiplication operator by $\mathcal{V}(x,t)$. The momentum and the kinetic energy are simultaneously measurable and so are the position and the potential energy. However, it is impossible to measure the position and the momentum simultaneously.

If A is a self-adjoint operator with discrete spectrum, then there exists an orthonormal basis of eigenvectors r_i in R for the operator A with corresponding eigenvalues a_i.

In the basis r_i, the matrices for the operators A and $e_\lambda(A)$ are diagonal, with the diagonal elements equal to a_i for the operator A and equal to $e_\lambda(a_i)$ for the operator $e_\lambda(A)$. Let us suppose that a normalized state vector is represented in the basis r_i as $\psi = \sum_i c_i r_i$, then

$$\langle e_\lambda(A)\psi, \psi \rangle = \sum_{ij} c_i c_j^* e_\lambda(a_i)\langle r_i, r_j \rangle = \sum_i e_\lambda(a_i)|c_i|^2.$$

It follows that when measuring the physical quantity corresponding to the operator A, we obtain only the eigenvalues a_i with non-zero probability. The probability of obtaining the value a in the state ψ is equal to $\sum_i |\langle \psi, r_i \rangle|^2$, where the sum is taken over those i where $Ar_i = ar_i$.

To each state vector ψ, we can assign the sequence $(c_1, \ldots, c_i, \ldots)$ of coefficients that arise from the decomposition of the vector ψ with respect to the basis r_i, consisting of the eigenvectors of operator A. This construction specifies an isomorphism of the space R and the space l^2. We call this isomorphism an *A-representation* and the sequence $(c_1, \ldots, c_i, \ldots)$, we call the A-representation of the vector ψ. The A-representation of the vector ψ allows us to easily calculate the probabilities of values of the observable a in the state ψ.

The notion of an A-representation can also be defined when the operator A has continuous spectrum. Moreover, it can be generalized to the case when we are dealing with a family of commutative self-adjoint operators. Namely, if $A_1, \ldots, A_k$ are self-adjoint, pairwise commuting operators, then an $(A_1, \ldots, A_k)$-representation is an isomorphism of Hilbert spaces R and $L^2(M)$ in which the operators $A_1, \ldots, A_k$ are transformed to the multiplication operators by the functions $a_1(m), \ldots, a_k(m)$ (here M is a measure space, with $m \in M$). In the $(A_1, \ldots, A_k)$-representation of the vector $\psi \in R$, one can easily calculate the probability distribution of the observables $A_1, \ldots, A_k$ in the state ψ. It is also possible to prove that the $(A_1, \ldots, A_k)$-presentation exists for any system of commuting self-adjoint operators $A_1, \ldots, A_k$. (If the operators $A_1, \ldots, A_k$ have discrete spectrum, then there exists an orthonormal basis of the space R consisting of the common eigenfunctions of the operators $A_1, \ldots, A_k$).

The $(A_1, \ldots, A_k)$-presentation is closely tied to the generalized eigenfunctions of the operators $A_1, \ldots, A_k$. That is, let r_m, where m belongs to the space with measure M, be a δ-normalized, generalized basis for the space R (for details, see A.8). If the basis is an eigenbasis for the operators $A_1, \ldots, A_k$ (i.e. $A_i r_m = a_i(m) r_m$), then the isomorphism of the spaces R and $L^2(M)$, given by the formula $a(m) = \langle x, r_m \rangle$, is a $(A_1, \ldots, A_k)$-presentation.

In what follows, we will identify physical quantities with their corresponding self-adjoint operators.

1.4 Heisenberg operators

Up to this point, we have considered our state vectors to be dependent on time, while the operators for the physical quantities were constant (Schrödinger's picture). However, it is possible to view this from a different but equivalent angle, where the operators depend on time, but the vectors stay constant (Heisenberg's picture).

Let us define for every operator A a time-dependent operator A_t (a Heisenberg operator) by the formula $A_t = U^*(t,0)AU(t,0)$.

We note the following relations:

$$(AB)_t = U^*(t,0)ABU(t,0)$$
$$= U^*(t,0)AU(t,)U^*(t,0)BU(t,0)$$
$$= A_t B_t,$$
$$(A^*)_t = U^*(t,0)A^*U(t,0)$$
$$= (U^*(t,0)AU(t,0))^*$$
$$= (A_t)^*,$$
$$f(A_t) = f(A)_t,$$
$$\langle A_t\psi, \psi \rangle = \langle A\psi_t, \psi_t \rangle.$$

Combining the last two relations, we can verify that calculating the probability distributions for the operator A and the state $\psi_t = U(t,0)\psi$, we obtain the same answer as when we calculate the probability distributions for the operator A_t and the state ψ. Hence, the Heisenberg and Schrödinger pictures are equivalent.

From (1.1) and the Hermitian conjugate equation $i\frac{\partial U^*(t,0)}{\partial t} = -U^*(t,0)H(t)$, we obtain the following equation for the operator A_t (*the Heisenberg equation of motion*):

$$\frac{dA_t}{dt} = i[H_t, A_t],$$

where $H_t = U^*(t,0)H(t)U(t,0)$. When the Hamiltonian H does not depend on time, $H_t = H$, $A_t = \exp(itH)A\exp(-itH)$, and the Heisenberg equation takes the form

$$\frac{dA_t}{dt} = i[H, A_t].$$

Let us calculate the Heisenberg equation of motion in a concrete example. Let R be the space of square-integrable functions of one variable x and $H = \frac{\hat{p}^2}{2m} + \mathcal{V}(\hat{x}, t)$, where $\hat{p} = \frac{1}{i}\frac{d}{dx}$ and $\hat{x}$ is the operator of multiplication by x. Using the fact that time evolution commutes with summation and multiplication of operators, as well as the operation of taking a function of an operator, we obtain

$$H_t = \frac{\hat{p}_t^2}{2m} + \mathcal{V}(\hat{x}_t, t).$$

(In the case when H does not depend on time, then $H = H_t$.)

From the relation $[\hat{p}, \mathcal{V}(\hat{x})] = \frac{1}{i}\mathcal{V}'(\hat{x})$, it follows that $[\hat{p}_t, \frac{\hat{p}_t^2}{2m} + \mathcal{V}(\hat{x}_t, t)] = \frac{1}{i}\mathcal{V}'(\hat{x}_t, t)$ and therefore $\frac{d\hat{p}}{dt} = -\mathcal{V}'(\hat{x}_t, t)$. (We are using the fact that evolution preserves commutation relations.) Analogously, we have that $\frac{d\hat{x}_t}{dt} = \frac{1}{m}\hat{p}_t$. Similar reasoning can be applied whenever it is necessary to write the Heisenberg equations for a specific operator.

1.5　Integrals of motion and stationary states

A physical quantity (an observable) A is called an *integral of motion* if for any state vector ψ, the probability distribution of the quantity A in the state $\psi_t = U(t,0)\psi$ does not depend on the time t.

In the Heisenberg picture, the quantity A is an integral of motion when the Heisenberg operator A_t does not depend on t. We will assume for the rest of this section that the Hamiltonian H does not depend on time. It is clear from the Heisenberg equations that the quantity A will be an integral of motion if and only if the operator A commutes with the Hamiltonian H. In particular, the physical quantity that corresponds to a time-independent operator H is an integral of motion; as we have noted, an observable of this form has the physical meaning of energy.

The state vectors corresponding to the eigenvectors of the operator H (states with definite energy E) are called *stationary states*. Indeed, if $H\psi = E\psi$, then $\psi_t = \exp(-iEt)\psi$ satisfies the Schrödinger equation. This means that ψ_t differs from ψ only by a numerical factor and consequently describes the same state; hence, the solution for the equation $H\psi = E\psi$ (the stationary Schrödinger equation) is "stationary". Conversely, when the vector $\psi_t = U(t,0)\psi$ for any t differs only by a factor $C(t)$ from the vector ψ, it follows from the Schrödinger equation that $iC'(t)\psi = C(t)H\psi$, hence $C(t) = \exp(-iEt)$ and $H\psi = E\psi$. Therefore, any state that doesn't change with time is stationary in the sense of the above definition.

A stationary state corresponding to the smallest energy value is called the ground state. It is easy to check that ground states can also be characterized as states with the smallest average energy; in other words, if Φ_0 is a ground state and Ψ is any other state, then $\langle H\Psi, \Psi \rangle \geq \langle H\Phi_0, \Phi_0 \rangle$ (we assume that Φ_0 and Ψ are normalized vectors). Indeed, let us consider for simplicity the case when the Hamiltonian H only has discrete spectrum. Decomposing the vector Ψ into the eigenvectors Φ_n of the operator H, we have

$$\langle H\Psi, \Psi \rangle = \left\langle H\sum c_n\Phi_n, \sum c_n\Phi_n \right\rangle$$

$$= \sum E_n|c_n|^2$$

$$\geq E_0 \sum |c_n|^2$$

$$\geq E_0 = \langle H\Phi_0, \Phi_0 \rangle$$

(here, E_n are the eigenvalues of the operator H corresponding to the eigenvectors Φ_n and $E_0 = \min E_n$ is the energy of the ground state).

Chapter 2

Quantum Mechanics of Single-Particle and Non-Identical Particle Systems

2.1 Quantum mechanics of a single scalar particle

To describe a quantum mechanical system, we specify the Hilbert space R of quantum states and the Hamiltonian operator H defining time evolution. Similarly, we need to specify the operators for the major physical quantities. In the following, we describe how to write simple quantum systems.

The state for a single scalar particle in three-dimensional space is described by a wave function — a square-integrable function $\psi(\mathbf{r})$ (i.e. the Hilbert space R in this case is $L^2(E^3)$). The Hamiltonian operator for a non-relativistic particle in a potential field with energy $\mathcal{V}(\mathbf{r})$ is written in the form $H = -\frac{1}{2m}\Delta + \hat{\mathcal{V}}(\mathbf{r})$, where m is the mass of the particle, Δ is the Laplace operator, and $\hat{\mathcal{V}}(\mathbf{r})$ is the operator of multiplication by the function $\mathcal{V}(\mathbf{r})$. The coordinate operator $\hat{x}$ for the coordinate x is defined as the multiplication operator by x (i.e. $\hat{\psi}(\mathbf{r}) = x\psi(\mathbf{r})$). Operators $\hat{y}$ and $\hat{z}$ are defined analogously. Since the operators $\hat{x}$, $\hat{y}$, $\hat{z}$ commute, we can obtain the joint probability distribution for the coordinates x, y, z in the state $\psi(\mathbf{r})$. From the previously formulated postulates, it follows that the density of this probability distribution is $|\psi(\mathbf{r})|^2$ (as always, we consider the state vectors $\psi(\mathbf{r})$ to be normalized). The operators for projections of the momentum $\mathbf{p}$ are $\hat{p}_x = \frac{1}{i}\frac{\partial}{\partial x}$, $\hat{p}_y = \frac{1}{i}\frac{\partial}{\partial y}$, $\hat{p}_z = \frac{1}{i}\frac{\partial}{\partial z}$, respectively. The components of angular momentum are operators defined by the formulas: $\hat{M}_x = \hat{y}\hat{p}_z - \hat{z}\hat{p}_y$; $\hat{M}_y = \hat{z}\hat{p}_x - \hat{x}\hat{p}_z$; $\hat{M}_z = \hat{x}\hat{p}_y - \hat{y}\hat{p}_x$.

For brevity, we will often use the vector operators: $\hat{\mathbf{r}} = (\hat{x}, \hat{y}, \hat{z})$, $\hat{\mathbf{p}} = (\hat{p}_x, \hat{p}_y, \hat{p}_z)$, $\hat{\mathbf{M}} = \hat{\mathbf{r}} \times \hat{\mathbf{p}} = (\hat{M}_x, \hat{M}_y, \hat{M}_z)$.

For the Heisenberg operators $\hat{r}_t$ and $\hat{p}_t$, the equations of motion are

$$\frac{d\hat{\mathbf{r}}_t}{dt} = \frac{\hat{\mathbf{p}}_t}{m}; \frac{d\hat{\mathbf{p}}_t}{dt} = -\frac{dV}{d\mathbf{r}}(\hat{\mathbf{r}}_t).$$

Up to this point, the states for one particle were written as $\psi(\mathbf{r})$ with $\mathbf{r} \in E^3$. If α is a unitary operator, transforming the space $L^2(E^3)$ to the space $L^2(M)$, where M is some measure space (an isomorphism of the space $L^2(E^3)$ and the space $L^2(M)$), then we can equivalently represent the state vectors $\psi \in L^2(E^3)$ as vectors $\tilde{\psi} = \alpha\psi \in L(M)$. For every operator A acting on the space $L^2(E^3)$, there will be a corresponding operator $\tilde{A} = \alpha A \alpha^{-1}$ acting on $L^2(M)$, and from $\langle f(\tilde{A})\tilde{\psi}, \tilde{\psi}\rangle = \langle f(A)\psi, \psi\rangle$, it follows that the probabilities obtained from the operator $\tilde{A}$ and the state vector $\tilde{\psi}$ will be the same as those obtained from A and ψ.

Changing the representation of a state vector is useful in many cases. For example, let α be the Fourier transform

$$\tilde{\psi}(\mathbf{p}) = (2\pi)^{-3/2} \int \exp(-i\mathbf{p}\mathbf{r})\psi(\mathbf{r})d\mathbf{r}$$

(in this case $M = E^3$). The Fourier transform of the operator $\hat{p}_x = \frac{1}{i}\frac{\partial}{\partial x}$ is the multiplication operator $\tilde{\hat{p}}_x\tilde{\psi}(p) = p_x\tilde{\psi}(p)$ (analogously for the other variables). It is clear that the joint probability density of $\hat{p}_x$, $\hat{p}_y$, $\hat{p}_z$ is equal to $|\tilde{\psi}(\mathbf{p})|^2$.

Let us describe the geometric origin of momentum and angular momentum operators. Note that to every translation T in the space E^3, there corresponds a unitary operator W_T on the space $L^2(E^3)$, that is, W_T transforms wave functions when the underlying space is transformed by T. W_T can be written in the form $W_T\psi(\mathbf{r}) = \psi(T^{-1}\mathbf{r})$. It is obvious that $W_{T_1}W_{T_2} = W_{T_1T_2}$ and therefore the operators W_T form a unitary representation of the translation group.

The component operator for the momentum along an axis can be written as the operator of infinitesimal translation along the axis, that is, the operator $\hat{p}_x$, for example, can be written as $i\lim_{a\to 0}\frac{W_{T_a}-1}{a}$, where T_a is translation by a along the axis x

$(T_a(x, y, z) = (x+a, y, z))$. Accordingly, the component of the angular momentum can be written as the operator of infinitesimal small rotation around the corresponding axis, that is, M_z can be written as $i \lim_{\psi \to 0} \frac{1}{\psi}(W_{S_\psi} - 1)$ where S_ψ denotes rotation by the angle ψ around the z axis (i.e. $S_\phi(x, y, z) = (x \cos \phi - y \sin \phi, x \sin \phi + y \cos \phi, z))$.

This geometric interpretation allows us to understand when components of momentum and angular momentum are integrals of motion. For example, if the Hamiltonian is invariant with respect to rotation around the z axis (i.e. if H commutes with W_{S_ϕ}), then the Hamiltonian will commute with the operator M_z and therefore M_z will be an integral of motion.

2.2 Quantum mechanics of particles with spin

The state of a particle with spin is written as a column vector ψ of k square-integrable functions defined on three-dimensional space

$$\psi = \begin{bmatrix} \psi_1(\mathbf{r}) \\ \psi_2(\mathbf{r}) \\ \vdots \\ \psi_k(\mathbf{r}) \end{bmatrix},$$

or, in other words, a function $\psi(\mathbf{r})$ with k-column outputs. Instead of notation $\psi_i(\mathbf{r})$, we will use $\psi(\mathbf{r}, i)$, where i is a parameter, that takes the values $1, 2, \ldots, k$. In other words, the space of states R can be taken to be the space of square-integrable functions $\psi(\xi)$, where $\xi = (\mathbf{r}, i) \in E^3 \times B, \mathbf{r} \in E^3, i \in B = \{1, \ldots, k\}$.

Integration over ξ has the same effect as integrating over $\mathbf{r}$ and summing over i, for example, taking two functions $\phi \in R$ and $\psi \in R$, we have

$$\langle \phi, \psi \rangle = \int \phi(\xi)\overline{\psi(\xi)}d\xi = \sum_{i=1}^{k} \int \phi(\mathbf{r}, i)\overline{\psi(\mathbf{r}, i)}d\mathbf{r}.$$

The operators of position, momentum, and the Hamiltonian, if the particle is in a potential field, are the same as in the case of the spinless particle. The discrete variable acts like a parameter under the action of the listed operators on the function $\psi(\xi) = \psi(\mathbf{r}, i)$. However, the operators of angular momentum are different. As we have seen, these operators are related to the action of rotations on the wave functions and, in general, the rotations act not only on space variables but also on discrete variables. Therefore, these operators are represented as sums of the orbital part and the part coming from discrete variables (the spin part). More precisely, the components of operators for the angular momentum $\hat{M}_x$, $\hat{M}_y$, $\hat{M}_z$ can be written in the form $\hat{M}_x = \hat{l}_x + \hat{s}_x$; $\hat{M}_y = \hat{l}_y + \hat{s}_y$; $\hat{M}_z = \hat{l}_z + \hat{s}_z$, where $\hat{\mathbf{l}} = (\hat{l}_x, \hat{l}_y, \hat{l}_z)$ is the operator of the orbital angular momentum and $\hat{\mathbf{s}} = (\hat{s}_x, \hat{s}_y, \hat{s}_z)$ is the operator for the spin angular momentum. The operator for the orbital angular momentum obeys the same formulas as the operator of angular momentum for spinless particles (i.e. $\hat{\mathbf{l}} = \hat{\mathbf{r}} \times \hat{\mathbf{p}}$); the operator for the spin angular momentum acts only on the discrete variable (we say that the operator A acts only on the discrete variable if there exist matrices a_{ij}, where $1 \leq i, j \leq k$, such that $A\psi(\mathbf{r}, i) = \sum_{j=1}^{k} a_{ij}\psi(\mathbf{r}, j)$).

Remark. We can verify these relationships by following the considerations in Section 2.1. To every rotation T in three-dimensional space, there corresponds a unitary transformation W_T of the space R, this transformation changes the argument $\mathbf{r}$ of the wave function (as in scalar case), but we should take into account discrete variables. For example, if $k = 3$, we can consider $\psi(\mathbf{r})$ as a function of $\mathbf{r}$, taking values in three-dimensional vectors. Then the operator W_T acts on the function $\psi(\mathbf{r})$ in the natural way, that is, the function $\psi' = W_T\psi$ is given by the equation $\psi'_i(\mathbf{r}) = \sum_{j=1}^{3} T_{ij}\psi_j(T^{-1}\mathbf{r})$, where T_{ij} is the matrix for the transformation T. In general, the operator W_T should be written in the form

$$(W_T\psi)(\mathbf{r}, i) = \sum_{j=1}^{k} D_{ij}^{T}\psi(T^{-1}\mathbf{r}, j).$$

This form of the operator W_T immediately leads to the formulas previously stated for the components of angular momentum, which are identified with operators of infinitesimally small rotation around an axis. Operators W_T must satisfy the relation $W_{T_1 T_2} = W_{T_1} W_{T_2}$; this means that they specify a unitary representation of the rotation group. It is clear that matrices D_T define k-dimensional unitary representation of the rotation group. Knowing the finite-dimensional unitary representation of the rotation group, it's not difficult to establish the form of spin projection operators. The correspondence between rotation T with the operator W_T and the matrices D^T is in general two-valued (the operator W_T and the matrix D_T are defined up to a sign). This does not affect our calculations since state vectors in quantum mechanics are defined up to multiplication by a number.

It is well known that for every value $k \geq 0$, there exists one (up to equivalence) k-dimensional unitary irreducible representation of the rotation group. For k odd, the representation is single-valued; for k even, it is two-valued. The functions

$$\psi(\mathbf{r}) = \begin{bmatrix} \psi_1(\mathbf{r}) \\ \psi_2(\mathbf{r}) \\ \vdots \\ \psi_k(\mathbf{r}) \end{bmatrix},$$

taking values in the state space, describe a particle with spin $s = \frac{k-1}{2}$. This terminology comes from the remark that the projection of the spin takes values between $-s$ and s (more precisely, every operator $\hat{s}_x$, $\hat{s}_y$, $\hat{s}_z$, constructed as described in this paragraph, has eigenvalues $-s, -s+1, \ldots, s-1, s$).

Especially important is the case of particles with spin $1/2$, due to many elementary particles, such as the electron, proton, and neutron, having spin $1/2$. Components of the spin can be written in this case as matrices

$$\hat{s}_x = \frac{1}{2} \begin{bmatrix} 0 & 1 \\ 1 & 0 \end{bmatrix}; \hat{s}_y = \frac{1}{2} \begin{bmatrix} 0 & -i \\ i & 0 \end{bmatrix}; \hat{s}_z = \frac{1}{2} \begin{bmatrix} 1 & 0 \\ 0 & -1 \end{bmatrix}.$$

The above apparatus is also useful when particles have various other quantum numbers, not just spin (e.g. isospin). The only difference is in that when there are no quantum numbers besides spin, the matrices D^T specify an irreducible representation; otherwise, they define a reducible representation.

2.3   Quantum description of a system with non-identical particles

Two electrons (or two protons) have the same mass, the same charge; their other properties are also identical. Therefore, we say that all electrons are identical particles. If two particles belonging to a system of particles are identical, then we must essentially change the quantum mechanical description of the system. In this section, we will consider only a system of particles where any two particles are not identical.

If we consider a system of n three-dimensional particles with spin zero, then the Hilbert space R of the states of such a system can be considered as a space $L^2(E^{3n})$ of square-integrable functions of n spatial variables.

Let us consider the case of a system with n particles with spin. We assume that the jth particle can be described by a function $\psi(\xi)$, defined on the set B_j (in coordinate representation B_j is the set of pairs $(\mathbf{r}, i)$, where $\mathbf{r}$ is a point in three-dimensional space and i runs over the discrete set $1, 2, \ldots, k_j$). Then the space of states of the whole system is the space $R = L^2(B_1 \times \cdots \times B_n)$ of square-integrable functions $\psi(\xi_1, \ldots, \xi_n)$ of variables $\xi_1, \ldots, \xi_n$ that run over the sets $B_1, \ldots, B_n$. In the coordinate representation, the wave function $\psi(x_1, y_1, z_1, i_1, \ldots, x_n, y_n, z_n, i_n)$ of the system of particles depends on the coordinates (x_j, y_j, z_j) and spin variables i_j of every particle.

Every operator corresponding to a physical quantity related to the jth particle can be considered in a natural way as an operator on the space R. For example, in the coordinate representation, the operators $\hat{x}_j, \hat{y}_j, \hat{z}_j$ of the coordinates of the jth particle act in R as operators of multiplication by x_j, y_j, z_j; the operators $\hat{p}_{jx}, \hat{p}_{jy}, \hat{p}_{jz}$ of the momentum components of the jth particle act in R as operators of coordinate differentiation of the jth particle: $\hat{p}_{jx} = \frac{1}{i}\frac{\partial}{\partial x_j}, \hat{p}_{jy} = \frac{1}{i}\frac{\partial}{\partial y_j}, \hat{p}_{jz} = \frac{1}{i}\frac{\partial}{\partial z_j}$ (which we can write more compactly as $\hat{\mathbf{p}}_j = \frac{1}{i}\frac{\partial}{\partial \mathbf{r}_j}$).

In general, if the operator A (acting on the space $L(B_j)$) transforms the function $\psi(\xi)$ into the function $\psi'(\xi) = \int A(\xi, \eta)\psi(\eta)d\eta$, then the corresponding operator in the space R transforms the function $\psi(\xi_1, \ldots, \xi_n)$ into the function

$$\psi'(\xi_1, \ldots, \xi_n) = \int A(\xi_j, \eta)\psi(\xi_1, \ldots, \xi_{j-1}, \eta, \xi_{j+1}, \ldots, \xi_n)d\eta.$$

The operator of total momentum $\mathbf{p}$ is given by the formula

$$\mathbf{p} = \sum_{j=1}^{n}\hat{\mathbf{p}}_j = \sum_{j=1}^{n}\frac{1}{i}\frac{\partial}{\partial \mathbf{r}_j}. \tag{2.1}$$

The Hamiltonian operator of a system of n particles, in the simplest case, can be written in the form

$$H = \sum_{j=1}^{n}\frac{\hat{\mathbf{p}}_j^2}{2m_j} + \mathcal{V}(\hat{\mathbf{r}}_1, \ldots, \hat{\mathbf{r}}_n) = \sum_{j=1}^{n}(-\frac{\Delta_j}{2m_j}) + \mathcal{V}(\hat{\mathbf{r}}_1, \ldots, \hat{\mathbf{r}}_n) \tag{2.2}$$

(here, the function $\mathcal{V}$ typically takes the form $\mathcal{V}(\mathbf{r}_1, \ldots, \mathbf{r}_n) = \sum_{j=1}^{n}\mathcal{V}_j(\mathbf{r}_j) + \sum_{j<l}\mathcal{V}_{jl}(|\mathbf{r}_j - \mathbf{r}_l|))$.

If $\mathcal{V}(\mathbf{r}_1, \ldots, \mathbf{r}_n) = \mathcal{V}_1(\mathbf{r}_1) + \cdots + \mathcal{V}_n(\mathbf{r}_n)$ (the case of non-interacting particles in an external field), then the problem of the motion of n particles can be reduced to n single-particle problems. Let us calculate, as an example, the stationary states of n non-interacting particles. Let us denote by $\psi_1^{(j)}(\xi_j), \ldots, \psi_r^{(j)}(\xi_j), \ldots$ the eigenfunctions of the operator $H_j = \frac{\hat{\mathbf{p}}_j^2}{2m_j} + \mathcal{V}_j(\hat{\mathbf{r}}_j)$ in the space $L^2(B_j)$ (the stationary states of the jth particle). The corresponding eigenvalues will be denoted by $E_r^{(j)}$. It is easy to check that the function $\phi_{r_1}^{(1)}(\xi_1)\phi_{r_2}^{(2)}(\xi_2)\ldots\phi_{r_n}^{(n)}(\xi_n)$ is a stationary state of the Hamiltonian $H = H_1 + \cdots + H_n$ in the space R with the corresponding eigenvalue $E_{r_1}^{(1)} + \cdots + E_{r_n}^{(n)}$. If, for every j, the functions $\phi_r^{(j)}(\xi_j)$ form a complete orthonormal system in the space $L^2(B_j)$, then the products $\phi_{r_1}^{(1)}(\xi_1)\phi_{r_2}^{(2)}(\xi_2)\ldots\phi_{r_n}^{(n)}(\xi_n)$ form a complete orthonormal system in the space R.

Until now, we have worked with coordinate representations, which means that we considered wave functions as functions of coordinates and spin variables. Let us use another representation for every particle; let us choose an isomorphism α_j of the space $L^2(B_j)$ and the space $L^2(M_j)$, where M_j is a measure space. Then it is easy to construct an isomorphism of the space $R = L^2(B_1 \times \cdots \times B_n)$ and the space $L^2(M_1 \times \cdots \times M_n)$ of square-integrable functions of n variables $(m_1, \ldots, m_n)$ that run over the sets $M_1, \ldots, M_n$. In other words, the state of a system of n particles can be represented by functions $\psi(m_1, \ldots, m_n)$, where $m_j \in M_j$. Denoting the space of states of the j-th particle by R_j, we can say that the space of states R of an n particle system is isomorphic to the tensor product $R_1 \otimes \cdots \otimes R_n$ (this follows immediately from the definition of the tensor product given in A.4).

In particular, it is very often convenient to use the momentum representation. In the momentum representation, the system of n particles is described as a function of momentum variables $\mathbf{p}_1, \ldots, \mathbf{p}_n$ and spin variables $i_1, \ldots, i_n$, given by the formula

$$\tilde{\psi}(\mathbf{p}_1, i_1, \ldots, \mathbf{p}_n, i_n) = (2\pi)^{-3n/2} \int \exp\left(-i \sum_{k=1}^{n} \mathbf{p}_k \mathbf{r}_k\right)$$
$$\times \, \psi(\mathbf{r}_1, i_1, \ldots, \mathbf{r}_n, i_n) d^n \mathbf{r},$$

where $\psi(\mathbf{r}_1, i_1, \ldots, \mathbf{r}_n, i_n)$ is the wave function in the coordinate representation.

The momentum operator of the ith particle $\hat{\mathbf{p}}_i$ in the momentum representation can be considered as the operator of multiplication by $\mathbf{p}_i$; the operator of total momentum, which is equal to $\mathbf{p} = \sum \hat{\mathbf{p}}_i$, can be considered as an operator of multiplication by $\mathbf{p}_1 + \cdots + \mathbf{p}_n$.

2.4 A particle in a box with periodic boundary conditions

Let us consider, as an example, a scalar particle moving freely in the cube Ω, specified by the inequalities $0 \leq x \leq L, 0 \leq y \leq L, 0 \leq z \leq L$. To make this problem precise, we will impose periodic boundary conditions on the wave functions of stationary states. The space of states R in this case can be identified with the space of all square-integrable functions on the cube Ω and the Hamiltonian is the operator $H = (-\frac{1}{2m})\Delta$ with periodic boundary conditions (more precise definition of this Hamiltonian will be given later).

Let us fix a vector $\mathbf{a}$ in three-dimensional space and consider the operation of parallel translation by the vector $\mathbf{a}$. The corresponding unitary shift operator $W_{\mathbf{a}}$ on the space R transforms state vectors as translation $(W_{\mathbf{a}}\psi)(\mathbf{r}) = \check{\psi}(\mathbf{r} - \mathbf{a})$, where $\check{\psi}(\mathbf{r})$ is a periodic extension of $\psi(\mathbf{r})$, given on the cube Ω, to the whole three-dimensional space. The momentum operators $\hat{p}_x$, $\hat{p}_y$, $\hat{p}_z$ can be defined as infinitesimal shift operators along the axes x, y, z (for example, $\hat{p}_x = i \lim_{a\to 0} \frac{W_{(a,0,0)} - 1}{a}$).

It is easy to check that $\hat{p}_x\psi = \frac{1}{i}\frac{\partial\psi}{\partial x}$. The operator $\hat{p}_x$ is defined on the functions $\psi \in L^2(\Omega)$, obeying the condition $\frac{\partial\psi}{\partial x} \in L^2(\Omega)$ and satisfying the periodic boundary conditions $\psi(0, y, z) = \psi(L, y, z)$. In terms of the operator $\hat{\mathbf{p}}$, the Hamiltonian can be written in the form $H = \frac{1}{2m}\hat{\mathbf{p}}^2$.

The functions $\phi_k = L^{-3/2}\exp(-i\mathbf{k}\mathbf{r})$, where $\mathbf{k}$ runs over the lattice T_Ω consisting of vectors $\frac{2\pi}{L}\mathbf{n}$ (where $\mathbf{n}$ is a vector of integer coordinates), form a complete orthonormal system of common eigenfunctions of the operators $H, \hat{p}_x, \hat{p}_y, \hat{p}_z$. Every function $\psi \in R$ can be represented in the form $\psi(\mathbf{r}) = \sum c_{\mathbf{k}}\phi_{\mathbf{k}}$ (where the sum is taken over the lattice T_Ω); the function $c_{\mathbf{k}}$ on the lattice T_Ω is called the wave function in the momentum representation.

2.5 One-dimensional harmonic oscillator

Let us consider a one-dimensional non-relativistic quantum particle with mass m and potential energy $V(q)$. We assume that $V(q)$ has one global minimum at the point q_0 and $V''(q_0)) = k > 0$. Then we can study the ground state and low-lying energy levels replacing $V(q)$ with $V(q_0) + \frac{k}{2}(q - q_0)^2$. This means that we should study a quantum harmonic oscillator. Taking into account higher order terms in the Taylor expansion of $V(q)$, we obtain the Hamiltonian of anharmonic oscillator. Without loss of generality, we can assume that $m = 1$, $k = 1$, $V(q_0) = 0$. Then in the variable $x = q - q_0$, the Hamiltonian of the harmonic oscillator takes the form

$$H = \frac{\hat{p}^2}{2} + \frac{\hat{x}^2}{2} = -\frac{1}{2}\frac{d}{dx^2} + \frac{1}{2}\hat{x}^2.$$

First, let us find the Heisenberg operators $\hat{p}_t$ and $\hat{x}_t$ from the Heisenberg equations

$$\frac{d\hat{p}_t}{dt} = -\hat{x}_t; \frac{d\hat{x}_t}{dt} = \hat{p}_t.$$

This system of operator equations is linear and therefore it can be solved precisely as the corresponding system of numerical equations. One of the possible ways to solve it is based on the introduction of auxiliary operators

$$\hat{a} = \frac{1}{\sqrt{2}}(\hat{x} + i\hat{p}); \hat{a}^+ = \frac{1}{\sqrt{2}}(\hat{x} - i\hat{p}).$$

The equations for the Heisenberg operators $\hat{a}_t$ and $\hat{a}_t^+$ have a very simple form

$$\frac{d\hat{a}_t}{dt} = -i\hat{a}_t; \frac{d\hat{a}_t^+}{dt} = i\hat{a}_t^+,$$

hence $\hat{a}_t^+ = \hat{a}^+ \exp(it); \hat{a}_t = \hat{a} \exp(-it)$. Using these relations, we can immediately obtain the formulae $\hat{x}_t = \frac{1}{\sqrt{2}}(\hat{a}_t + \hat{a}_t^+)$ and $\hat{p}_t = \frac{1}{\sqrt{2}}(\hat{a}_t^+ - \hat{a}_t)$.

The operators $\hat{a}^+$ and $\hat{a}$ are very convenient for solving problems related to the harmonic oscillator.

Note, first of all, that the Hamiltonian can be expressed in terms of these operators by the formula $H = \hat{a}^+\hat{a} + 1/2$. Second, note that the commutation relations with the Hamiltonian have the form $[H, \hat{a}^+] = \hat{a}^+$, $[H, \hat{a}] = -\hat{a}$, and their commutator is $[\hat{a}, \hat{a}^+] = 1$.

Let us find the stationary states of the Hamiltonian H. We use the following statement: If $H\phi = E\phi$, then $H(\hat{a}\phi) = (E-1)\hat{a}\phi$ (hence if ϕ is a stationary state and $\hat{a}\phi \neq 0$, then $\hat{a}\phi$ is also a stationary state). This statement follows from the relations $H(\hat{a}\phi) = \hat{a}(H - 1)\phi = (E - 1)\hat{a}\phi$.

Similarly, $H(\hat{a}^+\phi) = \hat{a}^+(H + 1)\phi = (E + 1)\hat{a}^+\phi$.

Note that the ground state ϕ_0 should satisfy the relation $\hat{a}\phi_0 = 0$ (otherwise, $\hat{a}\phi_0$ is a stationary state with lower energy). Solving the equation $\hat{a}\phi_0 = \frac{1}{\sqrt{2}}(x + \frac{d}{dx})\phi_0 = 0$, we obtain that $\phi_0 = \pi^{-1/4} \exp(-\frac{1}{2}x^2)$ (the constant $\pi^{-1/4}$ comes from the normalization condition) and that the corresponding eigenvalue is $E_0 = 1/2$. Starting with ϕ_0, we can get an infinite number of stationary states $\phi_n = c_n(\hat{a}^+)^n \phi_0 = c_n 2^{-n/2}(\hat{x} - \frac{d}{dx})^n(\pi^{-1/4}\exp(-\frac{x^2}{2}))$ (here c_n is the normalization constant). Noting that the operator $\hat{a}^+$ increases the energy by 1, we have $H\phi_n = (n + \frac{1}{2})\phi_n$. It is easy to see that the functions ϕ_n have the form $\phi_n(x) = H_n(x)\exp(-\frac{1}{2}x^2)$, where $H_n(x)$ are polynomials of n-th degree; one can check that $H_n(x)$ coincides with the n-th Hermite polynomial up to a constant factor. Let us calculate the normalization constant c_n assuming that all of them are taken to be real and positive. It is obvious that $\phi_n = \gamma_n \hat{a}^+ \phi_{n-1}$, where $\gamma_n = c_n/c_{n-1}$. Taking the scalar square of this equality, we obtain

$$
\begin{aligned}
1 &= \langle \phi_n, \phi_n \rangle \\
&= \gamma_n^2 \left\langle \hat{a}^+ \phi_{n-1}, \hat{a}^+ \phi_{n-1} \right\rangle \\
&= \gamma_n^2 \left\langle \hat{a}\hat{a}^+ \phi_{n-1}, \phi_{n-1} \right\rangle \\
&= \gamma_n^2 \left\langle \left(H + \frac{1}{2}\right) \phi_{n-1}, \phi_{n-1} \right\rangle \\
&= n\gamma_n^2,
\end{aligned}
$$

hence, $\gamma_n = 1/\sqrt{n}$ and $c_n = (n!)^{-1/2}$.

The orthonormal system of stationary states $\phi_n = \frac{1}{\sqrt{n!}}\hat{a}^{+n}\phi_0$, with energy levels $E_n = n + 1/2$, exhausts all stationary states. One can check this by proving the completeness of functions ϕ_n in the space $L^2(E^1)$. However, one can give a more direct proof. Let ϕ denote a stationary state of H. Let us denote by n then, the minimal number satisfying $\hat{a}^n\phi_0 = 0$ (such numbers necessarily exist because the eigenvalues of the Hamiltonian H are bounded from below). Then, $\hat{a}^{n-1}\phi = \lambda\phi_0$, where $\lambda \neq 0$. Applying the operator $(\hat{a}^+)^{n-1}$ to this equation, after some easy calculations, we obtain that ϕ is proportional to ϕ_{n-1}.

The more general Hamiltonian $H = \frac{\hat{p}^2}{2m} + \frac{m\omega^2\hat{x}^2}{2}$ can be reduced to the Hamiltonian we have considered by means of changing the system of units. One can, however, use the same consideration in this case, taking

$$\hat{a} = \frac{1}{\sqrt{2}}\left(\sqrt{m\omega}\,\hat{x} + \frac{i\hat{p}}{\sqrt{m\omega}}\right); \hat{a}^+ = \frac{1}{\sqrt{2}}\left(\sqrt{m\omega}\,\hat{x} - \frac{i\hat{p}}{\sqrt{m\omega}}\right).$$

These operators are also related by the formula $[\hat{a}, \hat{a}^+] = 1$, however, the expression of the Hamiltonian in terms of these operators has the form $H = \omega(\hat{a}^+\hat{a} + 1/2)$, hence $[H, \hat{a}] = -\omega\hat{a}, [H, \hat{a}^+] = \omega\hat{a}^+$. The dependence of operators $\hat{a}_t$ and $\hat{a}_t^+$ on time has the form $\hat{a}_t = \hat{a}\exp(-i\omega t), \hat{a}_t^+ = \hat{a}^+\exp(i\omega t)$. The stationary states can be written in terms of the ground state ϕ_0 by the formula $\phi_n = (n!)^{-1/2}(\hat{a}^+)^n\phi_0$, with the energies equal to $E_n = (n + 1/2)\omega$.

Adding to the quadratic potential energy some higher order terms with respect to $\hat{x}$, we obtain the Hamiltonian of anharmonic oscillators. It is convenient to express the Hamiltonian of anharmonic oscillators in terms of the operators $\hat{a}^+$ and $\hat{a}$. For example, if $H = \frac{\hat{p}^2}{2m} + \frac{m\omega^2\hat{x}^2}{2} + \alpha\hat{x}^3 + \beta\hat{x}^4$, then its expression in terms of $\hat{a}^+$ and $\hat{a}$ looks as follows:

$$H = \frac{\omega}{2} + \omega\hat{a}^+\hat{a} + \gamma(\hat{a}^+ + \hat{a})^3 + \delta(\hat{a}^+ + \hat{a})^4,$$

where $\gamma = \alpha(2m\omega)^{-3/2}, \delta = \beta(2m\omega)^{-2}$.

2.6   Multidimensional harmonic oscillator

Let us consider a quantum mechanical system with space of states $R = L^2(E^n)$ and the Hamiltonian

$$H = -\sum_{j=1}^{n} \alpha_j \frac{\partial^2}{\partial x_j^2} + \sum_{i,j} k_{i,j}\hat{x}_i\hat{x}_j$$

(here, $\alpha_j > 0$ and the quadratic form $\sum k_{ij}x_ix_j$ is positively definite). In this way, we can write down, for example, the Hamiltonian for a system of oscillators coupled by means of elastic forces. By means of a linear change of variables $\xi_i = \sum_j a_{ij}x_j$ (the coordinates ξ_i are called normal coordinates), the Hamiltonian can be written in the following form:

$$H = -\frac{1}{2}\sum_{i=1}^{n}\frac{\partial^2}{\partial\xi_i^2} + \frac{1}{2}\sum_{i=1}^{n}\omega_i^2\hat{\xi}_i^2.$$

(This statement follows immediately from the theorem that a symmetric matrix can be diagonalized by means of an orthogonal transformation, or from the equivalent theorem that for any two quadratic forms, one of which is positively definite, one can transform both into a sum of squares by means of a linear transformation.)

The Hamiltonian, written in normal coordinates, represents a system of n non-interacting one-dimensional oscillators. Using the results of Section 2.3, we can calculate its eigenvectors and eigenvalues. Note that the operators $\hat{a}_i^+ = \frac{1}{\sqrt{2}}(\sqrt{\omega_i}\hat{\xi}_i - \frac{1}{\sqrt{\omega_i}}\frac{\partial}{\partial\xi_i})$; $\hat{a}_i = \frac{1}{\sqrt{2}}(\sqrt{\omega_i}\hat{\xi}_i + \frac{1}{\sqrt{\omega_i}}\frac{\partial}{\partial\xi_i})$ allow us to write the Hamiltonian in the form $H = \sum_{i=1}^{n}\omega_i(\hat{a}_i^+\hat{a}_i + 1/2)$. The operators $\hat{a}_i^+, \hat{a}_i (1 \le i \le n)$ satisfy the relations

$$[\hat{a}_i^+, \hat{a}_i^+] = [\hat{a}_i, \hat{a}_j] = 0; \quad [\hat{a}_i, \hat{a}_j^+] = \delta_{ij}$$

(these relations, called the canonical commutation relations, will be explored further in a later section). The ground state Φ of the Hamiltonian H satisfies the relations $a_i\Phi = 0$ for all $i = 1,\ldots,n$. We can obtain the rest of the stationary states from Φ by means of the operators $\hat{a}_i^+$, i.e. the stationary states have the form $\hat{a}_1^{+v_1}a_2^{+v_2}\ldots a_n^{+v_n}\Phi$, where $v_1,\ldots,v_n$ are non-negative integers; the corresponding energies are $(v_1 + \frac{1}{2})\omega_1 + \cdots + (v_n + \frac{1}{2})\omega_n$.

The Hamiltonian H' for a multidimensional anharmonic oscillator has the form $H' = H + V$, where H is the Hamiltonian of harmonic oscillator and $V = \sum_{k_1,\ldots,k_n} c_{k_1,\ldots,k_n}\hat{x}_1^{k_1}\ldots\hat{x}_n^{k_n}$. Similar to H, the Hamiltonian H' is also conveniently represented in terms of the operators $\hat{a}_i^+, \hat{a}_i$. We obtain

$$H' = \mathrm{const} + \sum_{i=1}^{n}\omega_i\hat{a}_i^+\hat{a}_i$$

$$+ \sum_{k_1,\ldots,k_m,l_1,\ldots,l_n}\Gamma_{k_1,\ldots,k_m|l_1,\ldots,l_n}\hat{a}_1^{+k_1}\ldots\hat{a}_n^{+k_n}\hat{a}_1^{l_1}\ldots\hat{a}_n^{l_n},$$

where the coefficients $\Gamma_{k|l}$ can be expressed in terms of $c_{k_1,\ldots,k_n}$.

In general, one cannot calculate explicitly the stationary states of the Hamiltonian H' and other physical quantities associated with this Hamiltonian. However, in the case that the anharmonic terms V can be considered as a small perturbation, there exist techniques that allow us to calculate the expansions of physical quantities associated with H' as a power series with respect to the small parameter. These techniques (Feynman diagrams) are described in Section 6.4.

Chapter 3

Quantum Mechanics of a System
of Identical Particles

3.1 A system of n identical particles

In Section 2.3, we mentioned that two electrons are identical particles and that a system of two electrons cannot be described by means of the prescriptions of Section 2.3. If the system of two electrons were specified by a wave function $\psi(\xi_1, \xi_2)$, where $\xi_i = (\mathbf{r}_i, s_i)$ stands for the coordinate variable $\mathbf{r}_i$ and the spin variable s_i, then we could ask about the physical meaning of the number $\sum_{s_1, s_2} |\psi(\mathbf{r}_1, s_1, \mathbf{r}_2, s_2)|^2 = p(\mathbf{r}_1, \mathbf{r}_2)$. It would be natural to consider this number as the probability, or more precisely as the probability density, of the first electron occupying the point $\mathbf{r}_1$ and the second electron occupying the point $\mathbf{r}_2$. However, the first and second electrons are identical, therefore we can only talk about the probability of finding one of these electrons at the point $\mathbf{r}_1$ and the other electron at the point $\mathbf{r}_2$. So we should make some changes and the right way to make these changes is as follows. For the wave function of two electrons, we should consider only square-integrable functions $\psi(\xi_1, \xi_2) = \psi(\mathbf{r}_1, s_1, \mathbf{r}_2, s_2)$ that are anti-symmetric with respect to the variables ξ_1, ξ_2 (i.e. $\psi(\xi_2, \xi_1) = -\psi(\xi_1, \xi_2)$). Then, it is obvious that $p(\mathbf{r}_1, \mathbf{r}_2) = p(\mathbf{r}_2, \mathbf{r}_1)$ and that this quantity makes sense as the probability density measuring the probability of finding one of the electrons at the point $\mathbf{r}_1$ and the other at $\mathbf{r}_2$. For the system of two protons, two neutrons, or two μ-mesons, the situation is precisely the same. However, for the system of two π_0-mesons, we should consider as wave functions only functions $\psi(\mathbf{r}_1, \mathbf{r}_2)$ that are symmetric with respect to the variables $\mathbf{r}_1, \mathbf{r}_2$.

Let us now give precise definitions.

All particles are divided into two classes — bosons and fermions.

The space F_n of the states of the systems of n identical bosons (fermions) is a Hilbert space of square-integrable functions of n variables $\psi(\xi_1, \ldots, \xi_n)$ that are symmetric (antisymmetric) with respect to the variables $(\xi_1, \ldots, \xi_n)$ (here, $\xi_i \in X$, where X is a measure space). To emphasize that we are dealing with bosons, we use the notation F_n^s for the space F_n and in the case of fermions, we use the notation F_n^a. The space $L^2(X^n) = L^2(X \times \cdots \times X)$ will be denoted by $\mathcal{B}_n$. The spaces F_n^s and F_n^a are subspaces of this space.

Experiments show that particles with half-integer spin (electrons, protons, and so on) are fermions and particles with integer spin (π-mesons, photons, and so on) are bosons. In relativistic quantum theories, this statement can be derived from theoretical considerations.

It is easy to check that the space F_n is completely determined by the Hilbert space F_1 of states of one particle and the number n (i.e. it doesn't depend on the representation of the space F_1 of one particle in the form $L^2(X)$). Indeed, as we have mentioned already in Section 2.3, the isomorphism α of the spaces $L^2(X)$ and $L^2(Y)$ can be naturally extended to an isomorphism α_n of the spaces $L^2(X^n) = L^2(X \times \cdots \times X)$ and $L^2(Y^n) = L^2(Y \times \cdots \times Y)$. It is easy to check that the isomorphism α_n transforms symmetric functions into symmetric functions and antisymmetric functions into antisymmetric functions.

If $\mathcal{B}$ is a Hilbert space of states of one particle, then the space of states of n identical particles F_n can be described as an nth symmetric tensor power of the space $\mathcal{B}$ in the case of bosons and an nth antisymmetric tensor power of the space $\mathcal{B}$ in the case of fermions.

Let us define the operators of symmetrization P_s and antisymmetrization P_a in the space $\mathcal{B}_n = L^2(X^n)$ by means of the formulas

$$P_s\psi(\xi) = \frac{1}{n!}\sum_\pi \psi(\pi(\xi)),$$

$$P_a\psi(\xi) = \frac{1}{n!}\sum_\pi (-1)^{\gamma_\pi}\psi(\pi(\xi))$$

(here, $\xi = (\xi_1,\ldots,\xi_n) \in X^n$, π is a permutation $(i_1,\ldots,i_n)$ of the indices $1,\ldots,n$, with $\pi(\xi) = (\xi_{i_1},\ldots,\xi_{i_n}) \in X^n$, γ_π is the parity of the permutation π, and $\sum_\pi$ is a sum over all permutations π). In particular, for $n = 2$,

$$P_s\psi(\xi_1,\xi_2) = \frac{1}{2}(\psi(\xi_1,\xi_2) + \psi(\xi_2,\xi_1)),$$

$$P_a\psi(\xi_1,\xi_2) = \frac{1}{2}(\psi(\xi_1,\xi_2) - \psi(\xi_2,\xi_1)).$$

It is clear that $P_s\mathcal{B}_n = F_n^s$; $P_a\mathcal{B}_n = F_n^a$; $P_s\psi = \psi$ if $\psi \in F_n^s$, and $P_a\psi = \psi$ if $\psi \in F_n^a$.

It is easy to check that P_s and P_a are orthogonal projections of the space $\mathcal{B}_n$ onto the subspaces F_n^s and F_n^a, respectively.

Let us denote by H an operator on $\mathcal{B}_n$ that commutes with operators P_s and P_a. Then, F_n^s and F_n^a are H-invariant subspaces. If $\psi \in \mathcal{B}_n$ is an eigenfunction of the operator H, then $P_s\psi$ and $P_a\psi$ (if they are non-zero) are also eigenfunctions of the operator H that belong to subspaces F_n^s and F_n^a, respectively, with the same eigenvalue.

Let us consider a system of n non-interacting identical particles, i.e. a system having a Hamiltonian transforming the function $\psi(\xi_1,\ldots,\xi_n)$ into the function

$$\check{\psi}(\xi_1,\ldots,\xi_n) = \sum_{i=1}^n \int A(\xi_i,\eta_i)\psi(\xi_1,\ldots,\xi_{i-1},\eta_i,\xi_{i+1},\ldots,\xi_n)d\eta_i,$$

$$(3.1)$$

in particular, a single-particle Hamiltonian H_1 transforms the function $\psi(\xi)$ into the function $\check{\psi}(\xi) = \int A(\xi,\eta)\psi(\eta)d\eta$ (i.e.

the generalized function $A(\xi, \eta)$ is a kernel of the operator H_1). Formula (3.1) specifies an operator H_n on the space $\mathcal{B}_n$, commuting with the operators P_s and P_a, hence this operator transforms the subspaces F_n^s and F_n^a into themselves.[1] The operators on the spaces F_n^s and F_n^a, defined by relation (3.1), will be denoted by H_n^s and H_n^a, respectively. These operators are the Hamiltonians of systems of n non-interacting bosons or fermions.

Let us assume, for definiteness, that the single-particle Hamiltonian H_1 has discrete spectrum, and let us denote by ϕ_n the complete orthonormal system of its eigenvectors, with corresponding eigenvalues E_n. Then it is obvious that the functions $\phi_{k_1}(\xi_1) \ldots \phi_{k_n}(\xi_n)$ constitute a complete orthonormal system of eigenvectors of the operator H_n in the space $\mathcal{B}_n$ (see Section 2.3). It follows that $P_s \phi_{k_1}(\xi_1) \ldots \phi_{k_n}(\xi_n)$ are eigenfunctions of the Hamiltonian H_n^s and belong to the space F_n^s. Similarly, $P_a \phi_{k_1}(\xi_1) \ldots \phi_{k_n}(\xi_n)$ are eigenvectors of the Hamiltonian H_n^a and belong to F_n^a. The corresponding eigenvalues, in both cases, are equal to $E_{k_1} + \cdots + E_{k_n}$. In the case of fermions, the functions $P_a \phi_{k_1}(\xi_1) \ldots \phi_{k_n}(\xi_n)$ should only be considered in the case of distinct indices $k_1, \ldots, k_n$ because in the case when a pair of indices coincides, the antisymmetrization gives zero.[2] It is easy to check that the functions $P_s \phi_{k_1}(\xi_1) \ldots \phi_{k_n}(\xi_n)$ form a complete system of functions in F_n^s and the functions $P_a \phi_{k_1}(\xi_1) \ldots \phi_{k_n}(\xi_n)$ form a complete system in F_n^a. These functions are not normalized, however, one can check that two functions belonging to one of these systems are either orthogonal or proportional. In the coordinate representation, the space of states of one particle is realized as the space $L^2(E^3 \times B)$, where B is a finite set, therefore the space of states of n identical particles can be considered as a space of square-integrable symmetric (antisymmetric) functions $\psi(\mathbf{r}_1, i_1, \ldots, \mathbf{r}_n, i_n)$

---

[1]More precisely, the definition of the operator H_n can be given as follows: H_n is a self-adjoint operator on $\mathcal{B}_n$ that transforms functions $\phi_1(\xi_1) \ldots \phi_n(\xi_n)$, where the functions ϕ_i belong to the domain of the operator H_1, into the function $\sum_{i=1}^{n} \phi_1(\xi_1) \ldots \phi_{i-1}(\xi_{i-1})(H_1\phi_i)(\xi_i)\phi_{i+1}(\xi_{i+1}) \ldots \phi_n(\xi_n)$. Such an operator is unique.

[2]In other words, in the system of non-interacting identical fermions, two fermions cannot be found in the same state. This statement is called the Pauli exclusion principle.

depending on the coordinate variables $\mathbf{r}_1, \ldots, \mathbf{r}_n \in E^3$ and discrete variables $i_1, \ldots, i_n \in B$ (the functions ψ, in the case of bosons, are invariant with respect to simultaneous permutation of coordinates $\mathbf{r}_l$ and $\mathbf{r}_m$ and discrete variables i_l and i_m, while in the case of fermions, this function changes sign under such permutation of variables).

The operator of total momentum $\mathbf{P}_n$ and the Hamiltonian H_n of a system of n non-relativistic identical particles can be written precisely in the same way as specified in Section 2.3 ((2.1) and (2.2)); the only difference is that the masses of identical particles are equal and the potential energy $\mathcal{V}_n(\mathbf{r}_1, \ldots, \mathbf{r}_n)$ should be a symmetric function of the coordinates. We will always assume that

$$\mathcal{V}_n(\mathbf{r}_1, \ldots, \mathbf{r}_n) = \sum_{j=1}^{n} \mathcal{V}(\mathbf{r}_j) + \sum_{1 \leq j \leq l \leq n} \mathcal{W}(|\mathbf{r}_j - \mathbf{r}_l|)$$

(the first sum corresponds to the potential energy of particles in an external field and the second sum corresponds to the interaction). It is easy to see that the operators $\mathbf{P}_n$ and H_n transform symmetric functions and antisymmetric functions into antisymmetric functions, i.e. they can be considered as operators on the spaces F_n^s and F_n^a. The operators $\hat{\mathbf{P}}_j = \frac{1}{i}\frac{\partial}{\partial \mathbf{r}_j}$ do not have this property; this means that for the identical particles, we cannot speak of the momentum of the jth particle.

In the case of non-interacting particles, the operator of potential energy has the form $\mathcal{V}_n(\mathbf{r}_1, \ldots, \mathbf{r}_n) = \mathcal{V}(\mathbf{r}_1) + \cdots + \mathcal{V}(\mathbf{r}_n)$; it was shown earlier that the stationary states of n particles, in this case, can be expressed in terms of stationary states of a single particle.

3.2 Fock space

In Section 3.1, we fixed a measure space X. We have considered Hilbert spaces F_n^s and F_n^a of square-integrable symmetric and antisymmetric functions $\psi(\xi_1, \ldots, \xi_n)$, depending on n variables $\xi_1, \ldots, \xi_n \in X$. It is often convenient to consider the direct sums

$$F^s = F_0^s + F_1^s + \cdots + F_n^s + \cdots,$$

$$F^a = F_0^a + F_1^a + \cdots + F_n^a + \cdots.$$

Here, F_0^s and F_0^a are spaces of constant functions (i.e. one-dimensional spaces).

The spaces F^s and F^a are called Fock spaces. The elements of Fock spaces represent the states of a system of identical bosons or fermions in the case when the number of particles is not fixed. Speaking of bosons and fermions simultaneously, we will use the notation F for one of the spaces F^s or F^a. Vectors in the space F^s or F^a can be considered as sequences $(f_0, f_1, \ldots, f_n, \ldots)$, where $f_n \in F_n^s$ (or correspondingly $f_n \in F_n^a$), satisfying the condition $\sum_{n=0}^\infty \|f_n\|^2 < \infty$. In other words, the elements of the space F can be considered as column vectors

$$
f = \begin{bmatrix} f_0 \\ f_1(\xi_1) \\ \vdots \\ f_n(\xi_1, \ldots, \xi_n) \\ \vdots \end{bmatrix}
$$

of symmetric (antisymmetric) functions $f_n(\xi_1, \ldots, \xi_n)$ as entries; they should satisfy the condition[3]

$$
\sum_{n=0}^\infty \int |f_n(\xi_1, \ldots, \xi_n)|^2 d\xi_1 \ldots d\xi_n < \infty.
$$

Linear combination and scalar product of two such column vectors are defined in the natural way, in particular, the scalar product of f and g is

$$
\langle f, g \rangle = \sum_{n=0}^\infty \int f_n(\xi_1, \ldots, \xi_n)\overline{g_n(\xi_1, \ldots, \xi_n)} d\xi_1 \ldots d\xi_n.
$$

The spaces F_n are naturally embedded in the Fock space F (to the vector $f \in F_n$, there corresponds a sequence $(f_0, \ldots, f_n, \ldots) \in F$, where $f_n = f, f_k = 0$ for $k \neq n$).

Let us introduce the operator of the number of particles

$$
N = N_0 + N_1 + \cdots + N_n + \cdots,
$$

---

[3]Such column vectors are called *Fock states*.

where N_n is an operator on the space F_n, multiplying every vector by the number of particles n. The spaces F_n are eigenspaces of the operator N.

The operator of total momentum $\mathbf{P}$ and the Hamiltonian of the system of non-relativistic particles H in the Fock space are defined as direct sums of operators $\mathbf{P}_n$ and H_n introduced in Section 3.1

$$\mathbf{P} = \mathbf{P}_0 + \mathbf{P}_1 + \cdots + \mathbf{P}_n + \cdots ,$$

$$H = H_0 + H_1 + \cdots + H_n + \cdots$$

(the operators $\mathbf{P}$ and H act on Fock space, constructed starting with the measure space $E^3 \times B$).

In non-relativistic quantum mechanics, the operators of all observables leave the spaces $F_n \subset F$ invariant (i.e. they commute with the operator of the number of particles N). However, even in the non-relativistic case, it is very convenient to express the operators of physical quantities in terms of auxiliary operators that do not commute with N. Namely, we will introduce the operators $a(f)$ and $a^+(f)$ that can be considered as the operators of annihilation and creation of a particle with the wave function $f \in L^2(X)$.

First of all, let us define the operator $a_n(f)$, where $f \in L^2(X)$, acting on the space F_n into the space F_{n-1}. Namely, we will assume that the operator $a_n(f)$ transforms a function $\phi(\xi_1, \ldots, \xi_n) \in F_n$ into the function $\check{\phi}(\xi_1, \ldots, \xi_{n-1}) \in F_{n-1}$, defined by the formula

$$\check{\phi}(\xi_1, \ldots, \xi_{n-1}) = \sqrt{n} \int \phi(\xi_1, \ldots, \xi_{n-1}, \xi) f(\xi) d\xi$$

(the operator $a_0(f)$ transforms the space F_0 into zero).

The operator $a_n^+(f)$, defined on the space F_{n-1} and taking values in the space F_n, can be described as the operator adjoint to $a_n(f)$. It is easy to calculate that the operator $a_n^+(f)$ transforms the function $\phi(\xi_1, \ldots, \xi_{n-1})$ into the function $\sqrt{n}P^s(\phi(\xi_1, \ldots, \xi_{n-1})\overline{f(\xi_n)})$ in the bosonic case and into the function $\sqrt{n}P^a(\phi(\xi_1, \ldots, \xi_{n-1})\overline{f(\xi_n)})$ in the fermionic case.

Let us consider a subset D of the space F, consisting of finite sequences $(\phi_0, \ldots, \phi_n, \ldots)$ (in other words, D is the smallest linear subspace that contains all subspaces F_n). The operators $a(f)$ and

$a^+(f)$, where $f \in L^2(X)$, will be defined on the space D, namely, $a(f)$ transforms the sequence $(\phi_0, \ldots, \phi_n, \ldots) \in D$ into the sequence

$$(a_1(f)\phi_1, a_2(f)\phi_2, \ldots, a_{n+1}(f)\phi_{n+1}, \ldots)$$

and the operator $a^+(f)$, transforms the same sequence into

$$(0, a_1^*(f)\phi_0, \ldots, a_n^*(f)\phi_{n-1}, \ldots).$$

It is obvious that these operators transform the subspace D into itself and that they are Hermitian conjugate on D.

It is easy to check (see Section 6.1) that in the case of fermions, the operators $a(f), a^+(f)$ are bounded ($\|a(f)\| = \|f\|$) and therefore can be extended by continuity to the whole Fock space F; however, in the bosonic case, the operators $a(f), a^+(f)$ are not bounded ($\|a_n(f)\| = \sqrt{n}\|f\|$).

By means of straightforward calculations, it is easy to find the commutation relations for the operators $a(f), a^+(f)$. In the bosonic case, the operators satisfy the conditions

$$[a(f), a(g)] = [a^+(f), a^+(g)] = 0; [a(f), a^+(g)] = \langle f, g \rangle \qquad (3.2)$$

(these relations are called the *canonical commutation relations* or CCR). In the fermionic case, we have similar conditions for the anticommutators

$$[a(f), a(g)]_+ = [a^+(f), a^+(g)]_+ = 0; [a(f), a^+(g)]_+ = \langle f, g \rangle \qquad (3.3)$$

(these relations are called the *canonical anticommutation relations* or CAR).

The operators $a(f)$ depend linearly on $f \in L^2(X)$. This means that they can be considered as operator generalized functions on X. In other words, we can introduce the symbols $a(x), a^+(x)$, that are related to the operators $a(f), a^+(f)$ by the following formulas:

$$a(f) = \int f(x)a(x)dx,$$

$$a^+(f) = \int \overline{f(x)}a^+(x)dx.$$

The symbols $a(x), a^+(x)$ are not operators. The integral $\int f(x) a(x)dx$ is considered simply as another notation for the operators $a(f)$.

Note, however, that in principle, the symbol $a(x)$ can be considered as an operator on some subset of the space F and the symbol $a^+(x)$ can be considered as an operator acting from the space F into some larger space (see Appendix A.7).

In what follows, we will consider the expression of the form

$$A = \int f(x_1, \ldots, x_m | y_1, \ldots, y_n) a^+(x_1) \ldots a^+(x_m) a(y_1) \ldots$$

$$\times a(y_n) d^m x d^n y, \tag{3.4}$$

where f is a generalized function. Clearly, such expressions define an operator on D if the function f takes the form

$$f_1(x_1) \ldots f_m(x_m) g_1(y_1) \ldots g_n(y_n), \tag{3.5}$$

where $f_i, g_i \in L^2(X)$ (then one should assume that $A = a^+(\overline{f_1}) \ldots a^+(\overline{f_m}) a(g_1) \ldots a(g_n)$). This definition can be extended to the case when the function f is a finite sum of products of the form (3.5). One can prove easily that under this condition the operator A transforms a sequence $\phi = (\phi_0, \ldots, \phi_k, \ldots) \in D$, into the sequence $\psi = (\psi_0, \ldots, \psi_k, \ldots) \in D$, where

$$\psi_k(\xi_1, \ldots, \xi_k) = \sqrt{\frac{k!}{(k-m)!} \cdot \frac{(k-m+n)!}{(k-m)!}}$$

$$\times P \int \phi_{n-m+k}(\xi_1, \ldots, \xi_{k-m}, x_1, \ldots, x_n)$$

$$\times f(\xi_{k-m+1}, \ldots, \xi_k | x_1, \ldots, x_n) d^n x \tag{3.6}$$

(P denotes the operator of symmetrization in the bosonic case and antisymmetrization in the fermionic case).

We can use the formula (3.6) to define the operator A for an arbitrary function f. The domain of the operator A, in this case, consists of all sequences $\phi \in D$, for which the functions $\psi_k(\xi_1, \ldots, \xi_k)$, obtained by the formula (3.6), are square-integrable. Considering expressions of the form

$$A = \sum_{m,n}^{\infty} \int A_{m,n}(x_1, \ldots, x_m | y_1, \ldots, y_n) a^+(x_1)$$

$$\ldots a^+(x_m) a(y_1) \ldots a(y_n) d^m x d^n y, \tag{3.7}$$

we define every summand using the consideration above and the sum of the series we understand in terms of strong convergence. The representation of the operator A in the form (3.7) is called the *representation in normal form* (in the expression (3.7), the creation operators stand to the left of the operators of annihilation).

Every bounded operator can be represented in the form (3.7) if the convergence of the series is understood as weak convergence.

The vector θ, specified by the sequence $(1, 0, 0, \ldots)$, plays a special role in the Fock space F. This vector corresponds to the state that contains no particles ($\theta \in F_0$) and is called the *vacuum vector*; it satisfies the condition $a(f)\theta = 0$ for all functions $f \in L^2(X)$. It is easy to check that the vector $\theta \in F$ is a cyclic vector with respect to the family of operators $a^+(f)$ (in other words, linear combinations of the vectors of the form $a^+(f_1) \ldots a^+(f_s)\theta$ are dense in F). One can write down an explicit expression of every vector $\phi = (\phi_0, \ldots, \phi_n, \ldots) \in F$ in terms of the generalized operator functions $a^+(x)$ and the vector θ, namely, by applying the formula (3.6), we see that

$$f = \sum_{n=0}^{\infty} \frac{1}{\sqrt{n!}} \int f_n(\xi_1, \ldots, \xi_n) a^+(\xi_1) \ldots a^+(\xi_n)\theta d^n \xi. \tag{3.8}$$

Let us consider the simplest operators on the Fock space. Let us start with operators of the form

$$\int A(x, y) a^+(x) a(y) dx dy. \tag{3.9}$$

The operator (3.9) transforms the sequence $(\phi_0, \ldots, \phi_k, \ldots) \in D$ into the sequence $(\psi_0, \ldots, \psi_k, \ldots)$, where

$$\psi_k(\xi_1, \ldots, \xi_k) = kP \int \phi_k(\xi_1, \ldots, \xi_{k-1}, x) A(\xi_k, x) dx. \tag{3.10}$$

It is easy to check that in the case when the function $A(x, y)$ is a kernel of a self-adjoint operator on $L^2(X)$, the operator (3.9) is essentially self-adjoint on F. Let us denote by A the self-adjoint extension of the operator $\int A(x, y)a^+(x)a(y)dxdy$. The operator A commutes with the operator N, hence, it transforms the space F_n into itself. The operator induced by the operator A on the space F_n will be denoted by A_n. Operators of the form A_n were considered earlier, namely, the Hamiltonian of a system of n non-interacting identical particles can be written in this form. In Section 3.1, we have found eigenvectors and eigenvalues of the operator A_n in the assumption that the operator A_1 is a self-adjoint operator with discrete spectrum in $L^2(X)$. Using this result, we can obtain the description of eigenvectors and eigenvalues of the self-adjoint operator A. Namely, the vectors $\prod_{i=1}^{\infty}(a^+(\overline{\phi_i}))^{n_i}\theta$ constitute a complete system of eigenvectors of the operator A and the corresponding eigenvalues are equal to $\sum_{i=1}^{\infty} n_i E_i$ (here, $\phi_1, \ldots, \phi_i, \ldots$ is the complete system of eigenfunctions of the operator A_1 and $E_1, \ldots, E_n, \ldots$ are the corresponding eigenvalues; the numbers n_i constitute a finite sequence. In the fermionic case, the numbers n_i are equal to 0 or 1 and in the bosonic case $n = 0, 1, 2, \ldots$). The proof can be reduced to the remark that the vector $a^+(\overline{f_1}) \ldots a^+(\overline{f_n})\theta$ is equal to the vector that corresponds in Fock space to the function $Pf_1(\xi_1) \ldots f_n(\xi_n)\theta \in F_n$, up to a constant factor (recall that F_n is naturally embedded in F).

The operators

$$B = \int B(x_1, x_2|y_1, y_2)a^+(x_1)a^+(x_2)a(y_1)a(y_2)dx_1dx_2dy_1dy_2$$

$$(3.11)$$

also commute with the operator of the number of particles (as well as the operators of the form (3.4), obeying $m = n$). This means that the operator B transforms the subspace F_n into itself and it follows from formula (3.6) that after restricting the operator B to the subspace F_n, we obtain an operator B_n transforming the vector ϕ into function

$$\psi_k(\xi_1, \ldots, \xi_k) = k(k-1)P \int \phi_k(\xi_1, \ldots, \xi_{k-2}, x_1, x_2)$$

$$\times B(\xi_{k-1}, \xi_k|x_1, x_2)dx_1dx_2. \tag{3.12}$$

Now we can write down the operators $N, \mathbf{P}$, and H in the normal form.

The operator N can be represented as

$$N = \int a^+(\xi)a(\xi)d\xi \tag{3.13}$$

(or, more precisely, the operator N can be expressed by the formula (3.9), where $A(\xi, \eta) = \delta(\xi, \eta)$). This follows immediately from the relation (3.10), which implies that the operator N transforms the sequence $(\phi_0, \ldots, \phi_k, \ldots) \in D$ into the sequence $(\psi_0, \ldots, \psi_k, \ldots)$, where

$$\psi_k(\xi_1, \ldots, \xi_k) = kP \int \phi_k(\xi_1, \ldots, \xi_{k-1}, x)\delta(\xi_k, x)dx$$

$$= kP\phi_k(\xi_1, \ldots, \xi_k) = k\phi_k(\xi_1, \ldots, \xi_k).$$

The momentum operator $\mathbf{P}$ and the Hamiltonian H of a system of non-relativistic identical particles act on Fock space, constructed starting from the measure space $E^3 \times B$. The operators acting on this space can be expressed in terms of the operator generalized functions $a^+(\mathbf{x}, s), a(\mathbf{x}, s)$ in the coordinate representation and in terms of the operator generalized functions $a^+(\mathbf{k}, s), a(\mathbf{k}, s)$ in the momentum representation (here, $\mathbf{x}, \mathbf{k} \in E^3, s \in B$). In the following, we will sometimes use the notation $a_s^+(\mathbf{k}) = a^+(\mathbf{k}, s), a_s(\mathbf{k}) = a(\mathbf{k}, s)$. Again, using the relation (3.10), it is easy to derive that in the coordinate representation, the operator $\mathbf{P}$ is

$$\mathbf{P} = \sum_s \frac{1}{i} \int a^+(\mathbf{x}, s)\frac{\partial}{\partial \mathbf{x}}a(\mathbf{x}, s)d\mathbf{x}, \tag{3.14}$$

or, more precisely,

$$P = \sum_{s,s'} \int A(\mathbf{x}, s, \mathbf{y}, s')a^+(\mathbf{x}, s)a(\mathbf{x}, s')d\mathbf{x}d\mathbf{y},$$

where $A(\mathbf{x}, s, \mathbf{y}, s') = \delta_s^{s'}\frac{1}{i}\frac{\partial}{\partial \mathbf{x}}\delta(\mathbf{x} - \mathbf{y})$. In the momentum representation, the operator $\mathbf{P}$ takes the form

$$\mathbf{P} = \sum_s \int \mathbf{k}a^+(\mathbf{k}, s)a(\mathbf{k}, s)d\mathbf{k}. \tag{3.15}$$

It follows from the relations (3.10) and (3.12) that the Hamiltonian H of the system of non-relativistic identical particles can be represented as a sum of operators of the form (3.9) and (3.11). Namely, in the coordinate representation,

$$H = \sum_s \int a^+(\mathbf{x}, s) \left(-\frac{\Delta}{2m} \right) a(\mathbf{x}, s) dx$$

$$+ \sum_s \mathcal{V}(\mathbf{x}) a^+(\mathbf{x}, s) a(\mathbf{x}, s) dx + \sum_s \frac{1}{2} \int \mathcal{W}(|\mathbf{x}_1 - \mathbf{x}_2|) a^+(\mathbf{x}_1, s)$$

$$\times a^+(\mathbf{x}_2, s) a(\mathbf{x}_2, s) a(\mathbf{x}_1, s) d\mathbf{x}_1 d\mathbf{x}_2, \tag{3.16}$$

and in the momentum representation,

$$H = \sum_s \int \frac{k^2}{2m} a^+(\mathbf{k}, s) a(\mathbf{k}, s) dk$$

$$+ \sum_s \tilde{\mathcal{V}}(\mathbf{k}_1 - \mathbf{k}_2) a^+(\mathbf{k}_1, s) a(\mathbf{k}_2, s) d\mathbf{k}_1 d\mathbf{k}_2$$

$$+ \sum_s \frac{1}{2} \int \tilde{\mathcal{W}}(\mathbf{k}_1 - \mathbf{k}_4) \delta(\mathbf{k}_1 + \mathbf{k}_2 - \mathbf{k}_3 - \mathbf{k}_4)$$

$$\times a^+(\mathbf{k}_1, s) a^+(\mathbf{k}_2, s) a(\mathbf{k}_3, s) a(\mathbf{k}_4, s) d^4\mathbf{k}, \tag{3.17}$$

where $\tilde{\mathcal{V}}(\mathbf{k}), \tilde{\mathcal{W}}(\mathbf{k})$ are the Fourier transforms of the functions $\mathcal{V}(\mathbf{x}), \mathcal{W}(\mathbf{x})$.

It is easy to construct examples of expressions of the form (3.7) that do not define an operator in Fock space. For example, the expression

$$A = \int \alpha(x_1, \ldots, x_m) a^+(x_1) \ldots a^+(x_m) dx_1 \ldots dx_m \tag{3.18}$$

can define an operator on Fock space only in the case when the function α is square integrable. Namely, by the general definition, the operator defined by the expression (3.8) should transform the sequence $(\phi_0, \ldots, \phi_n, \ldots) \in D$ into the sequence $(\psi_0, \ldots, \psi_n, \ldots)$,

where

$$\psi_n(\xi_1,\ldots,\xi_n) = \sqrt{\frac{n!}{(n-m)!}}\, P\phi_{n-m}(\xi_1,\ldots,\xi_{n-m})\alpha(\xi_{n-m+1},\ldots,\xi_n).$$

If the function α is square integrable, then the functions ψ_n are also square integrable, hence the expression (3.18) specifies an operator A on the whole space D and this operator transforms D into itself. However, if the function α is not square integrable, then the function ψ_n may be square integrable only under the condition $\phi_{n-m} \equiv 0$. This means that the domain of the operator specified by the expression (3.18) contains only the zero vector; in other words, the expression (3.18) does not specify any operator because the domain of the operator should be dense in Fock space.

Let us consider the Fock space $F(L^2(E^3))$, constructed starting with the measure space E^3, and operators defined by expressions of the form

$$A = \sum_{m,n}^{m+n\leq s} \int A_{m,n}(\mathbf{k}_1,\ldots,\mathbf{k}_m|\mathbf{l}_1,\ldots,\mathbf{l}_n)$$

$$\times\, \delta(\mathbf{k}_1+\cdots+\mathbf{k}_m-\mathbf{l}_1-\cdots-\mathbf{l}_n)$$

$$\times\, a^+(\mathbf{k}_1)\ldots a^+(\mathbf{k}_m)a(\mathbf{l}_1)\ldots a(\mathbf{l}_n)d^m\mathbf{k}d^n\mathbf{l} \tag{3.19}$$

(operators of this form commute with the momentum operator). Let us suppose that the functions $A_{m,n}$ belong to the space $\mathcal{S}(E^{3(m+n)})$ of smooth, rapidly decreasing functions (faster than any power function). Let us single out the subspace $\mathcal{S}_\infty \subset F$ that consists of sequences $(\phi_0,\ldots,\phi_k,\ldots) \in D$, obeying $\phi_k \in \mathcal{S}(E^{3k})$. In what follows, it will be convenient to consider operators on the space $F(L^2(E^3))$ only on the set $\mathcal{S}_\infty$. By means of the relation (3.6), it is easy to check that in the case when $A_{m,0} \equiv 0$, the operator specified by the expression (3.19) is well defined on all elements of the set $\mathcal{S}_\infty$. This operator transforms every sequence in $\mathcal{S}_\infty$ into a sequence belonging to the same set. If one of the functions $A_{m,0}$ is non-zero, then the expression (3.19) cannot define an operator on Fock space because the function $A_{m,0}(\mathbf{k}_1,\ldots,\mathbf{k}_m)\delta(\mathbf{k}_1+\cdots+\mathbf{k}_m)$ is not square integrable.

Let us consider the fermionic case. In this case, we can introduce the notion of even vectors as vectors belonging to a direct sum of subspaces F_{2n} and odd vectors as vectors belonging to a direct sum of subspaces F_{2n+1}. An operator on Fock space is called parity-preserving if it transforms an even vector to an even vector and an odd vector to an odd vector. Introducing an involution τ, transforming a sequence $(\phi_0, \phi_1, \dots, \phi_{2n}, \phi_{2n+1}, \dots)$ into a sequence $(\phi_0, -\phi_1, \dots, \phi_{2n}, -\phi_{2n+1}, \dots)$, we can say that an even vector is invariant with respect to this involution, $\tau x = x$, and an odd vector satisfies the condition $\tau x = -x$. A parity-preserving operator commutes with this involution. An operator represented in the form (3.7) (normal form) is parity preserving if the functions $A_{m,n}$ do not vanish only in the case when the numbers m and n have the same parity.

Operators corresponding to physical quantities should be parity preserving. In particular, the Hamiltonian is always a parity-preserving operator.

Chapter 4

Operators of Time Evolution
$S(t, t_0)$ and $S_\alpha(t, t_0)$

4.1 Non-stationary perturbation theory

Let us consider a Hamiltonian $H(t)$ having the form $H(t) = H_0 + gV(t)$. Let us assume that we can calculate the operator of evolution $U_0(t, t_0) = \exp(-iH_0(t - t_0))$ corresponding to the Hamiltonian H_0. We will solve the problem of finding the evolution operator $U(t, t_0)$ from the Hamiltonian $H(t)$ as a series with respect to the parameter g. It is convenient to introduce the operator $S(t, t_0) = \exp(iH_0 t)U(t, t_0)\exp(-iH_0 t_0)$, expressed in terms of the operator $U(t, t_0)$, and calculate it. We will find out later that the operator $S(t, t_0)$ is very important on its own. In the case when we should emphasize that the operators $U(t, t_0)$ and $S(t, t_0)$ depend on the parameter g, we will use the notation $U(t, t_0|g)$ and $S(t, t_0|g)$.

The operator $U(t, t_0)$ satisfies equation (1.1). Using this equation, we can easily obtain the following equation for the operator $S(t, t_0)$:

$$i\frac{\partial S(t, t_0)}{\partial t} = g\tilde{V}(t)S(t, t_0), \tag{4.1}$$

where $\tilde{V}(t) = \exp(iH_0 t)V(t)\exp(-iH_0 t)$; the initial condition is specified in the form $S(t_0, t_0) = 1$.

In some physical situations, the operator H_0 can be considered as a *free Hamiltonian* (i.e. it describes non-interacting particles), and the operator $H - H_0 = gV(t)$ is the *interaction Hamiltonian*. This terminology is also commonly used in quantum field theory. However, in quantum field theory, as we will see later, there exists no natural partition of the Hamiltonian into the free and interaction parts. We will not use these terms in order to avoid misleading associations.

We will express $S(t, t_0)$ as a power series in the variable g as follows:

$$S(t, t_0) = \sum_{n=0}^{\infty} g^n S_n(t, t_0).$$

Substituting this series into the equation for the operator $S(t, t_0)$, we obtain the recurrence relation

$$i\frac{\partial S_n(t, t_0)}{\partial t} = \tilde{V}(t)S_{n-1}(t, t_0).$$

Using the initial condition $S(t_0, t_0) = 1$, we see that $S_0(t_0, t_0) = 1$ and $S_n(t_0, t_0) = 0$ for $n \geq 1$, hence

$$S_n(t, t_0) = \frac{1}{i} \int_{t_0}^{t} \tilde{V}(\tau)S_{n-1}(\tau, t_0)d\tau. \tag{4.2}$$

The differential equation for S with an initial condition is equivalent to the integral equation

$$S(t, t_0) = 1 + g \int_{t_0}^{t} \tilde{V}(\tau)S(\tau, t_0)d\tau,$$

one can solve the integral equation using the method of iterations.

From (4.2) or from the integral equation, we can conclude that

$$S_1(t, t_0) = \frac{1}{i} \int_{t_0}^{t} \tilde{V}(\tau)d\tau,$$

$$S_2(t, t_0) = \left(\frac{1}{i}\right)^2 \int_{t_0}^{t} d\tau_1 \int_{t_0}^{\tau_1} d\tau_2 \tilde{V}(\tau_1)\tilde{V}(\tau_2),$$

$$\vdots$$

$$S_n(t, t_0) = \left(\frac{1}{i}\right)^n \int_{t_0}^{t} d\tau_1 \int_{t_0}^{\tau_1} d\tau_2 \cdots \int_{t_0}^{\tau_{n-1}} d\tau_n \tilde{V}(\tau_1)\ldots\tilde{V}(\tau_n).$$

We can also write

$$S_n(t, t_0) = \left(\frac{1}{i}\right)^n \int_{\Gamma_n} d\tau_1 \ldots d\tau_n \tilde{V}(\tau_1) \ldots \tilde{V}(\tau_n),$$

where the domain of integration Γ_n is defined by $t \geq \tau_1 \geq \cdots \geq \tau_n \geq t_0$.

We introduce the following notation:

$$T(\tilde{V}(\tau_1) \ldots \tilde{V}(\tau_n)) = \tilde{V}(\tau_{i_1}) \ldots \tilde{V}(\tau_{i_n}),$$

where $i_1, \ldots, i_n$ is a permutation of the indices $1, \ldots, n$, satisfying the relation $\tau_{i_1} \geq \cdots \geq \tau_{i_n}$. In other words, $T(\tilde{V}(\tau_1) \ldots \tilde{V}(\tau_n))$ (*chronological* or a T-*product* of the operators $\tilde{V}(\tau_1) \ldots \tilde{V}(\tau_n)$) is defined as the product of the operators $\tilde{V}(\tau_1), \ldots, \tilde{V}(\tau_n)$ in order of decreasing time τ_i. If some of the times τ_i are identical, then the permutation for which $\tau_{i_1} \geq \cdots \geq \tau_{i_n}$ is not unique, however, it is easy to see that the T-product does not depend on this choice of permutation. Using the T-product, we can write $S_n(t, t_0)$ in the form

$$S_n(t, t_0) = \frac{1}{n!} \left(\frac{1}{i}\right)^n \int_{t_0}^{t} \cdots \int_{t_0}^{t} T(\tilde{V}(\tau_1) \ldots \tilde{V}(\tau_n)) d\tau_1 \ldots d\tau_n.$$

Indeed, the domain of integration in this integral can be partitioned into $n!$ regions G_P, with each region G_P corresponding to a permutation $P = (j_1, \ldots, j_n)$ of the indices $1, \ldots, n$. Here, G_P is the region singled out by the inequalities $t \geq \tau_{j_1} \geq \cdots \geq \tau_{j_n} \geq t_0$. It is clear that

$$\left(\frac{1}{i}\right)^n \int_{G_P} T(\tilde{V}(\tau_1) \ldots \tilde{V}(\tau_n)) d\tau_1 \ldots d\tau_n$$

$$= \left(\frac{1}{i}\right)^n \int_{G_P} \tilde{V}(\tau_{j_1}) \ldots \tilde{V}(\tau_{j_n}) d\tau_1 \ldots d\tau_n,$$

and that this integral, after a change of variables, is reduced to the integral $S_n(t, t_0)$. This means, that the integral $\frac{1}{n!}\left(\frac{1}{i}\right)^n \int_{t_0}^{t} \cdots \int_{t_0}^{t} T(\tilde{V}(\tau_1) \ldots \tilde{V}(\tau_n)) d\tau_1 \ldots d\tau_n$ splits into $n!$ identical integrals, equal to $S_n(t, t_0)$; this gives the needed formula. The full operator

$S(t, t_0)$ can be written as the series

$$S(t, t_0) = \sum_{n=0}^{\infty} \frac{1}{n!} \left(\frac{1}{i}\right)^n g^n \int_{t_0}^{t} \cdots \int_{t_0}^{t} T(\tilde{V}(\tau_1) \ldots \tilde{V}(\tau_n)) d\tau_1 \ldots d\tau_n.$$

$$(4.3)$$

This series can be compactly written in the form

$$S(t, t_0) = T \exp\left(\frac{1}{i} g \int_{t_0}^{t} \tilde{V}(\tau) d\tau\right), \qquad (4.4)$$

which is called T-*exponential*. If the operator $V(t)$ is bounded for every t and is continuous (in the sense of strong limits) in t, then the series (4.3) is convergent in norm (hence, in this case, the existence of solutions of equations (1.1) and (4.1) follows). Indeed, we have that $\sup_{t_0 \leq \tau \leq t} \|\tilde{V}(\tau)\| = \sup_{t_0 \leq \tau \leq t} \|V(\tau)\| < \infty$ and $\|S_n(t, t_0)\| \leq \frac{1}{n!} (\sup_{t_0 \leq \tau \leq t} \|V(\tau)\|)^n |t - t_0|^n$. In the case when it is not possible to prove the convergence of the series (4.3), the corresponding results are conditional (if a series expansion in g is possible, then the series takes the form (4.3)). Regardless of the possibility of the series expansion in g, one can prove the following relation:

$$\frac{\partial^n S(t, t_0 | g)}{\partial g^n}\Big|_{g=0} = \left(\frac{1}{i}\right)^n \int_{t_0}^{n} \cdots \int_{t_0}^{t} T(\tilde{V}(\tau_1) \ldots \tilde{V}(\tau_n)) d\tau_1 \ldots d\tau_n.$$

If the Hamiltonian H can be represented in the form $H_0 + V$, where H_0 and V do not depend on time, then it is convenient to consider the Hamiltonian $H_\alpha(t) = H_0 + \exp(-\alpha|t|)V$. We introduce the operator $S_\alpha(t, t_0) = \exp(iH_0 t) U_\alpha(t, t_0) \exp(-iH_0 t_0)$, where $U_\alpha(t, t_0)$ is the operator of evolution, constructed with $H_\alpha(t)$.

In a similar way, we can define the operator $S_\alpha(t, t_0)$ as the solution of the differential equation

$$i\frac{\partial S_\alpha(t, t_0)}{\partial t} = \exp(-\alpha|t|)\tilde{V}(t) S_\alpha(t, t_0)$$

with the initial condition $S_\alpha(t_0, t_0) = 1$ (here, $\tilde{V}(t) = \exp(iH_0 t) V \exp(-iH_0 t)$). Following the discussion above, we can write $S_\alpha(t, t_0)$

in the form of the T-exponent:

$$S_\alpha(t, t_0) = T \exp\left(\frac{1}{i} \int_{t_0}^t \exp(-\alpha|\tau|)\tilde{V}(\tau)d\tau\right).$$

The operator $S_\alpha(\infty, -\infty) = \operatorname{slim}_{\substack{t \to \infty \\ t_0 \to -\infty}} S_\alpha(t, t_0)$, denoted by S_α, plays a special role and is called the *adiabatic S-matrix*. Two other important operators $S_\alpha(0, \pm\infty) = \operatorname{slim}_{t \to \pm\infty} S_\alpha(0, t)$ will be denoted $S_{\alpha+}$, and $S_{\alpha-}$ and are called the *adiabatic Møller operators*.

In the case when H_0 can be considered a free Hamiltonian and V is the interaction, considering the Hamiltonian $H_\alpha(t)$ with $\alpha \to 0$ describes adiabatic (infinitesimally slow) turning on and off the interaction; this explains the above-introduced terminology. In place of the Hamiltonian $H = H_0 + V$, it is often useful to consider the family of Hamiltonians $H_g = H_0 + gV$, where g is a constant, which is typically called the coupling constant. The operators $S_\alpha(t, t_0), S_\alpha(\infty, -\infty), S_\alpha(0, \pm\infty)$, constructed with the Hamiltonian H_g, will be denoted by $S_\alpha(t, t_0|g), S_\alpha(\infty, -\infty|g) = S_\alpha(g)$, and $S_\alpha(0, \pm\infty|g) = S_{\alpha\pm}(g)$.

4.2 Stationary states of Hamiltonians depending on a parameter

Let us consider the family of Hamiltonians $H(g)$, where $0 \le g \le g_0$. We will assume that for every g, there exists a normalized eigenvector ϕ_g of the operator $H(g)$, with eigenvalue $E(g)$, which is differentiable with respect to the parameter g. Without loss of generality, we can assume that

$$\left\langle \phi_g, \frac{d\phi_g}{dg} \right\rangle = 0.$$

Indeed, if this condition is not satisfied, one can always find a phase factor $\exp(i\alpha(g))$, such that the vector $\tilde{\phi}_g = \exp(-i\alpha(g))\phi_g$ satisfies the condition: namely, we can choose $\alpha(g) = \frac{1}{i} \int_0^g \left\langle \phi_\lambda, \frac{d\phi_\lambda}{d\lambda} \right\rangle d\lambda$ (the function $\alpha(g)$ is real, since from the relation $\langle \phi_\lambda, \phi_\lambda \rangle = 1$, one can obtain that

$$\left\langle \phi_\lambda, \frac{d\phi_\lambda}{d\lambda} \right\rangle + \left\langle \frac{d\phi_\lambda}{d\lambda}, \phi_\lambda \right\rangle = \left\langle \phi_\lambda, \frac{d\phi_\lambda}{d\lambda} \right\rangle + \overline{\left\langle \phi_\lambda, \frac{d\phi_\lambda}{d\lambda} \right\rangle} = 0).$$

By differentiating in g the relation

$$H(g)\phi_g = E(g)\phi_g,$$

we obtain

$$(H(g) - E(g))\frac{d\phi_g}{dg} = -\frac{dH(g)}{dg}\phi_g + \frac{dE(g)}{dg}\phi_g. \qquad (4.5)$$

Taking the scalar product of this relation with ϕ_g, we obtain

$$\frac{dE(g)}{dg} = \left\langle \frac{dH(g)}{dg}\phi_g, \phi_g \right\rangle. \qquad (4.6)$$

Let us consider the commonly occurring case of $H(g) = H_0 + gV$; then, it is clear that

$$\frac{dE(g)}{dg} = \langle V\phi_g, \phi_g \rangle, \qquad (4.7)$$

$$(H_0 + gV - E(g))\frac{d\phi_g}{dg} = (\langle V\phi_g, \phi_g \rangle - V)\phi_g. \qquad (4.8)$$

We will sometimes assume that the vector ϕ_g is analytic with respect to g in the neighborhood around $g = 0$, and we will search for an expansion of the vector ϕ_g and the eigenvalue $E(g)$ in a series in g (stationary perturbation theory). Formulas (4.7) and (4.8), when $g = 0$, provide the linear terms in g (first-order terms), namely,

$$E(g) = E(0) + g\langle V\phi_0, \phi_0 \rangle + \cdots,$$

$$\phi_g = \phi_0 + g\psi + \cdots,$$

where ψ satisfies the relation

$$(H_0 - E(0))\psi = (\langle V\phi_0, \phi_0 \rangle - V)\phi_0. \qquad (4.9)$$

Given that the vector ϕ_g satisfies the equation $\left\langle \phi_g, \frac{d\phi_g}{dg} \right\rangle = 0$, we obtain the following condition on the vector ψ:

$$\langle \psi, \phi_0 \rangle = 0. \qquad (4.10)$$

In the case when $E(0)$ is a simple eigenvalue of the Hamiltonian H_0, then equations (4.9) and (4.10) give an unambiguous definition of the vector ψ. The derivation of the relations (4.7) and (4.8) is rigorous if all the operators $H_0 + gV$ have the same domain.

We did not consider the more subtle question of the existence of eigenvectors ϕ_g that depend on the parameter g smoothly. With mild assumptions, the answer to this question is given by the following theorem [Kato, 2013].

Let H_0 denote a self-adjoint operator and V a Hermitian operator with domain that contains the domain of H_0. Let us suppose that E_0 is a simple isolated eigenvalue of the operator H_0 and ϕ_0 is the corresponding eigenvector. Then for sufficiently small g: (1) the operator $H(g) = H_0 + gV$ is a self-adjoint operator with the same domain as the operator H_0; (2) there exists an eigenvector ϕ_g of the operator $H(g)$, which is analytic in g and is equal to ϕ_0 for $g = 0$; (3) the corresponding eigenvalue $E(g)$ also depends analytically on g and is simple; (4) the relations (4.7) and (4.8) are valid.

We do not wish to give a direct calculation of the higher terms with respect to g in the series for ϕ_g and $E(g)$. Instead, we will prove the formula that allows us to get the decomposition of eigenvectors and eigenvalues in a series in g from non-stationary perturbation theory. Let us suppose that for $0 \le g \le g_0$, the self-adjoint operators $H(g) = H_0 + gV$ have the same domain and $E(g)$ is an isolated eigenvalue of the operator $H(g)$ that depends continuously on the parameter g in the interval $0 \le g \le g_0$, and ϕ_g is the eigenvector corresponding to this eigenvalue, obeying $\left\langle \phi_g, \frac{d\phi_g}{dg} \right\rangle = 0$. Then

$$\phi_g = \lim_{a \to 0} \exp\left(i\frac{C(g)}{\alpha} \right) S_\alpha(0, -\infty|g)\phi_0,$$

$$\phi_g = \lim_{a \to 0} \exp\left(i\frac{C(g)}{\alpha} \right) S_\alpha(0, +\infty|g)\phi_0, \tag{4.11}$$

where $C(g) = \int_0^g \frac{E(\lambda)-E(0)}{\lambda}d\lambda$ and $S_\alpha(0, \mp\infty|g)$ are adiabatic Møller matrices, constructed with the pair of operators $H(g)$, H_0. This statement is proven in Section 4.3.

4.3   Adiabatic variation of stationary state

Let us suppose that the Hamiltonian $H(t)$ varies slowly (adiabatically) with time.

Let us consider a solution of the Schrödinger equation (1.2) with initial condition $\psi_{t_0} = \phi_0$, where ϕ_0 is a stationary state of the Hamiltonian $H(t_0)$ corresponding to a non-degenerate energy level. We will show that for every t this solution is very close to the stationary state of the Hamiltonian $H(t)$. (When we are talking about the stationary state of Hamiltonian $H(t)$, we always assume that we consider an eigenvector of $H(t)$ for a fixed t.) For definiteness, we restrict ourselves to the case when we consider a family of Hamiltonians H_g that depend on the parameter g, and the Hamiltonian $H(t)$ is defined by the formula $H(t) = H_{\alpha t}$, where α is a small positive number. Then equation (1.2) can be reduced to the equation

$$i\alpha \frac{d\psi(g)}{dg} = H_g \psi(g), \tag{4.12}$$

by means of the change of variables $g = \alpha t$. In what follows, we will consider equation (4.12).

Let us assume that H_g is a family of self-adjoint operators with the same domain D, smoothly depending on the parameter g in the interval $g_0 \leq g \leq g_1$ (i.e. for any $x \in D$, the vector $H_g x$ smoothly depends on g). We will assume that for every g in the interval $g_0 \leq g \leq g_1$, there exists a stationary state ϕ_g of the Hamiltonian H_g, smoothly depending on the parameter g. The energy level $E(g)$, corresponding to the state ϕ_g, will be supposed isolated and non-degenerate. The state ϕ_g is assumed to be normalized and obeying $\langle \phi_g, \frac{d\phi_g}{dg} \rangle = 0$. Finally, we will assume that $\frac{d^n \phi_g}{dg^n}|_{g=g_0} = 0$ for $n = 1, 2, \ldots$.

Lemma 4.1. *Assuming the conditions above, the solution of equation (4.12) that coincides with ϕ_{g_0} with $g = g_0$ can be written in the form*

$$\psi(g) = \exp\left[-\frac{i}{\alpha}C(g)\right](\phi_g + \alpha s_g + \alpha^2 r(g, \alpha)),$$

where $C(g) = \int_0^g E(\lambda)d\lambda$, s_g is defined by the relation

$$i\frac{d\phi_g}{dg} = (H_g - E(g))s_g, \left\langle \frac{ds_g}{dg}, \phi_g \right\rangle = 0, s_{g_0} = 0,$$

and the norm of the vector $r(g, \alpha)$ is bounded above by a constant that does not depend on g or α (here, g lies in the interval $g_0 \leq g \leq g_1$).

To prove this lemma, let us first use the change of variables $\psi(g) = \exp(\frac{i}{\alpha}C(g))\sigma(g)$ to transform equation (4.12) to the form

$$i\alpha\frac{d\sigma(g)}{dg} = (H_g - E(g))\sigma(g). \tag{4.13}$$

Let us assume that the solution of equation (4.13) has the form

$$\sigma(g) = \sum_{n=0}^{\infty} \alpha^n \sigma_n(g). \tag{4.14}$$

By comparing terms with equal powers of α, we obtain the relations

$$(H_g - E(g))\sigma_0(g) = 0, \tag{4.15}$$

$$i\frac{d\sigma_0(g)}{dg} = (H_g - E(g))\sigma_1(g),$$

$$\vdots$$

$$i\frac{d\sigma_{n-1}(g)}{dg} = (H_g - E(g))\sigma_n(g), \tag{4.16}$$

$$\sigma_n(g_0) = 0 \quad \text{for } n \geq 1. \tag{4.17}$$

Furthermore, taking the scalar product of these relations with ϕ_g, we obtain

$$\left\langle \frac{d\sigma_0(g)}{dg}, \phi_g \right\rangle = 0, \tag{4.18}$$

$$\left\langle \frac{d\sigma_1(g)}{dg}, \phi_g \right\rangle = 0, \tag{4.19}$$

$$\vdots$$

$$\left\langle \frac{d\sigma_n(g)}{dg}, \phi_g \right\rangle = 0. \tag{4.20}$$

Equation (4.15) and relation (4.18) are satisfied for $\sigma_0(g) = \phi_g$. It then follows that $\sigma_1(g)$ can be found from relations (4.16) and (4.19); we also obtain that $\sigma_1(g) = s_g$. By the recursive application of relations (4.17) and (4.20), we can find $\sigma_n(g)$. Note that the $\sigma_n(g)$ are specified by these relations uniquely (this follows from the fact that the eigenvalue $E(g)$ is non-degenerate and isolated) and are smooth in g. From equation $\frac{d^n \phi_g}{dg^n}\big|_{g=g_0} = 0$, it follows that all the above relations can be satisfied.

Let us show now that

$$\sigma(g) = \sum_{n=0}^{N} \alpha^n \sigma_n(g) + \alpha^{N+1} r_N(g, \alpha), \tag{4.21}$$

where $|r_N(g, \alpha)| \leq K$. Indeed, by inserting (4.21) into (4.14), we obtain the following equations for $r_N(g, \alpha)$:

$$i\alpha \frac{\partial r_N(g, \alpha)}{\partial g} = (H_g - E(g))r_N(g, \alpha) - i\frac{d\sigma_N(g)}{dg} \tag{4.22}$$

with initial condition $r_N(g_0, \alpha) = 0$. Let us denote by $V(g)$ the unitary operator defined by the relation

$$i\alpha \frac{dV(g)}{dg} = (H_g - E(g))V(g),$$

$$V(g_0) = 1.$$

Now, we can make a change of variables $r_N(g, \alpha) = V(g)\rho_N(g, \alpha)$ in the equation (4.22). We obtain

$$i\alpha \frac{\partial \rho_N(g, \alpha)}{\partial g} = -iV^{-1}(g)\frac{d\sigma_N(g)}{dg},$$

hence

$$\|r_N(g, \alpha)\| = \|\rho_N(g, \alpha)\| \leq \int_{g_0}^{g} \left\|\frac{d\rho_N(g', \alpha)}{dg'}\right\| dg'$$

$$\leq \frac{1}{\alpha} \int_{g_0}^{g} \left\|\frac{d\sigma_N(g')}{dg'}\right\| dg' \leq \frac{\text{const}}{\alpha}. \tag{4.23}$$

We can now see that

$$r_N(g, \alpha) = \sigma_N(g) + \alpha r_{N+1}(g, \alpha).$$

Applying inequality (4.23) to r_{N+1}, we can derive the following inequality for r_N:

$$\|r_N(g, \alpha)\| \le \|\sigma_N(g)\| + \alpha \frac{\text{const}}{\alpha} \le \text{const}.$$

Since $r(g, \alpha) = r_1(g, \alpha)$, the above estimate with $N = 1$ confirms Lemma 4.1.

Remark 4.1. In the above discussion, the condition of smooth dependence of H_g and ϕ_g on g can be relaxed to the requirement of three-fold differentiability; the condition $\frac{d^n \phi_g}{dg^n}\big|_{g=g_0} = 0$ can be relaxed to hold only for $n = 1$.

Remark 4.2. If we do not require that $\frac{d\phi_g}{dg}\big|_{g=g_0} = 0$, then it is clear, from the above proof, that the proof of Lemma 4.1 applies to the solution of equation (4.12), satisfying the initial condition $\psi_{g_0} = \phi_{g_0} + \alpha s_{g_0}$.

We will now show how to obtain the relation (4.11) using Lemma 4.1. Let us assume that the family of Hamiltonians H_g satisfy the conditions of Lemma 4.1, with the exception of the requirement that $\frac{d^n \phi_g}{dg^n}\big|_{g=g_0} = 0$ for $n \ge 1$. Let us consider the operator of evolution $U_\alpha(t, T)$ constructed with the Hamiltonian $H_\alpha(t) = H_{h(\alpha t)}$, where $h(\tau)$ is a smooth function defined on $\tau \le 0$ and taking values in the interval $[g_0, g_1]$. Let us further assume that $h(\tau) = g_0$ for $\tau \le a$. Then the family of Hamiltonians $\tilde{H}_\lambda = H_{h(\lambda)}$ with λ in $[a, 0]$ and the vectors $\tilde{\phi}_\lambda = \phi_{h(\lambda)}$ satisfy the conditions of Lemma 4.1, including the condition $\frac{d^n \tilde{\phi}_\lambda}{d\lambda^n}\big|_{\lambda=a} = 0$ (the latter condition follows from the relation $\frac{d^n h(\lambda)}{d\lambda^n}\big|_{\lambda=a} = 0$). Applying the statement of this lemma, we see that

$$\left\| U_\alpha\left(0, \frac{a}{\alpha}\right) \tilde{\phi}_a - \exp\left(-\frac{1}{\alpha} C\right) \tilde{\phi}_0 \right\| \to 0,$$

where $C = \int_a^0 E(h(\lambda))d\lambda$. We obtain

$$\phi_{h(0)} = \tilde{\phi}_0 = \lim_{a \to 0} \exp\left(\frac{i}{\alpha}C\right) U_\alpha\left(0, \frac{a}{\alpha}\right)\phi_{h(a)}. \qquad (4.24)$$

Introducing the operator

$$S_\alpha(t, T) = \exp(iH_{g_0}t)U_\alpha(t, T)\exp(-iH_{g_0}T),$$

we can rewrite the equation (4.14) in the form

$$\phi_{h(0)} = \lim_{a \to 0} \exp\left\{\frac{i}{\alpha}\int_a^0 [E(h(\lambda)) - E(h(a))]d\lambda\right\} S_\alpha\left(0, \frac{a}{\alpha}\right)\phi_{h(a)}.$$
$$(4.25)$$

Since $S_\alpha(\frac{a}{\alpha}, T) = 1$ for $T < \frac{a}{\alpha}$ and $h(\lambda) = h(a)$ for $\lambda < a$, using the relation (4.25), we can show that

$$\phi_{h(0)} = \lim_{a \to 0} \exp\left\{\frac{i}{\alpha}\int_{-\infty}^0 [E(h(\lambda)) - E(h(-\infty))]d\lambda\right\}$$
$$\times S_\alpha\left(0, -\infty\right)\phi_{h(-\infty)}. \qquad (4.26)$$

To obtain (4.11) from (4.26), we must choose for the family of Hamiltonians H_g the family $H_0 + gV$ with $0 \leq g \leq g_1$, and for $h(\tau)$ choose the function $g\exp(-\alpha|\tau|)$. The function $\exp(-\alpha|\tau|)$ does not vanish for $\tau \ll 0$ as required of the function $h(\tau)$, therefore, strictly speaking, we cannot use equation (4.26). However, a slight modification of the above proof, based on Remark 4.2, allows us to verify (4.11).

Remark 4.3. If the family of Hamiltonians H_g also depends on another parameter Ω, then it is not difficult to outline the conditions under which the limit in (4.26) is uniform in Ω (for this, it is necessary to give uniform in Ω estimates in the proof of the lemma). In particular, if $H_g^\Omega = H_0^\Omega + gV^\Omega$ $(0 \leq g \leq g_1, \Omega \in \mathcal{O})$, then the limit in (4.26) will be uniform in g and in Ω if the norm of the operators V^Ω is bounded by a constant not depending on Ω and it is possible to find a δ in such a way that the interval $(E^\Omega(g) - \delta, E^\Omega(g) + \delta)$ for any Ω and g contains no eigenvalues of the operator H_g^Ω except for $E^\Omega(g)$ (see Tyupkin and Shvarts, 1972).

Chapter 5

The Theory of Potential Scattering

5.1 Formal scattering theory

Let H and H_0 be two self-adjoint operators on the space R. The *Møller matrices* S_+ and S_- of the operator pair (H, H_0) are defined as strong limits

$$S_- = \operatorname*{slim}_{t \to -\infty} \exp(itH)\exp(-itH_0) = \operatorname*{slim}_{t \to -\infty} S(0, t), \qquad (5.1)$$

$$S_+ = \operatorname*{slim}_{t \to +\infty} \exp(itH)\exp(-itH_0) = \operatorname*{slim}_{t \to +\infty} S(0, t). \qquad (5.2)$$

The operators S_- and S_+ are isometries, as strong limits of unitary operators, however, they are not necessarily unitary. If they are unitary, one can check that

$$S_-^* = \operatorname*{slim}_{t \to -\infty} \exp(itH_0)\exp(-itH),$$

$$S_-^* = \operatorname*{slim}_{t \to +\infty} \exp(itH_0)\exp(-itH)$$

(see Appendix A.5).

The *S-matrix* of the pair of operators (H, H_0) is defined by the formula

$$S = S_+^* S_-.$$

If the operators S_+ and S_- are unitary, then the S-matrix is also unitary and can be written in the form

$$S = \operatorname*{slim}_{\substack{t \to \infty \\ t_0 \to -\infty}} S(t, t_0), \qquad (5.3)$$

51

where

$$S(t, t_0) = \exp(iH_0 t) \exp(-iH(t - t_0)) \exp(-iH_0 t_0).$$

However, the S-matrix can sometimes be unitary when the Møller matrices are not unitary. Namely, for the S-matrix to be unitary, it is sufficient (and necessary) to assume that the ranges of the operators S_- and S_+ coincide: $S_- R = S_+ R$.

In what follows, we will see that, under certain conditions, one can use the S-matrix to describe the process of scattering; in these conditions, it is usually possible to prove that the S-matrix is unitary.

Let us assume that the operators H and H_0 have the same domain; we will denote by the symbol V the difference $H - H_0$. Let us prove a simple sufficient condition for the existence of Møller matrices (Cook's condition).

If the integral $\int_0^\infty \|V \exp(-iH_0 t)x\| dt$ converges for all vectors x in a dense set $T \subset R$, then the Møller matrices S_+ and S_- exist.

Proof. Consider the vector $\Phi_x(t) = \exp(iHt) \exp(-iH_0 t)x$. Note that

$$\left\| \frac{d\Phi_x(t)}{dt} \right\| = \|i \exp(iHt) V \exp(-iH_0 t)x\| = \|V \exp(-iH_0 t)x\|.$$

Hence, we have that

$$\|\Phi_x(t_1) - \Phi_x(t_2)\| = \left\| \int_{t_1}^{t_2} \frac{d\Phi_x(t)}{dt} dt \right\| \leq \int_{t_1}^{t_2} \left\| \frac{d\Phi_x(t)}{dt} \right\| dt$$

$$= \int_{t_1}^{t_2} \|V \exp(-iH_0 t)x\| dt$$

and therefore, it follows from our assumptions that for $x \in T$, we have

$$\lim_{t_1, t_2 \to \pm\infty} \|\Phi_x(t_1) - \Phi_x(t_2)\| = 0.$$

This means that the limits of $\Phi_x(t)$ for $t \to \pm\infty$ exist for $x \in T$ and therefore, the limits exist for all $x \in R$ (see Appendix A.5).

Let us consider the relation of Møller matrices $S_\pm$ and the S-matrices S to the adiabatic Møller matrices $S_{\alpha\pm}$ and the adiabatic

S-matrices S_α defined in Section 4.1. Let us assume that for any t and t_0,

$$\mathop{\mathrm{slim}}_{\alpha \to 0} S_\alpha(t, t_0) = S(t, t_0).$$

It is easy to check that this condition is satisfied if the operator V is bounded (in this case, we even have convergence in norm); the condition is also satisfied in a much larger class of situations.

We will now show that the Cook condition implies

$$\mathop{\mathrm{slim}}_{\alpha \to 0} S_{\alpha+} = S_+, \tag{5.4}$$

$$\mathop{\mathrm{slim}}_{\alpha \to 0} S_{\alpha-} = S_-. \tag{5.5}$$

If the operator S_+ is unitary, then the relations (5.4) and (5.5) imply

$$\mathop{\mathrm{slim}}_{\alpha \to 0} S_\alpha = S. \tag{5.6}$$

To prove these statements, let us consider the vectors

$$\Phi_x^\alpha(t) = S_\alpha(0, t)x = S_\alpha^*(t, 0)x,$$

where $x \in T$. It is clear that

$$\left\| \frac{d\Phi_x^\alpha(t)}{dt} \right\| = \|iU_\alpha(0, t)V \exp(-\alpha|t|) \exp(-iH_0 t)x\|$$

$$= \exp(-\alpha|t|)\|V \exp(-iH_0 t)x\|,$$

from which it follows that

$$\|\Phi_x^\alpha(t_1) - \Phi_x^\alpha(t_2)\| \le \int_{t_1}^{t_2} \|V \exp(-iH_0 t)x\| \exp(-\alpha|t|)dt$$

$$\le \int_{t_1}^{t_2} \|V \exp(-iH_0 t)x\|dt.$$

It follows from this inequality that the limits $\Phi_x^\alpha(\pm\infty)$ exist, hence the operators $S_{\alpha\pm}$ also exist. Furthermore, it follows from this inequality that the limit

$$\lim_{t\to\pm\infty} \Phi_x^\alpha(t) = \Phi_x^\alpha(\pm\infty)$$

is uniform in α; this allows us to take the limit $\alpha \to 0$ under the limit sign. Taking this limit, we obtain, for $x \in T$, the relation

$$\lim_{\alpha\to 0} S_{\alpha\pm}x = \lim_{\alpha\to 0}\lim_{t\to\pm\infty} \Phi_x^\alpha(t) = \lim_{t\to\pm\infty}\lim_{\alpha\to 0} S_\alpha(0,t)x$$

$$= \lim_{t\to\pm\infty} S(0,t)x = S_\pm x,$$

implying that the strong limit of $S_{\alpha\pm}$ for $\alpha \to 0$ equals $S_\pm$.

If V is a bounded operator, then

$$\|\Phi_x^\alpha(t_1) - \Phi_x^\alpha(t_2)\| \le \|V\| \cdot \|x\| \cdot \left|\int_{t_1}^{t_2} \exp(-\alpha|t|)dt\right|.$$

This inequality implies that in the relations (5.4)–(5.6), one can talk about norm convergence (instead of strong convergence), hence the operators $S_{\alpha+}$, $S_{\alpha-}$, S_α are unitary.

Taking the limit of t to $\pm\infty$ in the identity,

$$\exp(iH\tau)S(0,t) = S(0,t+\tau)\exp(iH_0\tau),$$

we obtain the important relation

$$\exp(iH\tau)S_\pm = S_\pm \exp(iH_0\tau),$$

which implies

$$HS_\pm = S_\pm H_0, \quad H_0 S = SH_0. \tag{5.7}$$

If $S_\pm$ are unitary operators, then this relation establishes the *unitary equivalence* between H and H_0. It follows that if, for example, H_0 does not have a discrete spectrum but H does, then the operators S_+ and S_- are not unitary (it is easy to check that the ranges of the operators S_+ and S_- are orthogonal to eigenvectors in the discrete spectrum).

Let ϕ_λ be a complete system of δ-normalized, generalized eigenfunctions of the operator H_0 and E_λ the corresponding energy levels (for simplicity of notation, let us assume that the operator H_0 has no discrete spectrum). Then the functions $\psi_\lambda^\pm = S_\pm \phi_\lambda$ are generalized eigenfunctions of the operator H with the same eigenvalues E_λ. This follows from the relation

$$H\psi_\lambda^\pm = HS_\pm\phi_\lambda = S_\pm H_0\phi_\lambda = E_\lambda S_\pm\phi_\lambda = E_\lambda\psi_\lambda^\pm.$$

The matrix elements of S-matrices in the basis ϕ_λ can be easily expressed in terms of the functions $\psi_\lambda^\pm$, namely,

$$\langle S\phi_\lambda, \phi_\mu\rangle = \langle S_+^* S_-\phi_\lambda, \phi_\mu\rangle = \langle S_-\phi_\lambda, S_+\phi_\mu\rangle = \langle \psi_\lambda^-, \psi_\mu^+\rangle.$$

Therefore, it is convenient to obtain the equations for the functions $\psi_\lambda^\pm$. With this goal in mind, consider the operator

$$\Sigma_{\pm\epsilon} = \pm\epsilon \int_0^{\pm\infty} \exp(-\epsilon|t|) S(0,t) dt.$$

It is easy to check that in the case when the operators S_+ and S_- exist, we have

$$\lim_{\epsilon\to+0} \Sigma_{\pm\epsilon} = S_\pm.$$

For the proof, it is sufficient to use the relation

$$\lim_{\epsilon\to+0} \pm\epsilon \int_0^{\pm\infty} \exp(-\epsilon|t|) f(t) dt = f(\pm\infty), \qquad (5.8)$$

which holds if the vector function $f(t)$ is bounded and has the limit $f(\pm\infty) = \lim_{t\to\pm\infty} f(t)$.[1] It is clear that $\psi_\lambda^\pm = \lim_{\epsilon\to+0} \psi_\lambda^{\pm\epsilon}$, where

$$\psi_\lambda^{\pm\epsilon} = \Sigma_{\pm\epsilon}\phi_\lambda = \pm\epsilon \int_0^{\pm\infty} \exp(-\epsilon|t|) \exp(iHt) \exp(-iE_\lambda t)\phi_\lambda dt$$

$$= \frac{\pm i\epsilon}{H - E_\lambda \pm i\epsilon}\phi_\lambda.$$

---

[1] If $\|f(t)\| \le A$ and for $t \ge T$, we have that $\|f(t) - f(+\infty)\| \le \delta$, then

$$\left\| \epsilon \int_0^\infty \exp(-\epsilon t) f(t) dt - f(+\infty) \right\| = \epsilon \left\| \int_0^\infty \exp(-\epsilon t)(f(t) - f(+\infty)) dt \right\|$$

$$\le 2\epsilon T A + \delta.$$

We have obtained that $\psi_\lambda^{\pm\epsilon}$ satisfies the equation

$$(H - E_\lambda \pm i\epsilon)\psi_\lambda^{\pm\epsilon} = \pm i\epsilon\phi_\lambda,$$

which can be rewritten in the form

$$\phi_\lambda^{\pm\epsilon} = \phi_\lambda + (E_\lambda - H_0 \mp i\epsilon)^{-1}V\psi_\lambda^{\pm\epsilon}. \tag{5.9}$$

If the limit $\lim_{\epsilon\to 0}(E_\lambda - H_0 \pm i\epsilon)^{-1} = (E_\lambda - H_0 \pm i0)^{-1}$ exists in an appropriate sense, then one can take the limit $\epsilon \to 0$ in equation (5.9); we obtain the following equation for $\psi_\lambda^{\pm}$:

$$\psi_\lambda^{\pm} = \phi_\lambda + (E_\lambda - H_0 \mp i0)^{-1}V\psi_\lambda^{\pm}.$$

This equation is called the Lippman–Schwinger equation.[2] One can also express the matrix elements of the S-matrix in terms of the functions ψ_λ^{+} only (or of the functions ψ_λ^{-} only), namely,

$$\langle S\phi_\lambda, \phi_\mu \rangle = \langle \psi_\lambda^{-}, \psi_\mu^{+} \rangle = \delta(\lambda - \mu) - 2\pi i\delta(E_\lambda - E_\mu)\langle \phi_\lambda, V\psi_\mu^{+} \rangle$$

$$= \delta(\lambda - \mu) - 2\pi i\delta(E_\lambda - E_\mu)\langle V\psi_\lambda^{-}, \phi_\mu \rangle. \tag{5.10}$$

Indeed, using the Lippman–Schwinger equation, we can show that

$$\langle \psi_\lambda^{-}, \psi_\mu^{+} \rangle = \langle \phi_\lambda, \phi_\mu \rangle + \langle (E_\lambda - H_0 + i0)^{-1}V\psi_\lambda^{-}, \phi_\mu \rangle$$

$$+ \langle \phi_\lambda, (E_\mu - H_0 - i0)^{-1}V\psi_\mu^{+} \rangle$$

$$+ \langle V\psi_\lambda^{-}, (E_\lambda - H_0 - i0)^{-1}(E_\mu - H_0 - i0)^{-1}V\psi_\mu^{+} \rangle$$

$$= \delta(\lambda - \mu) + \langle V\psi_\lambda^{-}, (E_\lambda - iH_0 - i0)^{-1}\phi_\mu \rangle$$

$$+ \langle (E_\mu - H_0 + i0)^{-1}\phi_\lambda, V\psi_\mu^{+} \rangle$$

$$- \frac{1}{E_\lambda - E_\mu}(\langle \psi_\lambda^{-} - \phi_\lambda, V\psi_\mu^{+} \rangle - \langle V\psi_\lambda^{-}, \psi_\mu^{+} - \phi_\mu \rangle)$$

$$= \delta(\lambda - \mu) + \left(\frac{1}{E_\lambda - E_\mu + i0} - \frac{1}{E_\lambda - E_\mu} \right)\langle V\psi_\lambda^{-}, \phi_\mu \rangle$$

$$+ \left(\frac{1}{E_\mu - E_\lambda + i0} + \frac{1}{E_\lambda - E_\mu} \right)\langle \phi_\lambda, V\psi_\mu^{+} \rangle. \tag{5.11}$$

---

[2]We will not go further into the delicate question of specifying the precise meaning of the Lippman–Schwinger equation. The calculations at the end of this section are also informal.

In the above, we used the formula

$$(E_\lambda - H_0 - i0)^{-1}(E_\mu - H_0 - i0)^{-1} = -\frac{1}{E_\lambda - E_\mu}$$

$$\times [(E_\lambda - H_0 - i0)^{-1} - (E_\mu - H_0 - i0)^{-1}]. \tag{5.12}$$

Both parts of this formula can be seen as generalized operator functions of λ and μ. The numerical generalized function $\frac{1}{E_\lambda - E_\mu}$ can be understood as one of the functions $(E_\lambda - E_\mu + i0)^{-1}, (E_\lambda - E_\mu - i0)^{-1}$ or $P\frac{1}{E_\lambda - E_\mu}$ (all these functions coincide up to a summand of the form $c\delta(E_\lambda - E_\mu)$); formula (5.12) is correct in all cases, since

$$\delta(E_\lambda - E_\mu)[(E_\lambda - H_0 - i0)^{-1} - (E_\mu - H_0 - i0)^{-1}] = 0).$$

Replacing $\frac{1}{E_\lambda - E_\mu}$ with $\frac{1}{E_\lambda - E_\mu + i0}$ or $\frac{1}{E_\lambda - E_\mu - i0}$ in (5.11), we obtain the formula we wanted to prove. $\qquad\square$

5.2 Single-particle scattering

Let us consider the scattering of non-relativistic particles without spin in the potential field $\mathcal{V}(\mathbf{x})$ that decays to zero at infinity. The state space R in this case is the space $L^2(E^3)$ and the Hamiltonian has the form $H = \hat{\mathbf{p}}^2/2m + \mathcal{V}(\mathbf{x})$. When the particle is very far from the scattering center, we can describe the motion of the particle with the Hamiltonian $H_0 = \hat{\mathbf{p}}^2/2m$, assuming that the potential energy is equal to zero. Accordingly, the scattering of the particle in the field $\mathcal{V}(\mathbf{x})$ can be described by the scattering matrix S corresponding to the pair of operators (H, H_0).

Let us check that in the case when the potential $\mathcal{V}(\mathbf{x})$ is square-integrable, the Cook condition is satisfied, hence the Møller matrices exist. Note that for a dense set of functions $f \in R$, we have the inequality $|f_t(x)| \leq C|t|^{-3/2}$, where $f_t = \exp(-iH_0t)f$ (for the proof of this inequality, see Section 10.3, Lemma 10.5). This implies that the norm of the function $\psi_t = V \exp(-iH_0t)f$ does not exceed $C|t|^{-3/2}\sqrt{\int |\mathcal{V}(\mathbf{x})|^2 d\mathbf{x}}$, since $\psi_t(\mathbf{x}) = \mathcal{V}(\mathbf{x})f_t(\mathbf{x})$. This shows that the Cook condition is satisfied. One can also prove this condition with weaker assumptions.

If the potential $\mathcal{V}(\mathbf{x})$ is simultaneously square and absolutely integrable ($\mathcal{V} \in L^2 \cap L^1$), then one can show that the S-matrix is unitary. We will not give the proof of this non-trivial fact. We will only note that the unitarity of the S-matrix in different assumptions was considered in numerous papers. The proof of the theorem in the above formulation is given by Kuroda (1959).

Let us introduce the notation

$$S(\mathbf{p}, \mathbf{q}) = \langle S\phi_q, \phi_p \rangle,$$

where $\phi_{\mathbf{p}} = (2\pi)^{-3/2} \exp(i\mathbf{p}x)$ is a generalized eigenfunction of the momentum operator ($S(\mathbf{p}, \mathbf{q})$ can be considered as a kernel of the operator S in the momentum representation).

Since the operator S commutes with the operator H_0, the function $S(\mathbf{p}, \mathbf{q})$ takes the form

$$S(\mathbf{p}, \mathbf{q}) = S_1(\mathbf{p}, \mathbf{q})\delta(p^2 - q^2) = \frac{S_1(\mathbf{p}, \mathbf{q})}{2p}\delta(p - q).$$

The expression

$$d\sigma = \pi^2 |S_1(\mathbf{k}, \mathbf{p})|^2 d\omega, \tag{5.13}$$

where $d\omega = \sin\theta d\theta d\phi$ is the element of solid angle and $\mathbf{k}$ is a vector of the same length as $\mathbf{p}$ and directed at the angle $d\omega$ has the physical meaning of *differential cross-section*, it gives the number of outgoing particles in the solid angle $d\omega$ under the condition that the incoming particles have momentum $\mathbf{p}$ (we assume that the beam of incoming particles has unit flux).[3]

By the formula (5.10),

$$S(\mathbf{k}, \mathbf{p}) = \langle S\phi_{\mathbf{p}}, \phi_{\mathbf{k}} \rangle = \delta(\mathbf{k} - \mathbf{p}) - 2\pi i\delta\left(\frac{\mathbf{p}^2}{2m} - \frac{\mathbf{k}^2}{2m}\right) \langle V\psi_{\mathbf{p}}^-, \phi_{\mathbf{k}} \rangle,$$

and therefore the formula for the differential cross-section can be written in the form

$$d\sigma = 16m^2\pi^4 |\langle V\psi_{\mathbf{p}}^-, \phi_{\mathbf{k}} \rangle|^2 d\omega.$$

To understand the physical meaning of the quantity (5.13), let us assume that the incoming particles are described by the wave

---

[3] We consider only the scattering on non-zero angles.

function $\exp(-iH_0t)\phi$ with $t \to -\infty$ (in other words, we assume that the particles are described by the function ϕ at time $t = 0$ if the interaction with the potential field is neglected). From the definition of the S-matrix, it follows that for $t \to +\infty$, the wave function has the form $\exp(-iH_0t)S\phi$. In the momentum representation, we have

$$(S\phi)(\mathbf{p}) = \psi(\mathbf{p}) = \int S(\mathbf{p}, \mathbf{q})\phi(\mathbf{q})d\mathbf{q},$$

$$(\exp(-iH_0t)S\phi)(\mathbf{p}) = \exp\left(-i\frac{\mathbf{p}^2}{2m}t\right)\psi(\mathbf{p}),$$

hence the probability of having the momentum of outgoing particles directed in the solid angle Ω is equal to

$$\int_\Omega |\psi(\mathbf{p})|^2 d\mathbf{p} = \int_\Omega d\mathbf{p} \int d\mathbf{q}d\mathbf{q}' S(\mathbf{p}, \mathbf{q})\overline{S}(\mathbf{p}, \mathbf{q}')\phi(\mathbf{q})\phi(\mathbf{q}').$$

In actual scattering experiments, we never know the wave function of the incoming particle. In classical mechanics, this means that we know the initial momentum $\mathbf{p}_0$ of the incoming particle, however, we do not know its impact parameter (recall that the impact parameter is the distance of the particle trajectory from the scattering center in the case when we neglect the interaction and assume that the trajectory is a straight line). Therefore, in classical mechanics, one considers a particle beam with particles having the same initial momentum $\mathbf{p}_0$ but different impact parameters; we assume that the beam has a unit flux (i.e. the number of particles going through a unit area orthogonal to $\mathbf{p}_0$ in unit time is equal to one). The differential cross-section in the solid angle $d\Omega$ is the number of outgoing particles that are directed into the solid angle $d\Omega$.

In quantum mechanics, the notion of the impact parameter cannot be defined. However, we can consider a family of wave functions

$$\rho_{\mathbf{a}}(\mathbf{p}) = \alpha(\mathbf{p})\exp(i\mathbf{p}\mathbf{a}),$$

where $\alpha(\mathbf{p})$ is a normalized wave function that doesn't vanish only in a small neighborhood of $\mathbf{p}_0$ and the vector $\mathbf{a}$ is orthogonal to $\mathbf{p}_0$. Then one can say that the vector $\mathbf{a}$ is the analog of the impact parameter (recall that multiplication by $\exp(i\mathbf{p}\mathbf{a})$ in the momentum representation is equivalent to a shift by the vector $\mathbf{a}$ in the coordinate representation). The scattering cross-section in the angle Ω of the incoming particle with initial momentum $\mathbf{p}_0$ can be defined as

$$\sigma_\Omega = \int_{\mathbf{a}\perp\mathbf{p}_0} d\mathbf{a} \int_\Omega d\mathbf{p} |\psi_\mathbf{a}(\mathbf{p})|^2,$$

where $\psi_\mathbf{a} = S\rho_\mathbf{a}$. Using

$$\psi_\mathbf{a}(\mathbf{p}) = \int S(\mathbf{p},\mathbf{q})\rho_\mathbf{a}(\mathbf{q})d\mathbf{q} = \int S_1(\mathbf{p},\mathbf{q})\frac{\delta(\mathbf{p}-\mathbf{q})}{2\mathbf{p}}\alpha(\mathbf{q})\exp(i\mathbf{q}\mathbf{a})d\mathbf{q},$$

we can say that

$$\sigma_\Omega = \int_{\mathbf{a}\perp\mathbf{p}_0} d\mathbf{a} \int_\Omega d\mathbf{p} \int d\mathbf{q}d\mathbf{q}' S_1(\mathbf{p},\mathbf{q})\overline{S_1}(\mathbf{p},\mathbf{q}')$$

$$\times \frac{\delta(\mathbf{p}-\mathbf{q})}{2\mathbf{p}} \cdot \frac{\delta(\mathbf{p}-\mathbf{q}')}{2\mathbf{p}}\alpha(\mathbf{q})\overline{\alpha}(\mathbf{q}')\exp(i(\mathbf{q}-\mathbf{q}')\mathbf{a}),$$

Integrating over $\mathbf{a}$, we obtain

$$\sigma_\Omega = (2\pi)^2 \int_\Omega d\mathbf{p} \int d\mathbf{q}d\mathbf{q}' S_1(\mathbf{p},\mathbf{q})\overline{S_1}(\mathbf{p},\mathbf{q}')\frac{\delta(\mathbf{p}-\mathbf{q})}{4\mathbf{p}^2}$$

$$\times \alpha(\mathbf{q})\overline{\alpha}(\mathbf{q}')\delta(\mathbf{q}-\mathbf{q}')\delta(\mathbf{q}_T - \mathbf{q}'_T);$$

here $\mathbf{q}_T$, $\mathbf{q}'_T$ are projections of the vectors $\mathbf{q}$ and $\mathbf{q}'$ onto the plane orthogonal to the vector $\mathbf{p}_0$; we used the fact that

$$\int_{\mathbf{a}\perp\mathbf{p}_0} \exp(i(\mathbf{q}-\mathbf{q}')\mathbf{a})d\mathbf{a} = (2\pi)^2\delta(\mathbf{q}_T - \mathbf{q}'_T)].$$

It is easy to check that

$$\delta(q-q')\delta(\mathbf{q}_T - \mathbf{q}_{T'}) = 2q\delta(q^2 - q'^2)\delta(\mathbf{q}_T - \mathbf{q}_{T'})$$

$$= 2q\delta(q_n^2 - q_n'^2)\delta(\mathbf{q}_T - \mathbf{q}'_T)$$

$$= \frac{q}{q_n}\delta(q_n - q_n')\delta(\mathbf{q}_T - \mathbf{q}'_T)$$

$$+ \frac{q}{q_n}\delta(q_n + q_n')\delta(\mathbf{q}_T - \mathbf{q}'_T)$$

$$= \frac{q}{q_n}[\delta(\mathbf{q}-\mathbf{q}') + \delta(I\mathbf{q}-\mathbf{q}')],$$

where q_n, q_n' are the projections of the vectors $\mathbf{q}, \mathbf{q}'$ on the vector $\mathbf{p}_0$ and I is the symmetry with respect to the plane orthogonal to $\mathbf{p}_0$.

Using the δ-functions entering the integrand, we can do the integral over $d\mathbf{q}'$ and using spherical coordinates in the integral over $d\mathbf{p}$, we can instead integrate over dp. We obtain

$$\sigma_\Omega = \pi^2 \int_\omega \sin\theta d\theta d\phi \int d\mathbf{q} |S_1(q,\theta,\phi|\mathbf{q})|^2 \frac{q}{q_n} |\alpha(\mathbf{q})|^2$$

$$+ \pi^2 \int_\omega \sin\theta d\theta d\phi \int d\mathbf{q} S_1(q,\theta,\phi|\mathbf{q}) \overline{S}_1(q,\theta,\phi|I\mathbf{q})$$

$$\times \left(-\frac{q}{q_n}\right) \alpha(\mathbf{q})\overline{\alpha}(I\mathbf{q})$$

(when $\mathbf{p}$ runs over Ω, then the spherical coordinates θ, ϕ run over ω; if $\mathbf{p}$ is a vector with spherical coordinates (p, θ, ϕ), then we use the notation $S_1(\mathbf{p},\mathbf{q}) = S_1(p,\theta,\phi|\mathbf{q})$). If the function $S_1(\mathbf{p},\mathbf{q})$ is continuous when $\mathbf{q}$ changes in a neighborhood of the point $\mathbf{p}_0$ and $\mathbf{p}$ changes in the angle Ω, we obtain that

$$\sigma_\Omega \approx \pi^2 \int_\omega |S_1(p_0,\theta,\phi|\mathbf{p}_0)|^2 \sin\theta d\theta d\phi.$$

This agrees with the expression for the differential cross-section that was written earlier. We have used that for a normalized function $\alpha(\mathbf{q})$ with support in a small neighborhood of the point $\mathbf{p}_0$ and continuous function $f(\mathbf{q})$, we have that

$$\int f(\mathbf{q})|\alpha(\mathbf{q})|^2 d\mathbf{q} \approx f(\mathbf{p}_0).$$

If $\mathbf{p}_0 \neq 0$, then

$$\int f(\mathbf{q})\alpha(\mathbf{q})\overline{\alpha(I\mathbf{q})} d\mathbf{q} \approx 0.[4]$$

---

[4]It is not difficult to convert the above considerations into a rigorous proof. To be precise, one needs to consider a sequence of normalized functions α_n with support tending towards the point $\mathbf{p}_0 \neq 0$ and use the fact that for such a sequence and a continuous function $f(\mathbf{q})$, we have

$$\int f(\mathbf{q})|\alpha_n(\mathbf{q})|^2 d\mathbf{q} \to f(\mathbf{p}_0),$$

$$\int f(\mathbf{q})\alpha_n(\mathbf{q})\overline{\alpha}_n(I\mathbf{q}) d\mathbf{q} \to 0.$$

Let us write down the Lippman–Schwinger equations in this case. For the full system of generalized eigenfunctions of the Hamiltonian H_0, we will take the unctions $\phi_{\mathbf{p}}(\mathbf{x}) = (2\pi)^{-3/2}\exp(i\mathbf{px})$. In the momentum representation, the Hamiltonian H_0 can be considered as multiplication by $\frac{p^2}{2m}$, the operator $(E_{\mathbf{q}} - H_0 \pm i\epsilon)^{-1}$ can be identified with multiplication by $(\frac{q^2}{2m} - \frac{p^2}{2m} \pm i\epsilon)^{-1}$, and the operator $(E^{\mathbf{q}} - H_0 \pm i0)^{-1}$ can be identified with multiplication by the generalized function $(\frac{q^2}{2m} - \frac{p^2}{2m} \pm i0)^{-1}$. Hence, the momentum representation of the Lippman–Schwinger equation takes the form

$$\psi_{\mathbf{q}}^{\pm}(\mathbf{p}) = \delta(\mathbf{p} - \mathbf{q}) + \frac{\int \tilde{\mathcal{V}}(\mathbf{p} - \mathbf{p}')\psi_{\mathbf{q}}^{\pm}(\mathbf{p}')d\mathbf{p}'}{\frac{q^2}{2m} - \frac{p^2}{2m} \mp i0}, \qquad (5.14)$$

where

$$\tilde{\mathcal{V}}(\mathbf{p}) = (2\pi)^{-3/2}\int \mathcal{V}(\mathbf{x})\exp(-i\mathbf{px})d\mathbf{x}.$$

Taking into account that

$$\int \frac{\exp(i\mathbf{px})d\mathbf{p}}{\frac{q^2}{2m} - \frac{p^2}{2m} \pm i0} = -4m\pi^2\frac{\exp(\pm iqx)}{\mathbf{x}},$$

we can write the Lippman–Schwinger equation in the coordinate representation

$$\psi_{\mathbf{q}}^{\pm}(\mathbf{x}) = \phi_{\mathbf{q}}(\mathbf{x}) - \frac{m}{2\pi}\int d\mathbf{x}'\frac{\exp(\mp iq|\mathbf{x} - \mathbf{x}'|)}{|\mathbf{x} - \mathbf{x}'|}\mathcal{V}(\mathbf{x}')\psi_{\mathbf{q}}^{\pm}(\mathbf{x}'). \quad (5.15)$$

Let us calculate the asymptotic behavior of the function $\psi_{\mathbf{q}}^{\pm}(\mathbf{x})$ for $\mathbf{x} \to \infty$. We have assumed that the potential $\mathcal{V}(x)$ tends to zero sufficiently fast at infinity, hence we can assume that in the integral in the formula (5.15), we have $x \gg x'$, and hence in the exponential, we can replace the $|x - x'|$ in the numerator with $x - \frac{x}{x}x'$ and $|x - x'|$ in the denominator with x. We obtain that for large x,

$$\psi^{\pm}(\mathbf{x}) \approx \phi_{\mathbf{q}}(\mathbf{x}) + f_{\mathbf{q}}^{\pm}\left(\mp\frac{x}{x}\right)\frac{\exp(\mp iqx)}{x},$$

where

$$f_{\mathbf{q}}^{\pm}(e) = -\frac{m}{2\pi}\int d\mathbf{x}'\exp(-iqex')\mathcal{V}(x')\psi_q^{\pm}(x').$$

Hence, it follows that

$$f_{\mathbf{q}}^{\pm}(e) = -m\sqrt{2\pi}\,\langle V\psi_{\mathbf{q}}^{\pm}, \phi_{\mathbf{k}}\rangle, \tag{5.16}$$

where $\mathbf{k} = q\mathbf{e}$. Formula (5.16) together with the relations (5.10) and (5.13) allows us to express the matrix entries of the S-matrix and the differential cross-section in terms of the function $f_{\mathbf{q}}^{-}(e)$ that is specified by the asymptotic behavior at infinity of the function $\psi_{\mathbf{q}}^{-}(\mathbf{x})$. For example, the scattering cross-section in the solid angle $d\omega$ is equal to

$$d\sigma = (2\pi)^3 |f_{\mathbf{p}}^{-}(e)|^2 d\omega; \tag{5.17}$$

here $\mathbf{p}$ is the momentum of the incoming particles and e is a unit vector directed in the solid angle $d\omega$. The function $\psi_q^{+}(x)$ satisfies the equation

$$\left(-\frac{1}{2m}\Delta + V(\mathbf{x})\right)\psi_{\mathbf{q}}^{\pm}(x) = \frac{q^2}{2m}\psi_q^{\pm}(x) \tag{5.18}$$

and the boundary condition at infinity

$$\psi_q^{\pm}(x) \approx \phi_q(x) + f_q^{\pm}\left(\mp\frac{x}{x}\right)\frac{\exp(\mp iqx)}{x}. \tag{5.19}$$

It is easy to check that these two conditions characterize the function $\psi_q^{\pm}(x)$. In other words, the function $\psi_q^{-}(x)$ can be described as the solution of the stationary Schrödinger equation that can be represented at infinity as a sum of a plane wave and an outgoing spherical wave. The function $\psi_q^{+}(x)$ can be described similarly, but the outgoing spherical wave should be replaced by an incoming spherical wave.

Hence, we come to the stationary formulation of the scattering problem for non-relativistic particles. In this formulation, one can find the scattering cross-section by solving the stationary Schrödinger equation with the boundary condition (5.19). It is useful to note that in the case that $V(x) = V(-x)$,

$$\psi_q^{+}(x) = \overline{\psi_q^{-}(-x)}$$

(this follows from the fact that the complex conjugate of the incoming spherical wave is the outgoing spherical wave and equation (5.18) is invariant with respect to complex conjugation). Therefore, $\overline{f_q^+(e)} = f_q^-(e)$, and hence the scattering cross-section can be written in the form

$$d\sigma = (2\pi)^3 |f_{\mathbf{p}}^+(e)|^2 d\omega.$$

5.3 Multi-particle scattering

The multi-particle scattering problem can be considered on the Fock space F, corresponding to the measure space $E^3 \times B$, where B is a finite set. The momentum operator $\mathbf{P}$ on this space is defined by the formula (3.15). Let us consider a translation-invariant Hamiltonian H on the space F (i.e. a self-adjoint operator H, commuting with the momentum operator $\mathbf{P}$).

The definition of the scattering matrix given in Section 5.1 is not always applicable to the problem of scattering in a system of n particles. This is because composite particles can be generated in the process of collision. For example, if we start with protons and neutrons, then it is possible that after collision we will have a deuteron (a bound state of a proton and neutron) or an α-particle (a bound state of two protons and two neutrons). Therefore, we should first give a general definition of a particle that includes the objects we will call composite particles.

The vector $\Phi(\mathbf{k})$ that describes a single-particle state with momentum $\mathbf{k}$ should be an eigenvector of the operators $\mathbf{P}$ and H (in the single-particle state, having the momentum $\mathbf{k}$, the energy has definite value $\omega(\mathbf{k})$). However, the operator $\mathbf{P}$ has only one normalized eigenvector (the Fock vacuum), therefore, the vector function $\Phi(\mathbf{k})$ should be a generalized function. This consideration allows us to give the following definition of a particle.

We define a particle corresponding to the Hamiltonian H, as a generalized vector function $\Phi(\mathbf{k})$, obeying

$$H\Phi(\mathbf{k}) = \omega(\mathbf{k})\Phi(\mathbf{k}), \tag{5.20}$$

$$\mathbf{P}\Phi(\mathbf{k}) = \mathbf{k}\Phi(\mathbf{k}), \tag{5.21}$$

$$\langle \Phi(\mathbf{k}), \Phi(\mathbf{k}') \rangle = \delta(\mathbf{k} - \mathbf{k}'). \tag{5.22}$$

The function $\omega(\mathbf{k})$ is called *the energy of a single-particle state* or the dispersion law.

For example, the particles corresponding to the Hamiltonian

$$H_0 = \sum_{s=1}^{n} \int \epsilon_s(\mathbf{k}) a_s^+(\mathbf{k}) a_s(\mathbf{k}) d\mathbf{k} \qquad (5.23)$$

are generalized vector functions

$$\Phi_s(\mathbf{k}) = a_s^+(\mathbf{k})\theta. \qquad (5.24)$$

The same generalized vector functions can also be considered as particles corresponding to the Hamiltonian

$$H = H_0 + W, \qquad (5.25)$$

where

$$W = \sum_{m\geq 2, n\geq 2} \int W_{m,n}(\mathbf{k}_1, \ldots, \mathbf{k}_m | \mathbf{l}_1, \ldots, \mathbf{l}_n) a^+(\mathbf{k}_1)$$
$$\ldots a^+(\mathbf{k}_m) a(\mathbf{l}_1) \ldots a(\mathbf{l}_n) d^m \mathbf{k} d^n \mathbf{l}. \qquad (5.26)$$

These particles are called elementary particles. For translation-invariant Hamiltonians of the form $H_0 + V$, where V is an arbitrary self-adjoint operator, a particle is called an elementary particle if it can be obtained from the particle $a_s^+(\mathbf{k})\theta$, corresponding to the Hamiltonian H_0 by means of perturbation theory (in other words, if the Hamiltonian $H_0 + gV$ has a particle $\Psi_g(\mathbf{k})$ that depends continuously on the parameter g in the interval $[0, 1]$, and we also have $\Psi_0(\mathbf{k}) = a_s^+(\mathbf{k})\theta$, then the particle $\Psi_1(\mathbf{k})$ is called an elementary particle of the Hamiltonian $H_0 + V$).

The particles $\Phi(\mathbf{k}), \Phi'(\mathbf{k})$ are orthogonal if $\langle \Phi(\mathbf{k}), \Phi'(\mathbf{k}) \rangle = 0$; the system of particles $\{\Phi_1(\mathbf{k}), \ldots, \Phi_N(\mathbf{k})\}$ is called complete if there is no particle orthogonal to every particle $\Phi_i(\mathbf{k})$.

For example, for the Hamiltonian, $H_0 = \int \epsilon(\mathbf{k}) a^+(\mathbf{k}) a(\mathbf{k}) d\mathbf{k}$, where $\epsilon(\mathbf{k})$ is a strongly convex function, the system of particles consisting of a single particle $\Phi(\mathbf{k}) = a^+(\mathbf{k})\theta$ (elementary particle), is complete. To prove this, let us consider a particle

$$\Psi(\mathbf{k}) = \sum_n \int \psi_n(\mathbf{k}, \mathbf{k}_1, \ldots, \mathbf{k}_n) a^+(\mathbf{k}_1) \ldots a^+(\mathbf{k}_n)\theta$$

corresponding to the Hamiltonian H_0. From the condition $\mathbf{P}\Psi(\mathbf{k}) = \mathbf{k}\Psi(\mathbf{k})$, it follows that

$$(\mathbf{k}_1 + \cdots + \mathbf{k}_n)\psi_n(\mathbf{k}, \mathbf{k}_1, \ldots, \mathbf{k}_n) = \mathbf{k}\psi_n(\mathbf{k}, \mathbf{k}_1, \ldots, \mathbf{k}_n),$$

and hence

$$\psi_n(\mathbf{k}, \mathbf{k}_1, \ldots, \mathbf{k}_n) = \delta(\mathbf{k} - \mathbf{k}_1 - \cdots - \mathbf{k}_n)\phi_n(\mathbf{k}_1, \ldots, \mathbf{k}_n).$$

The condition $H\Psi(\mathbf{k}) = \omega(\mathbf{k})\Psi(\mathbf{k})$ gives the relation

$$(\epsilon(\mathbf{k}_1) + \cdots + \epsilon(\mathbf{k}_n))\psi_n(\mathbf{k}, \mathbf{k}_1, \ldots, \mathbf{k}_n) = \omega(\mathbf{k})\psi_n(\mathbf{k}, \mathbf{k}_1, \ldots, \mathbf{k}_n),$$

from which it is clear that

$$(\omega(\mathbf{k}_1 + \cdots + \mathbf{k}_n) - \epsilon(\mathbf{k}_1) - \cdots - \epsilon(\mathbf{k}_n))\phi_n(\mathbf{k}_1, \ldots, \mathbf{k}_n) = 0.$$

$$(5.27)$$

Equation (5.27) allows us to conclude that $\phi_n \equiv 0$ for $n > 1$ (to prove this, we should note that a strongly convex function cannot be constant on a set of positive measure and that the function $\epsilon(\mathbf{p}) + \epsilon(\mathbf{k} - \mathbf{p})$ is a strongly convex function of the variable $\mathbf{p}$). Hence, $\Psi(\mathbf{k}) = \phi_1(\mathbf{k})a^+(\mathbf{k})\theta$; this concludes the proof.

The complete system of particles does not always consist only of elementary particles. It is possible that there exist particles that are orthogonal to every elementary particle (such particles are called composite particles).

We will consider, for example, the Hamiltonian

$$H = \int \frac{\mathbf{k}^2}{2m} a^+(\mathbf{k})a(\mathbf{k})d\mathbf{k} + \frac{1}{2} \int \tilde{W}(\mathbf{k}_1 - \mathbf{k}_3)a^+(\mathbf{k}_1)$$

$$\times a^+(\mathbf{k}_2)a(\mathbf{k}_3)a(\mathbf{k}_4)\delta(\mathbf{k}_1 + \mathbf{k}_2 - \mathbf{k}_3 - \mathbf{k}_4)d\mathbf{k}_1 d\mathbf{k}_2 d\mathbf{k}_3 d\mathbf{k}_4$$

$$(5.28)$$

that describes a system of identical spinless bosons (see Section 3.2). This Hamiltonian commutes with N (the operator of number of particles), therefore, it is sufficient to find only particles that belong to the n-particle state F_n. Recall that F_n can be represented as the space of symmetric functions $\phi(\mathbf{k}_1, \ldots, \mathbf{k}_n)$, where $\mathbf{k}_i \in E^3$. Let us introduce new variables

$$\mathbf{p} = \frac{\mathbf{k}_1 + \cdots + \mathbf{k}_n}{n},$$

$$\mathbf{p}_i = \mathbf{k}_i - \mathbf{k}_n \quad (i = 1, \ldots, n-1).$$

The space F_n' will be defined as the space of functions $\psi(\mathbf{p}_1, \ldots, \mathbf{p}_{n-1})$ obeying the condition that the function $f(\mathbf{p})\psi(\mathbf{p}_1, \ldots, \mathbf{p}_{n-1})$, where $f(\mathbf{p}) \in L^2(E^3)$, belongs to the space F_n. Functions from the space F_n' can be considered as wave functions of relative motion of n particles. The consideration of motion of n particles in terms of functions from the space F_n' corresponds to the separation of the motion of the center of inertia. Mathematically, the possibility to separate the motion of the center of inertia for the Hamiltonian H_n means that there exists a Hamiltonian H_n' in the space F_n', that for every function,

$$f(\mathbf{p})\psi(\mathbf{p}_1, \ldots, \mathbf{p}_{n-1}) \in F_n,$$

where $f \in L^2(E^3), \psi \in F_n$, we have

$$H_n(f(\mathbf{p})\psi(\mathbf{p}_1, \ldots, \mathbf{p}_{n-1})) = \left(\frac{\mathbf{p}^2}{2nm} f(\mathbf{p}) \right) \psi(\mathbf{p}_1, \ldots, \mathbf{p}_{n-1})$$

$$+ f(\mathbf{p})(H_n'\psi)(\mathbf{p}_1, \ldots, \mathbf{p}_{n-1}).$$

Let us assume that $\psi \in F_n'$ is a normalized eigenvector of the operator H_n' with eigenvalue E. It is easy to check that the generalized vector function

$$\Phi(\mathbf{k}) = \delta(\mathbf{k} - \mathbf{p})\psi(\mathbf{p}_1, \ldots, \mathbf{p}_{n-1})$$

is a particle in the sense of the above definition. It is also easy to check that

$$H\Phi(\mathbf{k}) = \left(\frac{\mathbf{k}^2}{2nm} + E \right) \Phi(\mathbf{k}).$$

For $n > 1$, the particle $\Phi(\mathbf{k})$ is a composite particle (a bound state of a system of n particles).

Let us now ask what we should consider to be a scattering matrix for a Hamiltonian of the form (5.25) (for more general translation-invariant Hamiltonians, the definition of the scattering matrix will be given in Chapter 9). It is natural to try to define the scattering matrix of the Hamiltonian (5.25) as an S-matrix corresponding to the pair of operators (H, H_0). However, this works only in the case when elementary particles $\Phi_s(\mathbf{k}) = a_s^+(\mathbf{k})\theta$ constitute a complete system of particles (in other words, there exist no composite particles). From the physical viewpoint, it is clear that the S-matrix of the operators (H, H_0) cannot describe the scattering of composite particles. If composite particles are present, then the S-matrix corresponding to the pair of operators (H, H_0) will not be a unitary operator.

To formulate the definition of scattering matrix, we will restrict ourselves to the case when all particles are bosons.

Note, first of all, that every particle $\Phi(\mathbf{k})$ can be written in the form

$$\Phi(\mathbf{k}) = \sum_{n=1}^{\infty} \sum_{i_1,\ldots,i_n} \int \delta(\mathbf{k} - \mathbf{k}_1 - \cdots - \mathbf{k}_n)$$

$$\times \phi_n(\mathbf{k}_1, i_1, \ldots, \mathbf{k}_n, i_n) a_{i_1}^+(\mathbf{k}_1) \ldots a_{i_n}^+(\mathbf{k}_n)\theta d\mathbf{k}_1 \ldots d\mathbf{k}_n.$$

Let us assign to the particle $\Phi(\mathbf{k})$ an operator generalized function defined by the formula

$$A(\mathbf{k}) = \sum_{n=1}^{\infty} \sum_{i_1,\ldots,i_n} \int \delta(\mathbf{k} - \mathbf{k}_1 - \cdots - \mathbf{k}_n)$$

$$\times \overline{\phi}_n(\mathbf{k}_1, i_1, \ldots, \mathbf{k}_n, i_n) a_{i_1}(\mathbf{k}_1) \ldots a_{i_n}(\mathbf{k}_n)\theta d\mathbf{k}_1 \ldots d\mathbf{k}_n.$$

The operator generalized function $A(\mathbf{k})$ satisfies the conditions: (1) $A^+(\mathbf{k})\theta = \Phi(\mathbf{k})$ (the operator $A^+(\mathbf{k})$ creates the particle $\Phi(\mathbf{k})$ when applied to the vacuum) and, (2) $A(\mathbf{k})$ is a superposition of operators $a_{i_1}(\mathbf{k}_1) \ldots a_{i_n}(\mathbf{k}_n)$, i.e. it can be represented in the form

$$A(\mathbf{k}) = \sum_{n=1}^{\infty} \sum_{i_1,\ldots,i_n} \int \lambda_n(\mathbf{k}, \mathbf{k}_1, i_1, \ldots, \mathbf{k}_n, i_n)$$

$$\times a_{i_1}(\mathbf{k}_1) \ldots a_{i_n}(\mathbf{k}_n) d\mathbf{k}_1 \ldots d\mathbf{k}_n.$$

Moreover, these conditions specify $A(\mathbf{k})$ uniquely.

Let us now define the in- and out-operators $A_{\mathrm{in}}(\mathbf{k}, \boldsymbol{\tau})$ and $A_{\mathrm{out}}(\mathbf{k}, \boldsymbol{\tau})$, corresponding to the particle $\Phi(\mathbf{k})$, as the limits

$$A_{\mathrm{in}}(\mathbf{k}, \boldsymbol{\tau}) = \lim_{t \to -\infty} \exp(i\omega(\mathbf{k})(t - \tau))A(\mathbf{k}, t), \qquad (5.29)$$

$$A_{\mathrm{out}}(\mathbf{k}, \boldsymbol{\tau}) = \lim_{t \to +\infty} \exp(i\omega(\mathbf{k})(t - \tau))A(\mathbf{k}, t), \qquad (5.30)$$

where $A(\mathbf{k}, t) = \exp(iHt)A(\mathbf{k})\exp(-iHt)$ and $\omega(\mathbf{k})$ is the dispersion law for the particle $\Phi(\mathbf{k})$. The limit is understood in the sense of the limit of generalized functions (in other words, we assume that for every function $f(\mathbf{k}) \in \mathcal{S}(E^3)$, we have

$$\int f(\mathbf{k})A_{\substack{\mathrm{in}\\\mathrm{out}}}(\mathbf{k}, \tau)d\mathbf{k} = \lim_{t \to \mp\infty} \int f(\mathbf{k})\exp(i\omega(\mathbf{k})(t - \tau))A(\mathbf{k}, t)d\mathbf{k}$$

in the sense of strong operator limit on the linear subspace D, where D is defined as the smallest linear subspace containing all spaces F_n). When talking about the operators A_{in} and A_{out} at the same time, we will use the notation A_{ex}.

Under certain conditions on the Hamiltonian H (in particular, for the Hamiltonians of the form (5.28), under the assumption that interaction potential is square integrable), we can prove that

(1) the limits (5.29) and (5.30) exist;
(2) if the particles $\Phi_1(\mathbf{k}), \ldots, \Phi_m(\mathbf{k})$ are orthogonal, then the corresponding operators $A_{\mathrm{ex}}(\mathbf{k}, i, \tau), A_{\mathrm{ex}}^+(\mathbf{k}, i, \tau)$, where $i = 1, \ldots, m$, for fixed τ obey the canonical commutation relations (CCR)

$$[A_{\mathrm{ex}}^+(\mathbf{k}, i, \tau), A_{\mathrm{ex}}^+(\mathbf{k}', i', \tau)] = [A_{\mathrm{ex}}(\mathbf{k}, i, \tau), A_{\mathrm{ex}}(\mathbf{k}', i', \tau)] = 0;$$

$$[A_{\mathrm{ex}}(\mathbf{k}, i, \tau), A_{\mathrm{ex}}^+(\mathbf{k}', i', \tau)] = \delta_i^i \delta(\mathbf{k} - \mathbf{k}');$$

(3) $A_{\mathrm{ex}}(\mathbf{k}, i, \tau) = \exp(iH\tau)A_{\mathrm{ex}}(\mathbf{k}, i)\exp(-iH\tau) = \exp(-\omega_i(\mathbf{k})\tau)A_{\mathrm{ex}}(\mathbf{k}, i)$, where $A_{\mathrm{ex}}(\mathbf{k}, i) = A_{\mathrm{ex}}(\mathbf{k}, i, 0)$;
(4) $A_{\mathrm{ex}}(\mathbf{k}, i)\theta = 0$, $A_{\mathrm{ex}}^+(\mathbf{k}, i)\theta = \Phi_i(\mathbf{k})$.

The proof of these statements can be found, for example, in Hepp and Epstein (1971) (the last two statements are trivial).

The generalized vector functions

$$\Psi_{\mathrm{in}}(\mathbf{k}_1, i_1, \ldots, \mathbf{k}_n, i_n) = A_{\mathrm{in}}^+(\mathbf{k}_1, i_1) \ldots A_{\mathrm{in}}^+(\mathbf{k}_n, i_n)\theta,$$

$$\Psi_{\mathrm{out}}(\mathbf{k}_1, i_1, \ldots, \mathbf{k}_n, i_n) = A_{\mathrm{out}}^+(\mathbf{k}_1, i_1) \ldots A_{\mathrm{out}}^+(\mathbf{k}_n, i_n)\theta$$

are called in- and out-states.

Sometimes it is useful to note that the vectors Ψ_{in} and Ψ_{out} can also be represented in the form

$$\Psi_{\text{ex}}(\mathbf{k}_1, i_1, \ldots, \mathbf{k}_n, i_n) = \lim_{t \to \pm\infty} \exp(-i(\omega_{i_1}(\mathbf{k}_1) + \cdots + \omega_{i_n}(\mathbf{k}_n))t)$$
$$\times A^+(\mathbf{k}_1, i_1, t) \ldots A^+(\mathbf{k}_n, i_n, t)\theta. \qquad (5.31)$$

This follows immediately from the definition of the operators A_{in}^+ and A_{out}^+. Applying the formula (5.8), we can get from (5.31) the following representation of in- and out-states:

$$\Psi_{\substack{\text{in} \\ \text{out}}}(\mathbf{k}_1, i_1, \ldots, \mathbf{k}_n, i_n) = \pm \lim_{\alpha \to 0} i\alpha(H - (\omega_{i_1}(\mathbf{k}_1)$$
$$+ \cdots + \omega_{i_n}(\mathbf{k}_n)) \pm i\alpha)^{-1} A^+(\mathbf{k}_1, i_1) \ldots A^+(\mathbf{k}_n, i_n)\theta.$$

Let us fix a complete orthonormal system of particles $\Phi_1(\mathbf{k})$, $\ldots, \Phi_N(\mathbf{k})$ and the operators $A_{\text{ex}}(\mathbf{k}, 1), \ldots, A_{\text{ex}}(\mathbf{k}, N)$ corresponding to the particles of this system.

We define the matrix elements of the S-matrix (or scattering amplitudes) as functions

$$S_{m,n}(\mathbf{k}_1, i_1, \ldots, \mathbf{k}_m, i_m | \mathbf{l}_1, j_1, \ldots, \mathbf{l}_n, j_n)$$
$$= \left\langle A_{\text{in}}^+(\mathbf{l}_1, j_1) \ldots A_{\text{in}}^+(\mathbf{l}_n, j_n)\theta, A_{\text{out}}^+(\mathbf{k}_1, i_1) \ldots A_{\text{out}}^+(\mathbf{k}_m, i_m)\theta \right\rangle.$$

Knowing these matrix elements, we can express the probability that by collision of n particles with quantum numbers $\mathbf{l}_1, j_1, \ldots, \mathbf{l}_n, j_n$, we get particles with quantum numbers $\mathbf{k}_1, i_1, \ldots, \mathbf{k}_m, i_m$ (more general situations will be discussed in Chapter 10).

Let us consider the relation of the definition of the scattering matrix in terms of the in- and out-operators, with the definition of the S-matrix given in Section 5.1. Let us suppose that the Møller matrices $S_{\mp}$, corresponding to the pair of operators (H, H_0) are unitary (this assumption is always satisfied when there are no composite particles). The operators $A(\mathbf{k}, s)$ corresponding to elementary particles $\Phi_s(\mathbf{k}) = a_s^+(\mathbf{k})\theta$ are clearly equal to $a_s(\mathbf{k})$. Let us show that

$$A_{\text{in}}(\mathbf{k}, s) = a_{\text{in}}(\mathbf{k}, s) = S_- a_s(\mathbf{k}) S_-^*, \qquad (5.32)$$
$$A_{\text{out}}(\mathbf{k}, s) = a_{\text{out}}(\mathbf{k}, s) = S_+ a_s(\mathbf{k}) S_+^*. \qquad (5.33)$$

To check these equalities, we will use the relations (5.1) and (5.2). It follows from these relations that

$$S_{\mp}a_s(\mathbf{k})S_{\mp}^{+} = \operatorname*{slim}_{t\to\mp\infty} \exp(iHt)\exp(-iH_0t)a_s(\mathbf{k})$$

$$\times \exp(iH_0t)\exp(-iHt)$$

$$= \operatorname*{slim}_{t\to\mp\infty} \exp(i\epsilon_s(\mathbf{k})t)a_s(\mathbf{k},t) = a_{\substack{\text{in}\\\text{out}}}(\mathbf{k},s).$$

The relations (5.32) and (5.33) imply that

$$\left\langle a_{\text{in}}^{+}(\mathbf{l}_1,\sigma_1)\dots a_{\text{in}}^{+}(\mathbf{l}_n,\sigma_n)\theta, a_{\text{out}}^{+}(\mathbf{k}_1,s_1)\dots a_{\text{out}}^{+}(\mathbf{k}_m,s_m)\theta \right\rangle$$

$$= \left\langle S_-a_{\sigma_1}^{+}(\mathbf{l}_1)\dots a_{\sigma_n}^{+}(\mathbf{l}_n)\theta, S_+a_{s_1}^{+}(\mathbf{k}_1)\dots a_{s_m}^{+}(\mathbf{k}_m)\theta \right\rangle$$

$$= \left\langle Sa_{\sigma_1}^{+}(\mathbf{l}_1)\dots a_{\sigma_n}^{+}(\mathbf{l}_n)\theta, a_{s_1}^{+}(\mathbf{k}_1)\dots a_{s_m}^{+}(\mathbf{k}_m)\theta \right\rangle$$

(we have used $S_-^*\theta = S_+^*\theta = \theta$). In other words, the matrix elements of the S-matrix defined in the present section under the assumption we have made coincide with the matrix elements of the operator S in the generalized basis $a_{s_1}^{+}(\mathbf{k}_1)\dots a_{s_m}^{+}(\mathbf{k}_m)\theta$.

In the case when there exist bound states, we can also construct an operator S having the functions $S_{m,n}$ as matrix elements. This operator S acts on the space F_{as} that is called the space of asymptotic states. F_{as} can be defined as a Fock space constructed from the measure space $E^3 \times N$, where N is a set of types of particles. (Let us recall that we have fixed a complete orthonormal system of particles.) The operators of creation and annihilation in F_{as} will be denoted $b^{+}(\mathbf{k},i)$ and $b(\mathbf{k},i)$ (here, $i \in N$). The vectors from the space F_{as} can be considered as initial and final states of the scattering process (for example, the vector $b_{i_1}^{+}(f_1)\dots b_{i_n}^{+}(f_n)\theta$, where $b_i(f) = \int f(\mathbf{k})b(\mathbf{k},i)d\mathbf{k}$ corresponds to a state with the particles of the types $i_1,\dots,i_n$ with the wave functions $\overline{f}_1,\dots,\overline{f}_n$). Let us define the isometries S_- and S_+ acting from the space F_{as} into the space F by means of the relations

$$A_{\text{in}}(\mathbf{k},i)S_- = S_-b(\mathbf{k},i), \quad S_-\theta = \theta, \tag{5.34}$$

$$A_{\text{out}}(\mathbf{k},i)S_+ = S_+b(\mathbf{k},i), \quad S_+\theta = \theta. \tag{5.35}$$

(Such operators do exist and they are defined by the relations (5.34) and (5.35) in a unique way; the proof of this fact is given in

Section 6.1.) The scattering matrix is defined as the operator $S = S_+^* S_-$ acting on the space F_{as}. It is obvious that

$$S_{m,n}(\mathbf{k}_1, i_1, \ldots, \mathbf{k}_m, i_m | \mathbf{l}_1, j_1, \ldots, \mathbf{l}_n, j_n)$$
$$= \left\langle A_{\text{in}}^+(\mathbf{l}_1, j_1) \ldots A_{\text{in}}^+(\mathbf{l}_n, j_n)\theta, A_{\text{out}}^+(\mathbf{k}_1, i_1) \ldots A_{\text{out}}^+(\mathbf{k}_m, i_m)\theta \right\rangle$$
$$= \left\langle S_- b^+(\mathbf{l}_1, j_1) \ldots b^+(\mathbf{l}_n, j_n)\theta, S_+ b^+(\mathbf{k}_1, i_1) \ldots b^+(\mathbf{k}_m, i_m)\theta \right\rangle$$
$$= \left\langle S b^+(\mathbf{l}_1, j_1) \ldots b^+(\mathbf{l}_n, j_n)\theta, b^+(\mathbf{k}_1, i_1) \ldots b^+(\mathbf{k}_m, i_m)\theta \right\rangle.$$

In other words, the matrix elements of the operator S in the generalized basis $b^+(\mathbf{k}_1, i_1) \ldots b^+(\mathbf{k}_m, i_m)\theta$ coincide with the functions $S_{m,n}$ (the scattering amplitudes).

Hamiltonians of the form (5.28) commute with the operator of the number of particles, therefore, one can study separately the n-particle scattering for $n = 2, 3, \ldots$. For $n = 2$, this problem can be reduced to the problem of potential scattering considered in the preceding section. The proof of unitarity of the scattering matrix for the three-particle problem was obtained by Faddeev (1963). Faddeev's method is based on the construction of equations for the in- and out-states. Faddeev equations are also very useful for calculations. They were also generalized in different ways for the case of n-particle scattering problem for $n > 3$. The unitarity of the scattering matrix for any number of non-relativistic particles was proven in Sigal and Soffer (1987), see Hunziker and Sigal (2000) for review.

In conclusion, let us present the definition of the Møller matrices $S_\pm$ and the scattering matrix S in the form that is more convenient in general situations (Chapter 10). Let us say that the operator B is a good operator if it can be represented as

$$\sum_m \sum_{s_1, \ldots, s_m} \int f(\mathbf{k}_1, s_1, \ldots, \mathbf{k}_m, s_m)$$
$$\times a_{s_1}^+(\mathbf{k}_1) \ldots a_{s_m}^+(\mathbf{k}_m) d\mathbf{k}_1 \ldots d\mathbf{k}_m$$

and obeys the condition

$$B\theta = \int \phi(\mathbf{k})\Phi_i(\mathbf{k})d\mathbf{k},$$

where $\Phi_i(\mathbf{k})$ is one of the particles in the fixed complete system $\Phi_1(\mathbf{k}), \ldots, \Phi_n(\mathbf{k})$. The isometric operator $S_-(S_+)$ acting on the asymptotic space F_{as} into the space F is called the Møller matrix if for all good operators $B_1, \ldots, B_m$ and smooth functions with compact support $f_1(\mathbf{p}), \ldots, f_m(\mathbf{p})$, we have

$$\lim_{t \to \mp\infty} B_1(f_1, t) \ldots B_m(f_m, t)\theta$$

$$= S_{\mp} b^+(\overline{\phi}_1, \overline{f}_1, i_1) \ldots b^+(\overline{\phi}_m, \overline{f}_m, i_m)\theta.$$

Here, the operators $B_\alpha(f_\alpha, t)$ are defined by the formula

$$B_\alpha(f_\alpha, t) = \int \tilde{f}_\alpha(x, t) B_\alpha(\mathbf{x}, t) d\mathbf{x},$$

where

$$B_\alpha(\mathbf{x}, t) = \exp(iHt - i\mathbf{P}\mathbf{x}) B_\alpha \exp(-iHt + i\mathbf{P}\mathbf{x}),$$

$$\tilde{f}_\alpha(\mathbf{x}, t) = \int \exp(-i\omega_{i_\alpha}(\mathbf{p})t + i\mathbf{p}\mathbf{x}) f_\alpha(\mathbf{p}) \frac{d\mathbf{p}}{(2\pi)^3},$$

and the numbers i_α and the functions $\phi_\alpha(\mathbf{p})$, $\omega_{i_\alpha}(\mathbf{p})$ are defined by the relations

$$B_\alpha\theta = \int \phi_\alpha(\mathbf{p})\Phi_{i_\alpha}(\mathbf{p})d\mathbf{p},$$

$$H\Phi_{i_\alpha}(\mathbf{p}) = \omega_{i_\alpha}(\mathbf{p})\Phi_{i_\alpha}(\mathbf{p}).$$

The scattering matrix S, as always, is defined in terms of the Møller matrices $S = S_+^* S_-$.

To check that this definition of Møller matrices is equivalent to the above definition, we note that a good operator B_α can be represented in the form

$$B_\alpha = \int \phi_\alpha(\mathbf{p}) A_{i_\alpha}^+(\mathbf{p})d\mathbf{p},$$

where $A_i(\mathbf{p})$ is an operator generalized function corresponding to the particle $\Phi_i(\mathbf{p})$. Using the obvious relation

$$\exp(-i\mathbf{P}\mathbf{x}) A_i^+(\mathbf{k}) \exp(i\mathbf{P}\mathbf{x}) = \exp(-i\mathbf{k}\mathbf{x}) A_i^+(\mathbf{k}),$$

we see that

$$B_\alpha(f_\alpha, t) = \int \tilde{f}_\alpha(\mathbf{x}, t) \exp(-i\mathbf{k}\mathbf{x})\phi_\alpha(\mathbf{k})A^+_{i_\alpha}(\mathbf{k}, t)d\mathbf{k}$$

$$= \int \exp(-i\omega_{i_\alpha}(\mathbf{k})t)f_\alpha(\mathbf{k})\phi_\alpha(\mathbf{k})A^+_{i_\alpha}(\mathbf{k}, t)d\mathbf{k},$$

hence

$$\operatorname*{slim}_{t\to\mp\infty} B_\alpha(f_\alpha, t) = \int f_\alpha(\mathbf{k})\phi_\alpha(\mathbf{k})A^+_{\substack{\text{in}\\\text{out}}}(\mathbf{k}, i_\alpha)d\mathbf{k}. \tag{5.36}$$

To finish the proof of equivalence of the two definitions of the Møller matrices, we note that it follows from (5.36), (5.34) and (5.35) that

$$\lim_{t\to\mp\infty} B_1(f_1, t)\ldots B_n(f_n, t)\theta = \int f_1(\mathbf{k}_1)\phi_1(\mathbf{k}_1)A^+_{\substack{\text{in}\\\text{out}}}(\mathbf{k}_1, i_1)$$

$$\ldots f_n(\mathbf{k}_n)\phi_n(\mathbf{k}_n)A^+_{\substack{\text{in}\\\text{out}}}(\mathbf{k}_n, i_n)d^n\mathbf{k}\theta = S_\mp b^+(\overline{f}_1, \overline{\phi}_1, i_1)$$

$$\ldots b^+(\overline{f}_n\overline{\phi}_n, i_n)\theta.$$

Chapter 6

Operators on Fock Space

6.1 The representations of canonical and anticommutation relations: Fock representation

Let us assume that we have assigned to every vector f of a pre-Hilbert space $\mathcal{B}$ two conjugate operators $a(f)$ and $a^+(f)$ acting on a dense subspace of the Hilbert space $\mathcal{H}$, in such a way that $a(f)$ linearly depends on f (i.e. $a(\lambda f + \mu g) = \lambda a(f) + \mu a(g)$). If the operators satisfy the relations

$$[a(f), a(g)] = [a^+(f), a^+(g)] = 0, [a(f), a^+(g)] = \langle f, g \rangle \qquad (6.1)$$

(where $f, g \in \mathcal{B}$ and λ, μ are complex numbers), then we say that these operators specify a *representation of canonical commutation relations*, or CCR, on the space $\mathcal{H}$. If the operators satisfy the analogous relations

$$[a(f), a(g)]_+ = [a^+(f), a^+(g)]_+ = 0, [a(f), a^+(g)]_+ = \langle f, g \rangle, \quad (6.2)$$

then we say that these operators specify a *representation of canonical anticommutation relations* or CAR.[1]

We will also introduce the notation $a(f, -1) = a(f), a(f, +1) = a^+(f)$. Using this notation, we can write down CCR and CAR in the following form:

$$[a(f, \epsilon), a(g, \epsilon')]_{\mp} = A^{\epsilon}_{\epsilon'} \langle f, g \rangle, \qquad (6.3)$$

---

[1] In the definition of the representation of CCR and CAR, one should assume that the operators $a(f), a^+(f)$ are defined on the same linear subspace D that is dense in the space $\mathcal{H}$. They should transform D into itself. The operators $a(f)$ and $a^+(f)$ are conjugate (i.e. $\langle a(f)x, y \rangle = \langle x, a^+(f)y \rangle$ for all $x, y \in D$).

where $A^{\epsilon}_{\epsilon'}$ is a matrix with the elements $A^1_1 = A^{-1}_{-1} = 0, A^{-1}_1 = 1, A^1_{-1} = \mp 1$ (the upper sign is used in the case of CCR and the lower sign is used in the case of CAR).

In the case of CAR, the operators $a(f)$ and $a^+(f)$ are bounded (the equation $a(f)a^+(f) + a^+(f)a(f) = \langle f, f \rangle$ implies that $\langle a(f)x, a(f)x \rangle + \langle a^+(f)x, a^+(f)x \rangle = \langle f, f \rangle \langle x, x \rangle$, hence $\|a(f)\| \leq \|f\|, \|a^+(f)\| \leq \|f\|$). Therefore, one can assume that the operators $a(f), a^+(f)$ are defined on the whole space $\mathcal{H}$. In the case of CCR, the operators $a(f), a^+(f)$ are unbounded.

Speaking simultaneously of CCR and CAR, we will use the term "canonical relations" (CR). When it is necessary to stress that CR are constructed by means of a pre-Hilbert space $\mathcal{B}$, we will use the notation CR($\mathcal{B}$). If in the space $\mathcal{B}$ we have fixed an orthonormal basis ϕ_n, then the operators $a_n = a(\phi_n)$ and $a_n^+ = a^+(\phi_n)$ satisfy the relations

$$[a_m, a_n] = [a_m^+, a_n^+] = 0, [a_m, a_n^+] = \delta_{m,n} \tag{6.4}$$

in the case of CCR and the relations

$$[a_m, a_n]_+ = [a_m^+, a_n^+]_+ = 0, [a_m, a_n^+]_+ = \delta_{m,n} \tag{6.5}$$

in the case of CAR (these relations are also called CR).

The most important example of operators satisfying CR includes the operators of creation $a^+(f)$ and annihilation $a(f)$, defined in Section 3.2. (They specify a representation of CCR in the bosonic case and the representation of CAR in the fermionic case.) The operators $a^+(f), a(f)$ are called creation and annihilation operators in other cases as well; note, however, that these names don't always agree with their physical meaning.

One more important representation of CCR is given by the following operators:

$$a(f) = \sum_{n=1}^N f_n a_n,$$

$$a^+(f) = \sum_{n=1}^N \overline{f_n} a_n^+,$$

where a_n^+, a_n are operators satisfying the relations (6.4) that were constructed in Section 2.6 where we studied the system of coupled oscillators (here $f = (f_1, \ldots, f_N)$ runs over a complex N-dimensional space).

The representation of CR is called the Fock representation if one can find a cyclic vector θ in the space $\mathcal{H}$ that satisfies the condition $a(f)\theta = 0$. More precisely, we assume that the vector θ is a cyclic vector with respect to the operators $a^+(f), a(f)$ (this means that applying these operators to θ and taking linear combinations we get a subset of $\mathcal{H}$ that is dense in $\mathcal{H}$). The vector θ is called the *vacuum vector*[2] (see Section 3.2).

Let us prove that *two Fock representations are equivalent*. This means that there exists a unitary operator α mapping the space $\mathcal{H}_1$ onto the space $\mathcal{H}_2$ and satisfying the conditions $\alpha a_1(f) = a_2(f)\alpha, \alpha\theta_1 = \theta_2$ (here $a_i(f)$ is a Fock representation of CR in the space $\mathcal{H}_i$ with the vacuum vector θ_i). In the case of CCR, the equation $\alpha a_1(f) = a_2(f)\alpha$ should be satisfied on a dense subset of the space $\mathcal{H}_1$ (not necessarily on the whole domain of $a_1(f)$).

Let us construct an operator α. For vectors of the form

$$a_1(f_1, \epsilon_1) \ldots a_1(f_n, \epsilon_n)\theta_1, \tag{6.6}$$

we define the operator α, assuming that

$$\alpha(a_1(f_1, \epsilon_1) \ldots a_1(f_n, \epsilon_n)\theta_1) = a_2(f_1, \epsilon_1) \ldots a_2(f_n, \epsilon_n)\theta_2.$$

If x and y are two vectors of the form (6.6), then $\langle x, y \rangle = \langle \alpha x, \alpha y \rangle$. This follows from the fact that the scalar product of vectors of the form (6.6) can be calculated using only CR and the relation $a_i(f)\theta_i = 0$. Let us denote by S_1 the linear subspace consisting of linear combinations of vectors of the form (6.6). The operator α, by linearity, can be extended to S_1 and satisfies the condition $\langle x, y \rangle = \langle \alpha x, \alpha y \rangle$ for all $x, y \in S_1$. Using the assumption that the vector θ_i is cyclic with respect to the operators $a_i(f, \epsilon)$, we obtain

---

[2]Using CR, it is easy to check that every vector of the form $a(f_1, \epsilon_1) \ldots a(f_n, \epsilon_n)\theta$ can be represented as a linear combination of vectors of the form $a^+(\phi_1) \ldots a^+(\phi_m)\theta$. Hence, θ is also a cyclic vector with respect to the family of operators $a^+(f)$.

that the set S_1 is dense in $\mathcal{H}_2$ and the set $S_2 = \alpha S_1$ is dense in $\mathcal{H}_2$. The operator α can be extended to the closure $\overline{S}_1 = \mathcal{H}_1$ continuously; the density of the set S_2 in $\mathcal{H}_2$ implies the relation $\alpha \mathcal{H}_1 = \mathcal{H}_2$. It is clear that on the set S_1 we have $\alpha a_1(f, \epsilon) = a_2(f, \epsilon)\alpha$. This completes the proof of the equivalence of two Fock representations.[3]

Hence, it is sufficient to describe just one Fock representation; this was done in Section 3.2 (and will be done again in slightly different terms later).

The space of Fock representation $CR(\mathcal{B})$ will be called *Fock space* and denoted as $F(\mathcal{B})$.

Let us assume that the pre-Hilbert space $\mathcal{B}$ is realized as a dense subset of the space $L^2(E^r)$ (the space of square-integrable functions on the Euclidean space E^r). Then, together with the operators $a^+(f), a(f)$, it is convenient to introduce the operator generalized functions $a^+(x), a(x)$ on E^r, where

$$a(f) = \int f(x)a(x)dx, \quad a^+(\overline{f}) = \int f(x)a^+(x)dx.$$

The functions $a^+(x), a(x)$ satisfy the following relations:

$$[a(x), a(y)]_{\mp} = [a^+(x), a^+(y)]_{\mp} = 0,$$
$$[a(x), a^+(y)]_{\mp} = \delta(x, y); a(x)\theta = 0. \tag{6.7}$$

This definition of the generalized operator functions $a^+(x), a(x)$ can also be used in the case when $\mathcal{B}$ is a dense subset of the space $L^2(X)$, where $X = E^r \times B$ consists of pairs (e, s) with $e \in E^r$ a point in Eucliean space and $s \in B$ from a finite set B (then the integration over X is understood as integration over Eucliean space and summation over a finite set). We can also use this definition in the more general case when X is an arbitrary measure space; however, we do not need this level of generality. Later in this section,

---

[3]If the first representation of CR is a Fock representation, but the second representation has a vector $\theta_2 \in R_2$ satisfying the condition $a_2(f)\theta_2 = 0$ (not necessarily cyclic), then precisely in the same way, we can construct the operator α satisfying the relation $\alpha a_1(f) = a_2(f)\alpha, \alpha\theta_1 = \theta_2$. However, in general, α will be an isometry that is not necessarily unitary (this statement is used in Section 5.3 to construct the operators S_- and S_+).

we assume that X is any measure space; however, for simplicity, the reader can assume that $X = E^r$ or $X = E^r \times B$.

Operator generalized functions $a^+(x), a(x)$, in the case of the Fock representations of CR, were considered in Section 3.2. In particular, we have shown in this section that an expression of the form (3.4) can be understood as an operator acting on Fock space. For an arbitrary representation of CR, the expression (3.4) can be easily defined in the case when the function entering this expression has the form (3.5). If f is an arbitrary function, then it can be represented as a limit of functions f_k of the form (3.5) and the operator A corresponding to the function f should be defined, in this case, as a limit of operators A_k corresponding to the functions f_k. Of course, this definition of the operator A is not completely precise because we have not defined the notion of limit for the functions f_k and the operators A_k. We will not be concerned with these definitions; however, we will note that in the case when $\mathcal{B} = \mathcal{S}(E^3)$ and the functional $\langle a(f, \epsilon)u, v \rangle$ (for arbitrary $u, v \in D$) continuously depends on f in the topology of the space $\mathcal{S}(E^3)$, then using the operator analog of the kernel theorem (see Appendix A.5), one can define the operator (3.4) for the functions $f(\mathbf{x}_1, \ldots, \mathbf{x}_m | \mathbf{y}_1, \ldots, \mathbf{y}_n)$ from the space $\mathcal{S}(E^{3(m+n)})$.

In the case when $\mathcal{B} = L^2(X)$, every element Φ of Fock space can be written in a unique way in the form

$$\Phi = \sum_n \int f_n(x_1, \ldots, x_n) a^+(x_1) \ldots a^+(x_n) \theta dx_1 \ldots dx_n, \qquad (6.8)$$

where the functions $f_n(x_1, \ldots, x_n)$ are symmetric in the case of CCR and antisymmetric in the case of CAR in the arguments $x_1, \ldots, x_n \in X$. The norm of the vector Φ is equal to

$$\sum_n n! \int |f_n(x_1, \ldots, x_n)|^2 dx_1 \ldots dx_n.$$

Conversely, to every sequence $f_0, f_1(x), \ldots, f_n(x_1, \ldots, x_n), \ldots$ of symmetric (antisymmetric) functions obeying

$$\sum_n n! \int |f_n(x_1, \ldots, x_n)|^2 dx_1 \ldots dx_n < \infty,$$

we can assign a vector in Fock space. This description of the Fock space can be obtained from the formulation described in Section 3.2, namely, it is easy to check that the Fock state (Fock column)

$$\Phi = \begin{bmatrix} \phi_0 \\ \phi_1(x_1) \\ \vdots \\ \phi_n(x_1, \dots, x_n) \\ \vdots \end{bmatrix}$$

can be represented in the form

$$\Phi = \sum_n \frac{1}{\sqrt{n!}} \int \phi_n(x_1, \dots, x_n) a^+(x_1) \dots a^+(x_n)\theta dx_1 \dots dx_n.$$

This statement can also be formulated in slightly different terms. Namely, one can say that vector generalized functions $a^+(x_1) \dots a^+(x_n)\theta$, where $n = 0, 1, 2, \dots$, constitute a generalized basis of the space $F(L^2(X))$ in the following sense: every vector from $F(L^2(X))$ can be decomposed as a linear combination of these vector generalized functions (i.e. represented in the form (6.8)) and if the coefficient functions f_n are symmetric (antisymmetric), then this decomposition is unique. (Note that we used the term generalized basis not precisely in the form that is described in A.7.) Generalized functions

$$A_{kl}(x_1, \dots, x_k | y_1, \dots, y_l) = \left\langle Aa^+(y_1) \dots a^+(y_l)\theta, a^+(x_1) \dots a^+(x_k)\theta \right\rangle$$

are called matrix elements or matrix entries of the operator A in the generalized basis $a^+(x_1) \dots a^+(x_n)\theta$.

In conclusion, we will describe an example of a representation of CCR that is not a Fock representation. Let us consider, in the space $F(L^2(E^n))$, the operator generalized functions $b(x) = a(x) + \phi(x), b^+(x) = a^+(x) + \overline{\phi(x)}$, where $\phi \in \mathcal{S}'$ is a numerical generalized function. The operators $b(x), b^+(x)$ specify a representation of CCR (we assign the operators

$$b(f) = \int f(x)b(x)dx = a(f) + \int f(x)\phi(x)dx,$$

$$b^+(f) = \int \overline{f(x)}b^+(x)dx = a^+(f) + \int \overline{f(x)\phi(x)}dx$$

to every function $f \in \mathcal{S}$). It is easy to see that this representation of CCR can be a Fock representation only in the case when the function ϕ is square integrable. To check this, consider the vector $\xi \in F(L^2(E^n))$ obeying the condition

$$b(x)\xi = a(x)\xi + \phi(x)\xi = 0. \tag{6.9}$$

Then

$$\langle \xi, a^+(x)\theta \rangle = \langle a(x)\xi, \theta \rangle = -\phi(x)\langle \xi, \theta \rangle.$$

For every vector $\xi \in F(L^2(E^n))$, the function $\langle \xi, a^+(x)\theta \rangle$ is square integrable because this function enters the Fock column corresponding to the vector ξ.[4] This means that in the case when $\phi \notin L^2(E^n)$, equation (6.9) does not have a solution in the Fock space.

A broader class of representations of CCR is described by the following formulas:

$$b(f) = a(\Phi f) + a^+(\overline{\Psi f}) + \int f(x)\phi(x)dx,$$

$$b^+(f) = a(\overline{\Psi f}) + a^+(\Phi f) + \int \overline{f(x)\phi(x)}dx \tag{6.10}$$

(for concreteness, we assume that $f \in \mathcal{S}(E^n), \phi \in \mathcal{S}'(E^n)$, and Φ, Ψ are operators transforming the space $\mathcal{S}(E^n)$ into the space $L^2(E^n)$).

It is easy to check that the relations (6.10) specify a representation of CCR in the case when

$$\Phi^*\Phi - \Psi^*\Psi = 1,$$

$$\Phi^*\overline{\Psi} - \Psi^*\overline{\Phi} = 0,$$

where the operators $\overline{\Phi}, \overline{\Psi}$ are defined by the formulas $\overline{\Phi}f = \overline{\Phi\overline{f}}, \overline{\Psi}f = \overline{\Psi\overline{f}}$. The relation (6.10) can be written in the form

$$b(x) = \int \Phi(y, x)a(y)dy + \int \Psi(y, x)a^+(y)dy + f(x),$$

$$b^+(x) = \int \overline{\Psi(y, x)}a(y)dy + \int \overline{\Phi(y, x)}a^+(y)dy + \overline{f}(x), \tag{6.11}$$

---

[4]We have used the fact that $\langle \xi, \theta \rangle \neq 0$. If $\langle \xi, \theta \rangle = 0$, then $\langle \xi, a^+(x_1)a^+(x_2)\dots a^+(x_n)\theta \rangle = -\phi(x_1)\langle \xi, a^+(x_2), \dots, a^+(x_n)\theta \rangle$. By induction on n, we derive $\langle \xi, a^+(x_1)\dots a^+(x_n)\theta \rangle = 0$, hence $\xi = 0$.

where $\Phi(x,y)$ and $\Psi(x,y)$ are the kernels of the operators Φ and Ψ.

One can prove that the representation of CCR specified in (6.10) is equivalent to the Fock representation if and only if the functions $\Psi(x,y)$ and $f(x)$ are square integrable (see Berezin (2012), where this statement and the analogous statement for CAR were formulated and proved).

The transition from the operators $a^+(x), a(x)$ to the operators $b^+(x), b(x)$, described by the formulas (6.11), is called a canonical transformation in the case when the new operators also satisfy CR. If the operators $b^+(f), b(f)$ generate a Fock representation of CCR, then one can find a unitary operator U satisfying the condition

$$Ua(f)U^{-1} = b(f),$$

$$Ua^+(f)U^{-1} = b^+(f).$$

(This follows from the uniqueness of Fock representation that we have proven.) Otherwise, such a unitary operator doesn't exist.

6.2   The simplest operators on Fock space

Let us fix an orthonormal basis ϕ_k in the space $\mathcal{B}$, where the index k runs over the set M. The operators $a^+(\phi_k), a(\phi_k)$ in the Fock space $F(\mathcal{B})$ are denoted a_k^+, a_k. We will use the notation $a(k, \epsilon) = a_k^\epsilon$, where $\epsilon = \pm 1$, assuming that $a_k^1 = a(k, 1) = a_k^+, a_k^{-1} = a(k, -1) = a_k$.

Let us consider the operators $N_k = a_k^+ a_k$ in the Fock space $F(\mathcal{B})$.

It is easy to check that all the operators N_k commute. Let us find their common eigenvectors.

We consider first the case of CCR. Then the commutation relations of the operators N_k with the operators a_l^+, a_l have the form

$$[N_k, a_l^+] = \delta_{k,l} a_l^+, \quad [N_k, a_l] = -\delta_{k,l} a_l.$$

If ϕ is a common eigenvector of the operators N_k, in other words $N_k\phi = n_k\phi$, then $N_k a_l^+\phi = a_l^+ N_k\phi = n_k a_l^+\phi$ for $l \neq k$, $N_k a_k^+\phi = a_k^+(N_k + 1)\phi = (n_k + 1)a_k^+\phi$. Similarly, $N_k a_l\phi = n_k a_l\phi$ for $l \neq k$, $N_k a_k\phi = (n_k - 1)a_k\phi$. We conclude that vectors of the form

$a_{k_1}^{+n_1} \ldots a_{k_r}^{+n_r} \theta$, where n_i are non-negative integers and $k_i \neq k_j$, form a complete orthogonal[5] (not orthonormal) system of common eigenvectors for the operators N_k. The eigenvalue of the operator N_k in the state $a_{k_1}^{+n_1} \ldots a_{k_r}^{+n_r} \theta$ is equal to n_i, if $k = k_i$, and equal to zero, if k does not coincide with any k_i.

Let us consider the case of CAR. The operators N_k satisfy the relation $N_k^2 = N_k$ in this case, which implies that the eigenvalues of N_k are equal to 0 or 1. Using the equations $N_k a_l^+ = a_l^+ N_k, N_k a_l = a_l N_k$ for $l \neq k$, and $a_k^+ N_k = 0, N_k a_k = 0$ and $N_k a_k^+ = a_k^+(1 - N_k), a_k N_k = (1 - N_k) a_k$, we can analyze the action of the operators a_k^+, a_k on the common eigenvector ϕ of the operators N_k. Namely, if $N_k \phi = n_k \phi$ and $n_l = 1$, then $a_l^+ \phi = 0$; if $n_l = 0$, then $a_l^+ \phi$ is an eigenvector of the operators N_k; and $N_k a_l^+ \phi = n_k a_l^+ \phi$ for $l \neq k, N_k a_k^+ \phi = a_k^+ \phi$. Similarly, for $n_l = 0$ we have $a_l \phi = 0$, and for $n_l = 1$ we obtain that $a_l \phi$ is an eigenvector of the operators N_k, and we have $N_k a_l \phi = n_k a_l \phi$ for $l \neq k, N_k a_k \phi = 0$. We see that the vectors $a_{k_1}^+ \ldots a_{k_n}^+ \theta$, where all k_i are different, constitute a complete orthogonal[6] (but not orthonormal) system of common eigenvectors of the operators N_k, namely we have $N_{k_i} a_{k_1}^+ \ldots a_{k_r}^+ \theta = a_{k_1}^+ \ldots a_{k_r}^+ \theta$ and $N_k a_{k_1}^+ \ldots a_{k_r}^+ \theta = 0$, if k is not equal to any k_i. When we consider simultaneously the case CCR and CAR, we will denote the vectors of the complete system that we have constructed as $a_{k_1}^{+n_1} \ldots a_{k_s}^{+n_s} \theta$; however, we should remember that in the case of CAR the numbers n_i cannot be greater than one (i.e. $n_i = 0, 1$).

The operator N_k has the physical meaning of the operator of the number of particles in the state ϕ_k, and the operators a_k^+ and a_k are called the creation and annihilation operators of a particle in the state ϕ_k. The properties of the operators N_k, a_k^+, a_k agree with this terminology. The operator $N = \sum_k N_k = \sum_k a_k^+ a_k$ is called the operator of the number of particles and the state $x \in F(\mathcal{B})$ satisfying $Nx = nx$ is called an n-particle state.

---

[5] More precisely, two vectors of this kind are either proportional or orthogonal. The completeness of this system of eigenvectors follows from the cyclicity of the vector θ with respect to the family of operators $a^+(f)$.

[6] More precisely, two vectors of this kind are either orthogonal or proportional.

It is easy to see that these definitions agree with the definitions of Section 3.2. In particular, the set of n-particle states coincides with the subspace F_n. This implies that the operator of the number of particles N doesn't depend on the choice of orthonormal basis ϕ_k.

Let us consider the Hamiltonian

$$H_0 = \sum \omega_k a_k^+ a_k = \sum \omega_k N_k, \qquad (6.12)$$

where ω_k are real numbers.[7] Note that the Heisenberg operators $a_k(t) = \exp(-iH_0 t)a_k \exp(-iH_0 t)$ and $a_k^+(t) = \exp(iH_0 t)a_k^+ \exp(-iH_0 t)$ can be calculated; namely, it is clear that the operators $a_k(t) = a_k \exp(-i\omega_k t)$ and $a_k^+(t) = a_k^+ \exp(i\omega_k t)$ satisfy the following Heisenberg equations:

$$\frac{da_k(t)}{dt} = i[H_0, a_k(t)] = -i\omega_k a_k(t),$$

$$\frac{da_k^+(t)}{dt} = i[H_0, a_k^+(t)] = i\omega_k a_k^+(t).$$

The operator H_0 commutes with the operators N_k, therefore the system of vectors $a_{k_1}^{+n_1} \ldots a_{k_r}^{+n_r}\theta$ that we have constructed is a complete system of eigenvectors of the operator H_0 and the eigenvalues are equal to $\sum_{i=1}^{r} \omega_{k_i} n_i$.

Let us find the ground state of the Hamiltonian H_0. In the case of CCR, if at least one $\omega_k < 0$, then the spectrum of energies is unbounded from below and therefore the ground state does not exist. If all $\omega_k \geq 0$, then all energy levels are non-negative and the vacuum vector θ is the ground state. In the case of CAR, the lowest energy level is equal to $E_0 = \sum_{k \in L} \omega_k$, where the sum is taken over the set L consisting of k, for which $\omega_k < 0$. To obtain the ground state Φ from the vacuum vector θ, we should apply all operators a_k^+ with $k \in L$ to the vacuum vector (in other words, $\Phi = (\prod_{k \in L} a_k^+)\theta$). If the set L is infinite, then the ground state doesn't exist. Let's introduce the

---

[7]The operator H_0 is an essentially self-adjoint operator for any choice of real numbers ω_k (this follows from the fact that the operator H_0 has a complete system of eigenvectors). In particular, the operator of the number of particles N is essentially self-adjoint. As usual, we identify the essentially self-adjoint operator with its self-adjoint extension.

operators b_k in the following way: $b_k = a_k^+$ for $k \in L$ and $b_k = a_k$ for $k \notin L$. The operators b_k^+, b_k satisfy the relations (6.5). To check that these operators generate a representation of CAR, we should define the operators $b^+(f), b(f)$ for all $f \in \mathcal{B}$. It is easy to do this, setting $b(f) = a^+(Pf) + a((1 - P)f)$, where $Pf = \sum_{k \in L} \langle f, \phi_k \rangle \phi_k$. It is convenient to express the Hamiltonian H_0 in terms of the operators b_k^+, b_k. If the set L is finite, then

$$H_0 = \sum |\omega_k| b_k^+ b_k + \sum_{k \in L} \omega_k,$$

and then the ground state of the Hamiltonian H_0 is the vacuum vector for the operators b_k, i.e. it satisfies the conditions $b_k \Phi = 0$. If the set L is infinite, then the vector satisfying the condition $b_k \Phi = 0$ in the space $F(\mathcal{B})$ doesn't exist. In other words, the operators $b^+(f), b(f)$ specify a representation of CAR that is not equivalent to the Fock representation.

The operators acting in Fock space $F(\mathcal{B})$, where $\mathcal{B} = L^2(X)$, can be expressed in terms of operator generalized functions $a^+(x), a(x)$, satisfying the relations (6.7). The statements proven in this section can also be proven for the operators in the space $F(\mathcal{B})$ that are expressed in terms of $a^+(x), a(x)$. In particular, one can introduce the commuting operators (or more precisely operator generalized functions) $N(x) = a^+(x)a(x)$. The operator of the number of particles N can be expressed in terms of the operators $N(x)$ by the formula

$$N = \int N(x)dx$$

(more details about the operator of the number of particles can be found in Section 3.2). A complete system of generalized eigenvectors (generalized eigenbasis) of the operator

$$H_0 = \int \omega(x)a^+(x)a(x)dx = \int \omega(x)N(x)dx \qquad (6.13)$$

consists of vectors (vector generalized functions) $a^+(x_1) \ldots a^+(x_n)\theta$.

Operators of the form (6.12), (6.13) are among the simplest operators in Fock space. Slightly more general operators (3.9) are considered in Section 3.2. The analysis of Hamiltonians of the form

$$H = \int A(x,y)a^+(x)a(y)dxdy + \int B(x,y)a^+(x)a^+(y)dxdy$$

$$+ \int \overline{B(x,y)}a(y)a(x)dxdy$$

$$+ \int C(x)a^+(x)dx + \int \overline{C(x)}a(x)dx \tag{6.14}$$

(quadratic Hamiltonians) is more complicated. Under certain conditions, they can be reduced to the form (6.13) by means of a linear canonical transformation of the form (6.11). A complete analysis of the Hamiltonian (6.14) can be found in the book by Berezin (2012). Much harder is the situation for the Hamiltonians of degree higher than two, with respect to the operators of creation and annihilation. Very rarely one can find precisely the eigenvalues or the eigenvectors of these operators and hence it is necessary to apply other methods. In Chapter 11, we will explain a method of calculating the evolution operator and the eigenvalues of the Hamiltonian $H = H_0 + gV$, where $H_0 = \sum \omega_k a_k^+ a_k$, in the form of power series with respect to the parameter g (perturbation theory).

In this section, we will consider the conditions under which the formal expression

$$H = \sum_m \sum_{k_i,\epsilon_i} \Gamma^m_{k_1,\epsilon_1,\ldots,k_m,\epsilon_m} a_{k_1}^{\epsilon_1} \ldots a_{k_m}^{\epsilon_m} \tag{6.15}$$

defines an operator on the Fock space $F(\mathcal{B})$ (here, as always, $a_k^1 = a_k^+ = a^+(\phi_k), a_k^{-1} = a_k = a(\phi_k)$, where ϕ_k is an orthonormal basis for $\mathcal{B}$, k_i are taken from the set M, and $\epsilon_i = \pm 1$). In the case of CAR, one can prove the following statement. *If the numerical series*

$$\sum_m \sum_{k_i,\epsilon_i} \Gamma^m_{k_1,\epsilon_1,\ldots,k_m,\epsilon_m} \tag{6.16}$$

is absolutely convergent, then the operator series (6.15) *is also absolutely convergent (in the sense of norm convergence) and specifies a bounded operator in* $F(\mathcal{B})$. This follows from the remark that in the case of CAR we have that $\|a_k^\epsilon\| \leq \|\phi_k\| = 1$ and hence, $\|a_{k_1}^{\epsilon_1} \ldots a_{k_m}^{\epsilon_m}\| \leq 1$.

If, in addition, the expression (6.15) is formally Hermitian, i.e. obeys

$$\overline{\Gamma^m_{k_1,\epsilon_1,\ldots,k_m,\epsilon_m}} = \Gamma^m_{k_1,-\epsilon_1,\ldots,k_m,-\epsilon_m}, \tag{6.17}$$

then the operator H defined by the formula (6.15) is self-adjoint (since bounded Hermitian operators are self-adjoint).

In the case of CCR, we can use the following estimates for an n-particle state x:

$$\|a_k^\epsilon x\| \le \sqrt{n}\|x\|,$$

$$\|a_{k_1}^{\epsilon_1}\ldots a_{k_m}^{\epsilon_m} x\| \le \sqrt{n}\sqrt{n+1}\ldots\sqrt{n+m-1}\|x\| \tag{6.18}$$

(the first of which can be obtained from the relation

$$\|a_k x\|^2 = \langle a_k^+ a_k x, x\rangle \le \left\langle \sum_k a_k^+ a_k x, x\right\rangle = \langle Nx, x\rangle = n\|x\|^2,$$

and the second follows from the first).

Using the estimate (6.18), one can check that the expression (6.15) specifies an operator with everywhere dense domain in $F(\mathcal{B})$, if this expression is polynomial ($\Gamma^m_{k_1,\epsilon_1,\ldots,k_m,\epsilon_m} = 0$ for $m > s$) and the numerical series (6.16) is absolutely convergent. This follows from the remark that the series

$$\sum_{m=1}^{s}\sum_{k_i,\epsilon_i} \Gamma^m_{k_1,\epsilon_1,\ldots,k_m,\epsilon_m} a_{k_1}^{\epsilon_1}\ldots a_{k_m}^{\epsilon_m} x$$

converges for $x \in F_n$, and hence the domain of the operator H contains D — the union of all linear spaces F_n. If the operator H is formally Hermitian, i.e. the condition (6.17) is satisfied, then H will be Hermitian on D. However, it is not clear whether this operator is essentially self-adjoint on D; results in this direction can be found in the paper by Hepp (1966). The operators of interest for physics are usually bounded from below and hence they define a self-adjoint operator by means of Friedrich's extension (see Appendix A.5). The analysis of the operator H defined by the expression (6.15) is much easier when $\Gamma^m_{k_1,\epsilon_1,\ldots,k_m,\epsilon_m} \neq 0$ only for $\epsilon_1 + \cdots + \epsilon_m = 0$ (i.e.

if the operator H commutes with the operator of the number of particles N). Under this condition, one can consider the operator H separately on each subspace F_n. One can extend the class of operators of the form (6.15) that defines self-adjoint operators, by noting that adding to a self-adjoint operator (for example, to H_0) a bounded self-adjoint operator, we again obtain a self-adjoint operator.

We will not analyze the conditions under which expressions of the form

$$H = \sum_m \sum_{\epsilon_1,\ldots,\epsilon_m} \int \Gamma^m(x_1, \epsilon_1, \ldots, x_n, \epsilon_n) a(x_1, \epsilon_1)$$

$$\ldots a(x_n, \epsilon_n) dx_1 \ldots dx_n \tag{6.19}$$

define an operator on the space $F(\mathcal{B}) = F(L^2(X))$. Note that only for the case $X = E^r$ some results in this direction are proven in Section 3.2.

In the case of CAR an operator corresponding to a physical quantity (in particular, a Hamiltonian) should always contain an even number of creation and annihilation operators (i.e. $\Gamma^m = 0$ for odd m).

6.3   The normal form of an operator: Wick's theorem

Let us say that an operator C in the Fock space $F(\mathcal{B})$ is presented in normal form if the operators of creation are to the left of the annihilation operators:

$$C = \sum_{m,n} \Gamma_{m,n}(k_1, \ldots, k_m | l_1, \ldots, l_n) a_{k_1}^+ \ldots a_{k_m}^+ a_{l_1} \ldots a_{l_n} \tag{6.20}$$

(here $a_k = a(\phi_k)$, where ϕ_k is an orthonormal basis in $\mathcal{B}$. As usual, we sum over repeated indices k_i, l_j). The sum in the expression above can be infinite, in which case the convergence of the series is understood in the sense of strong operator convergence. The set of indices for the operators a_k will be denoted by M. Since M is countable, we can identify M with the set of natural numbers; however, this is not always convenient.

The product of an arbitrary number of operators a_k^+, a_l, taken in any order, as well as any finite sum of such operators, can be written in normal form by means of CR. It can be shown that any bounded operator can be written in normal form if the convergence in (6.20) is understood in the sense of weak convergence.

The representation of an operator in normal form is especially convenient by the calculation of the vacuum average (vacuum expectation value) $\langle C\theta, \theta \rangle$ of the operator C. It is clear that $\langle C\theta, \theta \rangle = \Gamma_{0,0}$ (the constant term in the representation of the operator C in normal form).

Let us assume that $C_1, \ldots, C_k$ are operators written in normal form. Let us consider the following question: How can we write the product $C_1 \ldots C_k$ of these operators in normal form?

Before we answer this question, let us consider a simpler (though, in some sense, more general) situation. Let $A_1, \ldots, A_m, B_1, \ldots, B_n$ be operators with the property that the commutators $[A_\alpha, B_\beta]$ are numbers (i.e. they can be written as product of the identity operator and a number). The number $B_\beta A_\alpha = [A_\alpha, B_\beta]$ is called the *contraction* of the operators A_α and B_β. The *product $B_1 \ldots B_n A_1 \ldots A_m$ with one contraction $B_\beta A_\alpha$* will be understood as this product with the operators B_β and A_α deleted and with the factor $B_\beta A_\alpha$ inserted instead of these operators; to denote the product with the contraction $B_\beta A_\alpha$ we will write $B_1 \ldots B_\beta \ldots B_n A_1 \ldots A_\alpha \ldots A_m$. The *product $B_1 \ldots B_n A_1 \ldots A_m$ with k contractions $B_{\beta_1} A_{\alpha_1}, \ldots, B_{\beta_k} A_{\alpha_k}$* will be understood as a product where the operators $B_{\beta_1}, \ldots, B_{\beta_k}, A_{\alpha_1}, \ldots, A_{\alpha_k}$ are deleted and the contractions $B_{\beta_1} A_{\alpha_1}, \ldots, B_{\beta_k} A_{\alpha_k}$ are included in their place. Hence, in order to define a product with k contractions, we should select k operators $B_{\beta_1}, \ldots, B_{\beta_k}$ (the order of the selection is unimportant), and to every one of these operators, we should assign one of the operators $A_1, \ldots, A_n$, assuming that different indices $\beta_1, \ldots, \beta_k$ correspond to different indices $\alpha_1, \ldots, \alpha_k$.

Let us now formulate the following statement, which we will call Wick's theorem, noting that it presents a simplified form of this theorem:

The product $A_1 \ldots A_m B_1 \ldots B_n$ is equal to the product $B_1 \ldots B_n A_1 \ldots A_m$ with additional summand terms obtained by all possible contractions.

For example,
$$A_1 A_2 B_1 B_2 = B_1 B_2 A_1 A_2 + B_1 B_2 \underline{A_1\, A_2} + B_1 \underline{B_2 A_1}\, A_2 + \underline{B_1 B_2 A_1 A_2}$$
$$+ B_1\, \underline{B_2 A_1}\, A_2 + B_1\, \underline{B_2 A_1 A_2} + \underline{B_1 B_2 A_1 A_2}$$
$$+ \underline{B_1\, B_2 A_1}\, A_2 \,.$$

This statement is almost obvious: when we transfer an operator B_β to the left, we use the relation $A_\alpha B_\beta = B_\beta A_\alpha + B_\beta \underline{A_\alpha}$, and therefore every time we change the order of B_β and A_α, we will add two summands: one with the same number of contractions as in the original summand and another where the number of contractions is greater by one.

In order to give a formal proof, we can start by demonstrating the following statement (either directly or by induction over m):

$$A_1 \ldots A_m B = B A_1 \ldots A_m + B \underline{A_1} \ldots A_m + B \underline{A_1 A_2 \ldots A_m}$$
$$+ \cdots + B \underline{A_1 \ldots A_m} \,. \tag{6.21}$$

In other words, we check Wick's theorem for $n = 1$. Then, we can use mathematical induction with respect to the number of operators B_β; at every induction step, we use the formula (6.21).

Let us now consider the reduction to normal form of the product of two operators K and L, both represented in normal form. Without loss of generality, we can assume that K and L have definite degree with respect to creation and annihilation operators:

$$K = K(k_1, \ldots, k_m | l_1, \ldots, l_n) a_{k_1}^+ \ldots a_{k_m}^+ a_{l_1} \ldots a_{l_n},$$

$$L = L(p_1, \ldots, p_r | q_1, \ldots, q_s) a_{p_1}^+ \ldots a_{p_r}^+ a_{q_1} \ldots a_{q_s}$$

(as always, we have in mind a summation over the repeated indices from the set M).

Let us begin with the case of CCR. In this case, applying Wick's theorem to the product $a_{l_1} \ldots a_{l_n} a_{p_1}^+ \ldots a_{p_r}^+$, one can show that KL is equal to

$$K(k_1, \ldots, k_m | l_1, \ldots, l_n) L(p_1, \ldots, p_r | q_1, \ldots, q_s)$$
$$\times a_{k_1}^+ \ldots a_{k_m}^+ a_{p_1}^+ \ldots a_{p_r}^+ a_{l_1} \ldots a_{l_n} a_{q_1} \ldots a_{q_s},$$

with the addition of summands obtained from this expression by means of arbitrary contractions between the operators $a_{p_i}^+$ and a_{l_j} (recall that $a_p^+ a_l = [a_l, a_p^+] = \delta_{pl}$).

For example, if $K = K(k|l)a_k^+ a_l$ and $L = L(p_1, p_2|q_1, q_2)a_{p_1}^+ a_{p_2}^+ a_{q_1} a_{q_2}$, then

$$KL = K(k|l)L(p_1, p_2|q_1, q_2)a_k^+ a_{p_1}^+ a_{p_2}^+ a_l a_{q_1} a_{q_2}$$

$$+ K(k|l)L(p_1, p_2|q_1, q_2)a_k^+ a_{p_1}^+ a_{p_2}^+ a_l\, a_{q_1} a_{q_2}$$

$$+ K(k|l)L(p_1, p_2|q_1, q_2)a_k^+ a_{p_1}^+\, a_{p_2}^+ a_l\, a_{q_1} a_{q_2}$$

$$= K(k|l)L(p_1, p_2|q_1, q_2)a_k^+ a_{p_1}^+ a_{p_2}^+ a_l a_{q_1} a_{q_2}$$

$$+ K(k|l)L(l, p_2|q_1, q_2)a_k^+ a_{p_2}^+ a_{q_1} a_{q_2}$$

$$+ K(k|l)L(p_1, l|q_1, q_2)a_k^+ a_{p_1}^+ a_{q_1} a_{q_2}.$$

The functions $K(k_1, \ldots, k_m | l_1, \ldots, l_n)$ and $L(p_1, \ldots, p_r | q_1, \ldots, q_s)$ can be assumed to be symmetric with respect to every group of indices. In what follows, we will assume that this symmetry condition is always satisfied. Using this assumption, we can considerably reduce the number of summands in the expression for KL, since the summands corresponding to the same number of contractions are identical. In our example, using the symmetry of coefficients, one notices that the second and third summands are equal.

The reduction to normal form can be represented graphically by means of diagrams suggested by Feynman. The operator

$$K = K(k_1, \ldots, k_m | l_1, \ldots, l_n)a_{k_1}^+ \ldots a_{k_m}^+ a_{l_1} \ldots a_{l_n}$$

will be denoted by a diagram consisting of one internal vertex with m dotted incoming lines and n dotted outgoing lines (we consider the lines to be directed topological intervals). We will say that such a diagram is a *star* (see Fig. 6.1). The beginning of an incoming line will be called an in-vertex and the end of an outgoing line will be called an out-vertex. We will assign the indices $k_1, \ldots, k_m \in M$ to in-vertices; to out-vertices, we will assign the indices $l_1, \ldots, l_n \in M$; and to the internal vertex, we will assign the function $K(k_1, \ldots, k_m | l_1, \ldots, l_n)$. The operator L is depicted by a similar diagram.

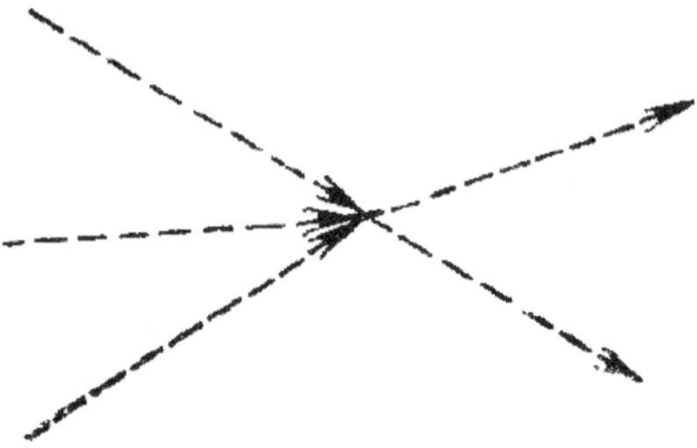

Fig. 6.1   Star: Feyman diagram of one internal vertex.

The normal form of the operator KL can be expressed now as a collection of all different diagrams, consisting of diagrams of the operators K and L, where some of the out-vertices of the diagram for the operator K are connected with solid lines with the in-vertices of the diagram for the operator L. These solid lines will be called edges of the diagram and we assign the contractions $a_p^+ a_l = [a_l, a_p^+] = \delta_{lp}$ to the edges. (Here, l is the index of the end of the edge and p is the index of the beginning of the edge.) The direction of the edge will always be fixed as the direction from the out-vertex of the diagram of the operator K to the in-vertex of the diagram of the operator L. For every diagram, we construct a product in which the internal vertices of the diagram contribute the factors $K(k_1, \ldots, k_m | l_1, \ldots, l_n)$ and $L(p_1, \ldots, p_r | q_1, \ldots, q_s)$, where $k_1, \ldots, k_m, p_1, \ldots, p_r$ are the indices of the ends of the lines entering the vertex and $l_1, \ldots, l_n, q_1, \ldots, q_s$ are the indices of the ends of the lines exiting the in-vertex. The factor $a_p^+ a_l = \delta_{lp}$ is assigned to every connecting edge. Finally, the lines with a free end contribute the operator a_k^+ or a_k, if the line is incoming or outgoing, respectively. We assume a summation over the set M for each index. The operators a_k^+, a_l are assumed to be written in normal order.

It is easy to check that the operator constructed by the diagram described above coincides with one of the summands in the product of the operators K and L written in normal form. Therefore, *the operator KL is equal to the sum of operators corresponding to all the possible diagrams* (more shortly, to the sum of diagrams).

By induction, it is easy to obtain diagrams for the reduction to normal form of a product of r operators $K_1, \ldots, K_r$, where

$$K_i = K_i(k_1, \ldots, k_{m_i} | l_1, \ldots, l_{n_i}) a_{k_1}^+ \ldots a_{k_{m_i}}^+ a_{l_1} \ldots a_{l_{n_i}}.$$

The normal form of the operator $K_1 \ldots K_r$ is depicted by a sum of diagrams consisting of the diagrams of the operators $K_1, \ldots, K_r$ and some number of lines connecting the outgoing vertices of the diagram for the operator K_i with the incoming vertices of the diagram of the operator K_j (here $i, j = 1, \ldots, r; i < j$). The prescription that assigns to a diagram an operator remains the same.

If the operator $K_i = \sum_{m,n} K_i^{m,n}$, where

$$K_i^{m,n} = K_i^{m,n}(k_1, \ldots, k_m | l_1, \ldots, l_n) a_{k_1}^+ \ldots a_{k_m}^+ a_{l_1} \ldots a_{l_n},$$

then the diagram of the operator K_i is a set of several stars, the diagrams of the operators $K_i^{m,n}$.

The normal form of the operator $K_1 \ldots K_r$ is represented then by the collection of diagrams consisting of r stars — a diagram of every operator $K_1, \ldots, K_r$ — and several lines connecting outgoing vertices of the star diagram of one of the operators K_i with incoming vertices of the star diagram of one of the operators $K_j, j > i$.

The diagram technique in Fig. 6.1 can be applied to reduce to the normal form a product of operators expressed in terms of creation and annihilation operators satisfying CCR. A similar diagram technique can also be constructed in the case of CAR; the only difference is in some signs in front of the diagrams. We do not describe explicitly the rules for these sign factors; however, we will formulate accurately the analog of Wick's theorem for the CAR case.

Let $A_1, \ldots, A_m, B_1, \ldots, B_n$ be operators having the property that the anticommutators $[A_\alpha, B_\beta]_+$ are numbers. The number $B_\beta A_\alpha = [A_\alpha, B_\beta]_+$ is called a *contraction* of the operators A_α and B_β. The product $B_1 \ldots B_n A_1 \ldots A_m$ with k contractions $B_{\beta_1} A_{\alpha_1}, \ldots, B_{\beta_k} A_{\alpha_k}$ is defined as the operator obtained by deleting from the product the operators $B_{\beta_1}, \ldots, B_{\beta_k}, A_{\alpha_1}, \ldots, A_{\alpha_k}$ and including instead of them the contractions $\underline{B_{\beta_1} A_{\alpha_1}}, \ldots, \underline{B_{\beta_k} A_{\alpha_k}}$ and the sign factor $(-1)^{mn+\lambda+\mu}$, where λ is the sum of the lengths of all contractions and μ is the number of pairs of overlapping contractions (the length of a contraction $\underline{B_\beta A_\alpha}$ is defined as $n - \beta + \alpha$; two contractions $\underline{B_\beta A_\alpha}$ and $\underline{B_{\beta'} A_{\alpha'}}$ are called overlapping if $(\beta' - \beta)(\alpha' - \alpha) > 0)$.

Given these definitions, one can formulate Wick's theorem in the form that is applicable in both situations.

The product $A_1 \ldots A_m B_1 \ldots B_n$ is equal to the product $B_1 \ldots B_n A_1 \ldots A_m$ with all possible contractions (we do not exclude the case where the number of contractions is equal to zero).

For example, in the case when the anticommutators $[A_\alpha, B_\beta]_+$ are numbers, we obtain

$$A_1 A_2 B_1 B_2 = (-1)^{2\cdot2} B_1 B_2 A_1 A_2 + (-1)^{2\cdot2+2}\, B_1 B_2 \underline{A_1}\, A_2$$

$$+ (-1)^{2\cdot2+3}\, B_1 B_2 \underline{A_1 A_2} + (-1)^{2\cdot2}\, B_1\, \underline{B_2 A_1}\, A_2$$

$$+ (-1)^{2\cdot2+2}\, B_1 \underline{B_2 A_1}\, A_2 + (-1)^{2\cdot2+4}\, B_1\, \underline{B_2 A_1}\, A_2$$

$$+ (-1)^{2\cdot2+4+1}\, B_1\, \underline{B_2 A_1\, A_2}\,.$$

An operator, acting on the Fock space $F(\mathcal{B})$, where $\mathcal{B} = L^2(X)$, is said to be represented in normal form if it is written in the form

$$A = \sum_{m,n} \int k_{m,n}(x_1, \ldots, x_m | y_1, \ldots, y_n) a^+(x_1) \ldots a^+(x_m)$$

$$\times\, a(y_1) \ldots a(y_n) d^m x\, d^n y \tag{6.22}$$

(here X is a measure space and $a^+(x), a(x)$ are operator generalized functions satisfying the relation (6.7)). The diagram techniques discussed above can also be used to represent in normal form the product of operators written in the form (6.22). The only modification that is necessary in the case in question is that the role of the indices $k \in M$ is played by the elements of the space

X, correspondingly the summation with respect to M is replaced by integration over the space X.[8]

Let us use the above results to prove some useful relations. For definiteness, we will consider the case of CCR.

First of all, we will calculate the inner product

$$\langle a^+(f_1)\dots a^+(f_m)\theta, a^+(g_1)\dots a^+(g_n)\theta\rangle$$
$$= \langle a(g_n)\dots a(g_1)a^+(f_1)\dots a^+(f_m)\theta, \theta\rangle.$$

The operator $a(g_n)\dots a(g_1)a^+(f_1)\dots a^+(f_m)$ can be reduced to normal form by means of Wick's theorem; it is equal to the sum where every summand is the product $a^+(f_1)\dots a^+(f_m) \times a(g_n)\dots a(g_1)$ with some contractions. Remembering that the vacuum average of an operator is equal to the constant term in the representation of the operator in normal form, we see that the quantity we would like to calculate is equal to the sum of the summands, where all operators are contracted (i.e. $m = n$ is equal to the number of contractions). Noting that

$$\underline{a^+(f_i)a(g_j)} = [a(g_j), a^+(f_i)] = \langle g_j, f_i\rangle = \overline{\langle f_i, g_j\rangle},$$

thus, we obtain the necessary formula

$$\langle a^+(f_1)\dots a^+(f_m)\theta, a^+(g_1)\dots a^+(g_n)\theta\rangle$$
$$= \delta_m^n \sum_P \overline{\langle f_1, g_{i_1}\rangle}\,\overline{\langle f_2, g_{i_2}\rangle}\dots\overline{\langle f_m, g_{i_m}\rangle} \qquad (6.23)$$

---

[8]It is useful to note that an orthonormal basis ϕ_k, with $k \in M$, for the space $\mathcal{B}$, specifies an isomorphism between the spaces $\mathcal{B}$ and $L^2(M)$. (The set M is considered as a measure space, equipped with the counting measure, where the measure of a finite subset is equal to the number of its elements; then the integral of $f(k)$ over the set M is equal to $\sum_{k\in M} f(k)$.) This remark allows us to say that the representation of an operator in the form (6.20) is a particular case of representation in the form (6.22); correspondingly, the diagram technique for the operators of the form (6.22) is a generalization of the techniques for operators of the form (6.20).

(where the sum is taken over all permutations $P = (i_1, \ldots, i_m)$). The formula (6.23) can also be expressed in the form

$$\langle a^+(x_1) \ldots a^+(x_m)\theta, a^+(y_1) \ldots a^+(y_n)\theta \rangle$$

$$= \delta_m^n \sum_P \delta(x_1, y_{i_1}) \ldots \delta(x_m, y_{i_m}), \qquad (6.24)$$

where $a^+(x), a(y)$ are operator generalized functions satisfying the relations (6.7).

Using the statements proved in this section, we can find the relation between the coefficient functions $K_{m,n}(x_1, \ldots, x_m | y_1, \ldots, y_n)$ in the representation of an operator A in the form (6.22) (normal form) and the matrix elements of the operator A in the generalized basis $a^+(x_1) \ldots a^+(x_m)\theta$. To find these relations, we can, for example, write the operator $Aa^+(y_1) \ldots a^+(y_n)\theta$ in normal form by means of Wick's theorem, and then we can calculate the matrix element

$$A_{m,n}(x_1, \ldots, x_m | y_1, \ldots, y_n)$$

$$= \langle Aa^+(y_1) \ldots a^+(y_n)\theta, a^+(x_1) \ldots a^+(x_m)\theta \rangle$$

using the formula (6.24). As a result, we obtain the following formula:

$$A_{m,n}(x_1, \ldots, x_m | y_1, \ldots, y_n) = K_{m,n}(x_1, \ldots, x_m | y_1, \ldots, y_n)m!n!$$

$$+ \sum_{i,j} K_{m-1,n-1}(x_1, \ldots, x_{i-1}, x_{i+1}, \ldots, x_m | y_1, \ldots, y_{j-1}, y_{j+1},$$

$$\ldots, y_n)\delta(x_i - y_j)(m-1)!(n-1)! + \cdots$$

$$+ \sum_{B,B',\alpha} K_{m-r,n-r}(x_i, i \overline{\in} B | y_j, j \overline{\in} B')$$

$$\times \prod_{i \in B} \delta(x_i - y_{\alpha(i)})(m-r)!(n-r)! + \cdots$$

(here B and B' consist of r elements, $B \subset \{1, \ldots, m\}, B' \subset \{1, \ldots, n\}$, α is a one-to-one correspondence between B and B', and we are summing over all possible B, B', α; $K_{m-r,n-r}(x_i, i \overline{\in} B | y_j, j \overline{\in} B')$ denotes the function $K_{m-r,n-r}$ with arguments x_i and y_j satisfying $i \overline{\in} B, j \overline{\in} B')$.

To conclude this section, we will formulate some useful definitions. Let K and L be operators represented in normal form as follows:

$$K = \sum_{m,n} \int K_{m,n}(k_1, \ldots, k_m | l_1, \ldots, l_n) a^+(k_1)$$

$$\ldots a^+(k_m) a(l_1) \ldots a(l_n) d^m k d^n l,$$

$$L = \sum_{m,n} \int L_{m,n}(k_1, \ldots, k_m | l_1, \ldots, l_n) a^+(k_1)$$

$$\ldots a^+(k_m) a(l_1) \ldots a(l_n) d^m k d^n l.$$

Then the normal product of operators K and L in the case of CCR is the operator

$$N(KL) = \sum_{m,n,r,s} \int K_{m,n}(k_1, \ldots, k_m | l_1, \ldots, l_n) L_{r,s}(p_1,$$

$$\ldots, p_r | q_1, \ldots, q_s) a^+(k_1) \ldots a^+(k_m) a^+(p_1) \ldots a^+(p_r) a(l_1)$$

$$\ldots a(l_n) a(q_1) \ldots a(q_s) d^m k d^n l d^r p d^s q,$$

and in the case of CAR is the operator

$$N(KL) = \sum_{m,n,r,s} (-1)^{nr} \int K_{m,n}(k_1, \ldots, k_m | l_1, \ldots, l_n) L_{r,s}(p_1,$$

$$\ldots, p_r | q_1, \ldots, q_s) a^+(k_1) \ldots a^+(k_m) a^+(p_1) \ldots a^+(p_r) a(l_1)$$

$$\ldots a(l_n) a(q_1) \ldots a(q_s) d^m k d^n l d^r p d^s q$$

(instead of using the symbol $N(KL)$ to denote normal product, one can also use the symbol $:KL:$). In other words, the normal product is obtained if we reduce the product KL to the normal form using the prescriptions above and delete all summands containing at least one contraction (i.e. we delete all diagrams having at least one non-dotted line). The normal product of n operators $K_1, \ldots, K_n$ can be defined by induction

$$N(K_1, \ldots, K_n) = N(N(K_1, \ldots, K_{n-1}) K_n).$$

The normal exponent of the operator K is the operator

$$N(\exp K) = \sum_{n=0}^{\infty} \frac{1}{n!} N(K^n).$$

6.4 Diagram techniques

Let us consider the Hamiltonian $H = H_0 + gV(t)$, where

$$H_0 = \sum \epsilon(k) a_k^+ a_k, \tag{6.25}$$

where

$$V(t) = \sum_{m,n} \sum_{k_1,\ldots,k_m} v_{m,n}(k_1,\ldots,k_m|p_1,\ldots,p_n|t) a_{k_1}^+ \ldots a_{k_m}^+$$
$$\times a_{p_1} \ldots a_{p_n}$$

(here, $a_k^+ = a^+(\phi_k)$, $a_k = a(\phi_k)$ are operators in the Fock space $F(\mathcal{B})$ corresponding to the orthonormal basis $\phi_k \in \mathcal{B}$ and the index k runs over the set M). For concreteness, we consider the case of CCR; however, the consideration of this section can also be performed in the case of CAR (in the case of CAR, the Hamiltonian H is assumed to be Fermi-even, i.e. it is assumed that every summand in the operator H contains an even number of operators a^+ and a ($m + n$ is even)).

Using the representation of the operator $S(t, t_0)$ in the form of T-exponent (4.4) and Wick's theorem (Section 6.3), we can construct diagram techniques for the calculation of the operator $S(t, t_0)$. More precisely, we obtain the decomposition of the normal form of $S(t, t_0)$ as a power series in g.

Let us recall that a star is a diagram consisting of a point with m incoming lines and n outgoing lines. The lines belonging to a star are depicted as dotted lines (Fig. 6.1 depicts a star with three incoming and two outgoing lines).

A diagram of the operator $S(t, t_0)$ is a collection of several stars and several edges that are depicted as directed lines. Every edge starts with an out-vertex of some star and ends in an in-vertex of another star. Let us assume that two lines belonging to a diagram cannot have common internal points and can have only one common vertex. Two vertices belonging to the same edge cannot belong to the same star.

The vertices belonging to edges are called internal vertices. Other vertices are called external vertices.

All vertices of a diagram should be numbered in such a way that the vertices belonging to the same star are numbered by neighboring numbers. Two diagrams are equivalent if there exists a topological equivalence between these diagrams transforming a vertex with some number to a vertex with the same number.

Some examples of diagrams are drawn in Fig. 6.2. The diagrams (a) and (b) are topologically equivalent, but are distinct, while the diagrams (a) and (c) are equivalent.

For every diagram, we can construct an operator in the following way. Let us assume that to every vertex of the diagram we have assigned an index $k_i \in M$ and a real number t_i (time). We assign the same time to every vertex belonging to one star. Every star contributes a factor $v_{m,n}(k_1, \ldots, k_m | p_1, \ldots, p_n | t)$, where t is the time of this star and $k_1, \ldots, k_m$ are indices of in-vertices of this star and $p_1, \ldots, p_n$ are indices of the out-vertices of the star. An edge contributes

$$\exp(-i\epsilon(k_1)(t_1 - t_2))\delta^{k_2}_{k_1}\theta(t_1 - t_2),$$

where the index $k_1 \in M$ and the time t_1 correspond to the beginning of the edge and $k_2 \in M$ and t_2 correspond to the end of the edge. To every free in-vertex, we assign an operator $\tilde{a}^+_k(t) = \exp(i\epsilon(k)t)a^+_k$, and to every free out-vertex, we assign the operator $\tilde{a}_k(t) = a_k \exp(-i\epsilon(k)t)$ (here k and t stand for the index and the time corresponding to the vertex, respectively).

To every diagram, we assign an operator obtained in the following way. First of all, we take the product of functions corresponding to the stars and edges of the diagram and the operators corresponding to the external in- and out-vertices. The operator is obtained from this product by means of summing over the indices of all vertices and integration over the times of all stars (calculating $S(t, t_0)$ we should integrate over the interval $[t_0, t]$). The operators corresponding to the external vertices are taken in normal order. We include the factor $\frac{1}{n!}g^n$ in the operator corresponding to the diagram with n stars.

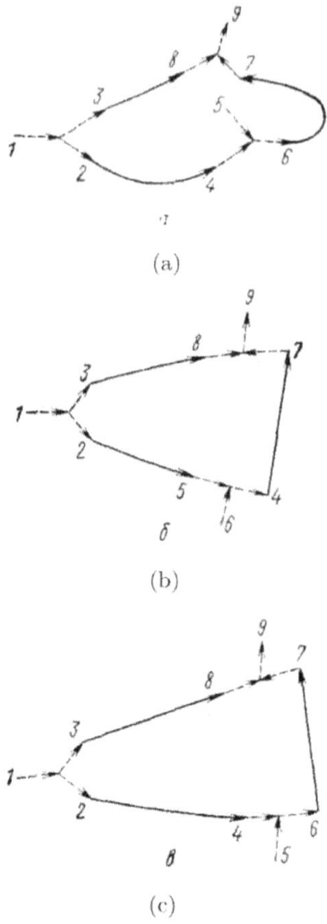

Fig. 6.2　Examples of diagrams.

Let us write down, for example, the operator corresponding to Fig. 6.2(a):

$$\frac{1}{3!}g^3 \sum_{k_1,\ldots,k_9} \int dt_1 dt_2 dt_3 v_{1,2}(k_1|k_2,k_3|t_1) v_{2,1}(k_4,k_5|k_6|t_2)$$

$$\times\, v_{2,1}(k_8,k_7|k_9|t_3) \exp(-i\epsilon(k_3)(t_1-t_3))\delta_{k_3}^{k_8}\theta(t_1-t_3)$$

$$\times\, \exp(-i\epsilon(k_2)(t_1-t_2))\delta_{k_2}^{k_4}\theta(t_1-t_2)\exp(-i\epsilon(k_6)(t_2-t_3))$$

$$\times\, \delta_{k_6}^{k_7}(t_2-t_3)\exp(i\epsilon(k_1)t_1+i\epsilon(k_5)t_2$$

$$-\,i\epsilon(k_9)t_3)a_{k_1}^{+}a_{k_5}^{+}a_{k_9}.$$

Let us prove that the operator $S(t, t_0)$ can be expressed as a sum of all different diagrams, or more precisely, as a sum of all the operators corresponding to these diagrams. Of course, this statement (and other statements of this kind) means only that the sum of the diagrams gives a decomposition $S(t, t_0)$ as a series in g; we cannot say anything about the convergence of this series.

Before giving the proof of this statement, let us make several remarks about the structure of the diagram representation of the operator $S(t, t_0)$ and consider an example. Let us note first that we can omit all diagrams containing n stars joined in a cyclic order by edges. (We say that n stars are connected in cyclic order by edges if they can be ordered in such a way that for every $i, 1 \leq i \leq n$, there exists an edge starting in a vertex of the ith star and ending in the vertex of the $(i+1)$-th star; we identify the $(n+1)$th star with the first star.) Let us denote the times of these stars by $\tau_1, \ldots, \tau_n$; then the edges introduce the number $\theta(\tau_1 - \tau_2) \ldots \theta(\tau_{n-1} - \tau_n)\theta(\tau_n - \tau_1)$, and hence in the integral defining our diagram, the integrand is equal to zero unless $\tau_1 = \tau_2 = \cdots = \tau_n$, which means that the integral vanishes.

It is easy to check that two topologically equivalent diagrams specify the same operator, therefore one usually draws only one of the class of topologically equivalent diagram and this diagram should be multiplied by the number of topologically equivalent diagrams.

Let us consider as an example the Hamiltonian of the system of non-relativist identical bosons in a box of the volume L^3. Here, $H = H_0 + V$, where

$$H_0 = \sum \frac{k^2}{2m} a_k^+ a_k,$$

$$V = \frac{1}{2} \left(\frac{2\pi}{L} \right)^3 \sum_{k_1, k_2, p_1, p_2} \tilde{\mathcal{W}}(\mathbf{k}_1 - \mathbf{p}_2)\delta_{p_1+p_2}^{k_1+k_2} \times a_{k_1}^+ a_{k_2}^+ a_{p_1} a_{p_2},$$

and k, k_1, k_2, p_1, p_2 run over a lattice in three-dimensional space with a spacing of $2\pi/L$.

Let us write down the summands corresponding to the diagrams of Fig. 6.3 in the decomposition of the adiabatic S-matrix S_α. The

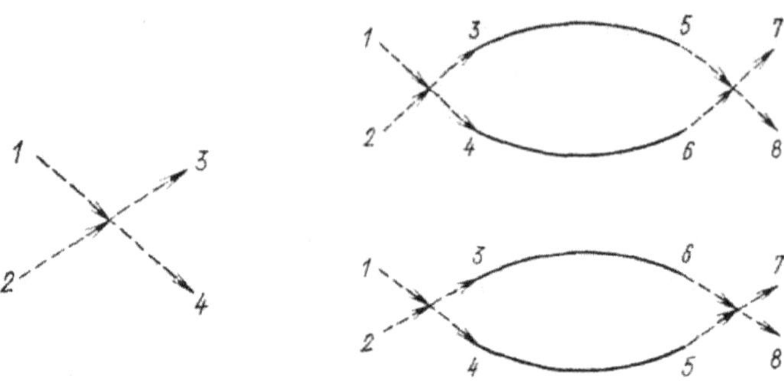

Fig. 6.3   More examples of diagrams.

first of these two diagrams represents the operator

$$\frac{1}{2}\left(\frac{2\pi}{L}\right)^3 \int_{-\infty}^{\infty} \sum_{k_1,k_2,p_1,p_2} \tilde{W}(\mathbf{k}_1 - \mathbf{p}_2)\delta_{p_1+p_2}^{k_1+k_2} \exp(-\alpha|\tau|)$$

$$\times \tilde{a}_{k_1}^+(\tau)\tilde{a}_{k_2}^+(\tau)\tilde{a}_{p_1}^+(\tau)\tilde{a}_{p_2}^+(\tau)d\tau$$

$$= \frac{1}{2}\left(\frac{2\pi}{L}\right)^3 \sum_{k_1,k_2,p_1,p_2} \tilde{W}(\mathbf{k}_1 - \mathbf{p}_2)\delta_{p_1+p_2}^{k_1+k_2}$$

$$\times \frac{2\alpha}{\left(\frac{k_1^2}{2m} + \frac{k_2^2}{2m} + \frac{p_1^2}{2m} + \frac{p_2^2}{2m}\right)^2 + \alpha^2} a_{k_1}^+ a_{k_2}^+ a_{p_1} a_{p_2}.$$

The operator corresponding to the second diagram has the form

$$\frac{1}{2l}\int d\tau_1 d\tau_2 \sum_{k_1,k_2,p_1,p_2} \sum_{k_1',k_2',p_1',p_2'} \frac{1}{4}\left(\frac{2\pi}{L}\right)^6 \tilde{W}(k_1 - p_2)$$

$$\times \delta_{p_1+p_2}^{k_1+k_2} \exp(-\alpha|\tau_1|)\exp\left(-i\frac{p_1^2}{2m}(\tau_1 - \tau_2)\right)\delta_{k_1'}^{p_1}\theta(\tau_1 - \tau_2)$$

$$\times \exp\left(-i\frac{p_2^2}{2m}(\tau_1 - \tau_2)\right)\theta(\tau_1 - \tau_2)\delta_{k_2'}^{p_2}\tilde{W}(k_1' - p_2')\delta_{p_1'+p_2'}^{k_1'+k_2'}$$

$$\times \exp(-\alpha|\tau_2|)\tilde{a}_{k_1}^+(\tau_1)\tilde{a}_{k_2}^+(\tau_1)\tilde{a}_{p_1'}^+(\tau_2)\tilde{a}_{p_2'}^+(\tau_2)$$

(where k_i, p_i and k_i', p_i' are summed over a lattice).

The third diagram is equivalent to the second; we take this into account by multiplying the second diagram by 2.

We can reduce the number of diagrams that should be calculated if we note that the operator corresponding to a disconnected diagram is equal to the normal product of the operators corresponding to each connected component.

Let us start now the derivation of the diagram representation of the operator $S(t, t_0)$. First of all, we note that for the representation of the operator $\tilde{V}(\tau_1)\ldots\tilde{V}(\tau_n)$ in normal form, we can use the technique developed in Section 6.3 with some modifications. (This follows from the formula

$$\tilde{V}(\tau) = \exp(iH_0\tau)V(\tau)\exp(-iH_0\tau)$$

$$= \sum_{m,n}\sum_{k_1,\ldots,k_m} v_{m,n}(k_1,\ldots,k_m|p_1,\ldots,p_n|\tau)\tilde{a}^+_{k_1}(\tau)$$

$$\ldots\tilde{a}^+_{k_m}(\tau)\tilde{a}_{p_1}(\tau)\ldots\tilde{a}_{p_n}(\tau)$$

$$= \sum_{m,n}\sum_{k_1,\ldots,k_m} \exp\left[i\left(\sum_{j=1}^{m}\epsilon(k_j) - \sum_{j=1}^{n}\epsilon(p_j)\right)\tau\right]$$

$$\times v_{m,n}(k_1,\ldots,k_m|p_1,\ldots,p_n|\tau)a^+_{k_1}\ldots a^+_{k_m}a_{p_1}\ldots a_{p_n}).$$

Let us describe the diagram representation of the operator $\tilde{V}(\tau_1)\ldots\tilde{V}(\tau_n)$ in more detail. The representation consists of diagrams that constitute the collection of n stars and some edges that point in the direction of increasing order (recall that the vertices of the diagram are ordered).

To every vertex of the diagram, we assign the index $k \in M$. To every free in-vertex (out-vertex), we assign the operator $a^+_k(\tau_i)$ (correspondingly the operator $a_k(\tau_i)$). To every edge, we assign the function $\exp(-i\epsilon(k)(\tau_i - \tau_j))\delta^k_l$, where k, l are indices of the vertices at the beginning and end of the edge and i, j are the numbers of stars that contain the beginning and end of the edge. The operator corresponding to the diagram is obtained from the product of the operators and functions that correspond to free vertices, stars, and edges. Namely, one should sum over the indices of all vertices (the operators are written in normal order).

It follows from the calculation of Section 6.3 that the operator $\tilde{V}(\tau_1)\ldots\tilde{V}(\tau_n)$ is the sum of operators corresponding to all the different diagrams.

Using the diagram techniques for the normal form of the operator $\tilde{V}(\tau_1)\ldots\tilde{V}(\tau_n)$, we can get the diagram technique for representing in normal form the operator $T(\tilde{V}(\tau_1)\ldots\tilde{V}(\tau_n))$, where all times are different. The only difference in the techniques is that we should consider arbitrary diagrams with n stars without restriction on the direction of the edges and to every edge we should assign the function $\exp(-i\epsilon(k)(\tau_i - \tau_j))\delta_l^k\theta(\tau_i - \tau_j)$. This follows from the remark that the factors $\theta(\tau_i - \tau_j)$ in the functions assigned to the edges are equal to 1 in the diagrams where the edges go in the direction of decreasing time and are equal to 0 in all other diagrams. This means that the diagrams that are not equal to zero coincide with the diagrams from the diagram representation for $\tilde{V}(\tau_{i_1})\ldots\tilde{V}(\tau_{i_n})$, where $i_1,\ldots,i_n$ is a permutation where $\tau_{i_1} > \cdots > \tau_{i_n}$. Noting that $\tilde{V}(\tau_{i_1})\ldots\tilde{V}(\tau_{i_n}) = T(\tilde{V}(\tau_1)\ldots\tilde{V}(\tau_n))$, we obtain the representation that we have claimed.

To get a diagram representation of the operator $S(t, t_0)$, it is sufficient to refer to the expression of this operator as a T-exponent (4.4) and to note that in the integral over $\tau_1,\ldots,\tau_n$ the value of the integrand on the set where at least two variables $\tau_1,\ldots,\tau_n$ coincide is irrelevant.

The diagram representation of the operator $S(t, t_0)$ can be easily obtained in the case when the index k runs over an arbitrary measure space X. More precisely, let us consider the Fock space $F(L^2(X))$, where X is a measure space, and in this space we define the operator generalized functions $a^+(k), a(k)$, obeying

$$[a(k), a(k')] = [a^+(k), a^+(k')] = 0,$$

$$[a(k), a^+(k')] = \delta(k, k'); a(k)\theta = 0.$$

Let us fix the Hamiltonian $H = H_0 + gV(t)$, where

$$H_0 = \int \epsilon(k)a^+(k)a(k)dk,$$

$$V(t) = \sum_{m,n} \int v_{m,n}(k_1, \ldots, k_m | p_1, \ldots, p_n | t)a^+(k_1)$$

$$\ldots a(p_n)d^m k d^n p.$$

We would like to find the normal form of the operator $S(t, t_0)$ for this Hamiltonian. The diagram technique to solve this problem is analogous to the technique described above. The only difference is that to every vertex of the diagram we should assign a point $k \in X$ instead of $k \in M$ and we should replace the summation with respect to M with integration with respect to X.

Chapter 7

Wightman and Green Functions

7.1 Wightman functions

Let us consider a Hilbert space $\mathcal{H}$, a Hamiltonian H with the ground state Φ, and a family of operators $\{a(\lambda)\}_{\lambda \in \Lambda}$ closed with respect to Hermitian conjugation. More precisely, let us assume that the operators $a(\lambda)$ are defined on a dense subspace $D \subset \mathcal{H}$ and for every operator $a(\lambda)$, there exists an operator $a(\hat{\lambda})$ that satisfies the relation $\langle a(\lambda)x, y \rangle = \langle x, a(\hat{\lambda})y \rangle$ for every $x, y \in D$. The symbol $a(\lambda, t)$ will denote the Heisenberg operators $a(\lambda, t) = \exp(iHt)a(\lambda)\exp(-iHt)$.

We will assume that the operators $a(\lambda)$ and $\exp(iHt)$ transform the set D into itself. Note that under these assumptions, the operators $a(\lambda_1, t_1) \ldots a(\lambda_n, t_n)$ also transform D into itself and that

$$(a(\lambda_1, t_1) \ldots a(\lambda_n, t_n))^+ = a^+(\lambda_n, t_n) \ldots a^+(\lambda_1, t_1)$$

$$= a(\hat{\lambda}_n, t_n) \ldots a(\hat{\lambda}_1, t_1).$$

We will assume that the ground state Φ belongs to the set D; its energy will be denoted by E_0.

Definition 7.1. The Wightman function $w_n(\lambda_1, t_1, \ldots, \lambda_n, t_n)$ of the Hamiltonian H is the expectation value of the operator

$$a(\lambda_1, t_1) \ldots a(\lambda_n, t_n)$$

in the ground state Φ of the Hamiltonian H:

$$w_n(\lambda_1, t_1, \ldots, \lambda_n, t_n) = \langle a(\lambda_1, t_1) \ldots a(\lambda_n, t_n)\Phi, \Phi \rangle.$$

Everything in this chapter can also be easily proved in the case when $a(\lambda)$ is a generalized operator function of the parameter λ. Naturally, in this case, the Wightman function will be a generalized function of the variables $\lambda_1, \ldots, \lambda_n$.

It is easy to prove the following properties of Wightman functions:

(1) *Invariance with respect to time translation:*

$$w_n(\lambda_1, t_1 + \tau, \ldots, \lambda_n, t_n + \tau) = w_n(\lambda_1, t_1, \ldots, \lambda_n, t_n).$$

$$(7.1)$$

(2) *Hermiticity:*

$$\overline{w_n}(\lambda_1, t_1, \ldots, \lambda_n, t_n) = w_n(\hat{\lambda}_n, t_n, \ldots, \hat{\lambda}_1, t_1). \qquad (7.2)$$

(3) *Positivity:* Let $f_n(\lambda_1, t_1, \ldots, \lambda_n, t_n)$ denote a sequence of functions that doesn't vanish only for finitely many values of n. Let us also assume that the functions belonging to this sequence do not vanish only on finitely many values of $\lambda_1, t_1, \ldots, \lambda_n, t_n$. Then

$$\sum_{m,n,\lambda_i,\mu_j,t_i,\tau_j} f_m(\lambda_1, t_1, \ldots, \lambda_m, t_m)\overline{f_n}(\mu_1, \tau_1, \ldots, \mu_n, \tau_n)$$

$$\times w_{m+n}(\lambda_1, t_1, \ldots, \lambda_m, t_m, \hat{\mu}_n, \tau_n, \ldots, \hat{\mu}_1, \tau_1) \geq 0$$

$$(7.3)$$

(the sum is taken over $\lambda_i, \mu_j \in \Lambda$ and $m, n = 0, 1, \ldots;\ -\infty < t_i, \tau_j < \infty$).

(4) *Spectral property:*

$$\int \exp(-i\omega a) w_n(\lambda_1, t_1, \ldots, \lambda_k, t_k, \lambda_{k+1}, t_{k+1}$$

$$+ a, \ldots, \lambda_n, t_n + a)da = 0 \qquad (7.4)$$

if the number $\omega + E_0$ does not belong to the spectrum of the Hamiltonian H (in particular, the relation (7.4) is true for every $\omega < 0$).

The proof of property (1) follows immediately from $\exp(-iHt)$ $\Phi = \exp(-iE_0 t)\Phi$. Property (2) follows from $\langle a(\lambda_1, t_1) \ldots$

$a(\lambda_n, t_n)\Phi, \Phi\rangle = \langle\Phi, a(\hat{\lambda}_n, t_n)\ldots a(\hat{\lambda}_1, t_1)\Phi\rangle$. Property (3) follows from applying the inequality $\langle AA^+\Phi, \Phi\rangle \geq 0$ to the operator

$$A = \sum_{m,\lambda_i,t_i} f_m(\lambda_1, t_1, \ldots, \lambda_m, t_m)a(\lambda_1, t_1)\ldots a(\lambda_m, t_m),$$

where the sum is taken over $m = 0, 1, \ldots$; $\lambda_i \in \Lambda$; $-\infty < t_i < \infty$.

The fourth property follows from the relation

$$\int \exp(-i\beta a)\,\langle\exp(iBa)\Psi_1, \Psi_2\rangle\,da = 0,$$

which is correct if the number β does not belong to the spectrum of the self-adjoint operator B (see Appendix A.5) (we should assume that $B = H$, $\Psi_1 = U\Phi$, $\Psi_2 = V^+\Phi$, where

$$U = a(\lambda_{k+1}, t_{k+1})\ldots a(\lambda_n, t_n), V = a(\lambda_1, t_1)\ldots a(\lambda_k, t_k)).$$

If the Hamiltonian H acts on a representation space of CR (on the Fock space $F(\mathcal{B})$, for example), we can consider as the family of operators $a(\lambda)$ the family $a(k, \epsilon)$ of annihilation and creation operators that correspond to some orthonormal basis $\phi_k \in \mathcal{B}$ (here, $k \in M$, $\epsilon = \pm 1$, $a(k, \epsilon) = a(\phi_k, \epsilon)$).

In other words, the Wightman functions of the Hamiltonian H acting on a representation space of CR are functions

$$w_n(k_1, \epsilon_1, t_1, \ldots, k_n, \epsilon_n, t_n)$$
$$= \langle a(k_1, \epsilon_1, t_1)\ldots a(k_n, \epsilon_n, t_n)\Phi, \Phi\rangle.$$

In the case of CAR, the operators $a(k, \epsilon, t)$ are bounded and hence the Wightman functions are well defined. In the case of CCR, we assume that the ground state Φ belongs to the domain D of operators $a(k, \epsilon)$ and that the domain D is invariant with respect to the operators $\exp(iHt)$.

For Hamiltonians acting on a representation space of CR, we can prove one more property of Wightman functions.

(5) *Permutation of arguments:* Let the functions $w_n^{(i)}$ be the functions obtained from w_n by swapping (k_i, ϵ_i, t_i) and $(k_{i+1}, \epsilon_{i+1}, t_{i+1})$

(with a change of sign in the case of CAR). If $t_i = t_{i+1}$, then

$$w_n^{(i)}(k_1, \epsilon_1, t_1, \ldots, k_n, \epsilon_n, t_n)$$

$$= w_n(k_1, \epsilon_1, t_1, \ldots, k_n, \epsilon_n, t_n) + A_{\epsilon_{i+1}}^{\epsilon_i} \delta_{k_i, k_{i+1}} w_{n-2}$$

$$\times (k_1, \epsilon_1, t_1, \ldots, k_{i-1}, \epsilon_{i-1}, t_{i-1}, k_{i+2}, \epsilon_{i+2}, t_{i+2}, \ldots, k_n, \epsilon_n, t_n)$$

$$(7.5)$$

(here, $A_{\epsilon'}^{\epsilon}$ is the matrix defined in formula (6.3)).

To check this equation, we note that from the commutation relations in the form (6.3), we can obtain

$$[a(k, \epsilon, t), a(k', \epsilon', t)]_{\mp} = [a(k, \epsilon), a(k', \epsilon')]_{\mp} = A_{\epsilon'}^{\epsilon} \delta_{kk'}.$$

Wightman functions can be easily calculated for the Hamiltonian $H_0 = \sum \omega(k) a^+(k) a(k)$ in Fock space because for this Hamiltonian, we have $a(k, \epsilon, t) = \exp(i \epsilon \omega(k) t) a(k, \epsilon)$, and hence,

$$w_n(k_1, \epsilon_1, t_1, \ldots, k_n, \epsilon_n, t_n)$$

$$= \exp(i(\epsilon_1 \omega(k_1) t_1 + \cdots + \epsilon_n \omega(k_n) t_n)) \langle a(k_1, \epsilon_1)$$

$$\ldots a(k_n, \epsilon_n) \Phi, \Phi \rangle.$$

In the case of CCR, the operator H_0 is bounded from below when $\omega_k \geq 0$; then $\Phi = \theta$ (see Section 6.2) and the functions $w_2(k_1, \epsilon_1, t_1, k_2, \epsilon_2, t_2)$ do not vanish only in the case when $\epsilon_1 = -1, \epsilon_2 = 1$; in this case, we have

$$w_2(k_1 - 1, t_1, k_2 - 1, t_2)$$

$$= \delta_{k_1, k_2} \exp(-i\omega(k_1) t_1 + i\omega(k_2) t_2).$$

In the case of CAR, the ground state is $\Phi = \prod_{\omega(k) < 0} a^+(k) \theta$. In this case, the function $w_2(k_1, \epsilon_1, t_1, k_2, \epsilon_2, t_2)$ does not vanish only if $\epsilon_2 \omega(k_2) > 0, \epsilon_1 \omega(k_1) < 0$; then

$$w_2(k_1, \epsilon_1, t_1, k_2, \epsilon_2, t_2) = \delta_{k_1, k_2} \exp(i\epsilon_1 \omega(k_1) t_1 + i\epsilon_2 \omega(k_2) t_2).$$

In other words,

$$w_2(k_1, \epsilon_1, t_1, k_2, \epsilon_2, t_2)$$

$$= \theta(\epsilon_2 \omega(k_2)) \theta(-\epsilon_1 \omega(k_1)) \delta_{k_1 k_2}$$

$$\times \exp(i(\epsilon_1 \omega(k_1) t_1 + \epsilon_2 \omega(k_2) t_2)).$$

For Hamiltonians acting on the space of representations $\mathrm{CR}(\mathcal{B})$, one can give a more invariant definition of Wightman functions that does not depend on the choice of basis in the space $\mathcal{B}$. Namely, for the Wightman function $w_n(f_1, \epsilon_1, t_1, \ldots, f_n, \epsilon_n, t_n)$, where $f_i \in \mathcal{B}, \epsilon_i = \pm 1, -\infty < t_i < \infty$, we can consider the expectation value of the operator $a(f_1, \epsilon_1, t_1) \ldots a(f_n, \epsilon_n, t_n)$ with respect to the ground state Φ of the Hamiltonian H (w_n is a linear functional of $f_1, \ldots, f_n$). If the space $\mathcal{B} = L^2(X)$, then we can write the operator $a(f, \epsilon)$ in the form $a(f, \epsilon) = \int f(x)a(x, \epsilon)dx$, where $a(x, \epsilon)$ is a generalized operator function ($x \in X, \epsilon = \pm 1$). It is easy to check that

$$w_n(f_1, \epsilon_1, t_1, \ldots, f_n, \epsilon_n, t_n)$$

$$= \int f_1(x_1) \ldots f_n(x_n) w_n(x_1, \epsilon_1, t_1, \ldots, x_n, \epsilon_n, t_n) dx_1 \ldots dx_n,$$

where $w_n(x_1, \epsilon_1, t_1, \ldots, x_n, \epsilon_n, t_n)$ is the Wightman function constructed by means of the generalized operator function $a(x, \epsilon)$ and the Hamiltonian H.

7.2   Green functions

Let us define Green functions for Hamiltonians acting on the space of representations of canonical commutation or anticommutation relations. As in Section 7.1, we use the notation $a(k, \epsilon) = a(\phi_k, \epsilon)$, where ϕ_k is an orthonormal basis in $\mathcal{B}$.

Let us define the T-product of Heisenberg operators $T(a(k_1, \epsilon_1, t_1) \ldots a(k_n, \epsilon_n, t_n))$ as the product of the operators $a(k_1, \epsilon_1, t_1) \ldots a(k_n, \epsilon_n, t_n)$ in chronological order (in order of decreasing times). In the case of CAR, this product is taken with a minus sign if we need an odd permutation to put the operators in chronological order; if we need an even permutation, we take it with a plus sign. In other words,

$$T(a(k_1, \epsilon_1, t_1) \ldots a(k_n, \epsilon_n, t_n))$$

$$= (-1)^\gamma a(k_{i_1}, \epsilon_{i_1}, t_{i_1}) \ldots a(k_{i_n}, \epsilon_{i_n}, t_{i_n}), \tag{7.6}$$

where $P = (i_1, \ldots, i_n)$ is the permutation obeying $t_{i_1} > \ldots t_{i_n}$, $\gamma = 0$ in the case of CCR and γ is equal to the parity of the permutation P

in the case of CAR (here, $k_i \in M$, $\epsilon_i = \pm 1$ and all the times t_i are distinct).

For the case of $n = 2$, we can write

$$T(a(k_1, \epsilon_1, t_1)a(k_2, \epsilon_2, t_2))$$

$$= \theta(t_1 - t_2)a(k_1, \epsilon_1, t_1)a(k_2, \epsilon_2, t_2)$$

$$\pm \theta(t_2 - t_1)a(k_2, \epsilon_2, t_2)a(k_1, \epsilon_1, t_1),$$

where we use the upper sign in the case of CCR and we use the lower sign in the case of CAR.

The Green function of the Hamiltonian H is defined as an expectation value of the T-product $T(a(k_1, \epsilon_1, t_1) \ldots a(k_n, \epsilon_n, t_n))$ with respect to the ground state Φ of the Hamiltonian H:

$$G_n(k_1, \epsilon_1, t_1, \ldots, k_n, \epsilon_n, t_n)$$

$$= \langle T(a(k_1, \epsilon_1, t_1) \ldots a(k_n, \epsilon_n, t_n))\Phi, \Phi \rangle.$$

Green functions can be easily expressed in terms of Wightman functions, e.g.,

$$G_2(k_1, \epsilon_1, t_1, k_2, \epsilon_2, t_2) = \theta(t_1 - t_2)w_2(k_1, \epsilon_1, t_1, k_2, \epsilon_2, t_2)$$

$$\pm \theta(t_2 - t_1)w_2(k_2, \epsilon_2, t_2, k_1, \epsilon_1, t_1).$$

The general expression of Green functions in terms of Wightman functions can be written in the following way:

$$G_n(k_1, \epsilon_1, t_1, \ldots, k_n, \epsilon_n, t_n)$$

$$= \sum_{\pi}(-1)^{\gamma(\pi)}\theta_{\pi}(t)w_n^{\pi}(k_1, \epsilon_1, t_1, \ldots, k_n, \epsilon_n, t_n), \qquad (7.7)$$

where the sum is taken over all permutations $\pi = (i_1, i_2, \ldots, i_n)$, $\theta_{\pi}(t) = \theta(t_{i_1} - t_{i_2})\theta(t_{i_2} - t_{i_3}) \ldots \theta(t_{i_{n-1}} - t_{i_n})$ and the functions w_n^{π} are defined by the relation

$$w_n^{\pi}(k_1, \epsilon_1, t_1, \ldots, k_n, \epsilon_n, t_n)$$

$$= w(k_{i_1}, \epsilon_{i_1}, t_{i_1}, \ldots, k_{i_n}, \epsilon_{i_n}, t_{i_n}),$$

$\gamma(\pi) = 0$ in the case of CCR, and γ is equal to the parity of the permutation π in the case of CAR.

The definition of the functions $G_n(k_1, \epsilon_1, t_1, \ldots, k_n, \epsilon_n, t_n)$ is meaningful only when all the times $t_1, \ldots, t_n$ are distinct. The set where a Green function is not defined has measure zero, therefore, the integral of a Green function multiplied by a conventional function of $t_1, \ldots, t_n$ is well defined. This means, for example, that the function G_n can be considered as a well-defined generalized function of the variables $t_1, \ldots, t_n$. We will need to specify the definition of a Green function in the case when some of the arguments $t_1, \ldots, t_n$ coincide only in one situation in Section 7.4. We will do this by requiring that when the time variables coincide, the corresponding operators in the T-product are written in normal order (i.e. the T-product is defined with respect to such a permutation of indices $(i_1, \ldots, i_n)$, such that $t_{i_1} \geq \cdots \geq t_{i_n}$ and in the case of $t_{i_k} = t_{i_l}$, we have $\epsilon_{i_k} \geq \epsilon_{i_l}$).

Let us formulate the following simple properties of Green functions:

(1) *Symmetry:* Green functions $G_n(k_1, \epsilon_1, t_1, \ldots, k_n, \epsilon_n, t_n)$ do not change under permutation of variables in the case of CCR. In the case of CAR, the Green function changes only by a sign under a permutation of (k_i, ϵ_i, t_i) and (k_j, ϵ_j, t_j).

(2) *Invariance with respect to time translation:*

$$G_n(k_1, \epsilon_1, t_1 + a, \ldots, k_n, \epsilon_n, t_n + a)$$
$$= G_n(k_1, \epsilon_1, t_1, \ldots, k_n, \epsilon_n, t_n).$$

(3) $G_n(k_1, \epsilon_1, t + 0, k_2, \epsilon_2, t, k_3, \epsilon_3, t_3, \ldots, k_n, \epsilon_n, t_n)$

$$\mp G_n(k_1, \epsilon_1, t_1, k_2, \epsilon_2, t + 0, k_3, \epsilon_3, t_3, \ldots, k_n, \epsilon_n, t_n)$$

$$= \delta_{k_1, k_2} A_{\epsilon_2}^{\epsilon_1} G_{n-2}(k_3, \epsilon_3, t_3, \ldots, k_n, \epsilon_n, t_n).$$

The first property follows from the fact that we can transpose factors under the sign of T-product. The third property follows from commutation relations. As for Wightman functions, one can give a definition of Green functions which does not depend on the choice of basis, namely, we should set

$$G_n(f_1, \epsilon_1, t_1, \ldots, f_n, \epsilon_n, t_n)$$
$$= \langle T(a(f_1, \epsilon_1, t_1) \ldots a(f_n, \epsilon_n, t_n)) \Phi, \Phi \rangle$$

(here, $f_i \in \mathcal{B}, \epsilon_i = \pm 1, -\infty < t_i < \infty$). If $\mathcal{B} = L^2(X)$, then

$$G_n(f_1, \epsilon_1, t_1, \ldots, f_n, \epsilon_n, t_n)$$

$$= \int f_1(x_1) \ldots f_n(x_n) G_n(x_1, \epsilon_1, t_1, \ldots, x_n, \epsilon_n, t_n) dx_1 \ldots dx_n,$$

where $G_n(x_1, \epsilon_1, t_1, \ldots, x_n, \epsilon_n, t_n)$ is a Green function constructed by means of the generalized operator function $a(x, \epsilon)$.

It is easy to construct a perturbation series for Green functions as well as a diagrammatic representation for this series. Let us describe this construction.

Suppose that the Hamiltonian H is written in the form $H = H_0 + gV$. The ground states of the Hamiltonians H and H_0 will be denoted by Φ and Φ_0 correspondingly. Let us use the relation

$$\Phi = \lim_{a \to 0} C_1(\alpha) S_\alpha(0, -\infty) \Phi_0 = \lim_{a \to 0} C_2(\alpha) S_\alpha(0, +\alpha) \Phi_0 \qquad (7.8)$$

that follows from (4.11) (our goal is to get the representation of Green functions in terms of perturbation series, therefore it is sufficient to know that the relation (7.8) is true in the framework of perturbation theory in the case when the ground states Φ_0 of the Hamiltonian H_0 is not degenerate). If $(i_1, \ldots, i_n)$ is a permutation obeying $t_{i_1} > \cdots > t_{i_n}$, then

$$G_n(k_1, \epsilon_1, t_1, \ldots, k_n, \epsilon_n, t_n)$$

$$= \lim_{\alpha \to 0} C_1(\alpha) \overline{C_2(\alpha)} \times \langle a(k_{i_1}, \epsilon_{i_1}, t_{i_1}) \ldots a(k_{i_n}, \epsilon_{i_n}, t_{i_n})$$

$$\times S_\alpha(0, -\infty) \Phi_0, S_\alpha(0, +\infty) \Phi_0 \rangle.$$

Let us introduce the operators

$$\tilde{a}(k, \epsilon, t) = \exp(iH_0 t) a(k, \epsilon) \exp(-iH_0 t).$$

It is easy to check that

$$a(k, \epsilon, t) = \exp(iHt) \exp(-iH_0 t) \tilde{a}(k, \epsilon, t) \exp(iH_0 t) \exp(-iHt)$$

$$= S(0, t) \tilde{a}(k, \epsilon, t) S(t, 0).$$

For fixed t and $\alpha \to 0$, we can write the approximate equations

$$S(t, 0) \approx S_\alpha(t, 0), \quad S(0, t) \approx S_\alpha(0, t),$$

$$a(k, \epsilon, t) \approx S_\alpha(0, t) \tilde{a}(k, \epsilon, t) S_\alpha(t, 0).$$

Using these formulas, we obtain

$$G_n(k_1, \epsilon_1, t_1, \ldots, k_n, \epsilon_n, t_n)$$

$$= \lim_{\alpha \to 0} C_1(\alpha)\overline{C_2(\alpha)}$$

$$\times \langle S_\alpha(0, t_{i_1})\tilde{a}(k_{i_1}, \epsilon_{i_1}, t_{i_1})S_\alpha(t_{i_1}, 0)S_\alpha(0, t_{i_2}) \ldots$$

$$\times \tilde{a}(k_{i_n}, \epsilon_{i_n}, t_{i_n})S_\alpha(t_{i_n}, 0)S_\alpha(0, -\infty)\Phi_0, S_\alpha(0, +\infty)\Phi_0 \rangle$$

$$= \lim_{\alpha \to 0} \frac{\langle A_\alpha \Phi_0, \Phi_0 \rangle}{\langle S_\alpha(\infty, -\infty)\Phi_0, \Phi_0 \rangle},$$

where

$$A_\alpha = S_\alpha(\infty, t_{i_1})\tilde{a}(k_{i_1}, \epsilon_{i_1}, t_{i_1})S_\alpha(t_{i_1}, t_{i_2})$$

$$\ldots \tilde{a}(k_{i_n}, \epsilon_{i_n}, t_{i_n})S_\alpha(t_{i_n}, -\infty). \tag{7.9}$$

We have used the group property

$$S_\alpha(t'', t) = S_\alpha(t'', t')S_\alpha(t', t),$$

and the relation

$$1 = \langle \Phi, \Phi \rangle = \lim_{\alpha \to 0} C_1(\alpha)\overline{C_2(\alpha)} \langle S_\alpha(0, -\infty)\Phi_0, S_\alpha(0, +\infty)\Phi_0 \rangle$$

$$= \lim_{\alpha \to 0} C_1(\alpha)\overline{C_2(\alpha)} \langle S_\alpha(\infty, -\infty)\Phi_0, \Phi_0 \rangle$$

in the derivation of (7.9). Equation (7.9) can be rewritten in the following way:

$$A_\alpha = T(\tilde{a}(k_1, \epsilon_1, t_1) \ldots \tilde{a}(k_n, \epsilon_n, t_n))$$

$$\times \exp\left(\frac{1}{i} \int_{-\infty}^{\infty} g \exp(-\alpha|\tau|)\tilde{V}(\tau)d\tau \right)$$

$$= \sum_{n=0}^{\infty} \frac{1}{n!} \left(\frac{g}{i} \right)^n$$

$$\times \int T(\tilde{a}(k_1, \epsilon_1, t_1) \ldots \tilde{a}(k_n, \epsilon_n, t_n)$$

$$\times \exp(-\alpha|\tau_1| - \cdots - \alpha|\tau_n|)\tilde{V}(\tau_1) \ldots \tilde{V}(\tau_n))d^n\tau.$$

One can use the short-hand notation

$$A_\alpha = T(\tilde{a}(k_1, \epsilon_1, t_1) \ldots \tilde{a}(k_n, \epsilon_n, t_n) S_\alpha(\infty, -\infty))$$

for the operator A_α.

Let us assume now that the operator $H = H_0 + gV$ is written in the form (6.26) with the functions $V_{m,n}$ that do not depend on time. For the operator $S_\alpha(\infty, -\infty)$ corresponding to this Hamiltonian, one can construct the diagram representation using the results of Section 6.3. Using this representation, we can also construct a diagram representation for the operator A_α. The diagram representation for the numbers $\langle A_\alpha \Phi_0, \Phi_0 \rangle$ and $\langle S_\alpha(\infty, -\infty) \Phi_0, \Phi_0 \rangle$ is obtained from the diagrams of the operators A_α and $S_\alpha(\infty, -\infty)$ if we exclude from the diagrams of these operators the diagrams where there are vertices that do not belong to any edges. Using the fact that disconnected diagrams decompose as a product of diagrams corresponding to the connected components, it is easy to construct the diagram representation for the expression

$$\frac{\langle A_\alpha \Phi_0, \Phi_0 \rangle}{\langle S_\alpha(\infty, -\infty) \Phi_0, \Phi_0 \rangle}. \tag{7.10}$$

It consists of the diagrams $\langle A_\alpha \Phi_0, \Phi_0 \rangle$ that do not contain as a component a diagram for $\langle S_\alpha(\infty, -\infty) \Phi_0, \Phi_0 \rangle$. To get a diagram representation for a Green function $G_n(k_1, \epsilon_1, t_1, \ldots, k_n, \epsilon_n, t_n)$, we should take the limits α tending to zero in the diagrams for the expression (7.10).

7.3 Källén–Lehmann representation

Let us consider the Wightman function w_n of the Hamiltonian H with respect to the family of operators $a(\lambda)$. It is easy to express this function in terms of matrix entries of the operator $a(\lambda)$ in the basis of the eigenvectors of the Hamiltonian H. Let us prove such an expression (called the Källén–Lehmann representation) for the function w_2. Let us assume first that the operator H has discrete spectrum, i.e. there exists an orthonormal basis Φ_n of eigenvectors of the operator H with the corresponding eigenvalues denoted by E_n.

Then

$$w_2(\lambda_1, t_1, \lambda_2, t_2) = \langle a(\lambda_1, t_1)a(\lambda_2, t_2)\Phi, \Phi \rangle$$

$$= \left\langle a(\lambda_2, t_2)\Phi, a(\tilde{\lambda}_1, t_1)\Phi \right\rangle$$

$$= \sum_n \langle a(\lambda_2, t_2)\Phi, \Phi_n \rangle \left\langle \Phi_n, a(\tilde{\lambda}_1, t_1)\Phi \right\rangle$$

$$= \sum_n \langle \exp(iHt_2)a(\lambda_2)\exp(-iHt_2)\Phi, \Phi_n \rangle$$

$$\times \left\langle \Phi_n, \exp(iHt_1)a(\tilde{\lambda}_1)\exp(-iHt_1)\Phi \right\rangle$$

$$= \sum_n \exp(i(E_n - E_0)t_2) \langle a(\lambda_2)\Phi, \Phi_n \rangle$$

$$\times \exp(i(E_0 - E_n)t_1) \left\langle \Phi_n, a(\tilde{\lambda}_1)\Phi \right\rangle.$$

Introducing the notation $\rho_n^\lambda = \langle \Phi_n, a(\lambda)\Phi \rangle$, we obtain the representation

$$w_2(\lambda_1, t_1, \lambda_2, t_2) = \sum_n \exp[-i(E_n - E_0)(t_1 - t_2)]\rho_n^{\tilde{\lambda}_1}\bar{\rho}_n^{\lambda_2}.$$

If the operator H additionally has continuous spectrum, then the complete system of eigenvectors consists of normalized eigenvectors Φ_n and generalized eigenvectors Φ_γ (we assume that the vectors Φ_n are orthonormal and Φ_γ are δ-normalized; the corresponding eigenvalues are denoted by E_n and E_γ). Repeating the above calculations and using the relation

$$\langle x, y \rangle = \sum_n \langle x, \Phi_n \rangle \langle \Phi_n, y \rangle + \int \langle x, \Phi_\gamma \rangle \langle \Phi_\gamma, y \rangle \, d\gamma,$$

we see that

$$w_2(\lambda_1, t_1, \lambda_2, t_2) = \sum_n \exp[-i(E_n - E_0)(t_1 - t_2)]\rho_n^{\tilde{\lambda}_1}\bar{\rho}_n^{\lambda_2}$$

$$+ \int \exp[-(E_\gamma - E_0)(t_1 - t_2)]\rho_\gamma^{\tilde{\lambda}_1}\bar{\rho}_\gamma^{-\lambda_2}d\gamma,$$

$$(7.11)$$

where

$$\rho_n^\lambda = \langle \Phi_n, a(\lambda)\Phi \rangle, \quad \rho_\gamma^\lambda = \langle \Phi_\gamma, a(\lambda)\Phi \rangle.$$

It is clear from the above formulas that the asymptotic behavior of the function w_2 as $t_1 - t_2 \to \infty$ is determined only by the discrete spectrum because the contribution of the continuous spectrum for $t_1 - t_2 \to \infty$ is an integral of quickly oscillating functions.[1]

Let us note that the statement about the asymptotic behavior of the function w_2 should be modified in the case when $a(\lambda)$ is a generalized operator function.

The function w_2 can also be expressed in terms of

$$\sigma^{\lambda_1, \lambda_2}(\mu) = \langle a(\lambda_1 E_\mu a(\lambda)\Phi, \Phi)\rangle,$$

where $E_\mu = e_\mu(H)$ is the spectral decomposition of the operator H (recall that $e_\mu(x) = 1$ for $x \leq \mu$, $e_\mu(x) = 0$ for $x > \mu$), namely,

$$w_2(\lambda_1, t_1, \lambda_2, t_2) = \int \exp(i(\mu - E_0)(t_2 - t_1))d\sigma^{\lambda_1, \lambda_2}(\mu).$$

Indeed,

$$
\begin{aligned}
w_2(\lambda_1, t_1, \lambda_2, t_2) &= \langle a(\lambda_2, t_2)\Phi, a(\hat{\lambda}_1, t_1)\Phi\rangle \\
&= \langle \exp(iHt_2)a(\lambda_2)\exp(-iE_0 t_2)\Phi, \exp(iHt_1) \\
&\quad \times a(\hat{\lambda}_1)\exp(-iE_0 t_1)\Phi\rangle \\
&= \langle \exp[i(H - E_0)(t_2 - t_1)]a(\lambda_2)\Phi, a(\hat{\lambda}_1)\Phi\rangle \\
&= \int \exp[i(\mu - E_0)(t_2 - t_1)]d\langle E_\mu a(\lambda_2)\Phi, a(\hat{\lambda}_1)\Phi\rangle \\
&= \int \exp[i(\mu - E_0)(t_2 - t_1)]d\sigma^{\lambda_1, \lambda_2}(\mu).
\end{aligned}
$$

The Green function G_2 of the operator acting on the representation space of CR can be expressed in terms of the function w_2 by means of the simple formula

$$G_2(k_1, \epsilon_1, t_1, k_2, \epsilon_2, t_2) = \theta(t_1 - t_2)w_2(k_1, \epsilon_1, t_1, k_2, \epsilon_2, t_2)$$

$$\pm \theta(t_2 - t_1)w_2(k_2, \epsilon_2, t_2, k_1, \epsilon_1, t_1).$$

---

[1] A rigorous proof of this statement can be given under the assumption of absolute continuity of the continuous spectrum.

Representations for the function G_2 follow immediately from representations of w_2. For example, for operators with a discrete spectrum, we have

$$G_2(k_1, \epsilon_1, t_1, k_2, \epsilon_2, t_2)$$
$$= \theta(t_1 - t_2) \sum_n \exp[-i(E_n - E_0)(t_1 - t_2)] \rho_n^{k_1, -\epsilon_1} \bar{\rho}_n^{k_2, \epsilon_2}$$
$$\pm \theta(t_2 - t_1) \sum_n \exp[-i(E_n - E_0)(t_2 - t_1)] \rho_n^{k_2, -\epsilon_2} \bar{\rho}_n^{k_1, \epsilon_1}$$

(here, $\rho_n^{k, \epsilon} = \langle \Phi_n, a(k, \epsilon) \Phi \rangle$ and Φ_n are the stationary states of the Hamiltonian H and E_n are the corresponding eigenvalues).

Let us also consider the functions

$$\tilde{w}_2(k_1, \epsilon_1, \omega_1, k_2, \epsilon_2, \omega_2)$$
$$= (2\pi)^{-1} \int \exp[i(\epsilon_1 \omega_1 t_1 + \epsilon_2 \omega_2 t_2)]$$
$$w_2(k_1, \epsilon_1, t_1, k_2, \epsilon_2, t_2) \times dt_1 dt_2;$$
$$\tilde{G}_2(k_1, \epsilon_1, \omega_1, k_2, \epsilon_2, \omega_2)$$
$$= (2\pi)^{-1} \int \exp[i(\epsilon_1 \omega_1 t_1 + \epsilon_2 \omega_2 t_2)]$$
$$\times G_2(k_1, \epsilon_1, t_1, k_2, \epsilon_2, t_2) dt_1 dt_2$$

(the Wightman and Green functions in energy representations).

The representations of w_2 and G_2 imply representations of the functions $\tilde{w}_2$ and $\tilde{G}_2$. In particular, for operators with discrete spectrum, we have

$$\tilde{G}_2(k_1, 1, \omega_1, k_2, -1, \omega_2) = G(k_1, k_2, \omega_1) \delta(\omega_1 - \omega_2),$$

where

$$G(k_1, k_2, \omega) = \int \exp(i\omega\tau) G_2(k_1, 1, \tau, k_2, -1, 0) d\tau$$
$$= i \sum_n \frac{\rho_n^{k_1, -1} \bar{\rho}_n^{k_2, -1}}{\omega - (E_n - E_0) + i0} \mp \sum_n \frac{\rho_n^{k_2, 1} \bar{\rho}_n^{k_1, -1}}{\omega + (E_n - E_0) - i0}.$$

$$(7.12)$$

Hence, the poles of the function G with respect to the variable ω correspond to points in the discrete spectrum. If H has continuous spectrum, we should add an integral over the continuous spectrum to the expression (7.12). In all physical examples, the contribution of the continuous spectrum is a continuous function of ω.

7.4　The equations for Wightman and Green functions

Let us consider the Hamiltonian H acting on the representation space of CR ($\mathcal{B}$). We will represent this Hamiltonian in normal form:

$$H = \sum_{m,n} \sum_{k_i, l_j} \Gamma^{m,n}(k_1, \ldots, k_m | l_1, \ldots, l_n) a_{k_1}^+ \ldots a_{k_m}^+ a_{l_1} \ldots a_{l_n},$$

where $a_k^+ = a^+(\phi_k), a_k = a(\phi_k)$, ϕ_k is an orthonormal basis in $\mathcal{B}$, the functions $\Gamma^{m,n}$ are symmetric with respect to variables k_i and l_j in the case of CCR and antisymmetric in the case of CAR. Then the Heisenberg equations for the operators $a_k(t) = \exp(iHt)a_k \exp(-iHt)$, $a_k^+(t) = \exp(iHt)a_k^+ \exp(-iHt)$ can be written in the form

$$\frac{1}{i}\frac{da_k(t)}{dt} = [H, a_k(t)]$$

$$= - \sum_{m,n,k_1,\ldots,k_{m-1},l_1,\ldots,l_n} m\Gamma^{m,n}(k, k_1, \ldots, k_{m-1} | l_1, \ldots, l_n)$$

$$\times a_{k_1}^+(t) \ldots a_{k_{m-1}}^+(t) a_{l_1}(t) \ldots a_{l_n}(t);$$

$$\frac{1}{i}\frac{da_k^+(t)}{dt} = [H, a_k^+(t)]$$

$$= \sum_{m,n,k_1,\ldots,k_{m-1},l_1,\ldots,l_n} n\Gamma^{m,n}(k, k_1, \ldots, k_m | l_1, \ldots, l_{n-1}, k)$$

$$\times a_{k_1}^+(t) \ldots a_{k_m}^+(t) a_{l_1}(t) \ldots a_{l_{n-1}}(t).$$

These equations immediately imply equations for Wightman functions

$$w_n(k_1, \epsilon_1, t_1, \ldots, k_n, \epsilon_n, t_n).$$

Let us calculate, for example, the expression of the derivative of the function w_n with respect to the variable t_1 in terms of Wightman functions (the expressions for the derivatives of w_n with respect to t_i have similar form). It is easy to see that

$$\frac{1}{i}\frac{\partial}{\partial t_1}w_r(k_1,\epsilon_1,t_1,\ldots,k_r,\epsilon_r,t_r)$$

$$=-\delta_{-1}^{\epsilon_1}\sum_{m,n,p_i,q_j}m\Gamma^{m,n}(k_1,p_1,\ldots,p_{m-1}|q_1,\ldots,q_n)$$

$$\times w_{r+m+n-1}(p_1,1,t_1,\ldots,p_{m-1},1,t_1,q_1,-1,t_1,$$

$$\ldots,q_n,-1,t_1,k_2,\epsilon_2,t_2,\ldots,k_r,\epsilon_r,t_r)$$

$$+\delta_{+1}^{\epsilon_1}\sum_{m,n,p_i,q_j}n\Gamma^{m,n}(p_1,\ldots,p_m|q_1,\ldots,q_{n-1},k_1)$$

$$\times w_{r+m+n-1}(p_1,1,t_1,\ldots,p_m,1,t_1,q_1,-1,t_1,$$

$$\ldots,q_{n-1},-1,t_1,k_2,\epsilon_2,t_2,\ldots,k_r,\epsilon_r,t_r). \qquad (7.13)$$

It is easy to derive the equations for Green functions from (7.13) or directly from Heisenberg equations. Let us calculate, for example,

$$\frac{\partial}{\partial t_1}G_2(k_1,\epsilon_1,t_1,k_2,\epsilon_2,t_2)$$

$$=\frac{\partial}{\partial t_1}[w_2(k_1,\epsilon_1,t_1,k_2,\epsilon_2,t_2)\theta(t_1-t_2)$$

$$\pm w_2(k_2,\epsilon_2,t_2,k_1,\epsilon_1,t_1)\theta(t_2-t_1)]$$

$$=\theta(t_1-t_2)\frac{\partial}{\partial t_1}w_2(k_1,\epsilon_1,t_1,k_2,\epsilon_2,t_2)$$

$$\pm\theta(t_2-t_1)\frac{\partial}{\partial t_1}w_2(k_2,\epsilon_2,t_2,k_1,\epsilon_1,t_1)$$

$$+\delta(t_1-t_2)[w_2(k_1,\epsilon_1,t_1,k_2,\epsilon_2,t_2)$$

$$\mp w_2(k_2,\epsilon_2,t_2,k_1,\epsilon_1,t_1)].$$

Using CR (or property (5) of Wightman functions), we can simplify the last summand:

$$\delta(t_1-t_2)(w_2(k_1,\epsilon_1,t_1,k_2,\epsilon_2,t_2)\mp w_2(k_2,\epsilon_2,t_2,k_1,\epsilon_1,t_1))$$

$$=A_{\epsilon_2}^{\epsilon_1}\delta(t_1-t_2)\delta_{k_1,k_2}.$$

The first two summands in the formula for $\frac{\partial G_2}{\partial t_1}$ can be expressed in terms of Green functions (one should use (7.13)). As a result, we obtain the equation

$$\frac{1}{i}\frac{\partial}{\partial t_1}G_2(k_1,\epsilon_1,t_1,k_2,\epsilon_2,t_2) = \frac{1}{i}\delta(t_1 - t_2)A^{\epsilon_1}_{\epsilon_2}\delta_{k_1,k_2}$$

$$-\delta^{\epsilon_1}_{-1}\sum_{m,n,p_i,q_j} m\Gamma^{m,n}(k_1,p_1,\ldots,p_{m-1}|q_1,\ldots,q_n)$$

$$\times G_{m+n}(p_1,1,t_1,\ldots,p_{m-1},1,t_1,q_1,-1,t_1,$$

$$\ldots,q_n,-1,t_1,k_2,\epsilon_2,t_2)$$

$$+\delta^{\epsilon_1}_{+1}\sum_{m,n,p_i,q_j} n\Gamma^{m,n}(p_1,\ldots,p_m|q_1,\ldots,q_{n-1},k_1)$$

$$\times G_{m+n}(p_1,1,t_1,\ldots,p_m,1,t_1,q_1,-1,t_1,$$

$$\ldots,q_{n-1},-1,t_1,k_2,\epsilon_2,t_2).$$

Similarly, one can derive the equations for Green functions G_n. We will not write down the form of these equations since, in more general situations, they are very complicated.

Chapter 8

Translation-Invariant Hamiltonians

8.1 Translation-invariant Hamiltonians in Fock space

Let us consider operators acting on the Fock space $F(\mathcal{B})$, where $\mathcal{B} = L^2(E^3)$ is the space of square-integrable functions $\psi(\mathbf{k})$ on the Euclidean space E^3 with variable $\mathbf{k}$ denoting the momentum.[1]

It is convenient to express these operators in terms of operator generalized functions $a^+(\mathbf{k}), a(\mathbf{k})$ satisfying the commutation relations

$$[a(\mathbf{k}), a(\mathbf{k}')]_{\mp} = [a^+(\mathbf{k}), a^+(\mathbf{k}')]_{\mp} = 0,$$

$$[a(\mathbf{k}), a^+(\mathbf{k}')]_{\mp} = \delta(\mathbf{k} - \mathbf{k}').$$

These operator generalized functions were defined in Section 3.2.

The Hamiltonian H is called *translation-invariant* if it commutes with the momentum operators $\mathbf{P} = \int \mathbf{k} a^+(\mathbf{k}) a(\mathbf{k}) d\mathbf{k}$. An example of a translation-invariant Hamiltonian is the Hamiltonian for a system of interacting non-relativistic identical particles considered in Section 3.2.

We point out the following easy-to-check statement: the vacuum vector θ is always an eigenvector of a translation-invariant Hamiltonian H.

---

[1] The case $\mathcal{B} = L^2(E^3)$ corresponds to identical spinless particles. If we consider particles with spin, or if we have several types of particles, we should consider $\mathcal{B} = L^2(E^3 \times B)$, where B is a finite set. It is easy to generalize all the results of the present chapter and the following chapters to this case, hence there is no need to describe the generalization in detail. It is important to note, however, that fermions always have spin, therefore, in the case of CAR, the situation $\mathcal{B} = L^2(E^3)$ that we consider does not correspond to any particle existing in nature.

123

To prove this statement, we note that the momentum operator $\mathbf{P}$ has precisely one (up to a factor) eigenvector θ that corresponds to a zero eigenvalue (here, we consider conventional — normalized — eigenvectors; of course, $\mathbf{P}$ has generalized eigenvectors). Since we have assumed that the operator H commutes with the operator $\mathbf{P}$, it transforms the single eigenvector of the operator $\mathbf{P}$ into an eigenvector of $\mathbf{P}$. Therefore, we see that $H\theta = \lambda\theta$, which proves our statement.

Let us assume that the Hamiltonian H is written in the normal form

$$H = \sum_{m,n} \int \Gamma_{m,n}(\mathbf{p}_1, \ldots, \mathbf{p}_m | \mathbf{q}_1, \ldots, \mathbf{q}_n) a^+(\mathbf{p}_1) \ldots a^+(\mathbf{p}_m) a(\mathbf{q}_1)$$

$$\ldots a(\mathbf{q}_n) d^m \mathbf{p} d^n \mathbf{q}. \tag{8.1}$$

It is easy to check that it is translation-invariant only in the case when the functions $\Gamma_{m,n}$ have the form

$$\Gamma_{m,n}(\mathbf{p}_1, \ldots, \mathbf{p}_m | \mathbf{q}_1, \ldots, \mathbf{q}_n) = \Lambda_{m,n}(\mathbf{p}_1, \ldots, \mathbf{p}_m | \mathbf{q}_1, \ldots, \mathbf{q}_n)$$

$$\times \, \delta(\mathbf{p}_1 + \cdots + \mathbf{p}_m - \mathbf{q}_1 - \cdots - \mathbf{q}_n).$$

The statement we have proven implies that expression (8.1) cannot define a self-adjoint operator on Fock space if for some m the function $\Gamma_{m,0} \not\equiv 0$ (in physical terms, the Hamiltonian H generates vacuum polarization in this case).[2]

The above statement implies that we should consider translation-invariant operators with vacuum polarization outside the Fock space framework. It is not that we should consider these Hamiltonians to be bad, but that Fock space is too narrow for these Hamiltonians.[3]

We see that we should consider translation-invariant Hamiltonian (8.1) as a formal expression, composed of the symbols $a^+(\mathbf{k}), a(\mathbf{k})$. We will show how we can construct various physical

---

[2]Indeed, in the case of vacuum polarization, the expression (8.1) does not determine an operator on Fock space at all (see Section 3.2 for details).

[3]To construct an operator on Fock space corresponding to a translation-invariant Hamiltonian with vacuum polarization, we should make the volume or, in other words, the infrared cutoff (see below).

quantities for such Hamiltonians. In particular, in Chapter 9, we will give a definition for the scattering matrix of a translation-invariant Hamiltonian.

We will use the following construction. Let H be a translation-invariant Hamiltonian, i.e. a formal expression of the form

$$H = \sum_{m,n} \int \Lambda_{m,n}(\mathbf{k}_1, \ldots, \mathbf{k}_m | \mathbf{l}_1, \ldots, \mathbf{l}_n)$$

$$\times \, \delta(\mathbf{k}_1 + \cdots + \mathbf{k}_m - \mathbf{l}_1 - \cdots - \mathbf{l}_n)$$

$$\times \, a^+(\mathbf{k}_1) \ldots a^+(\mathbf{k}_m) a(\mathbf{l}_1) \ldots a(\mathbf{l}_n) d^m \mathbf{k} d^n \mathbf{l}. \qquad (8.2)$$

We will assume that the expression for H is formally Hermitian (i.e. we assume that $\Lambda_{m,n}(\mathbf{k}_1, \ldots, \mathbf{k}_m | \mathbf{l}_1, \ldots, \mathbf{l}_n) = \overline{\Lambda}_{n,m}(\mathbf{l}_n, \ldots, \mathbf{l}_1 | \mathbf{k}_m, \ldots, \mathbf{k}_1)$). Let us define a Hamiltonian H with volume cutoff Ω. We will denote by Ω the cube with edge length L in coordinate space ($0 \leq x \leq L, 0 \leq y \leq L, 0 \leq z \leq L$). Then, we denote by $B_\Omega = L^2(\Omega)$ the space of square-integrable functions $\psi(\mathbf{r})$, where $\mathbf{r} \in \Omega$, and by $F_\Omega = F(B_\Omega)$, we denote the Fock space constructed with the Hilbert space B_Ω. In the space B_Ω, we select an orthonormal basis of functions $\phi_\mathbf{k} = L^{-3/2} \exp(-i\mathbf{k}\mathbf{r})$, where $\mathbf{k}$ runs over the lattice T_Ω with step size $\frac{2\pi}{L}$ (i.e. $\mathbf{k} = \frac{2\pi}{L}\mathbf{n}$, where $\mathbf{n}$ is an integer vector). The operators $a^+(\phi_\mathbf{k}), a(\phi_\mathbf{k})$ in F_Ω corresponding to this basis will be denoted by $a_\mathbf{k}^+, a_\mathbf{k}$.

The operator H_Ω on the space F_Ω will be defined by the formula

$$H_\Omega = \sum_{m,n} \sum_{k_i, l_j} \left(\frac{2\pi}{L}\right)^{\frac{3}{2}(m+n-2)} \Lambda_{m,n}(\mathbf{k}_1, \ldots, \mathbf{k}_m | \mathbf{l}_1, \ldots, \mathbf{l}_n)$$

$$\times \, \delta_{\mathbf{k}_1 + \cdots + \mathbf{k}_m, \mathbf{l}_1 + \cdots + \mathbf{l}_n} a_{\mathbf{k}_1}^+ \ldots a_{\mathbf{k}_m}^+ a_{\mathbf{l}_1}^+ \ldots a_{\mathbf{l}_n}^+ \qquad (8.3)$$

(in other words, we obtain the expression for H_Ω from the expression for H by replacing $a^+(\mathbf{k}), a(\mathbf{k})$ with $(\frac{L}{2\pi})^{3/2} a_\mathbf{k}^+, (\frac{L}{2\pi})^{3/2} a_\mathbf{k}$, replacing the integration with summation over the lattice T_Ω and multiplying by $(\frac{2\pi}{L})^3$, and replacing the function $\delta(\mathbf{k}_1 + \cdots + \mathbf{k}_m - \mathbf{l}_1 - \cdots - \mathbf{l}_n)$ by $(\frac{L}{2\pi})^3 \delta_{\mathbf{k}_1 + \cdots + \mathbf{k}_m, \mathbf{l}_1 + \cdots + \mathbf{l}_n}$).

Let us assume that the formula (8.3) specifies a self-adjoint operator on the space F_Ω. If the functions $\Lambda_{m,n}$ are "good" enough, then this assumption holds. For example, in the case of CAR, one can prove the following statements.

If the coefficient functions $\Lambda_{m,n}$ tend to zero at infinity faster than $q^{-3(m+n)-\epsilon}$ where $q = (k_1^2 + \cdots + k_m^2 + l_1^2 + \cdots + l_n^2)^{1/2}, \epsilon > 0$, and $\Lambda_{m,n} = 0$ for $m + n \geq s$, then the formula (8.3) specifies a bounded self-adjoint operator in the space F_Ω. If the Hamiltonian H is represented in the form $H_0 + V$, where $H_0 = \int \omega(\mathbf{k})a^+(\mathbf{k})a(\mathbf{k})d\mathbf{k}$, and the coefficient functions of the expression V tend to zero at infinity sufficiently fast, then the formula (8.3) specifies a self-adjoint operator[4] in the space F_Ω.

The proof of these statements can be obtained from the considerations at the end of Section 6.2.

Note that the operator H_Ω commutes with the operator $\mathbf{P}_\Omega = \sum_{\mathbf{k} \in T_\Omega} \mathbf{k} a_\mathbf{k}^+ a_\mathbf{k}$ (the momentum operator).

We will define the physical quantities corresponding to the formal Hamiltonian H by taking the limit $\Omega \to \infty$.

For example, let us say that the number E belongs to the spectrum (it is an energy level) of the formal Hamiltonian H, if one can find such eigenvalues E_Ω of the Hamiltonian H_Ω, such that $\lim_{\Omega \to \infty}(E_\Omega - E_\Omega^0) = E$ (here, E_Ω^0 denotes the energy of the ground state Φ_Ω of the Hamiltonian H). We will say that the Hamiltonian H has an energy level E with momentum $\mathbf{k}$, if there exist vectors $\Psi_\Omega \in F_\Omega$ satisfying the conditions $H_\Omega \Psi_\Omega = E_\Omega \Psi_\Omega; \mathbf{P}_\Omega \Psi_\Omega = \mathbf{k}_\Omega \Psi_\Omega; \lim_{\Omega \to \infty}(E_\Omega - E_\Omega^0) = E; \lim_{\Omega \to \infty} \mathbf{k}_\Omega = \mathbf{k}$.

Another way to analyze a translation-invariant Hamiltonian is based on the construction of an operator realization of the Hamiltonian.

Let us suppose that commuting self-adjoint operators $\hat{H}$ and $\hat{\mathbf{P}} = (\hat{P}_1, \hat{P}_2, \hat{P}_3)$ (energy operator and momentum operator) act on the Hilbert space $\mathcal{H}$. Let us assume further that the operator functions $a(\mathbf{k}, \epsilon, t)$ are generalized functions with respect to the variable $\mathbf{k}$, obey

---

[4]In the case of CCR, with the given assumptions on the coefficient functions, one can prove only that the formula (8.3) defines a Hermitian operator.

CR and act on the same space (here, $\mathbf{k} \in E^3, \epsilon = \pm 1, a(\mathbf{k}, 1, t) = a^+(\mathbf{k}, -1, t)$).

We will say that the operators $\hat{H}$, $\hat{\mathbf{P}}$, the operator generalized functions $a(\mathbf{k}, \epsilon, t)$, and the vector $\Phi \in \mathcal{H}$ constitute an *operator realization* of the formal Hamiltonian (8.2), if

(1) The operator generalized functions $a(\mathbf{k}, \epsilon, t)$ satisfy the Heisenberg equations that formally correspond to the Hamiltonian (8.2):

$$i\frac{\partial a(\mathbf{k}, -1, t)}{\partial t} = \sum_{m,n} m \int \Lambda_{m,n}(\mathbf{k}, \mathbf{k}_1, \ldots, \mathbf{k}_{m-1} | \mathbf{l}_1, \ldots, \mathbf{l}_n)$$

$$\times \; \delta(\mathbf{k} + \mathbf{k}_1 + \cdots + \mathbf{k}_{m-1} - \mathbf{l}_1 - \cdots - \mathbf{l}_n)$$

$$\times \; a(\mathbf{k}_1, 1, t) \ldots a(\mathbf{k}_{m-1}, 1, t) a(\mathbf{l}_1, -1, t)$$

$$\ldots a(\mathbf{l}_n, -1, t) d^{m-1}\mathbf{k} d^n \mathbf{l}; \tag{8.4}$$

$$i\frac{\partial a(\mathbf{k}, 1, t)}{\partial t} = -\sum_{m,n} n \int \Lambda_{m,n}(\mathbf{k}_1, \ldots, \mathbf{k}_m | \mathbf{l}_1, \ldots, \mathbf{l}_{n-1}, \mathbf{k})$$

$$\times \; \delta(\mathbf{k}_1 + \cdots + \mathbf{k}_m - \mathbf{l}_1 - \cdots - \mathbf{l}_{n-1} - \mathbf{k})$$

$$\times \; a(\mathbf{k}_1, 1, t) \ldots a(\mathbf{k}_m, 1, t) a(\mathbf{l}_1, -1, t)$$

$$\ldots a(\mathbf{l}_{n-1}, 1, t) d^m \mathbf{k} d^{n-1} \mathbf{l}. \tag{8.5}$$

(2) $\exp(i\tau\hat{H}) a(\mathbf{k}, \epsilon, t) \exp(-i\tau\hat{H}) = a(\mathbf{k}, \epsilon, t + \tau)$;
$\exp(i\boldsymbol{\alpha}\hat{\mathbf{P}}) a(\mathbf{k}, \epsilon, t) \exp(-i\boldsymbol{\alpha}\hat{\mathbf{P}}) = \exp(i\boldsymbol{\alpha}\mathbf{k}\epsilon) a(\mathbf{k}, \epsilon, t)$.

(3) The operators $a(f, \epsilon, t) = \int f(\mathbf{k}) a(\mathbf{k}, \epsilon, t) d\mathbf{k}$, where $f \in \mathcal{S}(E^3)$ are defined on a dense subset D of the space $\mathcal{H}$ and transform this subset into itself; if $\Psi_1, \Psi_2 \in D$, then $\langle a(f, \epsilon, t)\Psi_1, \Psi_2 \rangle$ continuously depends on $f \in \mathcal{S}(E^3)$. The operators $a(f, \epsilon, t)$ also satisfy CCR for fixed t.

(4) The vector Φ is a ground state of the operator $\hat{H}$ and satisfies the condition $\hat{H}\Phi = \hat{\mathbf{P}}\Phi = 0$.

(5) The vector Φ is a cyclic vector of the family of operators $a(f, \epsilon, t)$.

Note that it follows from condition (3) that expressions of the form

$$\int f(\mathbf{k}_1, \ldots, \mathbf{k}_n) a(\mathbf{k}_1, \epsilon_1, t_1) \ldots a(\mathbf{k}_n, \epsilon_n, t_n) d\mathbf{k}_1 \ldots d\mathbf{k}_n,$$

where $f \in \mathcal{S}(E^{3n})$ make sense as operators defined on the set D. This can be proven by means of an operator analog of the kernel theorem (see Appendix A.5). This remark allows us to define rigorously the right-hand side of (8.5) in the case when the function $\Lambda_{1,1}(\mathbf{k})$ is smooth and all of its derivatives grow no faster than a power of $\mathbf{k}$ and the other functions $\Lambda_{m,n}$ belong to the space $\mathcal{S}$. The derivative $\frac{\partial a(\mathbf{k}, \epsilon, t)}{\partial t}$ on the left-hand side is understood as a weak derivative.

The Wightman and Green functions for the operator realizations of translation-invariant Hamiltonians can be defined by the formulas

$$\breve{w}_n(\mathbf{k}_1, \epsilon_1, t_1, \ldots, \mathbf{k}_n, \epsilon_n, t_n) = \langle a(\mathbf{k}_1, \epsilon_1, t_1) \ldots a(\mathbf{k}_n, \epsilon_n, t_n)\Phi, \Phi \rangle;$$

$$\breve{G}_n(\mathbf{k}_1, \epsilon_1, t_1, \ldots, \mathbf{k}_n, \epsilon_n, t_n) = \langle T(a(\mathbf{k}_1, \epsilon_1, t_1) \ldots a(\mathbf{k}_n, \epsilon_n, t_n))\Phi, \Phi \rangle.$$

It would be more precise to say that the functions $\breve{w}_n$ are Wightman functions in $(\mathbf{k}, t)$-representations. Their Fourier transforms with respect to the variables $\mathbf{k}_1, \ldots, \mathbf{k}_n$ are the functions

$$w_n(\mathbf{x}_1, \epsilon_1, t_1, \ldots, \mathbf{x}_n, \epsilon_n, t_n) = (2\pi)^{-\frac{3}{2}n} \int \exp\left(i \sum \epsilon_j \mathbf{x}_j \mathbf{k}_j \right)$$
$$\times \breve{w}_n(\mathbf{k}_1, \epsilon_1, t_1, \ldots, \mathbf{k}_n, \epsilon_n, t_n) d^n\mathbf{k}$$

which are Wightman functions in $(\mathbf{x}, t)$-representations. Their Fourier transforms with respect to the variables $t_1, \ldots, t_n$ are given by

$$\tilde{w}_n(\mathbf{k}_1, \epsilon_1, \omega_1, \ldots, \mathbf{k}_n, \epsilon_n, \omega_n) = (2\pi)^{-\frac{1}{2}n} \int \exp\left(i \sum \epsilon_j \omega_j t_j \right)$$
$$\times \breve{w}_n(\mathbf{k}_1, \epsilon_1, t_1, \ldots, \mathbf{k}_n, \epsilon_n, t_n) d^n t$$

which are Wightman functions in the $(\mathbf{k}, \omega)$-representation. Analogously, we can define Green functions in $(\mathbf{x}, t)$ and $(\mathbf{k}, \omega)$-representations; we denote them by G_n and $\tilde{G}_n$.

In the following sections of this chapter, we define the Wightman and Green functions of translation-invariant Hamiltonians by means of taking the limit of Wightman and Green functions in finite volume. We will relate these definitions to the definitions of Wightman and Green functions by means of operator realization of the translation-invariant Hamiltonian. We will show also how one can construct the operator realization by taking the limit when the volume tends to infinity. In Chapter 11, we consider the construction of operator realization in perturbation theory.

Note that for translation-invariant Hamiltonians, we can write down the Källén–Lehmann representation for functions $\breve{w}_2(\mathbf{k}_1, \epsilon_1, t_1, \mathbf{k}_2, \epsilon_2, t_2)$ and $\breve{G}_2(\mathbf{k}_1, \epsilon_1, t_1, \mathbf{k}_2, \epsilon_2, t_2)$.

To derive this representation, we fix a system of generalized eigenfunctions Ψ_λ of the operators $\hat{H}$ and $\hat{\mathbf{P}}$:

$$\hat{H}\Psi_\lambda = E(\lambda)\Psi_\lambda;$$

$$\hat{\mathbf{P}}\Psi_\lambda = \mathbf{k}(\lambda)\Psi_\lambda.$$

Let us prove, first, that the generalized function $\langle a(\mathbf{k}, \epsilon, t)\Phi, \Psi_\lambda \rangle$ can be represented by the formula

$$\langle a(\mathbf{k}, \epsilon, t)\Phi, \Psi_\lambda \rangle = \exp(iE(\lambda)t)\delta(\epsilon\mathbf{k} - \mathbf{k}(\lambda))\rho(\epsilon, \lambda). \qquad (8.6)$$

Indeed,

$$\langle a(\mathbf{k}, \epsilon, t)\Phi, \Psi_\lambda \rangle = \langle \exp(i\hat{H}t)a(\mathbf{k}, \epsilon)\Phi, \Psi_\lambda \rangle$$

$$= \langle a(\mathbf{k}, \epsilon)\Phi, \exp(-i\hat{H}t)\Psi_\lambda \rangle$$

$$= \exp(iE(\lambda)t) \langle a(\mathbf{k}, \epsilon)\Phi, \Psi_\lambda \rangle.$$

The function $\langle a(\mathbf{k}, \epsilon)\Phi, \Psi_\lambda \rangle$ satisfies the relation

$$\exp(i\epsilon\boldsymbol{\alpha}\mathbf{k}) \langle a(\mathbf{k}, \epsilon)\Phi, \Psi_\lambda \rangle = \exp(i\boldsymbol{\alpha}\mathbf{k}(\lambda)) \langle a(\mathbf{k}, \epsilon)\Phi, \Psi_\lambda \rangle \qquad (8.7)$$

that can be derived using the transformations

$$\langle \exp(i\epsilon\boldsymbol{\alpha}\mathbf{k})a(\mathbf{k}, \epsilon)\Phi, \Psi_\lambda \rangle = \langle \exp(i\boldsymbol{\alpha}\hat{\mathbf{p}})a(\mathbf{k}, \epsilon)\exp(-i\boldsymbol{\alpha}\hat{\mathbf{p}})\Phi, \Psi_\lambda \rangle$$

$$= \langle a(\mathbf{k}, \epsilon)\Phi, \exp(-i\boldsymbol{\alpha}\hat{\mathbf{p}}\Psi_\lambda) \rangle$$

$$= \exp(i\boldsymbol{\alpha}\mathbf{k}(\lambda)) \langle a(\mathbf{k}, \epsilon)\Phi, \Psi_\lambda \rangle.$$

It follows from (8.7) that the function $\langle a(\mathbf{k}, \epsilon)\Phi, \Psi_\lambda \rangle$ has the form $\rho(\epsilon, \lambda)\delta(\epsilon\mathbf{k} - \mathbf{k}(\lambda))$; this gives the proof of the statement we need.

Using (8.6), we can write down the following representation of Wightman functions:

$$\check{w}_2(\mathbf{k}_1, \epsilon_1, t_1, \mathbf{k}_2, \epsilon_2, t_2) = \langle a(\mathbf{k}_1, \epsilon_1, t_1)a(\mathbf{k}_2, \epsilon_2, t_2)\Phi, \Phi\rangle$$

$$= \langle a(\mathbf{k}_2, \epsilon_2, t_2)\Phi, a(\mathbf{k}_1, -\epsilon_1, t_1)\Phi\rangle$$

$$= \int \langle a(\mathbf{k}_2, \epsilon_2, t_2)\Phi, \Psi_\lambda\rangle \overline{\langle a(\mathbf{k}_1, -\epsilon_1, t_1)\Phi, \Psi_\lambda\rangle} d\lambda$$

$$= \int \exp(iE(\lambda)(t_2 - t_1))\delta(\epsilon_1\mathbf{k}_1 + \epsilon_2\mathbf{k}_2)$$

$$\times \ \delta(\epsilon_1\mathbf{k}_1 + \mathbf{k}(\lambda))\rho(\epsilon_2, \lambda)\overline{\rho(-\epsilon_1, \lambda)}d\lambda. \quad (8.8)$$

The relation

$$\check{G}_2(\mathbf{k}_1, \epsilon_1, t_1, \mathbf{k}_2, \epsilon_2, t_2) = \theta(t_1 - t_2)\check{w}_2(\mathbf{k}_1, \epsilon_1, t_1, \mathbf{k}_2, \epsilon_2, t_2)$$

$$\pm\theta(t_2 - t_1)\check{w}(\mathbf{k}_2, \epsilon_2, t_2, \mathbf{k}_1, \epsilon_1, t_1),$$

and (8.8) implies the representation of the function G_2 in terms of the functions $\rho(\epsilon, \lambda)$ (the Källén–Lehmann representation). In particular,

$$\check{G}_2(\mathbf{k}_1, 1, t_1, \mathbf{k}_2, -1, t_2) = \check{G}(\mathbf{k}_1, t_1 - t_2)\delta(\mathbf{k}_1 - \mathbf{k}_2),$$

where

$$\check{G}(\mathbf{k}, t) = \theta(t) \int \exp(-iE(\lambda)t)\delta(\mathbf{k} + \mathbf{k}(\lambda))|\rho(-1, \lambda)|^2 d\lambda$$

$$\pm\theta(-t) \int \exp(iE(\lambda)t)\delta(-\mathbf{k} + \mathbf{k}(\lambda))|\rho(1, \lambda)|^2 d\lambda. \quad (8.9)$$

In the $(\mathbf{k}, \omega)$-representation, we have

$$\tilde{G}_2(\mathbf{k}_1, 1, \omega_1, \mathbf{k}_2, -1, \omega_2) = G(\mathbf{k}_1, \omega_1)\delta(\omega_1 - \omega_2)\delta(\mathbf{k}_1 - \mathbf{k}_2),$$

where

$$G(\mathbf{k}, \omega) = \int \exp(i\omega t)\check{G}(\mathbf{k}, t)dt$$

$$= i \int \frac{|\rho(-1, \lambda)|^2}{\omega - E(\lambda) + i0}\delta(\mathbf{k} + \mathbf{k}(\lambda))d\lambda$$

$$\mp i \int \frac{|\rho(+1, \lambda)|^2}{\omega + E(\lambda) - i0}\delta(-\mathbf{k} + \mathbf{k}(\lambda))d\lambda. \quad (8.10)$$

8.2   Reconstruction theorem

The Wightman functions $\breve{w}_n(\mathbf{k}_1, \epsilon_1, t_1, \ldots, \mathbf{k}_n, \epsilon_n, t_n)$ of a translation-invariant Hamiltonian H can be defined by the following relation:

$$\breve{w}_n(\mathbf{k}_1, \epsilon_1, t_1, \ldots, \mathbf{k}_n, \epsilon, t_n)$$

$$= \lim_{\Omega \to \infty} \left(\frac{L}{2\pi}\right)^{\frac{3}{2}n} w_n^{\Omega}(\mathbf{k}_1, \epsilon_1, t_1, \ldots, \mathbf{k}_n, \epsilon_n, t_n) \qquad (8.11)$$

(here, $w_n^{\Omega}(\mathbf{k}_1, \epsilon_1, t_1, \ldots, \mathbf{k}_n, \epsilon_n, t_n) = \langle a_{\mathbf{k}_1}^{\epsilon_1}(t_1) \ldots a_{\mathbf{k}_n}^{\epsilon_n}(t_n)\Phi_{\Omega}, \Phi_{\Omega}\rangle$ are Wightman functions of the Hamiltonian H_{Ω}, constructed with the basis $\phi_{\mathbf{k}}$, where $\mathbf{k} \in T_{\Omega}$). The relation (8.11) requires some explanation because the functions of continuous argument are defined as the limit of functions of arguments running over a lattice. We will understand this limit in the sense of generalized functions: for every test function $\phi(\mathbf{k}_1, \ldots, \mathbf{k}_n)$, we require

$$\int \phi(\mathbf{k}_1, \ldots, \mathbf{k}_n)\breve{w}_n(\mathbf{k}_1, \epsilon_1, t_1, \ldots, \mathbf{k}_n, \epsilon_n, t_n)d\mathbf{k}_1 \ldots d\mathbf{k}_n$$

$$= \lim_{\Omega \to \infty} \left(\frac{2\pi}{L}\right)^{\frac{3n}{2}} \sum_{\mathbf{k}_1, \ldots, \mathbf{k}_n \in T_{\Omega}} \phi(\mathbf{k}_1, \ldots, \mathbf{k}_n)$$

$$\times w_n^{\Omega}(\mathbf{k}_1, \epsilon_1, t_1, \ldots, bk_n, \epsilon_n, t_n). \qquad (8.12)$$

We assume that the limit (8.11) exists (the existence of this limit can be checked in the framework of perturbation theory).

The functions $\breve{w}_n(\mathbf{k}_1, \epsilon_1, t_1, \ldots, \mathbf{k}_n, \epsilon_n, t_n)$ will be considered as generalized functions with respect to the variables $\mathbf{k}_1, \ldots, \mathbf{k}_n$ and conventional functions with respect to the variables $t_1, \ldots, t_n$.

Green functions $\breve{G}_n(\mathbf{k}_1, \epsilon_1, t_1, \ldots, \mathbf{k}_n, \epsilon_n, t_n)$ of the translation-invariant Hamiltonian H are defined in the same way, as limits of finite-volume Green functions. Namely, we use the following definition:

$$\breve{G}_n(\mathbf{k}_1, \epsilon_1, t_1, \ldots, \mathbf{k}_n, \epsilon_n, t_n)$$

$$= \lim_{\Omega \to \infty} \left(\frac{L}{2\pi}\right)^{\frac{3n}{2}} G_n^{\Omega}(\mathbf{k}_1, \epsilon_1, t_1, \ldots, \mathbf{k}_n, \epsilon_n, t_n),$$

where $G_n^{\Omega}(\mathbf{k}_1, \epsilon_1, t_1, \ldots, \mathbf{k}_n, \epsilon_n, t_n) = \langle T(a_{\mathbf{k}_1}^{\epsilon_1}(t_1) \ldots a_{\mathbf{k}_n}^{\epsilon_n}(t_n))\Phi_{\Omega}, \Phi_{\Omega}\rangle$ are Green functions of the Hamiltonian H_{Ω} constructed with the basis $\phi_{\mathbf{k}}$, where $\mathbf{k} \in T_{\Omega}$. The limit is understood in the same way as the limit in the definition of Wightman functions.

Green functions $\check{G}_n$ can be expressed in terms of Wightman functions $\check{w}_n$. Namely, taking the limit of $\Omega \to \infty$ in the relation (7.7) applied to the functions G_n^{Ω} and w_n^{Ω}, we obtain

$$\check{G}_n(\mathbf{k}_1, \epsilon_1, t_1, \ldots, \mathbf{k}_n, \epsilon_n, t_n)$$
$$= \sum_{\pi} (-1)^{\gamma(\pi)} \theta_{\pi}(t) \check{w}_n^{\pi}(\mathbf{k}_1, \epsilon_1, t_1, \ldots, \mathbf{k}_n, \epsilon_n, t_n) \quad (8.13)$$

(the notation is the same as in Section 7.2).

The definition of Green functions in $(\mathbf{x}, t)$-representation G_n and in the $(\mathbf{k}, \omega)$-representation $\tilde{G}_n$ is analogous to similar definitions for Wightman functions.

It is easy to check that the Wightman functions $\check{w}_n(\mathbf{k}_1, \epsilon_1, t_1, \ldots, \mathbf{k}_n, \epsilon_n, t_n)$ of a translation-invariant Hamiltonian H have the following properties:

(1) *Invariance with respect to time translation:*

$$\check{w}_n(\mathbf{k}_1, \epsilon_1, t_1, \ldots, \mathbf{k}_n, \epsilon_n, t_n) = \check{w}_n(\mathbf{k}_1, \epsilon_1, t_1+\tau, \ldots, \mathbf{k}_n, \epsilon_n, t_n+\tau).$$

(2) *Hermiticity:*

$$\overline{\check{w}}_n(\mathbf{k}_1, \epsilon_1, t_1, \ldots, \mathbf{k}_n, \epsilon_n, t_n) = \check{w}_n(\mathbf{k}_n, -\epsilon_n, t_n, \ldots, \mathbf{k}_1, -\epsilon_1, t_1).$$

(3) *Positive definiteness:* For every sequence $f_n(\mathbf{k}_1, \epsilon_1, t_1, \ldots, \mathbf{k}_n, \epsilon_n, t_n)$ of test functions that do not vanish for only a finite number of indices n, we have

$$\sum_{m,n} \sum_{\epsilon_\alpha, \sigma_\beta} \int f_m(\mathbf{k}_1, \epsilon_1, t_1, \ldots, \mathbf{k}_m, \epsilon_m, t_m)\overline{f_n}(\mathbf{q}_n, -\sigma_n, \tau_n,$$
$$\ldots, \mathbf{q}_1, -\sigma_1, \tau_1)\check{w}_{m+n}(\mathbf{k}_1, \epsilon_1, t_1, \ldots, \mathbf{k}_m, \epsilon_m, t_m,$$
$$\mathbf{q}_1, \sigma_1, \tau_1, \ldots, \mathbf{q}_n, \sigma_n, \tau_n)d^m\mathbf{k}d^n\mathbf{q}d^mtd^n\tau \geq 0.$$

(4) *Spectral property*:

$$\int \exp(-i\omega a)\breve{w}_n(\mathbf{k}_1, \epsilon_1, t_1, \ldots, \mathbf{k}_r, \epsilon_r, t_r, \mathbf{k}_{r+1}, \epsilon_{r+1}, t_{r+1} + a,$$

$$\ldots, \mathbf{k}_n, \epsilon_n, t_n + a)da = 0, \tag{8.14}$$

if $\omega < 0$. The relation (8.14) is also valid under a weaker condition — it is sufficient to assume that ω does not belong to the spectrum of the Hamiltonian H.

(5) *Symmetry with respect to the permutation of arguments*: Let us introduce the notation

$$\breve{w}_n^{(i)}(\mathbf{k}_1, \epsilon_1, t_1, \ldots, \mathbf{k}_i, \epsilon_i, t_i, \mathbf{k}_{i+1}, \epsilon_{i+1}, t_{i+1}, \ldots, \mathbf{k}_n, \epsilon_n, t_n)$$

$$= \pm \breve{w}_n(\mathbf{k}_1, \epsilon_1, t_1, \ldots, \mathbf{k}_{i+1}, \epsilon_{i+1}, t_{i+1}, \mathbf{k}_i, \epsilon_i, t_i,$$

$$\ldots, \mathbf{k}_n, \epsilon_n, t_n)$$

(here, the plus sign corresponds to the case of CCR and the minus sign corresponds to the case of CAR). In other words, $\breve{w}_n^{(i)}$ is obtained from $\breve{w}_n$ by permuting $\mathbf{k}_i, \epsilon_i, t_i$ with $\mathbf{k}_{i+1}, \epsilon_{i+1}, t_{i+1}$ (with a change of sign in the case of CAR).
If $t_i = t_{i+1}$, then

$$\breve{w}_n^{(i)}(\mathbf{k}_1, \epsilon_1, t_1, \ldots, \mathbf{k}_n, \epsilon_n, t_n)$$

$$= \breve{w}_n(\mathbf{k}_1, \epsilon_1, t_1, \ldots, \mathbf{k}_n, \epsilon_n, t_n) + A_{\epsilon'}^{\epsilon}\delta(\mathbf{k}_i - \mathbf{k}_i')$$

$$\times \breve{w}_{n-2}(\mathbf{k}_1, \epsilon_1, t_1, \ldots, \mathbf{k}_{i-1}, \epsilon_{i-1}, t_{i-1}, \mathbf{k}_{i+2}, \epsilon_{i+2}, t_{i+2},$$

$$\ldots, \mathbf{k}_n, \epsilon_n, t_n) \tag{8.15}$$

(the definition of the matrix $A_{\epsilon}^{\epsilon'}$ can be found in Section 6.1).

(6) *Translation-invariance*:

$$\breve{w}_n(\mathbf{k}_1, \epsilon_1, t_1, \ldots, \mathbf{k}_n, \epsilon_n, t_n)$$

$$= v_n(\mathbf{k}_1, \epsilon_1, t_1, \ldots, \mathbf{k}_n, \epsilon_n, t_n)\delta\left(\sum \epsilon_j \mathbf{k}_j\right)$$

(in the $(\mathbf{x}, t)$-representation this property takes the form $w_n(\mathbf{x}_1, \epsilon_1, t_1, \ldots, \mathbf{x}_n, \epsilon_n, t_n) = w_n(\mathbf{x}_1 + a, \epsilon_1, t_1, \ldots, \mathbf{x}_n + a, \epsilon_n, t_n))$.

All the above listed properties (with the exception of (6)) can be easily proven if we recall that $\breve{w}_n$ are obtained as limits of the functions w_n^Ω, which have similar properties proven in Section 7.1.

Theorem 8.1. (Reconstruction theorem.)

Existence: *Given a family of functions $\breve{w}_n(\mathbf{k}_1, \epsilon_1, t_1, \ldots, \mathbf{k}_n, \epsilon_n, t_n)$ obeying properties $(1) - (6)$, one can construct a Hilbert space $\mathcal{H}$, four commuting self-adjoint operators $\hat{H}, \hat{\mathbf{P}}$, operator functions $a(\mathbf{k}, \epsilon, t)$ acting on $\mathcal{H}$ that are generalized[5] functions with respect to the variable $\mathbf{k}$, and the vector $\Phi \in \mathcal{H}$ in such a way that the following requirements are met:*

(a) $\breve{w}_n(\mathbf{k}_1, \epsilon_1, t_1, \ldots, \mathbf{k}_n, \epsilon_n, t_n) = \langle a(\mathbf{k}_1, \epsilon_1, t_1) \ldots a(\mathbf{k}_n, \epsilon_n, t_n) \, \Phi, \Phi \rangle$.

(b) $\exp(i\tau H) a(\mathbf{k}, \epsilon, t) \exp(-i\tau H) = a(\mathbf{k}, \epsilon, t + \tau)$,
$\exp(i\boldsymbol{\alpha}\mathbf{p}) a(\mathbf{k}, \epsilon, t) \exp(-i\boldsymbol{\alpha}\mathbf{p}) = \exp(i\epsilon\boldsymbol{\alpha}\mathbf{k}) a(\mathbf{k}, \epsilon, t)$;

(c) *The operators $a(f, \epsilon, t) = \int f(\mathbf{k}) a(\mathbf{k}, \epsilon, t) d\mathbf{k}$ are defined on the dense set D of the space $\mathcal{H}$ and transform the subset into itself; the operators $a(f, \epsilon, t)$ for fixed t specify a representation of CR; the expression $\langle a(f, \epsilon, t)\Psi_1, \Psi_2 \rangle$ continuously depends on f in the topology of the space $\mathcal{S}$ for all $\Psi_1, \Psi_2 \in D$.*

(d) *The vector Φ is the ground state of the operator $\hat{H}$ and satisfies the conditions $H\Phi = 0$ and $\mathbf{P}\Phi = 0$.*

(e) *The vector Φ is a cyclic vector with respect to the operators $a(f, \epsilon, t)$.*

Uniqueness: *If $\mathcal{H}_i, H_i, \mathbf{P}_i, a_i(\mathbf{k}, e\, t)$ and Φ_i are two sets of objects $(i = 1, 2)$ satisfying the conditions $(a), (b),$ and (c) of the reconstruction theorem, then there exists a unitary operator U that maps the space $\mathcal{H}_1$ onto the space $\mathcal{H}_2$ and obeys the conditions $U\Phi_1 = \Phi_2, U\hat{H}_1 = \hat{H}_2 U, U\hat{\mathbf{P}}_1 = \hat{\mathbf{P}}_2 U,$ as well as the relations $U a_1(f, \epsilon, t) = a_2(f, \epsilon, t) U$ on some dense subset of the space $\mathcal{H}_1$ (in other words, these two sets are isomorphic).*

Let us start the proof from the second part of the theorem. Let us consider the set $\mathcal{N}$ of sequences of functions $f =$

---

[5] We will consider the space of test functions to be $\mathcal{S}(E^3)$ (i.e. all functions f in the formulation of the theorem satisfy $f \in \mathcal{S}(E^3)$).

$\{f_n(\mathbf{k}_1, \epsilon_1, t_1, \ldots, \mathbf{k}_n, \epsilon_n, t_n)\}$ such that (1) every function f_n is a linear combination of functions of the form

$$\lambda_1(\mathbf{k}_1, \epsilon_1)\delta(t_1 - \tau_1) \ldots \lambda_m(\mathbf{k}_m, \epsilon_m)\delta(t_m - \tau_m),$$

where $\lambda_i(\mathbf{k}_i, \epsilon_i)$ are test functions and (2) only a finite number of functions f_n do not vanish. Let us assign the vectors Ψ_f^i, where $i = 1, 2$, to every sequence $f \in \mathcal{N}$ by the formula

$$\Psi_f^i = \sum_{n=1}^{\infty} \sum_{\epsilon_1, \ldots, \epsilon_n} \int f_n(\mathbf{k}_1, \epsilon_1, t_1, \ldots, \mathbf{k}_n, \epsilon_n, t_n)$$

$$\times\, a_i(\mathbf{k}_1, \epsilon_1, t_1) \ldots a_i(\mathbf{k}_n, \epsilon_n, t_n) d^n \mathbf{k} d^n t.$$

The set of vectors Ψ_f^i, where $f \in \mathcal{N}$, will be denoted by D_i. The inner product $\langle \Psi_f^i, \Psi_g^i \rangle$ of two vectors from the set D_i can be easily expressed in terms of the sequences f, g and Wightman functions:

$$\langle \Psi_f^i, \Psi_g^i \rangle = \sum_{m,n} \sum_{\epsilon_\alpha, \sigma_\alpha} \int f_m(\mathbf{k}_1, \epsilon_1, t_1, \ldots, \mathbf{k}_m, \epsilon_m, L_m) \overline{f_n}(\mathbf{q}_1, \sigma_1, \tau_1, \ldots,$$

$$\mathbf{q}_n, \sigma_n, \tau_n) \breve{w}_{m+n}(\mathbf{k}_1, \epsilon_1, t_1, \ldots, \mathbf{k}_m, \epsilon_m, t_m, \mathbf{q}_n, -\sigma_n,$$

$$\tau_n, \ldots, \mathbf{q}_1, -\sigma_1, \tau_1) d^m \mathbf{k} d^n \mathbf{q} d^m t d^n \tau. \tag{8.16}$$

It is easy to check that

$$a_i(\phi)\Psi_f^i = \Psi_{\phi f}^i \tag{8.17}$$

(here, $\phi = \phi(\mathbf{k}, \epsilon, t) = \lambda(\mathbf{k}, \epsilon)\delta(t - \tau)$, where $\lambda(\mathbf{k}, \epsilon)$ is a test function, $a_i(\phi) = \sum_\epsilon \int \phi(\mathbf{k}, \epsilon, t) a_i(\mathbf{k}, \epsilon, t) dk dt = \sum_\epsilon \int \lambda(\mathbf{k}, \epsilon) a(\mathbf{k}, \epsilon, \tau)) dk$, and $\phi f \in \mathcal{N}$ is a sequence whose nth entry is equal to $\phi(\mathbf{k}_1, \epsilon_1, t_1) f_{n-1}(\mathbf{k}_2, \epsilon_2, t_2, \ldots, \mathbf{k}_n, \epsilon_n, t_n))$. Condition (b) implies that

$$\exp(-i\hat{H}_i\tau)\Psi_f^i = \Psi_{V_\tau f}^i; \quad \exp(-i\alpha\hat{\mathbf{P}})\Psi_f^i = \Psi_{W_\alpha f}^i, \tag{8.18}$$

where $V_\tau f \in \mathcal{N}$ and $W_\alpha f \in \mathcal{N}$ are sequences of functions with the nth entry given by

$$f_n(\mathbf{k}_1, \epsilon_1, t_1 - \tau, \ldots, \mathbf{k}_n, \epsilon_n, t_n - \tau),$$

$$\exp\left(-i\sum \alpha \mathbf{k}_j \epsilon_j\right) f_n(\mathbf{k}_1, \epsilon_1, t_1, \ldots, \mathbf{k}_n, \epsilon_n, t_n).$$

Let us now construct the operator U on the set D_1 assuming that

$$U\Psi_f^1 = \Psi_f^2.$$

It follows from (8.16) that this operator preserves inner products, therefore it can be extended by continuity to the unitary operator mapping $\mathcal{H}_1$ into $\mathcal{H}_2$. It follows from the relations (8.17) and (8.18) that the constructed operator obeys the conditions we need.

The above proof of the second part of the theorem gives us a path to proving the first part of the theorem.

Let us start with the sequence of Wightman functions $\breve{w}$. Let us introduce the inner product $\langle f, g \rangle$ on the set $\mathcal{N}$ by means of the formula

$$\langle f, g \rangle = \sum_{m,n} \sum_{\epsilon_\alpha, \sigma_\beta} \int f_m(\mathbf{k}_1, \epsilon_1, t_1, \ldots, \mathbf{k}_n, \epsilon_n, t_n) \overline{f_n}(\mathbf{q}_1, \sigma_1, \tau_1, \ldots,$$

$$\mathbf{q}_n, \sigma_n, \tau_n) \breve{w}_{m+n}(\mathbf{k}_1, \epsilon_1, t_1, \ldots, \mathbf{k}_m, \epsilon_m, t_m, \mathbf{q}_n, -\sigma_n, \tau_n, \ldots,$$

$$\mathbf{q}_1, -\sigma_1, \tau_1) d^m\mathbf{k} d^n\mathbf{q} d^m t d^n \tau.$$

Then, it follows from (3) in Section 8.2 that $\langle f, f \rangle \geq 0$. The elements $f, g \in \mathcal{N}$ we consider to be equivalent ($f \sim g$), if $\langle f - g, f - g \rangle = 0$. The set of equivalence classes will be denoted by D and the equivalence class of the element $f \in \mathcal{N}$ will be denoted by Ψ_f. In other words, the set D consists of symbols Ψ_f, where $f \in \mathcal{N}$, and two symbols Ψ_f, Ψ_g specify the same element of the set D if $f \sim g$. The inner product of the elements $\Psi_f, \Psi_g \in D$ is defined by the formula $\langle \Psi_f, \Psi_g \rangle = \langle f, g \rangle$. An element of the set D can be represented in different ways in the form Ψ_f, however the inner product in D does not depend on the choice of the representative because from the relation $f \sim f', g \sim g'$, it follows that $\langle f, g \rangle = \langle f', g' \rangle$. (Similar considerations can be applied to other operations in D.) A linear combination of elements in D is defined by the formula

$$\lambda\Psi_f + \mu\Psi_g = \Psi_{\lambda f + \mu g},$$

where the linear combination $\lambda f + \mu g$ of the sequences f, g is defined in the usual way. Hence, the set D can be considered a pre-Hilbert space.

In the space D, we can define the operator generalized functions $a(f, \epsilon, t) = \int f(\mathbf{k})a(\mathbf{k}, \epsilon, t)d\mathbf{k}$, assuming that $a(f, \epsilon, t)\Psi_g = \Psi_{\phi g}$, where $\phi(\mathbf{k}, \sigma, \tau) = f(\mathbf{k})\delta_{\sigma,\epsilon}\delta(t - \tau)$, and we can define the family of operators $\tilde{V}_\tau, \tilde{W}_\alpha$, satisfying $\tilde{V}_\tau\Psi_g = \Psi_{V_\tau g}, \tilde{W}_\alpha\Psi_g = \Psi_{W_\alpha g}$.

Let us now define the Hilbert space $\mathcal{H}$ as the completion of the pre-Hilbert space D.

The operators $\tilde{V}_\tau, \tilde{W}_\alpha$ map the set D into itself and preserve inner products. Therefore, they can be extended by continuity to unitary operators on the space $\mathcal{H}$. In this way, we obtain a one-parameter and a three-parameter group of unitary operators on $\mathcal{H}$; the generators of these groups will be denoted by $\hat{H}$ and $\hat{\mathbf{P}}$ (in other words, $\tilde{V}_\tau = \exp(-iH\tau), \tilde{W}_\alpha = \exp(-i\boldsymbol{\alpha}\hat{\mathbf{P}})$). The symbol Φ will denote the vector Ψ_θ, where θ is a function sequence with $f_0 = 1, f_n = 0$, for $n > 0$. The operator generalized functions $a(f, \epsilon, t)$ are defined on the dense subset $D \subset \mathcal{H}$.

Hence, starting with the Wightman functions, we have constructed the objects that were described in the reconstruction theorem. It is easy to check that they have all the necessary properties. The only point we will consider in detail is the proof that the vector Φ is the ground state of the Hamiltonian $\hat{H}$. We derive this fact from the following lemma.

Lemma 8.1. *The number ω does not belong to the spectrum of the operator $\hat{H}$ if and only if for all Wightman functions we have*

$$\int \exp(-i\omega\tau)\breve{w}_n(\mathbf{k}_1, \epsilon_1, t_1, \ldots, \mathbf{k}_i, \epsilon_i, t_i, \mathbf{k}_{i+1}, \epsilon_{i+1}, t_{i+1} + \tau,$$

$$\ldots, \mathbf{k}_n, \epsilon_n, t_n + \tau)d\tau = 0. \tag{8.19}$$

Then, by property (4) of Wightman functions and the lemma, it follows that the operator $\hat{H}$ is non-negative. Taking into account that $\hat{H}\Phi = 0$, we see that Φ is the ground state.

To prove (2), we first note that the functions $\breve{w}_n$ are Wightman functions of the operator $\hat{H}$ with respect to the operator generalized function $a(\mathbf{k}, \epsilon, t)$ (in the sense of the definition in Section 7.1). Therefore, it follows from property (4) (Section 7.1) that for every ω that does not belong to the spectrum of the operator $\hat{H}$, we have (8.19). To prove the inverse statement, it is sufficient to check that for a smooth finite function $\chi(\omega)$ that does not vanish only for ω satisfying (8.19), we have $\int \tilde{\chi}(t) \exp(i\hat{H}t)dt = 0$, where $\tilde{\chi}(t) = \int \exp(-i\omega t)\chi(w)dw$ (see Appendix A.5). It is easy to check that

$$\int \tilde{\chi}(t)\langle\exp(i\hat{H}t)\Psi_1, \Psi_2\rangle dt = 0,$$

if $\Psi_i = a(f_1^{(i)}, \epsilon_1^{(i)}, t_1^{(i)}) \ldots a(f_{n_i}^{(i)}, \epsilon_{n_i}^{(i)}, t_{n_i}^{(i)})\Phi$; to prove this, we should express

$$\int \tilde{\chi}(t) \langle\exp(iHt)\Psi_1, \Psi_2\rangle \, dt = \int \chi(\omega)\langle\exp(i(\hat{H} - \omega)t)\Psi_1, \Psi_2\rangle d\omega dt$$

in terms of Wightman functions. By using the cyclicity of the vector Φ, we can see that the relation we have proven implies $\int \chi(t) \exp(i\hat{H}t)dt = 0$.

This finishes the proof of the reconstruction theorem.

It follows from the lemma that every point of the spectrum of the operator $\hat{H}$ belongs to the spectrum of the translation-invariant Hamiltonian H in the sense of Section 8.1. Analogously, one can prove the following statement: if the point $(\mathbf{k}, E)$ belongs to the joint spectrum of a family of commuting operators $(\hat{\mathbf{P}}, \hat{H})$, then the Hamiltonian H has an energy level E with momentum $\mathbf{k}$.

In conclusion, we will check that under certain conditions, the space $\mathcal{H}$, the operators $\hat{H}, \hat{\mathbf{P}}$, the operator generalized functions $a(\mathbf{k}, \epsilon, t)$, and the vector Φ, constructed as in the reconstruction theorem, can be considered as an operator realization of the translation-invariant Hamiltonian H in the sense of Section 8.1. Namely, we will show that this statement is correct if in formula (8.2), that specifies the Hamiltonian H, the function $\Lambda_{1,1}(\mathbf{k})$ is smooth, all derivatives of this function do not grow faster than a polynomial, and the remaining functions $\Lambda_{m,n}$ belong to the space $\mathcal{S}$ (as noted in Section 8.1, equations (8.4) and (8.5) have precise meaning in this case). To give the proof, it is sufficient to check that the operator generalized functions $a(\mathbf{k}, \epsilon, t)$ in the reconstruction theorem satisfy the Heisenberg equations (8.4) and (8.5) that correspond formally to the Hamiltonian H (all other conditions in the definition of an operator realization follow from the reconstruction theorem). It is easy to give this proof, using the remark that the Wightman functions of the Hamiltonian H satisfy

$$\frac{1}{i}\frac{\partial}{\partial t_1}\breve{w}_r(\mathbf{k}_1,\epsilon_1,t_1,\ldots,\mathbf{k}_r,\epsilon_r,t_r)$$

$$= \delta_1^{\epsilon_1}\sum_{m,n} n\int \Lambda_{m,n}(\mathbf{p}_1,\ldots,\mathbf{p}_m|\mathbf{q}_1,\ldots,\mathbf{q}_{n-1},\mathbf{k}_1)\delta(\mathbf{p}_1+\cdots+\mathbf{p}_m$$

$$-\mathbf{q}_1-\cdots-\mathbf{q}_{n-1}-\mathbf{k}_1)\breve{w}_{r+m+n-2}(\mathbf{p}_1,1,t_1,\ldots,\mathbf{p}_m,1,t_1,\mathbf{q}_1,$$

$$-1,t_1,\ldots,\mathbf{q}_{n-1},-1,t_1,\mathbf{k}_2,\epsilon_2,t_2,\ldots,\mathbf{k}_r,\epsilon_r,t_r)d^m\mathbf{p}d^{n-1}\mathbf{q}$$

$$-\delta_{-1}^{\epsilon_1}\sum_{m,n} m\int \Lambda_{m,n}(\mathbf{k}_1,\mathbf{p}_1,\ldots,\mathbf{p}_{m-1}|\mathbf{q}_1,\ldots,\mathbf{q}_n)$$

$$\times\ \delta(\mathbf{k}_1+\mathbf{p}_1+\cdots+\mathbf{p}_{m-1}-\mathbf{q}_1-\cdots-\mathbf{q}_n)\breve{w}_{r+m+n-2}(\mathbf{p}_1,1,t_1,$$

$$\ldots,\mathbf{p}_{m-1},1,t_1,\mathbf{q}_1,-1,t_1,\ldots,\mathbf{q}_n,-1,t_1,\mathbf{k}_2,\epsilon_2,t_2,$$

$$\ldots,\mathbf{k}_r,\epsilon_r,t_r)d^{m-1}\mathbf{p}d^n\mathbf{q} \tag{8.20}$$

and satisfy similar conditions for derivatives with respect to other time variables. (Equation (8.20) can be obtained if we use the equations for the functions w_r^Ω that follow from the considerations in Section 7.4 and take the limit $\Omega\to\infty$ in these equations. The conditions we have imposed on the functions $\Lambda_{m,n}$ imply that we can take this limit because we have assumed that the functions w_r^Ω tend to $\breve{w}_r$ in the sense of generalized functions.) Using (8.20), we can derive the Heisenberg equations (8.4) and (8.5); it is sufficient to express in terms of Wightman functions the quantity

$$\left\langle \frac{d}{dt}a(f,t)\Psi_\alpha,\Psi_\beta\right\rangle,$$

where $\alpha,\beta\in\mathcal{N}$, using equations similar to (8.20).

It follows from the above statement that the Wightman functions for a translation-invariant Hamiltonian defined in Section 8.2 are Wightman functions of the operator realizations of this Hamiltonian in the sense of Section 8.1. A similar statement can be proven for Green functions. In the following, we will use the definition in Section 8.1, but use the terminology from Section 8.2.

Let us define Schwinger functions that, together with Wightman and Green functions, play an important role in quantum field theory.

Let us consider the operators $\exp(iH\tau)$, where τ is a complex number. It follows from the non-negativity of the operator H that these operators are bounded for τ in the upper half plane $(\operatorname{Im}\tau \geq 0)$. Therefore, it is natural to assume that under the condition $\operatorname{Im}\tau \geq 0$, the operators $\exp(iH\tau)$ transform the set D into itself; then, in particular, the operator $\exp(-H\sigma)$, where $\sigma \geq 0$ has this property. Under this assumption, we can define Schwinger functions using the formula

$$
\begin{aligned}
S_n(\mathbf{k}_1, \epsilon_1, t_1, \ldots, \mathbf{k}_n, \epsilon_n, t_n) = \langle a(\mathbf{k}_1, \epsilon_1, 0)\exp(-H(t_1 - t_2)) \\
\times\, a(\mathbf{k}_2, \epsilon_2, 0)\ldots\exp(-H(t_{n-1} - t_n)) \\
\times\, a(\mathbf{k}_n, \epsilon_n, 0)\Phi, \Phi\rangle
\end{aligned}
$$

(we assume that $t_1 \geq \cdots \geq t_n$). Let us note that Schwinger functions can also be defined without any additional assumptions. Indeed by the property of time-translation invariance, the Wightman function $\breve{w}_n$ can be written in the following form:

$$
\begin{aligned}
\breve{w}_n(\mathbf{k}_1, \epsilon_1, t_1, \ldots, \mathbf{k}_n, \epsilon_n, t_n) \\
= \breve{v}_n(\mathbf{k}_1, \epsilon_1, \ldots, \mathbf{k}_n, \epsilon_n, t_2 - t_1, \ldots, t_n - t_{n-1}).
\end{aligned}
$$

The functions $\breve{v}_n(\mathbf{k}_1, \epsilon_1, \ldots, \mathbf{k}_n, \epsilon_n, \tau_1, \ldots, \tau_{n-1})$ can be analytically continued with respect to the variables $\tau_1, \ldots, \tau_{n-1}$ in the domain $\operatorname{Im}\tau_1 \geq 0, \ldots, \operatorname{Im}\tau_{n-1} \geq 0$ (to prove this fact, we should represent $\breve{v}_n$ in the form

$$
\begin{aligned}
\breve{v}_n(\mathbf{k}_1, \epsilon_1, \ldots, \mathbf{k}_n, \epsilon_n, \tau_1, \ldots, \tau_{n-1}) \\
= \int \exp(i\sum_j \omega_j\tau_j)\tilde{v}_n(\mathbf{k}_1, \ldots, \epsilon_n, \omega_1, \ldots, \omega_{n-1})d^{n-1}\omega
\end{aligned}
$$

and note that by the spectrum condition the support of the function v_n is contained in the set $\omega_1 \geq 0, \ldots, \omega_{n-1} \geq 0$). The analytic continuation of the function $\tilde{v}_n$ will be denoted by the same symbol.

Schwinger functions can then be defined by the formula

$$S_n(\mathbf{k}_1, \epsilon_1, t_1, \ldots, \mathbf{k}_n, \epsilon_n, t_n)$$

$$= \breve{v}_n(\mathbf{k}_1, \epsilon_1, \ldots, \mathbf{k}_n, \epsilon_n, i(t_1 - t_2), \ldots, i(t_{n-1} - t_n))$$

(where the definition assumes $t_1 \geq \cdots \geq t_n$).

It is easy to formulate properties of Schwinger functions that are analogous to the properties of Wightman functions and to prove the analog of the reconstruction theorem for Schwinger functions; see Osterwalder and Schrader (1973).

8.3 Interactions of the form $V(\phi)$

In this section, we will consider an important class of translation-invariant Hamiltonians. The Hamiltonians of this class can be obtained by the quantization of classical systems with an infinite number of degrees of freedom; they can be written in the form (8.2), which allows us to apply the results of Section 8.1.

In the present section, it will be convenient not to use the assumption that $\hbar = 1$, made in the rest of the book.

Let us recall that by quantizing a classical mechanical system with the Hamiltonian

$$\mathcal{H}(p, q) = \sum_{k=1}^{n} \frac{p_k^2}{2} + U(q_1, \ldots, q_n), \tag{8.21}$$

where p_k are generalized momenta and q_k are generalized coordinates, we obtain a quantum mechanical system described by the Hamiltonian

$$H = \sum_{k=1}^{n} \frac{1}{2}\hat{p}_k^2 + U(\hat{q}_1, \ldots, \hat{q}_n), \tag{8.22}$$

where $\hat{p}_k, \hat{q}_k$ are self-adjoint operators satisfying the canonical commutation relations (CCR)

$$[\hat{p}_k, \hat{p}_l] = [\hat{q}_k, \hat{q}_l] = 0; \quad [\hat{p}_k, \hat{q}_l] = \frac{\hbar}{i}\delta_{kl}$$

(the operators $\hat{p}_k, \hat{q}_k$ can be realized in the space of square-integrable functions $\psi(q_1, \ldots, q_n)$ and defined by the formulas $\hat{p}_k \psi = \frac{\hbar}{i}\frac{\partial}{\partial q_k}\psi$; $\hat{q}_k \psi = q_k \psi$).

The Heisenberg operators $\hat{p}_k(t) = \exp(iHt)\hat{p}_k \exp(-iHt), \hat{q}_k(t) = \exp(iHt)\hat{q}_k \exp(-iHt)$ satisfy the equations

$$\frac{d\hat{p}_k(t)}{dt} = -\frac{\partial U}{\partial q_i}(\hat{q}_1(t), \ldots, \hat{q}_n(t));$$

$$\frac{d\hat{q}_k(t)}{dt} = \hat{p}_k(t).$$

Let us now consider the analog of a classical system with the Hamiltonian (8.21) in the case of an infinite number of degrees of freedom. Namely, we assume that the classical system is described by the Hamiltonian functional

$$\mathcal{H}(\pi, \phi) = \frac{1}{2}\int \pi^2(\mathbf{x})d\mathbf{x} + V(\phi),$$

$$V(\phi) = \sum_n \int V_n(\mathbf{x}_1, \ldots, \mathbf{x}_n)\phi(\mathbf{x}_1) \ldots \phi(\mathbf{x}_n)d\mathbf{x}_1 \ldots d\mathbf{x}_n,$$

where $\pi(\mathbf{x})$ are the generalized momentum variables and $\phi(\mathbf{x})$ are the generalized coordinates. For definiteness, we assume that $\mathbf{x}$ runs over three-dimensional Euclidean space. We consider only translation-invariant functionals (i.e. we assume that the function $V_n(\mathbf{x}_1, \ldots, \mathbf{x}_n)$ has the form $v_n(\mathbf{x}_1 - \mathbf{x}_n, \ldots, \mathbf{x}_{n-1} - \mathbf{x}_n)$). It is natural to conjecture that by quantizing such a system we will obtain a quantum system described by the Hamiltonian

$$H = \frac{1}{2}\int \hat{\pi}^2(\mathbf{x})d\mathbf{x} + \sum_n \int V_n(\mathbf{x}_1, \ldots, \mathbf{x}_n)\hat{\phi}(\mathbf{x}_1) \ldots \hat{\phi}(\mathbf{x}_n)d\mathbf{x}_1 \ldots d\mathbf{x}_n,$$

$$(8.23)$$

where $\hat{\pi}(\mathbf{x}), \hat{\phi}(\mathbf{x})$ are Hermitian operators (more precisely, operator generalized functions) that obey the commutational relations

$$[\hat{\pi}(\mathbf{x}), \hat{\pi}(\mathbf{x}')] = [\hat{\phi}(\mathbf{x}), \hat{\phi}(\mathbf{x}')] = 0,$$

$$[\hat{\pi}(\mathbf{x}), \hat{\phi}(\mathbf{x}')] = \frac{\hbar}{i}\delta(\mathbf{x} - \mathbf{x}').$$
$$(8.24)$$

However, in trying to correctly define this quantum system, we encounter some difficulties that we have already encountered in considering a Hamiltonian of the form (8.2). Namely, there exist many essentially different systems of operators satisfying (8.24). For simple constructions of the operators $\hat{\pi}(\mathbf{x}), \hat{\phi}(\mathbf{x})$, the expression (8.23) does not specify a well-defined operator. These difficulties can be overcome by means of the approach in Section 8.1. Here, we will consider only the method of solving the Heisenberg equations formally written using the Hamiltonian (8.23).

Namely, we will consider the Hamiltonian H of the form (8.23) as a formal expression. We define the operator realization of the Hamiltonian H as a Hilbert space $\mathcal{H}$ with the energy operator $\hat{H}$, the momentum operator $\hat{\mathbf{P}}$, the vector ϕ, and the operator functions $\hat{\phi}(\mathbf{x}, t)$, that are generalized functions with respect to the variable $\mathbf{x}$ and conventional functions with respect to the variable t, satisfying the following conditions:

(1)

$$\frac{\partial^2 \hat{\phi}(\mathbf{x}, t)}{\partial t^2} = -\sum_n n \int V_n(\mathbf{x}, \mathbf{x}_1, \ldots, \mathbf{x}_{n-1})$$

$$\times \hat{\phi}(\mathbf{x}_1, t) \ldots \hat{\phi}(\mathbf{x}_{n-1}, t) d\mathbf{x}_1 \ldots d\mathbf{x}_{n-1}.$$

$$(8.25)$$

(2)

$$\exp\left(\frac{i}{\hbar}\tau\hat{H}\right) \hat{\phi}(\mathbf{x}, t) \exp\left(\frac{-i}{\hbar}\tau\hat{H}\right) = \hat{\phi}(\mathbf{x}, t + \tau),$$

$$\exp\left(\frac{-i}{\hbar}\mathbf{a}\hat{\mathbf{p}}\right) \hat{\phi}(\mathbf{x}, t) \exp\left(\frac{i}{\hbar}\mathbf{a}\hat{\mathbf{p}}\right) = \hat{\phi}(\hat{x} + a, t).$$

(3) The operators $\hat{\phi}(f, t) = \int f(\mathbf{x})\hat{\phi}(\mathbf{x}, t)d\mathbf{x}$ and $\hat{\pi}(f, t) = \frac{d}{dt}\hat{\phi}(f, t) = \int f(\mathbf{x})\frac{\partial}{\partial t}\hat{\phi}(\mathbf{x}, t)d\mathbf{x}$, where $f \in \mathcal{S}(E^3)$, are defined on a dense subset D of the space $\mathcal{H}$ and transform the subset into itself; if the function f is real, then these operators are Hermitian. The expressions $\langle\hat{\phi}(f, t)\Psi_1, \Psi_2\rangle$ and $\langle\hat{\pi}(f, t)\Psi_1, \Psi_2\rangle$ should depend

continuously on the functions $f \in \mathcal{S}(E^3)$ in the topology of the space $\mathcal{S}(E^3)$ for all $\Psi_1, \Psi_2 \in D$. We assume further that for all t, the conditions

$$[\hat{\pi}(f,t), \hat{\pi}(f',t)] = [\hat{\phi}(f,t), \hat{\phi}(f',t)] = 0,$$

$$[\hat{\pi}(f,t), \hat{\phi}(f',t)] = \frac{\hbar}{i} \int f(\mathbf{x}) f'(\mathbf{x}) d\mathbf{x}$$

are satisfied.

(4) The operators $\hat{H}, \hat{P}_1, \hat{P}_2, \hat{P}_3$ commute. The vector Φ is the ground state of the energy operator $\hat{H}$ and satisfies the conditions $H\Phi = 0, \hat{\mathbf{P}}\Phi = 0$.

(5) The vector Φ is a cyclic vector of the family of operators $\hat{\phi}(f,t)$.

In some simple cases, one can define the precise meaning for equation (8.25) by means of the operator analog of the kernel theorem (as in Section 8.1).

For Hamiltonians of the form

$$H_0 = \frac{1}{2} \int \hat{\pi}^2(\mathbf{x}) d\mathbf{x} + \frac{1}{2} \int v(\mathbf{x} - \mathbf{y}) \hat{\phi}(\mathbf{x}) \hat{\phi}(\mathbf{y}) d\mathbf{x} d\mathbf{y} \qquad (8.26)$$

(free Hamiltonians), it is easy to construct an operator realization. Let us assume that the function $\check{v}(\mathbf{k}) = \int \exp(-i\mathbf{k}\mathbf{x}) v(\mathbf{x}) d\mathbf{x}$ is positive almost everywhere (if this condition is not satisfied, then an operator realization of the Hamiltonian H_0 does not exist). Let $\mathcal{H}, \hat{H}, \hat{\mathbf{P}}, \Phi, \hat{\phi}(\mathbf{x}, t)$ be an operator realization of the Hamiltonian H_0. Let us construct the operator generalized functions

$$a(\mathbf{k}, \epsilon, t) = \hbar^{-1/2} \left(\frac{1}{\sqrt{2}} \sqrt{\omega(\mathbf{k})} \check{\phi}(\epsilon\mathbf{k}, t) + \frac{i\epsilon}{\sqrt{\omega(\mathbf{k})}} \hat{\pi}(\epsilon\mathbf{k}, t) \right),$$

where $\omega(\mathbf{k}) = \sqrt{\check{v}(\mathbf{k})}$, $\check{\phi}(\mathbf{k}, t) = (2\pi)^{-3/2} \int \exp(-i\mathbf{k}\mathbf{x}) \hat{\phi}(\mathbf{x}, t) d\mathbf{x}$, and $\check{\pi}(\mathbf{k}, t) = \frac{\partial}{\partial t} \check{\phi}(\mathbf{k}, t)$. It is easy to check that these operator generalized functions, together with the operators $\hat{H}, \hat{\mathbf{P}}$ and the vector Φ, specify an operator realization of the Hamiltonian $\hbar \int \omega(\mathbf{k}) a^+(\mathbf{k}) a(\mathbf{k}) d\mathbf{k}$ in the sense of Section 8.1. This statement prompts the following construction of an operator realization of the Hamiltonian (8.26).

As the space $\mathcal{H}$, we should take the Fock space $F(L^2(E^3))$, and the energy operator H and the momentum operator $\hat{\mathbf{P}}$ should be defined by the formulas

$$\hat{H} = \hbar \int \omega(\mathbf{k}) a^+(\mathbf{k}) a(\mathbf{k}) d\mathbf{k},$$

$$\hat{\mathbf{P}} = \hbar \int \mathbf{k} a^+(\mathbf{k}) a(\mathbf{k}) d\mathbf{k},$$

and the operator generalized functions $\hat{\phi}(\mathbf{x}, t)$ should be defined by the relation

$$\hat{\phi}(\mathbf{x}, t) = (2\pi)^{-3/2} \hbar^{1/2} \int (a^+(\mathbf{k}) \exp(i\omega(\mathbf{k})t - i\mathbf{k}\mathbf{x})$$

$$+ a(\mathbf{k}) \exp(-i\omega(\mathbf{k})t + i\mathbf{k}\mathbf{x})) \frac{d\mathbf{k}}{\sqrt{2\omega(\mathbf{k})}}.$$

The ground state Φ coincides with Fock vacuum θ. It is easy to check that the objects we have constructed satisfy the conditions for an operator realization and that any other operator realization is unitarily equivalent to the realization we have described.

Let us now consider an arbitrary Hamiltonian H of the form (8.23). Let us express it in terms of the symbols $a^+(\mathbf{k}), a(\mathbf{k})$ satisfying CCR, assuming that

$$\hat{\phi}(\mathbf{x}) = (2\pi)^{-3/2} \hbar^{1/2} \int (a^+(\mathbf{k}) \exp(-i\mathbf{k}\mathbf{x}) + a(\mathbf{k}) \exp(i\mathbf{k}\mathbf{x})) \frac{d\mathbf{k}}{\sqrt{2\omega(\mathbf{k})}},$$

$$\hat{\pi}(\mathbf{x}) = (2\pi)^{-3/2} \hbar^{1/2} \int \frac{i\sqrt{\omega(\mathbf{k})}}{\sqrt{2}} (a^+(\mathbf{k}) \exp(-i\mathbf{k}\mathbf{x})$$

$$- a(\mathbf{k}) \exp(i\mathbf{k}\mathbf{x})) d\mathbf{k}, \tag{8.27}$$

where $\omega(\mathbf{k})$ is an almost everywhere positive function. This expression can be written in normal form by means of CCR; we obtain a quadratic expression plus a (possibly infinite) constant. We will discard this constant and as a result we will obtain a formal expression for $\tilde{H}$ of the form (8.2).

It is easy to see that the problem of constructing an operator realization of the Hamiltonian H is equivalent to the same problem for $\tilde{H}$. For example, if $\mathcal{H}, \hat{H}, \hat{\mathbf{P}}, \Phi, a(\mathbf{k}, \epsilon, t)$ is an operator realization of the Hamiltonian $\tilde{H}$, then the operator realization of the Hamiltonian H can be obtained if we take the same Hilbert space $\mathcal{H}$, the same operators $\hat{H}, \hat{\mathbf{P}}$, and vector Φ, and the operator generalized functions $\hat{\phi}(\mathbf{x}, t)$ can be written in the form

$$\hat{\phi}(\mathbf{x}, t) = (2\pi)^{-3/2} \hbar^{1/2} \int \big(a(\mathbf{k}, 1, t) \exp(-i\mathbf{k}\mathbf{x})$$

$$+ a(\mathbf{k}, -1, t) \exp(i\mathbf{k}\mathbf{x}) \big) \frac{d\mathbf{k}}{\sqrt{2\omega(\mathbf{k})}}.$$

This remark allows us to transfer to Hamiltonians of the form (8.23) everything we know for Hamiltonians of the form (8.2).

Chapter 9

The Scattering Matrix for Translation-Invariant Hamiltonians

9.1 The scattering matrix for translation-invariant Hamiltonians in Fock space

In this chapter, we review the basic facts of scattering theory for translation-invariant Hamiltonians. Some of these facts are proved, however, the proofs are not fully rigorous (Sections 9.2, 9.5), while other facts are conveyed almost without proof (Sections 9.1, 9.3 and 9.4). Most of the results in this chapter will be proved later on the basis of axiomatic scattering theory (see Chapter 11).

To construct a scattering matrix for a translation-invariant Hamiltonian, one cannot directly use the general construction of formal scattering theory (Section 5.1). The first obstacle for applying this construction is the fact that a translation-invariant Hamiltonian can define a self-adjoint operator on Fock space only in the case when vacuum polarization is absent (Section 8.1). However, even translation-invariant Hamiltonians that specify an operator on Fock space still have problems. These problems are related to the fact that in the case under consideration, a natural representation of the Hamiltonian H as a sum of "free" Hamiltonian H_0 and "interaction" V usually does not exist.

It is reasonable to take the Hamiltonian H_0 to be of the form $\int \epsilon(\mathbf{k}) a^+(\mathbf{k}) a(\mathbf{k}) d\mathbf{k}$ because such a Hamiltonian describes a system of non-interacting identical particles. We assume, therefore, that the translation-invariant Hamiltonian H specifying an operator on

Fock space $F(L^2(E^3))$ is represented in the form $H = H_0 + V$, where

$$H_0 = \int \epsilon(\mathbf{k}) a^+(\mathbf{k}) a(\mathbf{k}) d\mathbf{k}, \tag{9.1}$$

$$V = \sum_{m,n \geq 1} \int V_{m,n}(\mathbf{k}_1, \ldots, \mathbf{k}_m | \mathbf{p}_1, \ldots, \mathbf{p}_n) a^+(\mathbf{k}_1)$$
$$\ldots a^+(\mathbf{k}_m) a(\mathbf{p}_1) \ldots a(\mathbf{p}_n) d^m \mathbf{k} d^n \mathbf{p} \tag{9.2}$$

(here,

$$V_{m,n}(\mathbf{k}_1, \ldots, \mathbf{k}_m | \mathbf{p}_1, \ldots, \mathbf{p}_n) = v_{m,n}(\mathbf{k}_1, \ldots, \mathbf{k}_m | \mathbf{p}_1, \ldots, \mathbf{p}_n)$$
$$\times \delta(\mathbf{k}_1 + \cdots + \mathbf{k}_m - \mathbf{p}_1 - \cdots - \mathbf{p}_n)). \tag{9.3}$$

However, such a representation of the Hamiltonian H has physical meaning only under the condition $v_{m,1} \equiv 0$. Otherwise, the choice of free and interaction parts can be performed in different ways and is dictated only by convenience.

One can prove the following statement: if $V_{m,1} \not\equiv 0$ (i.e. the "interaction" contains summands with one annihilation operator), then the Møller matrices $S_\pm$ cannot be defined by means of the definitions of formal scattering theory (5.1) and (5.2). Indeed, if there exists a limit entering the relations (5.1) and (5.2), then for every vector $x \in F$, we have

$$\lim_{\substack{t_1, t_2 \to -\infty; \\ (t_1, t_2 \to +\infty)}} \left\| \int_{t_1}^{t_2} \exp(iHt) V \exp(-iH_0 t) x \, dt \right\| = 0.$$

$\Bigg($ We can check this using the relation

$$\exp(itH) V \exp(-itH_0) x = \frac{1}{i} \frac{d\xi(t)}{dt},$$

where $\xi(t) = \exp(iHt) \exp(iH_0 t) x$ and hence

$$\int_{t_1}^{t_2} \exp(iHt) V \exp(-iH_0 t) x \, dt = \frac{1}{i}(\xi(t_2) - \xi(t_1)). \Bigg)$$

However, if $x = \int f(\mathbf{k})a^+(\mathbf{k})d\mathbf{k}\theta$, then

$$\left\| \frac{d\xi}{dt} \right\| = \| \exp(iHt)V\exp(-iH_0t)x \| = \| V\exp(-iH_0t)x \|$$

does not depend on t. This is clear because in this case

$$\exp(iH_0t)x = \int \exp(i\epsilon(\mathbf{k})t)f(\mathbf{k})a^+(\mathbf{k})d\mathbf{k}\theta,$$

$$V\exp(-iH_0t)x = \sum_n \int f_n(\mathbf{k}_1,\ldots,\mathbf{k}_n|t)a^+(\mathbf{k}_1)\ldots a^+(\mathbf{k}_n)d^n\mathbf{k}\theta,$$

where

$$f_n(\mathbf{k}_1,\ldots,\mathbf{k}_n|t) = v_{n,1}(\mathbf{k}_1,\ldots,\mathbf{k}_n|\mathbf{k}_1+\cdots+\mathbf{k}_n)\int(\mathbf{k}_1+\cdots+\mathbf{k}_n)$$

$$\times\ \exp(-it\epsilon(\mathbf{k}_1+\cdots+\mathbf{k}_n)).$$

Noting that

$$\| V\exp(-iH_0t)x \| = \sum_n n!\int |f_n(\mathbf{k}_1,\ldots,\mathbf{k}_n|t)|^2 d^n\mathbf{k},$$

we see that $\left\|\frac{d\xi}{dt}\right\|$ does not depend on t. Similar considerations show that $\left\|\frac{d^2\xi}{dt^2}\right\| = \left\|\frac{d}{dt}(\exp(iHt)\times V\exp(-iH_0t)x)\right\| = \| \exp(-iHt)(HV - VH_0)\times\exp(-iH_0t)x \| = \| (HV - VH_0)\exp(-iH_0t)x \|$ does not depend on t. Now, to finish the proof, we should apply the following lemma to the vector $\eta(t) = \frac{d\xi}{dt}$: if $\|\eta(t)\|$ does not depend on t and $\left\|\frac{d\eta}{dt}\right\|$ is bounded above, then $\int_{t_1}^{t_2}\eta(t)dt$ cannot tend to zero as $t_1, t_2 \to \infty$.

This mathematical statement — the fact that it is impossible to give the definition of a Møller matrix in the same way as in the theory of potential scattering — has a clear physical background. The problem is that the states $a^+(\mathbf{k})\theta$ (one particle states) are eigenstates of the Hamiltonian H_0 but are not eigenstates of the full Hamiltonian H. In Section 5.3, we introduced the notion of a particle (single-particle state) for the Hamiltonian H as a generalized vector function $\Phi(\mathbf{k})$ satisfying the conditions (5.20)–(5.22).

Using this notion, we can reformulate the above statement in the following way: the generalized vector function $a^+(\mathbf{k})\theta$ is a single-particle state of the Hamiltonian H_0 (bare single-particle state), however, it is not a single-particle state of the Hamiltonian H ("dressed" single-particle state). The fact that the notions of bare and dressed particles do not coincide explains the necessity to modify the definition of a scattering matrix.

In the case when $v_{m,0} \equiv v_{m,1} \equiv 0$, bare particles coincide with dressed ones, and therefore, it is not necessary to modify the definition of a scattering matrix.

Let us sketch how to perform the necessary modification of the definition of a scattering matrix for a translation-invariant Hamiltonian that specifies an operator on Fock space.

We will give three definitions of Møller matrices $S_\pm$ that can be used in the situation when dressed particles coincide with bare particles (i.e. the Hamiltonian has the form (9.1) and $v_{m,1} \equiv 0$). We will show how to modify these definitions for the case at hand (recall that knowing the Møller matrices $S_\pm$, we can define the scattering matrix by the formula $S = S_+^* S_-$).

The first definition is the same definition that was used in the formal scattering theory (see (5.1) and (5.2)). One can also define Møller matrices $S_\pm$ as strong limits of the adiabatic Møller matrices $S_{\alpha\pm}$ with $\alpha \to 0$. In Section 5.1, we described the conditions that guaranteed that the second definition is equivalent to the first one (see (5.4) and (5.5)). Finally, the third definition is based on the consideration of the in- and out-operators

$$a_{\substack{\text{in}\\\text{out}}}(\mathbf{k}) = \operatorname*{slim}_{t\to\mp\infty} \exp(i\epsilon(\mathbf{k})t)a(\mathbf{k},t),$$

$$a^+_{\substack{\text{in}\\\text{out}}}(\mathbf{k}) = \operatorname*{slim}_{t\to\mp\infty} \exp(-i\epsilon(\mathbf{k})t)a^+(\mathbf{k},t). \tag{9.4}$$

If we know the in- and out-operators, then the Møller matrices can be defined by the relations

$$S_- a_{\text{in}}(\mathbf{k}) = a(\mathbf{k})S_-, \quad S_-\theta = \theta,$$

$$S_+ a_{\text{out}}(\mathbf{k}) = a(\mathbf{k})S_+, \quad S_+\theta = \theta \tag{9.5}$$

(the equivalence of this definition with the first one was discussed in Section 5.3).

Let us start by generalizing the third definition. The limit in (9.4) does not exist for $v_{m,1} \not\equiv 0$; this is clear from the fact that the expression

$$\exp(iHt)\exp(-iH_0t)\int f(\mathbf{k})a^+(\mathbf{k})\theta d\mathbf{k}$$

$$= \int f(\mathbf{k})\exp(-i\epsilon(\mathbf{k})t)a^+(\mathbf{k},t)\theta d\mathbf{k}$$

does not have a limit as $t \to \pm\infty$ (as was proven already).

However, this fact does not mean that the expression $\exp(i\omega(\mathbf{k})t)a(\mathbf{k},t)$ does not have a weak limit. Therefore, the in- and out-operators (operator generalized functions) $a_{\text{in}}(\mathbf{k}), a_{\text{in}}^+(\mathbf{k}), a_{\text{out}}(\mathbf{k}), a_{\text{out}}^+(\mathbf{k})$ will be defined by means of the following relations:

$$a_{\text{in}}(\mathbf{k}) = \underset{t\to-\infty}{\text{wlim}}\ \Lambda(\mathbf{k})\exp(i\omega(\mathbf{k})t)a(\mathbf{k},t),$$

$$a_{\text{in}}^+(\mathbf{k}) = \underset{t\to-\infty}{\text{wlim}}\ \overline{\Lambda(\mathbf{k})}\exp(-i\omega(\mathbf{k})t)a^+(\mathbf{k},t),$$

$$a_{\text{out}}(\mathbf{k}) = \underset{t\to+\infty}{\text{wlim}}\ \Lambda(\mathbf{k})\exp(i\omega(\mathbf{k})t)a(\mathbf{k},t),$$

$$a_{\text{out}}^+(\mathbf{k}) = \underset{t\to+\infty}{\text{wlim}}\ \overline{\Lambda(\mathbf{k})}\exp(-i\omega(\mathbf{k})t)a^+(\mathbf{k},t),$$

where a positive function $\omega(\mathbf{k})$ should be found from the condition of the existence of the limit and the function $\Lambda(\mathbf{k})$ from the condition that the operators $a_{\text{in}}(\mathbf{k}), a_{\text{in}}^+(\mathbf{k})$ and $a_{\text{out}}(\mathbf{k}), a_{\text{out}}^+(\mathbf{k})$ satisfy CR:

$$[a_{\substack{\text{in}\\\text{out}}}(\mathbf{k}), a_{\substack{\text{in}\\\text{out}}}(\mathbf{k}')]_\mp = [a_{\substack{\text{in}\\\text{out}}}^+(\mathbf{k}), a_{\substack{\text{in}\\\text{out}}}^+(\mathbf{k}')] = 0,$$

$$[a_{\substack{\text{in}\\\text{out}}}(\mathbf{k}), a_{\substack{\text{in}\\\text{out}}}^+(\mathbf{k}')]_\mp = \delta(\mathbf{k}-\mathbf{k}').$$

It is easy to check (see Section 9.2) that the vector generalized functions $a_{\text{in}}^+(\mathbf{k})\theta$ and $a_{\text{out}}^+(\mathbf{k})\theta$ are single-particle states of the Hamiltonian H and therefore conclude that the functions $\omega(\mathbf{k})$ has meaning of the energy of a single-particle state:

$$Ha_{\substack{\text{in}\\\text{out}}}^+(\mathbf{k})\theta = \omega(\mathbf{k})a_{\substack{\text{in}\\\text{out}}}^+(\mathbf{k})\theta.$$

Møller matrices are defined as earlier in terms of in- and out-operators by (9.5).

The second definition can be generalized in the following way. The Møller matrices $S_\pm$ are defined as unitary operators that can be represented in the form

$$S_+ = \operatorname*{slim}_{\alpha \to 0} S_{\alpha+} U_\alpha^*, \tag{9.6}$$

$$S_- = \operatorname*{slim}_{\alpha \to 0} S_{\alpha-} U_\alpha, \tag{9.7}$$

where U_α is an operator of the form

$$\exp\left(\frac{i}{\alpha} \int r(\mathbf{k}) a^+(\mathbf{k}) a(\mathbf{k}) d\mathbf{k}\right), \tag{9.8}$$

and the function $r(\mathbf{k})$ is chosen from the condition of the existence of the limits (9.6) and (9.7).

If we want to define directly the scattering matrix, we can use the relation

$$S = \operatorname*{slim}_{\alpha \to 0} U_\alpha S_\alpha U_\alpha, \tag{9.9}$$

where S_α is the adiabatic S matrix, the operator U_α has the form (9.8) and the function $r(\mathbf{k})$ is chosen from the condition of existence of the limit (9.9).

Finally, we give the following modification for the first definition: Møller matrices $S_\pm$ are operators of the form

$$S_\pm = \operatorname*{slim}_{t \to \pm} \exp(iHt) T \exp(-iH_{\text{as}}t), \tag{9.10}$$

where T is an operator satisfying the conditions

$$T\theta = \theta,$$

$$Ta^+(\mathbf{k})\theta = \Phi(\mathbf{k})$$

and having the form $T = N(\exp B)$, where

$$B = \sum_n \int b_n(\mathbf{k}_1, \ldots, \mathbf{k}_n) \delta\left(\sum_i \mathbf{k}_i - \mathbf{k}\right) a^+(\mathbf{k}_1) \ldots$$

$$a^+(\mathbf{k}_n) a(\mathbf{k}) d\mathbf{k} d\mathbf{k}_1 \ldots d\mathbf{k}_n,$$

($\Phi(\mathbf{k})$ denotes a singe-particle state of the Hamiltonian H, $H_{\text{as}} = \int \omega(\mathbf{k}) a^+(\mathbf{k}) a(\mathbf{k}) d\mathbf{k}$, where $\omega(\mathbf{k})$ is the energy of the single-particle state: $H\Phi(\mathbf{k}) = \omega(\mathbf{k})\Phi(\mathbf{k})$).[1]

In the relation (9.10), one can replace the operator T described above by other operators. All operators D such that

$$S_\pm = \operatorname*{slim}_{t \to \pm\infty} \exp(itH) D \exp(-itH_{\text{as}})$$

will be called *dressing operators* (this term is related to the fact that they transform the bare single-particle state $a^+(\mathbf{k})\theta$ into the dressed single-particle state $\Phi(\mathbf{k})$). One can construct a broad class of dressing operators (see Sections 9.4 and 9.5).

All three definitions of Møller matrices are equivalent (under certain conditions). They can be generalized further to translation-invariant Hamiltonians generating the polarization of vacuum. Naturally, there are additional complications related to the fact that the Hamiltonian does not specify an operator in Fock space.

For the definition by means of in- and out-operators, these complications can be overcome by considering an operator realization of the Hamiltonian H. Other definitions should be modified by considering the volume cutoff Ω of the Hamiltonian H and taking the limit $\Omega \to \infty$ (in this case one should define the scattering matrix directly because Møller matrices cannot be defined in this case). In more detail, we will study various definitions of the scattering matrix for a translation-invariant Hamiltonian in other sections of this chapter and in Chapter 11.

In Section 5.3, we noted that the definition of scattering matrix should be modified even for the simplest Hamiltonians if there exist bound states. Similar modifications are necessary in the situation at hand if there exist bound states. The definition of scattering matrix that can be used in the case when bound states exist is given in Section 11.1.

---

[1] The definition of Møller matrices by means of (9.10) was suggested by I. Ya. Arefieva.

9.2 The definition of scattering matrix by means of operator realization of a translation-invariant Hamiltonian

Let H denote a translation-invariant Hamiltonian and let us consider its operator realization $(\mathcal{H}, \hat{H}, \hat{\mathbf{P}}, a(\mathbf{k}, \epsilon, t), \Phi)$ (see Section 8.1).

Let us define the in-operators $a_{\text{in}}(\mathbf{k}, \epsilon, t)$ and out-operators $a_{\text{out}}(\mathbf{k}, \epsilon, t)$ as limits

$$a_{\substack{\text{in}\\\text{out}}}(\mathbf{k}, t) = a_{\substack{\text{in}\\\text{out}}}(\mathbf{k}, -1, t)$$

$$= \operatorname*{wlim}_{\tau \to \mp\infty} \Lambda_{\mp}(\mathbf{k}) \exp(i\omega(\mathbf{k})\tau) a(\mathbf{k}, -1, t + \tau),$$

$$a^{+}_{\substack{\text{in}\\\text{out}}}(\mathbf{k}, t) = a_{\substack{\text{in}\\\text{out}}}(\mathbf{k}, 1, t)$$

$$= \operatorname*{wlim}_{\tau \to \mp\infty} \overline{\Lambda_{\mp}(\mathbf{k})} \exp(-i\omega(\mathbf{k})\tau) a(\mathbf{k}, 1, t + \tau),$$

$$(9.11)$$

where $\omega(\mathbf{k})$ is found from the condition of the existence of limits and $\Lambda_{\mp}(\mathbf{k})$ is found from the condition that the operators $a^{+}_{\text{in}}(\mathbf{k}, t)$ and $a_{\text{in}}(\mathbf{k}, t)$ (and, correspondingly, $a^{+}_{\text{out}}(\mathbf{k}, t), a_{\text{out}}(\mathbf{k}, t)$) satisfy CR for fixed t. (Here, $\omega(\mathbf{k})$ is an almost everywhere positive function and the limit is understood as a weak limit of operator generalized functions, $\epsilon = \pm 1$. Introducing the notation $\Lambda_{\mp}(\mathbf{k}, -1) = \Lambda_{\mp}(\mathbf{k}), \Lambda_{\mp}(\mathbf{k}, 1) = \overline{\Lambda_{\mp}(\mathbf{k})}$, we can say that for every function $f \in \mathcal{S}$, we have

$$a_{\substack{\text{in}\\\text{out}}}(f, \epsilon, t) = \int f(\mathbf{k}) a_{\substack{\text{in}\\\text{out}}}(\mathbf{k}, \epsilon, t) d\mathbf{k}$$

$$= \operatorname*{wlim}_{\tau \to \mp\infty} \int f(\mathbf{k}) \Lambda_{\mp}(\mathbf{k}, \epsilon) \exp(-i\epsilon\omega(\mathbf{k})\tau) a(\mathbf{k}, \epsilon, t + \tau) d\mathbf{k}$$

in the sense of weak limits of operators.)

This definition of in- and out-operators differs from the definition accepted in the theory of potential scattering (Section 5.3) by replacing strong limits with weak limits. The factor $\Lambda_{\mp}(\mathbf{k})$ is related to this modification: strong limits preserve CR, but weak limits do not.

The question of the existence of in- and out-operators (i.e. the question of the existence of the functions $\omega(\mathbf{k})$ and $\Lambda_{\mp}(\mathbf{k})$ such that the limit in (9.11) exists and satisfies CR) will be discussed in Chapters 10 and 11. To denote the in- and out-operators simultaneously, we will use the notation $a_{\mathrm{ex}}(\mathbf{k}, \epsilon, t)$.

Let us establish a few simple properties of in- and out-operators:

(1) $\hat{a}_{\mathrm{ex}}(\mathbf{k}, \epsilon, t) = \exp(iHt)a_{\mathrm{ex}}(\mathbf{k}, \epsilon, 0)\exp(-i\hat{H}t) = \exp(i\epsilon\omega(\mathbf{k})t)a_{\mathrm{ex}}(\mathbf{k}, \epsilon, 0)$;

(2) $\exp(i\hat{\mathbf{P}}\boldsymbol{\alpha})a_{\mathrm{ex}}(\mathbf{k}, \epsilon, t)\exp(-i\hat{\mathbf{P}}\boldsymbol{\alpha}) = \exp(i\epsilon\mathbf{k}\boldsymbol{\alpha})a_{\mathrm{ex}}(\mathbf{k}, \epsilon, t)$;

(3) $a_{\mathrm{ex}}(\mathbf{k}, t)\Phi = 0$;

(4) the generalized vector function $\Phi_{\mp}(\mathbf{k}) = a_{\mathrm{ex}}^{+}(\mathbf{k}, 0)\Phi$ is δ-normalized and is an eigenfunction of the operators H and $\hat{\mathbf{P}}$:

$$\hat{H}\Phi_{\mp}(\mathbf{k}) = \omega(\mathbf{k})\Phi_{\mp}(\mathbf{k}),$$

$$\hat{\mathbf{P}}\Phi_{\mp}(\mathbf{k}) = \mathbf{k}\Phi_{\mp}(\mathbf{k})$$

(the functions $\Phi_{\mp}(\mathbf{k})$ satisfying this condition describe a single-particle state; see Section 5.3).

We first note that (2) follows from the relation

$$\exp(i\hat{\mathbf{P}}\boldsymbol{\alpha})a(\mathbf{k}, \epsilon, t)\exp(i\hat{\mathbf{P}}\boldsymbol{\alpha}) = \exp(i\epsilon\mathbf{k}\boldsymbol{\alpha})a(\mathbf{k}, \epsilon, t).$$

Then,

$$a_{\mathrm{ex}}(\mathbf{k}, \epsilon, t) = \operatorname*{wlim}_{\tau\to\mp\infty} \Lambda_{\mp}(\mathbf{k}, \epsilon)\exp(i\epsilon\omega(\mathbf{k})\tau)a(\mathbf{k}, \epsilon, t+\tau)$$

$$= \operatorname*{wlim}_{\tau\to\mp\infty} \exp(i\hat{H}t)\Lambda_{\mp}(\mathbf{k}, \epsilon)\exp(i\epsilon\omega(\mathbf{k})\tau)$$

$$\times\ a(\mathbf{k}, \epsilon, \tau)\exp(-i\hat{H}t)$$

$$= \exp(i\hat{H}t)\big(\operatorname*{wlim}_{\tau\to\mp\infty} \Lambda_{\mp}(\mathbf{k}, \epsilon)\exp(i\epsilon\omega(\mathbf{k})\tau)a(\mathbf{k}, \epsilon, \tau)\big)$$

$$\times\exp(-iHt) = \exp(i\hat{H}t)a_{\mathrm{ex}}(\mathbf{k}, \epsilon, 0)\exp(-i\hat{H}t).$$

From the other side, substituting ρ instead of $t + \tau$ in (9.11), we obtain

$$a_{\mathrm{ex}}(\mathbf{k}, \epsilon, t) = \operatorname*{wlim}_{\rho\to\mp\infty} \Lambda_{\mp}(\mathbf{k}, \epsilon)\exp(i\epsilon\omega(\mathbf{k})(\rho - t))a(\mathbf{k}, \epsilon, \rho)$$

$$= \exp(-i\epsilon\omega(\mathbf{k})t)a_{\mathrm{ex}}(\mathbf{k}, \epsilon, 0).$$

It follows from (1) that

$$\exp(i\hat{H}t)a_{\mathrm{ex}}(\mathbf{k}, \epsilon)\Phi = \exp(i\hat{H}t)a_{\mathrm{ex}}(\mathbf{k}, \epsilon)\exp(-i\hat{H}t)\Phi$$

$$= a_{\mathrm{ex}}(\mathbf{k}, \epsilon, t)\Phi = \exp(i\epsilon\omega(\mathbf{k})t)\Phi,$$

hence,

$$\hat{H}a_{\mathrm{ex}}(\mathbf{k}, \epsilon)\Phi = \epsilon\omega(\mathbf{k})a_{\mathrm{ex}}(\mathbf{k}, \epsilon)\Phi. \tag{9.12}$$

From (9.12), it follows that $a_{\mathrm{ex}}(\mathbf{k}, -1)\Phi = 0$ (if this condition is not satisfied for $\mathbf{k}$ in the set K having non-zero measure, then the generalized vector function $a_{\mathrm{ex}}(\mathbf{k}, -1)\Phi$ is a generalized eigenfunction of the operator $\hat{H}$, hence the number $-\omega(\mathbf{k})$, where $\mathbf{k} \in K$, belongs to the spectrum of the operator $\hat{H}$; this is impossible because the operator $\hat{H}$ is positive).

Applying CR and (3), we see that the generalized function $\Phi_{\mp}(\mathbf{k}) = a_{\mathrm{ex}}(\mathbf{k}, 1)\Phi$ is δ-normalized ($\langle\Phi_{\mp}(\mathbf{k}), \Phi_{\mp}(\mathbf{k}')\rangle = \langle a_{\mathrm{ex}}(\mathbf{k}', -1)a_{\mathrm{ex}}(\mathbf{k}, 1)\Phi, \Phi\rangle = \langle[a_{\mathrm{ex}}(\mathbf{k}'), a_{\mathrm{ex}}^{+}(\mathbf{k})]_{\mp}\Phi, \Phi\rangle = \delta(\mathbf{k} - \mathbf{k}')$). Formula (9.12) implies that $\Phi_{\mp}(\mathbf{k})$ is a generalized eigenfunction of the operator $\hat{H}$. In order to prove that $\Phi_{\mp}(\mathbf{k})$ is a generalized eigenfunction of the operator $\hat{\mathbf{P}}$, we should recall that by (2), we have

$$\exp(i\hat{\mathbf{P}}\alpha)a_{\mathrm{ex}}(\mathbf{k}, \epsilon, t)\exp(-i\hat{\mathbf{P}}\alpha) = \exp(i\epsilon\mathbf{k}\alpha)a_{\mathrm{ex}}(\mathbf{k}, \epsilon, t),$$

which implies that

$$\exp(i\hat{\mathbf{P}}\alpha)\Phi_{\mp}(\mathbf{k}) = \exp(i\hat{\mathbf{P}}\alpha)a_{\mathrm{ex}}^{+}(\mathbf{k}, 0)\exp(-i\hat{\mathbf{P}}\alpha)\Phi = \exp(\epsilon\mathbf{k}\alpha)\Phi_{\mp}(\mathbf{k})$$

and therefore $\hat{\mathbf{P}}\Phi_{\mp}(\mathbf{k}) = \mathbf{k}\Phi_{\mp}(\mathbf{k})$.

Remark 9.1. We have assumed that the function $\omega(\mathbf{k})$ in the definition of in- and out-operators is almost everywhere positive. One can replace this condition by the condition that the operator $\hat{H}$ has a unique ground state. Then, in the case of CCR, we can modify the above considerations to check that the function $\omega(\mathbf{k})$ is automatically almost everywhere positive. In the case of CAR, one should introduce new operators $\tilde{a}^{+}(\mathbf{k}, t), \tilde{a}(\mathbf{k}, t), \tilde{a}_{\mathrm{ex}}^{+}(\mathbf{k}, t), \tilde{a}_{\mathrm{ex}}(\mathbf{k}, t)$ also satisfying CAR, using the formulas

$$\tilde{a}(\mathbf{k}, t) = \theta(\omega(\mathbf{k}))a(\mathbf{k}, t) + \theta(-\omega(\mathbf{k}))a^{+}(-\mathbf{k}, t),$$

$$\tilde{a}_{\mathrm{ex}}(\mathbf{k}, t) = \theta(\omega(\mathbf{k}))a_{\mathrm{ex}}(\mathbf{k}, t) + \theta(-\omega(\mathbf{k}))a_{\mathrm{ex}}^{+}(-\mathbf{k}, t).$$

Then, it is clear that

$$\tilde{a}_{\mathrm{ex}}(\mathbf{k}, t) = \underset{\tau \to \mp\infty}{\mathrm{wlim}} \, \Lambda'_{\mp}(\mathbf{k}) \exp(i|\omega(\mathbf{k})|\tau)\tilde{a}(\mathbf{k}, t + \tau),$$

where the function $|\omega(\mathbf{k})|$ is almost everywhere positive.

Let us introduce the space of asymptotic states $\mathcal{H}_{\mathrm{as}}$ as the space $F(L^2(E^3))$ of Fock representation of CR. The symbol $b(\mathbf{k}, \epsilon)$ denotes the operator generalized functions in this space that satisfy the conditions

$$[b(\mathbf{k}, \epsilon), b(\mathbf{k}', \epsilon')]_{\mp} = A_{\epsilon'}^{\epsilon}\delta(\mathbf{k} - \mathbf{k}'),$$

$$b(\mathbf{k}, -1)\theta = 0, \quad b^{+}(\mathbf{k}, -1) = b(\mathbf{k}, 1).$$

Møller matrices S_- and S_+ are defined as isometric operators mapping the space $\mathcal{H}_{\mathrm{as}}$ into the space $\mathcal{H}$ and satisfying the equations

$$a_{\mathrm{in}}(\mathbf{k}, \epsilon)S_- = S_-b(\mathbf{k}, \epsilon), \quad S_-\theta = \Phi, \tag{9.13}$$

$$a_{\mathrm{out}}(\mathbf{k}, \epsilon)S_+ = S_+b(\mathbf{k}, \epsilon), \quad S_+\theta = \Phi \tag{9.14}$$

(it follows from the results of Section 6.1 that such operators exist and are defined by conditions (9.13) and (9.14) uniquely).

The scattering matrix of a translation-invariant Hamiltonian H is defined as the operator $S = S_+^* S_-$.

It is easy to verify that the scattering matrix S will be unitary if and only if the spaces $\mathcal{H}_{\mathrm{in}} = S_-\mathcal{H}_{\mathrm{as}}$ and $\mathcal{H}_{\mathrm{out}} = S_+\mathcal{H}_{\mathrm{as}}$ coincide.

We will assume that $\mathcal{H} = \mathcal{H}_{\mathrm{in}} = \mathcal{H}_{\mathrm{out}}$ (in other words, not only is the S-matrix unitary, but so are the Møller matrices S_- and S_+).

Some of the relations proven later, in particular (9.22), are correct without this assumption.

Let us define the operator $\hat{H}_{as}$ (asymptotic Hamiltonian) and the operator $\hat{\mathbf{P}}_{as}$ in the space $\mathcal{H}_{as}$ using the formulas

$$\hat{H}_{as} = \int \omega(\mathbf{k})b^+(\mathbf{k})b(\mathbf{k})dk,$$

$$\hat{\mathbf{P}}_{as} = \int \mathbf{k}b^+(\mathbf{k})b(\mathbf{k})dk$$

(as usual, $b^+(\mathbf{k}) = b(\mathbf{k}, 1), b(\mathbf{k}) = b(\mathbf{k}, -1)$). It is easy to see that

$$b(\mathbf{k}, \epsilon, t) = \exp(i\hat{H}_{as}t)b(\mathbf{k}, \epsilon)\exp(-i\hat{H}_{as}t) = \exp(i\epsilon\omega(\mathbf{k})t)b(\mathbf{k}, \epsilon),$$

$$\exp(i\hat{\mathbf{P}}_{as}\boldsymbol{\alpha})b(\mathbf{k}, \epsilon, t)\exp(-i\hat{\mathbf{P}}_{as}\boldsymbol{\alpha}) = \exp(i\epsilon\boldsymbol{\alpha}\mathbf{k})b(\mathbf{k}, \epsilon, t).$$

Using this relation and the properties (1) and (2) of the operators a_{ex}, we can show that

$$\hat{H}S_{\pm} = S_{\pm}\hat{H}_{as}, \quad \hat{\mathbf{P}}S_{\pm} = S_{\pm}\hat{\mathbf{P}}_{as} \tag{9.15}$$

(i.e. the operators S_+ and S_- specify unitary equivalences between the operators $\hat{H}$ and $\hat{H}_{as}$, $\hat{\mathbf{P}}$ and $\hat{\mathbf{P}}_{as}$). From (9.15), it follows that the scattering matrix commutes with the operators $\hat{H}_{as}$ and $\hat{\mathbf{P}}_{as}$:

$$S\hat{H}_{as} = \hat{H}_{as}S,$$

$$S\hat{\mathbf{P}}_{as} = \hat{\mathbf{P}}_{as}S.$$

Let us now prove that the scattering matrix defined above has the following properties:

(1) $S\theta = \theta$ (the vacuum is stable),
(2) $Sb^+(\mathbf{k})\theta = c(\mathbf{k})b^+(\mathbf{k})\theta$,

where $|c(\mathbf{k})| = 1$ (single-particle states are stable). The second of these statements will be proved only under certain restrictions on the function $\omega(\mathbf{k})$; it is sufficient to assume that it is strongly convex.

The stability of the vacuum follows immediately from the relations $S_-\theta = \Phi, S_+\theta = \Phi$. To prove that single-particle states are stable, we note that

$$\hat{H}_{as}Sb^+(\mathbf{k})\theta = S\hat{H}_{as}b^+(\mathbf{k})\theta = \omega(\mathbf{k})Sb^+(\mathbf{k})\theta,$$

$$\hat{\mathbf{P}}_{as}Sb^+(\mathbf{k})\theta = S\hat{\mathbf{P}}_{as}b^+(\mathbf{k})\theta = \mathbf{k}Sb^+(\mathbf{k})\theta.$$

Hence, the generalized vector function $\Psi(\mathbf{k}) = Sb^+(\mathbf{k})\theta$ satisfies the condition $\hat{H}_{\mathrm{as}}\Psi(\mathbf{k}) = \omega(\mathbf{k})\Psi(\mathbf{k}), \hat{\mathbf{P}}_{\mathrm{as}}\Psi(\mathbf{k}) = \mathbf{k}\Psi(\mathbf{k})$; it is easy to see that this generalized vector function is δ-normalized. It follows that $\Psi(\mathbf{k}) = c(\mathbf{k})b^+(\mathbf{k})\theta$, where $|c(\mathbf{k})| = 1$ (see Section 5.3).

We will now consider an ambiguity in our definition of in- and out-operators. It is not difficult to check that the functions $\Lambda_-(\mathbf{k})$ and $\Lambda_+(\mathbf{k})$ are not specified uniquely by the requirement the operators $a_{\mathrm{ex}}(\mathbf{k}, t)$ obey CR. Namely, we can replace the functions $\Lambda_-(\mathbf{k})$ and $\Lambda_+(\mathbf{k})$ by functions $\Lambda'_-(\mathbf{k})$ and $\Lambda'_+(\mathbf{k})$ that have the same absolute value as the functions $\Lambda_-(\mathbf{k})$ and $\Lambda_+(\mathbf{k})$ and obtain the new in- and out-operators $a'_{\mathrm{in}}(\mathbf{k}, t)$ and $a'_{\mathrm{out}}(\mathbf{k}, t)$:

$$a'_{\mathrm{in}}(\mathbf{k}, t) = \exp(i\phi_-(\mathbf{k}))a_{\mathrm{in}}(\mathbf{k}, t),$$

$$a'_{\mathrm{out}}(\mathbf{k}, t) = \exp(i\phi_+(\mathbf{k}))a_{\mathrm{out}}(\mathbf{k}, t) \tag{9.16}$$

(here, $\exp(i\phi_\mp(\mathbf{k})) = \Lambda'_\mp(\mathbf{k})\Lambda_\pm^{-1}(\mathbf{k})$ and $\phi_\mp(\mathbf{k})$ are real-valued functions).

One can check that under the above assumptions, all possible in- and out-operators can be represented as operators $a'_{\mathrm{ex}}(\mathbf{k}, t)$.

It is easy to check that the Møller matrices S'_- and S'_+ constructed by means of the operators $a'_{\mathrm{ex}}(\mathbf{k}, t)$ are related to the Møller matrices S_- and S_+ corresponding to the operators $a_{\mathrm{ex}}(\mathbf{k}, t)$ by the formula

$$S'_\mp = S_\mp U_\mp,$$

where $U_\mp = \exp(i \int \phi_\mp(\mathbf{k})b^+(\mathbf{k})b(\mathbf{k})d\mathbf{k})$ (this follows from the relation $U_\mp b(\mathbf{k})U_\mp^{-1} = \exp(i\phi_\mp(\mathbf{k}))b(\mathbf{k})$).

Hence, the scattering matrix $S' = S'^*_+ S'_-$ corresponding to the new ex-operators is related to the old scattering matrix $S = S^*_+ S_-$ by the relation $S' = U_+^{-1}SU_-$.

Using this ambiguity in the definition of in- and out-operators, one can strengthen the stability condition of single-particle states, namely, we can require

$$Sb^+(\mathbf{k})\theta = b^+(\mathbf{k})\theta. \tag{9.17}$$

In what follows, we always assume that the in- and out-operators are chosen in such a way that condition (9.17) is satisfied.

There is less ambiguity in the definition of in- and out-operators if we make the following assumption: the relation between the old and new ex-operators is given by the formulas (9.16), where $\phi_-(\mathbf{k}) = \phi_+(\mathbf{k})$; correspondingly, in the formulas relating the new and old Møller matrices and new and old scattering matrices, we have $U_- = U_+$.

Let us check that the condition (9.17) is satisfied if and only if the functions $\Lambda_-(\mathbf{k})$ and $\Lambda_+(\mathbf{k})$ are equal: $\Lambda_-(\mathbf{k}) = \Lambda_+(\mathbf{k}) = \Lambda(\mathbf{k})$. To check this, we consider the decomposition of generalized vector functions $a^+(\mathbf{k})\Phi$ with respect to the generalized basis $a_{\mathrm{in}}^+(\mathbf{k}_1)\dots a_{\mathrm{in}}^+(\mathbf{k}_n)\Phi = \Phi_-(\mathbf{k}_1,\dots,\mathbf{k}_n)$. This decomposition has the form

$$a_+(\mathbf{k})\Phi = \rho(1,\mathbf{k})\Phi_-(\mathbf{k}) + \sum_{n=2}^{\infty} \frac{1}{\sqrt{n!}} \int \rho(1,\mathbf{k}_1,\dots,\mathbf{k}_n)$$

$$\times \delta\left(\mathbf{k} - \sum_{i=1}^{n} \mathbf{k}_i\right) \Phi_-(\mathbf{k}_1,\dots,\mathbf{k}_n)d^n\mathbf{k}, \qquad (9.18)$$

where the function $\rho(\epsilon,\mathbf{k}_1,\dots,\mathbf{k}_n)$ is defined by the equation

$$\frac{1}{\sqrt{n!}} \langle a(\mathbf{k},\epsilon)\Phi, \Phi_-(\mathbf{k}_1,\dots,\mathbf{k}_n)\rangle = \rho(\epsilon,\mathbf{k}_1,\dots,\mathbf{k}_n)\delta\left(\epsilon\mathbf{k} - \sum_i \mathbf{k}_i\right).$$

From (9.18), it follows that

$$\int f(\mathbf{k})\exp(-i\omega(\mathbf{k})t)a^+(\mathbf{k},t)\Phi d\mathbf{k}$$

$$= \int f(\mathbf{k})\exp(-i\omega(\mathbf{k})t)\exp(i\hat{H}t)a^+(\mathbf{k})\Phi d\mathbf{k}$$

$$= \int f(\mathbf{k})\rho(1,\mathbf{k})\Phi_-(\mathbf{k})d\mathbf{k} + \sum_{n\geq 2} \frac{1}{\sqrt{n!}} \int \rho(1,\mathbf{k}_1,\dots,\mathbf{k}_n)$$

$$\times \delta\left(\mathbf{k} - \sum_i \mathbf{k}_i\right)\exp(i(\omega(\mathbf{k}_1) + \cdots + \omega(\mathbf{k}_n)$$

$$- \omega(\mathbf{k}))t)\Phi_-(\mathbf{k}_1,\dots,\mathbf{k}_n)d^n\mathbf{k}.$$

For large t, the second summand contains a fast-oscillating factor and therefore it weakly tends to zero as $t \to \pm\infty$. Hence, we see that

$$\underset{t\to\pm\infty}{\mathrm{wlim}} \int f(\mathbf{k}) \exp(-i\omega(\mathbf{k})t) a^+(\mathbf{k}, t)\Phi d\mathbf{k}$$

$$= \int f(\mathbf{k})\rho(1, \mathbf{k})\Phi_+(\mathbf{k})d\mathbf{k}. \tag{9.19}$$

From the other side, from (9.11), we obtain

$$\underset{t\to\mp\infty}{\mathrm{wlim}} \int f(\mathbf{k}) \exp(-i\omega(\mathbf{k})t)\overline{\Lambda_\mp(\mathbf{k})} a^+(\mathbf{k}, t)\Phi d\mathbf{k}$$

$$= \int f(\mathbf{k})a^+_{\substack{\mathrm{in} \\ \mathrm{out}}}(\mathbf{k})\Phi d\mathbf{k} = \int f(\mathbf{k})\Phi_\mp(\mathbf{k})d\mathbf{k}. \tag{9.20}$$

Comparing Equations (9.19) and (9.20) and noting that by (9.17), $\Phi_-(\mathbf{k}) = \Phi_+(\mathbf{k})$, we obtain

$$\Lambda_-(\mathbf{k}) = \Lambda_+(\mathbf{k}) = \Lambda(\mathbf{k}) = (\overline{\rho(1, \mathbf{k})})^{-1}. \tag{9.21}$$

Hence, we have proven the equality $\Lambda_-(\mathbf{k}) = \Lambda_+(\mathbf{k})$ and also related these quantities with the function $\rho(1, \mathbf{k})$. Using this relation and the Källén–Lehmann representation, we will prove that the functions $\omega(\mathbf{k})$ and $|\Lambda(\mathbf{k})|$ can be expressed in terms of the Green function

$$\tilde{G}_2(\mathbf{p}_1, 1, \omega_1, \mathbf{p}_2, -1, \omega_2) = G(\mathbf{p}_1, \omega_1)\delta(\omega_1 - \omega_2)\delta(\mathbf{p}_1 - \mathbf{p}_2).$$

Namely, the function $\omega(\mathbf{p})$ specifies the location of poles of the functions $G(\mathbf{p}, \omega)$ (i.e. the poles of the function $G(\mathbf{p}, \omega)$ with respect to the variable ω, for fixed $\mathbf{p}$, are located at the points $\omega(\mathbf{p})$ and $-\omega(\mathbf{p})$). The function $|\Lambda(\mathbf{p})|$ is equal to $Z(\mathbf{p})^{-1/2}$, where $iZ(\mathbf{p})$ is the residue of the function $G(\mathbf{p}, \omega)$ at the pole $\omega(\mathbf{p})$.

To prove this statement, we note that we can take as a generalized eigenbasis of the operators $\hat{H}$ and $\hat{\mathbf{P}}$ the basis of vectors $a^+_{\mathrm{in}}(\mathbf{k}_1)\ldots a^+_{\mathrm{in}}(\mathbf{k}_n)\Phi = \Phi_-(\mathbf{k}_1, \ldots, \mathbf{k}_n)$. Let us write down the Källén–Lehmann representation for $G(\mathbf{p}, w)$ (8.10) in this basis.

We obtain

$$G(\mathbf{p}, \omega) = \frac{i|\rho(-1, -\mathbf{p})|^2}{\omega - \omega(-\mathbf{p}) + i0} \mp \frac{i|\rho(+1, \mathbf{p})|^2}{\omega + \omega(\mathbf{p}) - i0}$$

$$+ i \sum_{n \geq 2} \int \frac{|\rho(-1, \mathbf{k}_1, \dots, \mathbf{k}_n)|^2}{\omega - (\omega(\mathbf{k}_1) + \cdots + \omega(\mathbf{k}_n)) + i0}$$

$$\times \, \delta(\mathbf{p} + \mathbf{k}_1 + \cdots + \mathbf{k}_n) d\mathbf{k}_1 \dots d\mathbf{k}_n$$

$$\mp i \sum_{n \geq 2} \int \frac{|\rho(+1, \mathbf{k}_1, \dots, \mathbf{k}_n)|^2}{\omega + \omega(\mathbf{k}_1) + \cdots + \omega(\mathbf{k}_n) - i0}$$

$$\times \, \delta(-\mathbf{p} + \mathbf{k}_1 + \cdots + \mathbf{k}_n) d\mathbf{k}_1 \dots d\mathbf{k}_n.$$

In combination with (9.21), this relation gives a proof of the statements we need.

The generalized functions

$$S_{m,n}(\mathbf{p}_1, \dots, \mathbf{p}_m | \mathbf{q}_1, \dots, \mathbf{q}_n)$$

$$= \left\langle S b^+(\mathbf{q}_1) \dots b^+(\mathbf{q}_n)\theta, b^+(\mathbf{p}_1) \dots b^+(\mathbf{p}_m)\theta \right\rangle$$

(the matrix entries of the operator S in the generalized basis $b^+(\mathbf{p}_1) \dots b^+(\mathbf{p}_m)\theta$) are called the scattering amplitudes. They can be easily expressed in terms of the generalized vector functions $a_{\mathrm{ex}}^+(\mathbf{p}_1) \dots a_{\mathrm{ex}}^+(\mathbf{p}_m)\theta$ (that are called the in- and out-states), namely,

$$S_{m,n}(\mathbf{p}_1, \dots, \mathbf{p}_m | \mathbf{q}_1, \dots, \mathbf{q}_n)$$

$$= \left\langle S_- b^+(\mathbf{q}_1) \dots b^+(\mathbf{q}_n)\theta, S_+ b^+(\mathbf{p}_1) \dots b^+(\mathbf{p}_m)\theta \right\rangle$$

$$= \left\langle a_{\mathrm{in}}^+(\mathbf{q}_1) \dots a_{\mathrm{in}}^+(\mathbf{q}_n)\Phi, a_{\mathrm{out}}^+(\mathbf{p}_1) \dots a_{\mathrm{out}}^+(\mathbf{p}_n)\Phi \right\rangle.$$

Knowing the scattering amplitudes $S_{m,n}$, we can calculate the differential collision cross-section (see Section 10.5). It is often convenient to represent the operator S in normal form

$$S = \sum_{m,n} \frac{1}{m!n!} \int \sigma_{m,n}(\mathbf{p}_1, \dots, \mathbf{p}_m | \mathbf{q}_1, \dots, \mathbf{q}_n) b^+(\mathbf{p}_1) \dots b(\mathbf{q}_n) d^m \mathbf{p} \, d^n \mathbf{q}.$$

The functions $\sigma_{m,n}$ are closely related to the scattering amplitudes. This can be easily checked with Wick's theorem (see Section 6.3); we note here that under the condition $\mathbf{p}_i \neq \mathbf{q}_j (1 \leq i \leq m, 1 \leq j \leq n)$, we have

$$S_{m,n}(\mathbf{p}_1, \ldots, \mathbf{p}_m | \mathbf{q}_1, \ldots, \mathbf{q}_n) = \sigma_{m,n}(\mathbf{p}_1, \ldots, \mathbf{p}_m | \mathbf{q}_1, \ldots, \mathbf{q}_n).$$

We will now show how to express the functions $\sigma_{m,n}$ in terms of the Green functions of a translation-invariant Hamiltonian H.

We will prove the following *Lehmann–Symanzik–Zimmermann formula* (LSZ):

$$\sigma_{m,n}(\mathbf{p}_1, \ldots, \mathbf{p}_m | \mathbf{q}_1, \ldots, \mathbf{q}_n)$$

$$= (i\sqrt{2\pi})^{m+n} \prod_{i=1}^{m} \Lambda(\mathbf{p}_i) \prod_{j=1}^{n} \bar{\Lambda}(\mathbf{q}_j) \lim_{\omega_i \to \omega(\mathbf{p}_i)} \lim_{\sigma_j \to \omega(\mathbf{q}_j)} \prod_{i=1}^{m} (\omega_i - \omega(\mathbf{p}_i))$$

$$\times \prod_{j=1}^{n} (\sigma_j - \omega(\mathbf{q}_j)) \tilde{G}_{m+n}(\mathbf{q}_1, 1, \sigma_1, \ldots, \mathbf{q}_n, 1, \sigma_n | \mathbf{p}_1, -1,$$

$$\omega_1, \ldots, \mathbf{p}_m, -1, \omega_m), \tag{9.22}$$

where $\tilde{G}_{m+n}$ are the Green functions of the translation-invariant Hamiltonian H in the $(\mathbf{k}, \omega)$ representation.

We will give the proof of the LSZ formula in the CCR case (the case of CAR differs only by signs). First, let us note that

$$a_{\text{out}}(\mathbf{k}_1, \epsilon_1) T(a(\mathbf{k}_2, \epsilon_2, t_2) \ldots a(\mathbf{k}_n, \epsilon_n, t_n))$$

$$= \underset{t_1 \to +\infty}{\text{wlim}} \, a(\mathbf{k}_1, \epsilon_1, t_1) \exp(-i\epsilon_1 \omega(\mathbf{k}_1) t_1) \Lambda(\mathbf{k}_1, \epsilon_1)$$

$$\times T(a(\mathbf{k}_2, \epsilon_2, t_2) \ldots a(\mathbf{k}_n, \epsilon_n, t_n))$$

$$= \underset{t_1 \to +\infty}{\text{wlim}} \, \exp(-i\epsilon_2 w(\mathbf{k}_1) t_1)$$

$$\times \Lambda(\mathbf{k}_1, \epsilon_1) T(a(\mathbf{k}_1, \epsilon_1, t_1) \ldots a(\mathbf{k}_n, \epsilon_n, t_n)). \tag{9.23}$$

Indeed, this follows from the remark that if $t_1 > t_2, \ldots, > t_n$, then

$$a(\mathbf{k}_1, \epsilon_1, t_1) T(a(\mathbf{k}_2, \epsilon_2, t_2) \ldots a(\mathbf{k}_n, \epsilon_n, t_n))$$

$$= T(a(\mathbf{k}_1, \epsilon_1, t_1) \ldots a(\mathbf{k}_n, \epsilon_n, t_n))).$$

Similarly,

$$T(a(\mathbf{k}_2, \epsilon_2, t_2)\ldots a(\mathbf{k}_n, \epsilon_n, t_n))a_{\mathrm{in}}(\mathbf{k}_1, \epsilon_1)$$

$$= \lim_{t_1 \to -\infty} \Lambda(\mathbf{k}_1, \epsilon_1)\exp(-i\epsilon_1\omega(\mathbf{k}_1)t_1)$$

$$\times\, T(a(\mathbf{k}_1, \epsilon_1, t_1)\ldots a(\mathbf{k}_n, \epsilon_n, t_n)). \qquad (9.24)$$

Combining the relations (9.23) and (9.24), we obtain

$$a_{\mathrm{out}}(\mathbf{k}_1, \epsilon_1)T(a(\mathbf{k}_2, \epsilon_2, t_2)\ldots a(\mathbf{k}_n, \epsilon_n, t_n))$$

$$-\, T(a(\mathbf{k}_2, \epsilon_2, t_2)\ldots a(\mathbf{k}_n, \epsilon_n, t_n))a_{\mathrm{in}}(\mathbf{k}_1, \epsilon_1)$$

$$= \Lambda(\mathbf{k}_1, \epsilon_1)\int_{-\infty}^{\infty}\frac{\partial}{\partial t_1}(\exp(-i\epsilon_1\omega(\mathbf{k}_1)t_1)T(a(\mathbf{k}_1, \epsilon_1, t_1)$$

$$\ldots a(\mathbf{k}_n, \epsilon_n, t_n)))dt_1$$

$$= \int_{-\infty}^{\infty} L_1 T(a(\mathbf{k}_1, \epsilon_1, t_1)\ldots a(\mathbf{k}_n, \epsilon_n, t_n))dt_1. \qquad (9.25)$$

Here, we used the notation

$$L_i f(\mathbf{k}_1, \epsilon_1, t_1, \ldots, \mathbf{k}_n, \epsilon_n, t_n) = \Lambda(\mathbf{k}_i, \epsilon_i)\frac{\partial}{\partial t_i}(\exp(-i\epsilon_i\omega(\mathbf{k}_i)t)$$

$$\times\, f(\mathbf{k}_1, \epsilon_1, t_1, \ldots, \mathbf{k}_n, \epsilon_n, t_n)).$$

It is useful to note the simplest form of (9.25)

$$a_{\mathrm{out}}(\mathbf{k}, \epsilon) - a_{\mathrm{in}}(\mathbf{k}, \epsilon) = \int_{-\infty}^{\infty}\Lambda(\mathbf{k}, \epsilon)\frac{\partial}{\partial t}(\exp(-i\epsilon\omega(\mathbf{k})t)a(\mathbf{k}, \epsilon, t))dt$$

$$= \int_{-\infty}^{\infty} La(\mathbf{k}, \epsilon, t)dt$$

(Yang–Feldman equation).

In the case of CAR, we analogously have

$$a_{\mathrm{out}}(\mathbf{k}_1, \epsilon_1)T(a(\mathbf{k}_2, \epsilon_2, t_2)\ldots a(\mathbf{k}_n, \epsilon_n, t_n))$$

$$-(-1)^{n-1}T(a(\mathbf{k}_2, \epsilon_2, t_2)\ldots a(\mathbf{k}_n, \epsilon_n, t_n))a_{\mathrm{in}}(\mathbf{k}_1, \epsilon_1)$$

$$= \int_{-\infty}^{\infty}\Lambda(\mathbf{k}_1, \epsilon_1)\frac{\partial}{\partial t_1}(\exp(-i\epsilon_1\omega(\mathbf{k}_1)t_1)T(a(\mathbf{k}_1, \epsilon_1, t_1)$$

$$\ldots a(\mathbf{k}_n, \epsilon_n, t_n)))dt_1$$

$$= \int_{-\infty}^{\infty} L_1 T(a(\mathbf{k}_1, \epsilon_1, t_1)\ldots a(\mathbf{k}_n, \epsilon_n, t_n))dt_1.$$

Applying the relation (9.25) several times, we can express the matrix entries of the scattering matrix in terms of Green functions.

We will give another, shorter, proof. Let us introduce the operators $c(\mathbf{k}, \epsilon, t) = S_-^* a(\mathbf{k}, \epsilon, t) S_-$. We can easily express the Green function of the Hamiltonian H in terms of these operators:

$$\check{G}_n(\mathbf{k}_1, \epsilon_1, t_1, \ldots, \mathbf{k}_n, \epsilon_n, t_n)$$

$$= \langle T(a(\mathbf{k}_1, \epsilon_1, t_1) \ldots a(\mathbf{k}_n, \epsilon_n, t_n)) S_- \theta, S_- \theta \rangle$$

$$= \langle T(c(\mathbf{k}_1, \epsilon_1, t_1) \ldots c(\mathbf{k}_n, \epsilon_n, t_n)) \theta, \theta \rangle .$$

Multiplying (9.25) on the left by $S_+^* = S S_-^*$ and on the right by S_- and using the relations (9.13) and (9.14), we obtain

$$[S \cdot T(c(\mathbf{k}_2, \epsilon_2, t_2) \ldots c(\mathbf{k}_n, \epsilon_n, t_n)), b(\mathbf{k}_1, \epsilon_1)]$$

$$= - \int_{-\infty}^{\infty} L_1 S \cdot T(c(\mathbf{k}_1, \epsilon_1, t_1) \ldots c(\mathbf{k}_n, \epsilon_n, t_n)) dt_1. \quad (9.26)$$

Applying this formula several times, we see that

$$[\ldots [S, b(\mathbf{k}_1, \epsilon_1)] \ldots b(\mathbf{k}_n, \epsilon_n)]$$

$$= (-1)^n \int L_n \ldots L_1(S \cdot T(c(\mathbf{k}_1, \epsilon_1, t_1) \ldots c(\mathbf{k}_n, \epsilon_n, t_n))) dt_1 \ldots dt_n.$$

$$(9.27)$$

The expression for the functions $\sigma_{m,n}$ can be obtained from the remark that

$$\sigma_{m,n}(\mathbf{p}_1, \ldots, \mathbf{p}_m | \mathbf{q}_1, \ldots, \mathbf{q}_n)$$

$$= (-1)^n \langle [\ldots [[\ldots [S, b^+(\mathbf{q}_1)] \ldots b^+(\mathbf{q}_n)] b(\mathbf{p}_1)] \ldots b(\mathbf{p}_m)] \theta, \theta \rangle .$$

$$(9.28)$$

Namely, from (9.27) and (9.28) and the relation $S\theta = \theta$, it follows that

$$\sigma_{m,n}(\mathbf{p}_1, \ldots, \mathbf{p}_m | \mathbf{q}_1, \ldots, \mathbf{q}_n)$$

$$= (-1)^m \int L_{m+n} \ldots L_1 \check{G}_{m+n}(\mathbf{q}_1, 1, t_1, \ldots, \mathbf{q}_n, 1, t_n, \mathbf{p}_1, -1,$$

$$t_{n+1}, \ldots, \mathbf{p}_m, -1, t_{m+n}) dt_1 \ldots dt_{m+n}.$$

In order to obtain (9.22) from the above formula, we should take the Fourier transform over the variables $t_1, \ldots, t_{m+n}$, using that

$$\int L f(\mathbf{k}, \epsilon, t) dt = \int_{-\infty}^{\infty} \Lambda(\mathbf{k}, \epsilon) \frac{d}{dt} (\exp(-i\epsilon\omega(\mathbf{k})t) f(\mathbf{k}, \epsilon, t)) dt$$

$$= i\epsilon \Lambda(\mathbf{k}, \epsilon) \lim_{\omega \to \omega(\mathbf{k})} (\omega - \omega(\mathbf{k}))$$

$$\times \int_{-\infty}^{\infty} \exp(-i\epsilon\omega t) f(\mathbf{k}, \epsilon, t) dt.$$

For a more rigorous proof of (9.22), see, for example, Robertson *et al.* (1980). Such a proof can be obtained also from considerations of Chapter 13.

9.3 The adiabatic definition of scattering matrix

Let us now provide a definition of scattering matrix of a translation-invariant Hamiltonian that does not use the notion of operator realization. Let H be a translation-invariant Hamiltonian of the form $H = H_0 + V$, where $H_0 = \int \epsilon(\mathbf{k}) a^+(\mathbf{k}) a(\mathbf{k}) d\mathbf{k}$.

We will consider the adiabatic S-matrix $S_\alpha^\Omega = S_\alpha^\Omega(\infty, -\infty)$ corresponding to the pair of operators $(H_\Omega, H_{0\Omega})$ (the Hamiltonian H_Ω is defined as the Hamiltonian H with volume cutoff, see Section 8.1).

Definition 9.1. The scattering matrix for a Hamiltonian H is an operator in the space $\mathcal{H}_{\mathrm{as}}$ that has the matrix entries

$$\langle \mathbf{p}_1, \ldots, \mathbf{p}_m | S | \mathbf{q}_1, \ldots, \mathbf{q}_m \rangle = \lim_{\alpha \to 0} \lim_{\Omega \to \infty} \left(\frac{L}{2\pi} \right)^{\frac{3}{2}(m+n)}$$

$$\times \frac{\langle \mathbf{p}_1, \ldots, \mathbf{p}_m | S_\alpha^\Omega | \mathbf{q}_1, \ldots, \mathbf{q}_m \rangle \langle \theta | S_\alpha^\Omega | \theta \rangle^{\frac{m+n}{2} - 1}}{\sqrt{\prod_{i=1}^{m} \langle \mathbf{p}_i | S_\alpha^\Omega | \mathbf{p}_i \rangle \prod_{j=1}^{n} \langle \mathbf{q}_j | S_\alpha^\Omega | \mathbf{q}_j \rangle}} \tag{9.29}$$

in the generalized basis $b^+(\mathbf{p}_1) \ldots b^+(\mathbf{p}_n)\theta$. Here $\mathcal{H}_{\mathrm{as}}$, as in Section 9.2, denotes the space of Fock representations of CR (the operator generalized functions $b^+(\mathbf{k}), b(\mathbf{k})$ acting on $\mathcal{H}_{\mathrm{as}}$ satisfy CR and $\mathbf{k}$ runs over E^3). We use the notation

$$\langle \mathbf{p}_1, \ldots, \mathbf{p}_m | S | \mathbf{q}_1, \ldots, \mathbf{q}_n \rangle = \langle Sb^+(\mathbf{q}_1) \ldots b^+(\mathbf{q}_n)\theta, b^+(\mathbf{p}_1)$$

$$\ldots b^+(\mathbf{p}_m)\theta \rangle,$$

$$\langle \mathbf{p}_1, \ldots, \mathbf{p}_m | S_\alpha^\Omega | \mathbf{q}_1, \ldots, \mathbf{q}_n \rangle = \left\langle S_\alpha^\Omega a_{\mathbf{q}_1}^+ \ldots a_{b q_n}^+ \theta, a_{\mathbf{p}_1}^+ \ldots a_{\mathbf{p}_m}^+ \theta \right\rangle$$

(recall that the operator S_α^Ω, like the operator H_Ω, acts on the Fock space F_Ω and the symbols $a_{\mathbf{p}}^+$, $a_{\mathbf{p}}$ denote the operators $a^+(\phi_{\mathbf{p}}), a(\phi_{\mathbf{p}})$, where $\phi_{\mathbf{p}}(\mathbf{x}) = L^{-3/2} \exp(i\mathbf{px})$ and $\mathbf{p}$ runs over the lattice T_Ω, consisting of the vectors $\frac{2\pi}{L}\mathbf{n}$).

We should explain the meaning of convergence in (9.29), since the functions of continuous argument are defined as limits of functions with arguments in a lattice. We dealt with a similar situation in Section 8.2; here and in the rest of the book, we will deal with similar limits analogously (i.e. in the sense of generalized functions). Namely, the relation (9.29) means that for any test function $\phi(\mathbf{p}_1, \ldots, \mathbf{p}_m | \mathbf{q}_1, \ldots, \mathbf{q}_n)$

$$\int \phi(\mathbf{p}_1, \ldots, \mathbf{p}_m | \mathbf{q}_1, \ldots, \mathbf{q}_n) \langle \mathbf{p}_1, \ldots, \mathbf{p}_m | S | \mathbf{q}_1, \ldots, \mathbf{q}_n \rangle d^m \mathbf{p} d^n \mathbf{q}$$

$$= \lim_{\alpha \to \infty} \lim_{\Omega \to \infty} \left(\frac{2\pi}{L} \right)^{\frac{3}{2}(m+n)} \sum_{\mathbf{p}_i, \mathbf{q}_j \in T_\Omega} \phi(\mathbf{p}_1, \ldots, \mathbf{p}_m | \mathbf{q}_1, \ldots, \mathbf{q}_n)$$

$$\times \frac{\langle \mathbf{p}_1, \ldots, \mathbf{p}_m | S_\alpha^\Omega | \mathbf{q}_1, \ldots, \mathbf{q}_m \rangle \langle \theta | S_\alpha^\Omega | \theta \rangle^{\frac{m+n}{2}-1}}{\sqrt{\prod_{i=1}^m \langle \mathbf{p}_i | S_\alpha^\Omega | \mathbf{p}_i \rangle \prod_{j=1}^n \langle \mathbf{q}_j | S_\alpha^\Omega | \mathbf{q}_j \rangle}}.$$

We can represent the definition of scattering matrix in a different form, noting that

$$\frac{\langle \mathbf{p}_1, \ldots, \mathbf{p}_m | S_\alpha^\Omega | \mathbf{q}_1, \ldots, \mathbf{q}_m \rangle \langle \theta | S_\alpha^\Omega | \theta \rangle^{\frac{m+n}{2}-1}}{\sqrt{\prod_{i=1}^m \langle \mathbf{p}_i | S_\alpha^\Omega | \mathbf{p}_i \rangle \prod_{j=1}^n \langle \mathbf{q}_j | S_\alpha^\Omega | \mathbf{q}_j \rangle}}$$

$$= \langle \mathbf{p}_1, \ldots, \mathbf{p}_m | U_\alpha^\Omega S_\alpha^\Omega U_\alpha^\Omega | \mathbf{q}_1, \ldots, \mathbf{q}_m \rangle, \tag{9.30}$$

where $U_\alpha^\Omega = \exp[i(C + \sum_{\mathbf{k}} r_\alpha^\Omega(\mathbf{k}) a_{\mathbf{k}}^+ a_{\mathbf{k}})]$ (the sum is taken over the lattice T_Ω), $C = \frac{i}{2} \ln \langle \theta | S_\alpha^\Omega | \theta \rangle$, $r_\alpha^\Omega(\mathbf{k}) = \frac{i}{2} \ln \frac{\langle \mathbf{k} | S_\alpha^\Omega | \mathbf{k} \rangle}{\langle \theta | S_\alpha^\Omega | \theta \rangle}$ (the relation (9.30) follows from $U_\alpha^\Omega a_{\mathbf{k}_1}^+ \ldots a_{\mathbf{k}_n}^+ \theta = \exp[i(C + r_\alpha^\Omega(\mathbf{k}_1) + \cdots + r_\alpha^\Omega(\mathbf{k}_n))] a_{\mathbf{k}_1}^+ \ldots a_{\mathbf{k}_n}^+ \theta)$. The relation (9.30) follows from the fact that

$$S = \lim_{\alpha \to 0} \lim_{\Omega \to \infty} U_\alpha^\Omega S_\alpha^\Omega U_\alpha^\Omega,$$

where the convergence of operators is understood in the sense of the convergence of the matrix entries in the basis $a_{\mathbf{k}_1}^+ \ldots a_{\mathbf{k}_n}^+ \theta$ to the matrix elements in the generalized basis $b^+(\mathbf{k}_1) \ldots b^+(\mathbf{k}_n)\theta$ (further in the book, we will understand analogous situations concerning the convergence of operators acting on the space F_Ω to operators on the space $\mathcal{H}_{as}$ in the same way).

Using the results of Chapter 4 (see Sections 4.2 and 4.3), one can prove that the S-matrix defined by the relation (9.29) satisfies the conditions $S\theta = \theta$, $Sb^+(\mathbf{k})\theta = b^+(\mathbf{k})\theta$ (the stability of the vacuum and single-particle states).

Indeed, let $\phi^\Omega(\lambda)$ (correspondingly, $\phi_{\mathbf{k}}^\Omega(\lambda)$) be a stationary state of the Hamiltonian $H_\Omega(\lambda) = H_{0\Omega} + \lambda V_\Omega$ continuously depending on the parameter λ and converging to θ for $\lambda = 0$ (correspondingly $a_{\mathbf{k}}^+\theta$). Furthermore, we let $\epsilon_\Omega(\lambda)$ and $\epsilon_\Omega(\mathbf{k}, \lambda)$ be the energy levels of the Hamiltonian $H_\Omega(\lambda)$

$$\rho_\Omega = \frac{1}{\alpha} \int_0^1 \frac{1}{\mu} (\epsilon_\Omega(\mu) - \epsilon_\Omega(0)) d\mu,$$

$$\rho_\Omega(\mathbf{k}) = \frac{1}{\alpha} \int_0^1 \frac{1}{\lambda} (\epsilon_\Omega(\mathbf{k}, \lambda) - \epsilon_\Omega(\mathbf{k}, 0)) d\lambda.$$

It follows from (4.11) that

$$\theta = \lim_{\alpha \to 0} \exp(2i\rho_\Omega) S_\alpha^\Omega \theta,$$

$$a_{\mathbf{k}}^+\theta = \lim_{\alpha \to 0} \exp(2i\rho_\Omega(\mathbf{k})) S_\alpha^\Omega a_{\mathbf{k}}^+\theta.$$

We conclude that for $\alpha \to 0$,

$$\langle \mathbf{k}|S_\alpha^\Omega|\mathbf{k}'\rangle \approx \exp(2i\rho_\Omega(\mathbf{k}))\delta_{\mathbf{k}'}^{\mathbf{k}}, \tag{9.31}$$

$$\langle \theta|S_\alpha^\Omega|\theta\rangle \approx \exp(2i\rho_\Omega), \tag{9.32}$$

and therefore,

$$U_\alpha^\Omega \approx \tilde{U}_\alpha^\Omega = \exp\left(-i\rho_\alpha - i\sum_{\mathbf{k}}(\rho_\Omega(\mathbf{k}) - \rho_\Omega)a_{\mathbf{k}}^+ a_{\mathbf{k}}\right). \tag{9.33}$$

Using Equations (9.31) and (9.32), we see that

$$\lim_{\alpha \to 0} U_\alpha^\Omega S_\alpha^\Omega U_\alpha^\Omega \theta = \theta, \tag{9.34}$$

$$\lim_{\alpha \to 0} U_\alpha^\Omega S_\alpha^\Omega U_\alpha^\Omega a_{\mathbf{k}}^+ \theta = a_{\mathbf{k}}^+ \theta. \tag{9.35}$$

Multiplying the relations (9.34) and (9.35) by the vectors $a_{\mathbf{k}_1}^+ \ldots a_{\mathbf{k}_n}^+ \theta$ and taking the limit $\Omega \to \infty$ in the resulting matrix elements of the operator $U_\alpha^\Omega S_\alpha^\Omega U_\alpha^\Omega$, we obtain a proof of the stability of the vacuum and single-particle states.

Let us also note that from the previous reasoning, we can obtain the "almost unitarity" of the operators U_α^Ω when $\alpha \to 0$. More precisely, the scattering matrix can be represented in the form

$$S = \lim_{\alpha \to 0} \lim_{\Omega \to \infty} \tilde{U}_\alpha^\Omega S_\alpha^\Omega \tilde{U}_\alpha^\Omega, \tag{9.36}$$

where $\tilde{U}_\alpha^\Omega = \exp(-i\rho_\Omega - i \sum_{\mathbf{k}}(\rho_\Omega(\mathbf{k}) - \rho_\Omega)a_{\mathbf{k}}^+ a_{\mathbf{k}})$ are unitary operators.

Note that the proofs above are clearly not rigorous.

We can modify the scattering matrix definition given above by introducing the notion of N-equivalent and S-equivalent operators, acting in Fock space.

Two operators S and S' on Fock space T_{as} are N-*equivalent* if there exist unitary operators U and V of the form $\exp[i(\int \nu(\mathbf{k})b^+(\mathbf{k})b(\mathbf{k})d\mathbf{k} + C)]$, such that $S' = USV$. If $U = V^{-1}$, then we say that the operators S and S' are S-*equivalent*.

The definitions of N-equivalence and S-equivalence for operators acting on the Fock space F_Ω only differ by assuming that the operators U and V take the form $\exp[i(\sum_{\mathbf{k}} \nu_{\mathbf{k}} a_{\mathbf{k}}^+ a_{\mathbf{k}} + C)]$.

Definition 9.2. The operator S on the space $\mathcal{H}_{\text{as}}$ is called the *scattering matrix of the Hamiltonian* H if it satisfies the conditions $S\theta = \theta, Sb^+(\mathbf{k})\theta = b^+(\mathbf{k})\theta$ and it can be represented in the form

$$S = \lim_{\alpha \to 0} \lim_{\Omega \to \infty} \tilde{S}_\alpha^\Omega, \tag{9.37}$$

where $\tilde{S}_\alpha^\Omega$ are operators that are N-equivalent to the operators S_α^Ω.

Let us recall that the limit in the expression of the form (9.37) is understood in the sense of the convergence of the matrix entries of $\tilde{S}_\alpha^\Omega$ in the basis $a_{\mathbf{k}_1}^+ \ldots a_{\mathbf{k}_n}^+ \theta$ to the matrix entries of the operator S in the generalized basis $b^+(\mathbf{k}_1) \ldots b^+(\mathbf{k}_n)\theta$.

It is clear from the relation (9.36) that a scattering matrix specified by Definition 7.1 is also a scattering matrix in the sense of Definition 9.1. However, unlike Definition 7.1, Definition 9.1 does not specify a scattering matrix unambiguously: using the ambiguity in the choice of the operators S_α^Ω, it is easy to see that an operator that is S-equivalent to the scattering matrix is also a scattering matrix in the sense of Definition 9.1. The converse of this observation holds as well: two scattering matrices of a Hamiltonian H are S-equivalent. Note that in Section 9.2 the scattering matrix was also defined up to S-equivalence.

In the framework of perturbation theory, one can prove that *Definition 9.1 is equivalent to the definition of the scattering matrix given in Section 9.2.*[2] However, to show this equivalence, we need to make a few assumptions about the Hamiltonian H (it is sufficient to require that the Hamiltonian belongs to the class $\mathcal{M}$ defined in Section 11.1 and the function $\epsilon(\mathbf{k})$, entering the definition of the Hamiltonian H_0, satisfies the condition $\epsilon(\mathbf{k}_1 + \mathbf{k}_2) < \epsilon(\mathbf{k}_1) + \epsilon(\mathbf{k}_2))$.

We consider the connection between adiabatic S-matrices and scattering matrices in Section 10.6 (in the framework of axiomatic scattering theory) and in Section 11.5 (for Hamiltonians that do not generate vacuum polarization). The proof of the equivalence of Definition 9.1 with other definitions of scattering matrices in the framework of perturbation theory with partial summation can be obtained by modifying the considerations in Section 10.6. Another less rigorous proof can be found in Likhachev *et al.* (1972).

9.4 Faddeev's transformation and equivalence theorems

Let us consider a Hamiltonian $H = H_0 + V$ of the form (9.29). In the case of $v_{m,0} \equiv v_{m,1} \equiv 0$, the scattering matrix of H can be defined

---

[2] More precisely, to show this equivalence, we should perform a partial summation of the series in perturbation theory for S_α^Ω.

by using the standard relations of formal scattering theory (5.3) (see Section 9.1). If the Hamiltonian H is arbitrary, then under certain conditions, we may replace H with an equivalent, in a certain sense, Hamiltonian $H' = H'_0 + V'$ for which the scattering matrix can be defined as

$$S = \operatorname*{slim}_{\substack{t \to \infty \\ t_0 \to -\infty}} \exp(iH'_0 t) \exp(-iH'(t - t_0)) \exp(-iH'_0 t_0). \tag{9.38}$$

This scattering matrix can be regarded as the scattering matrix for the original Hamiltonian H.

The following discussion will proceed in the framework of perturbation theory. This means that the Hamiltonian $H_0 + V$ will be included in the family of Hamiltonians $H_0 + gV$ depending on a parameter g (coupling constant) and the Hamiltonian H' will be constructed as a series in powers of g.

To begin, let us prove a few helpful statements. Consider the equation

$$[h_0, x] = r, \tag{9.39}$$

where h_0, x, r are operators on the Fock space $F_\Omega = F(L^2(\Omega))$ corresponding to the finite-volume Ω:

$$h_0 = \sum_{\mathbf{k} \in T_\Omega} \epsilon(\mathbf{k}) a_{\mathbf{k}}^+ a_{\mathbf{k}}, \tag{9.40}$$

where the positive function $\epsilon(\mathbf{k})$ satisfies

$$\epsilon(\mathbf{k}_1 + \mathbf{k}_2) < \epsilon(\mathbf{k}_1) + \epsilon(\mathbf{k}_2). \tag{9.41}$$

Let us write down the unknown operator x and the known operator r in normal form:

$$x = \sum_{m,n} \sum_{\mathbf{k}_i, \mathbf{p}_j} \xi_{m,n}(\mathbf{k}_1, \ldots, \mathbf{k}_m | \mathbf{p}_1, \ldots, \mathbf{p}_n) a_{\mathbf{k}_1}^+$$
$$\ldots a_{\mathbf{k}_m}^+ a_{\mathbf{p}_1} \ldots a_{\mathbf{p}_n}; \tag{9.42}$$
$$r = \sum_{m,n} \sum_{\mathbf{k}_i, \mathbf{p}_j} \rho_{m,n}(\mathbf{k}_1, \ldots, \mathbf{k}_m | \mathbf{p}_1, \ldots, \mathbf{p}_n) a_{\mathbf{k}_1}^+$$
$$\ldots a_{\mathbf{k}_m}^+ a_{\mathbf{p}_1} \ldots a_{\mathbf{p}_n}$$

(in these formulas, just as in the following analogous formulas, the sums are taken over $\mathbf{k}_i, \mathbf{p}_j \in T_\Omega$). If we calculate the commutator

$[h_0, x]$, we obtain the relation between the unknown functions $\xi_{m,n}$ and the known functions $\rho_{m,n}$:

$$(\epsilon(\mathbf{k}_1) + \cdots + \epsilon(\mathbf{k}_m) - \epsilon(\mathbf{p}_1) - \cdots - \epsilon(\mathbf{p}_n))$$

$$\times \xi_{m,n}(\mathbf{k}_1, \ldots, \mathbf{k}_m | \mathbf{p}_1, \ldots, \mathbf{p}_n)$$

$$= \rho_{m,n}(\mathbf{k}_1, \ldots, \mathbf{k}_m | \mathbf{p}_1, \ldots, \mathbf{p}_n). \tag{9.43}$$

If the operator is presented in normal form, then the summands containing m creation operators and n annihilation operators will be called summands of type (m, n). Summands of type $(m, 0), (0, m), (n, 1)$, and $(1, n)$, where $m \geq 1, n \geq 2$, will be called bad summands.

Let us prove the following lemma.

Lemma 9.1. *If an operator r commutes with the momentum operator $\mathbf{P}_\Omega = \sum_{\mathbf{k}} \mathbf{k} a_{\mathbf{k}}^+ a_{\mathbf{k}}$ and only contains bad summands, then equation (9.39) is solvable. If we also assume that the operator x commutes with the momentum operator and contains only bad summands, then equation (9.39) has a unique solution which will be denoted by $\Gamma(r)$. If the operator r is Hermitian, then so is the operator $\Gamma(r)$.*

In order to prove this lemma, we first note that from the condition (9.43), we can find the functions $\xi_{m,n}$ in terms of the functions $\rho_{m,n}$ if and only if for all values of $\mathbf{k}_1, \ldots, \mathbf{k}_m, \mathbf{p}_1, \ldots, \mathbf{p}_n$ satisfying $\epsilon(\mathbf{k}_1) + \cdots + \epsilon(\mathbf{k}_m) = \epsilon(\mathbf{p}_1) + \cdots + \epsilon(\mathbf{p}_n)$, we have $\rho_{m,n}(\mathbf{k}_1, \ldots, \mathbf{k}_m | \mathbf{p}_1, \ldots, \mathbf{p}_n) = 0$. If we assume $\epsilon(\mathbf{k}_1) + \ldots + \epsilon(\mathbf{k}_m) > 0$, then for the functions $\rho_{m,0}$ and $\rho_{0,m}$, we can find the functions $\xi_{m,0}$ and $\xi_{0,m}$. Furthermore, from the assumption that r commutes with the momentum operator, it follows that the quantity $\rho_{m,n}(\mathbf{k}_1, \ldots, \mathbf{k}_m | \mathbf{p}_1, \ldots, \mathbf{p}_n)$ is non-zero only if $\mathbf{k}_1 + \cdots + \mathbf{k}_m = \mathbf{p}_1 + \cdots + \mathbf{p}_n$. This allows us to say that the inequality

$$\epsilon(\mathbf{k}_1) + \cdots + \epsilon(\mathbf{k}_m) > \epsilon(\mathbf{p}),$$

which holds in the case of $\mathbf{k}_1 + \cdots + \mathbf{k}_m = \mathbf{p}, m > 1$, guarantees the solvability of (9.43) for $m > 1, n = 1$ and $n > 1, m = 1$ (this follows from (9.41)). Therefore, under the conditions of the lemma, equation (9.39) is solvable. The proofs of the rest of the statements in the lemma are trivial (for example, the uniqueness follows from the fact that under the condition $\epsilon(\mathbf{k}_1) + \cdots + \epsilon(\mathbf{k}_m) \neq \epsilon(\mathbf{p}_1) + \cdots + \epsilon(\mathbf{p}_n)$ we can find a unique $\xi_{m,n}(\mathbf{k}_1, \ldots, \mathbf{k}_m | \mathbf{p}_1, \ldots, \mathbf{p}_n)$ for each $\rho_{m,n}(\mathbf{k}_1, \ldots, \mathbf{k}_m | \mathbf{p}_1, \ldots, \mathbf{p}_n)$).

Let us take a Hermitian operator h, acting on the Fock space F_Ω, that commutes with the momentum operator $\mathbf{P}_\Omega$, and consider the question of finding a unitary operator w such that the operator $h' = whw^{-1}$ contains no bad summands. We will show that in the framework of perturbation theory, the operator w can be constructed. More precisely, we will show that for an operator of the form $h = h_0 + gv$, where h_0 takes the form (9.40) considered above and g is a scalar parameter (coupling constant), the operators h' and w can be constructed as formal series in powers of g. The constructed operators h' and w will commute with the momentum operator $\mathbf{P}_\Omega$; this implies that the operator h' can be represented in the form $h' = c + h'_0 + v'$, where c is a constant [a summand of type $(0,0)$], $h'_0 = \sum \omega_\mathbf{k} a_\mathbf{k}^+ a_\mathbf{k}$ [a summand of type $(1,1)$], $v' = \sum_{m,n \geq 2} \sum v'_{m,n}(\mathbf{k}_1, \ldots, \mathbf{k}_m | \mathbf{p}_1, \ldots, \mathbf{p}_n) \delta_{\mathbf{k}_1 + \cdots + \mathbf{k}_m}^{\mathbf{p}_1 + \cdots + \mathbf{p}_n} a_{\mathbf{k}_1}^+ \cdots a_{\mathbf{k}_m}^+ a_{\mathbf{p}_1} \ldots a_{\mathbf{p}_n}$ (a sum of summands of type (m,n) with $m \geq 2, n \geq 2$). We will assume that the operator w takes the form

$$w = \exp(-i\alpha),$$

where α is a Hermitian operator that can be represented as a series in powers of g:

$$\alpha = \sum_{n=1}^{\infty} g^n \alpha_n.$$

We will use the formula[3]

$$h' = \exp(-i\alpha)h\exp(i\alpha) = \sum_{n=0}^{\infty} \frac{i^n}{n!}[\ldots[h, \alpha]\ldots, \alpha].$$

---

[3]It is easy to see that the operators $C(t) = \exp(-it\alpha)b\exp(it\alpha)$ and $D(t) = \sum_{n=0}^{\infty} \frac{(it)^n}{n!}[\ldots[b, \alpha]\ldots, \alpha]$ satisfy identical equations $i\frac{dC(t)}{dt} = [\alpha, C(t)], i\frac{dD(t)}{dt} = [\alpha, D(t)]$ with the same initial condition $C(0) = D(0) = b$; hence $C(t) = D(t)$.

Equating the terms of the same degree with respect to g, we obtain the equation

$$h'_n = i[h_0, \alpha_n] + q_n,$$

where q_n denotes the operator that can be expressed in terms of h_0, v, and α_k with $k \leq n - 1$, and $h'_n g^n$ denotes an order-n term in the expansion of the operator h' in powers of g. We will show by induction with respect to n that we can find operators α_n and h'_n such that the operator h'_n contains no bad summands. This can be done if we take as h'_n the sum of good summands in the normal form of the operator q_n. Then, the operator $h'_n - q_n$ contains only bad terms and satisfies the conditions imposed on the operator r in the lemma. Therefore, the operator α_n satisfying the above equation exists; let us define it by the formula $\alpha_n = \Gamma(h'_n - q_n)$. It is easy to check that the operators q_n are Hermitian; it follows that h'_n and α_n are Hermitian as well.

By this method, we can inductively construct the operators h'_n and α_n such that the operator $h' = \sum h'_n g^n$ does not contain bad summands and the operator $w = \exp(-i\alpha) = \exp(-i \sum \alpha_n g^n)$ establishes a unitary equivalence between the operators h and h' (i.e. $h' = whw^{-1}$).

Note that our construction for the operators h' and w is unambiguous; we will call this construction the *Faddeev construction* and the unitary operator w the *Faddeev transformation*.

It is worth noting that the operators h' and w were constructed in the framework of perturbation theory; in other words, these operators were constructed in the form of formal series in powers of g. It is not clear whether these series converge.

Moreover, as was done in the lemma and in the construction of the Faddeev transformation, we used the term "operator" for expression (9.42), but we did not prove that this formal expression specifies an operator on Fock space. In the situations where we use Faddeev construction, it is easy to check that the formal expressions h_n and α_n specify Hermitian operators; the proof is based on the considerations in Section 6.2.

Let us consider the Faddeev transformation from another point of view.

Consider the operators

$$b_{\mathbf{k}}^{+} = w^{-1} a_{\mathbf{k}}^{+} w; \quad b_{\mathbf{k}} = w^{-1} a_{\mathbf{k}} w. \tag{9.44}$$

It is clear that these operators satisfy the same commutation relations as the operators $a_{\mathbf{k}}^{+}, a_{\mathbf{k}}$:

$$[b_{\mathbf{k}}, b_{\mathbf{k}'}]_{\mp} = [b_{\mathbf{k}}^{+}, b_{\mathbf{k}'}^{+}]_{\mp} = 0, \quad [b_{\mathbf{k}}, b_{\mathbf{k}'}^{+}]_{\mp} = \delta_{\mathbf{k},\mathbf{k}'}.$$

A transformation from the operators $a_{\mathbf{k}}, a_{\mathbf{k}}^{+}$ to the operators $b_{\mathbf{k}}, b_{\mathbf{k}}^{+}$ satisfying the same commutation relations is called a *canonical transformation*. From the relation $h = w^{-1}h'w$, it follows that the operator h can be expressed in terms of the operators $b_{\mathbf{k}}^{+}, b_{\mathbf{k}}$ in the same way that h' can be expressed with $a_{\mathbf{k}}^{+}, a_{\mathbf{k}}$; in this way, representing the operator h with the operators $b_{\mathbf{k}}^{+}, b_{\mathbf{k}}$ in normal form, we obtain an expression without bad summands.

Let us now return to the translation-invariant Hamiltonian $H = H_0 + gV$, where H_0 and V satisfy the formulas (9.1) and (9.2). Suppose that $\epsilon(\mathbf{k})$ is a smooth function, whose derivatives do not grow faster than a power, and that $w_{m,n}$ belongs to the space $\mathcal{S}$. Let us also assume that the function $\epsilon(\mathbf{k})$ satisfies equation (9.41).

Let us now apply the Faddeev construction to the Hamiltonian H_Ω obtained from H by means of a volume cutoff. The Hamiltonian obtained through this construction contains no bad summands and will be denoted by H'^{Ω}. The operator establishing a unitary equivalence between H'^{Ω} and H_Ω will be denoted by $W^{\Omega} = \exp(-iA^{\Omega})$. It is easy to understand the behavior of the operators H'^{Ω} and A^{Ω} as $\Omega \to \infty$. Namely, the operator H'^{Ω} can be written in the form $H'^{\Omega} = C^{\Omega} + H_0'^{\Omega} + V'^{\Omega}$, where C^{Ω} is a constant that grows linearly as $\Omega \to \infty$. We obtain

$$H_0'^{\Omega} = \sum \omega_{\mathbf{k}}^{\Omega} a_{\mathbf{k}}^{+} a_{\mathbf{k}},$$

$$V'^{\Omega} = \sum_{m,n} \sum \left(\frac{L}{2\pi}\right)^{-3\left(\frac{m+n}{2}-1\right)} v_{m,n}'^{\Omega}(\mathbf{k}_1,\dots,\mathbf{k}_m|\mathbf{p}_1,\dots,\mathbf{p}_n)$$

$$\times \delta_{\mathbf{k}_1+\dots+\mathbf{k}_m}^{\mathbf{p}_1+\dots+\mathbf{p}_n} a_{\mathbf{k}_1}^{+}\dots a_{\mathbf{k}_m}^{+} a_{\mathbf{p}_1}\dots a_{\mathbf{p}_n},$$

the functions $\omega_{\mathbf{k}}^{\Omega}, v_{m,n}^{\prime\Omega}$ converge to a limit as $\Omega \to \infty$.[4] Let us introduce the notation

$$\lim \omega_{\mathbf{k}}^{\Omega} = \omega(\mathbf{k});$$

$$\lim v_{m,n}^{\prime\Omega}(\mathbf{k}_1, \ldots, \mathbf{k}_m | \mathbf{p}_1, \ldots, \mathbf{p}_n)$$
$$= v_{m,n}^{\prime}(\mathbf{k}_1, \ldots, \mathbf{k}_m | \mathbf{p}_1, \ldots, \mathbf{p}_n), \tag{9.45}$$

$$H_0^{\prime} = \int \omega(\mathbf{k}) a^{+}(\mathbf{k}) a(\mathbf{k}) d\mathbf{k};$$

$$V^{\prime} = \sum_{m,n} \int v_{m,n}^{\prime}(\mathbf{k}_1, \ldots, \mathbf{k}_m | \mathbf{p}_1, \ldots, \mathbf{p}_n)$$

$$\times \, \delta(\mathbf{k}_1 + \cdots + \mathbf{k}_m - \mathbf{p}_1 - \cdots - \mathbf{p}_n) a^{+}(\mathbf{k}_1)$$

$$\ldots a^{+}(\mathbf{k}_m) a(\mathbf{p}_1) \ldots a(\mathbf{p}_n) d^m \mathbf{k} d^n \mathbf{p}. \tag{9.46}$$

The translation-invariant Hamiltonian $H^{\prime} = H_0^{\prime} + V^{\prime}$ does not contain bad summands; we will say that this Hamiltonian is obtained from the Hamiltonian H by means of Faddeev transformation. One can say that

$$H^{\prime} = \lim_{\Omega \to \infty} (H^{\prime\Omega} - C^{\Omega}) \tag{9.47}$$

(the relation (9.47) can be understood as a shorthand version of (9.46)).

The operator A^{Ω} behaves the same way in the limit $\Omega \to \infty$, namely, it can be written in the form

$$A^{\Omega} = \sum_{m,n} \left(\frac{L}{2\pi}\right)^{-3\left(\frac{m+n}{2}-1\right)} \sum \alpha_{m,n}^{\Omega}(\mathbf{k}_1, \ldots, \mathbf{k}_m | \mathbf{p}_1, \ldots, \mathbf{p}_n)$$

$$\times \, \delta_{\mathbf{k}_1 + \cdots + \mathbf{k}_m}^{\mathbf{p}_1 + \cdots + \mathbf{p}_n} a_{\mathbf{k}_1}^{+} \ldots a_{\mathbf{k}_m}^{+} a_{\mathbf{p}_1} \ldots a_{\mathbf{p}_n},$$

---

[4]The functions $\omega_{\mathbf{k}}^{\Omega}, v_{m,n}^{\prime\Omega}$ are expressed in the form of a series in powers of g; we understand the convergence as $\Omega \to \infty$ in the sense of convergence for every power of g. It is useful to note that we can find such functions $v_{m,n}^{r} \in \mathcal{S}$ and $\sigma^{r}(\mathbf{k})$ that do not grow faster than a power function and that satisfy the conditions $|(v_{m,n}^{\prime\Omega})^{r}| \leq v_{m,n}^{r}, |(\omega_{\mathbf{k}}^{\Omega})^{r}| \leq \sigma^{r}(\mathbf{k})$ (by the symbol f^{r} here, we denote a term of order r in g in the expansion of the function f).

where the functions $\alpha_{m,n}^{\Omega}(\mathbf{k}_1,\ldots,\mathbf{k}_m|\mathbf{p}_1,\ldots,\mathbf{p}_n)$ have a limit as $\Omega \to \infty$,

$$\alpha_{m,n}(\mathbf{k}_1,\ldots,\mathbf{k}_m|\mathbf{p}_1,\ldots,\mathbf{p}_n) = \lim_{\Omega \to \infty} \alpha_{m,n}^{\Omega}(\mathbf{k}_1,\ldots,\mathbf{k}_m|\mathbf{p}_1,\ldots,\mathbf{p}_n)$$

(9.48)

in each term in the series with respect to g. Using the formal translation-invariant expression

$$A = \sum_{m,n} \int \alpha_{m,n}(\mathbf{k}_1,\ldots,\mathbf{k}_m|\mathbf{p}_1,\ldots,\mathbf{p}_n)$$
$$\times \; \delta(\mathbf{k}_1 + \cdots + \mathbf{k}_m - \mathbf{p}_1 - \cdots - \mathbf{p}_n) a^+(\mathbf{k}_1)$$
$$\ldots a^+(\mathbf{k}_m) a(\mathbf{p}_1) \ldots a(\mathbf{p}_n) d^m \mathbf{k} d^n \mathbf{p},$$

we can write the relation (9.49) in the form $A = \lim_{\Omega\to\infty} A^{\Omega}$. The behavior of the operator W^{Ω} for large Ω is more complicated; one cannot find a formal translation-invariant expression for this operator in the limit $\Omega \to \infty$. By construction, the expression for V' has no summands of type $(m,0)$ or $(m,1)$ and therefore the formal expression for H' can specify an operator $\hat{H}'$ in Fock space. In this case, the scattering matrix for the Hamiltonian H' can be constructed like an S-matrix for the pair of operators $(\hat{H}', \hat{H}_0')$ (see Section 9.1). In Chapter 11, we will show that the S-matrix for the pair of operators $(\hat{H}', \hat{H}_0')$ can be constructed in the framework of perturbation theory.

Faddeev proposed to define the *scattering matrix for the Hamiltonian H* as the S-matrix constructed for the pair of operators $(\hat{H}', \hat{H}_0')$. It can be proved that Faddeev's definition of the scattering matrix is equivalent to the other definitions we have considered (the proof can only be given in perturbation theory, since Faddeev's construction is based on perturbation theory). The proof of this statement is provided in Section 11.3, while here we will instead turn our attention to some generalizations of Faddeev's construction. First, we will represent this construction in a slightly different form.

It is easy to check that the S-matrix (9.38) can be written in the form

$$S = \lim_{t\to\infty, t_0\to-\infty} \lim_{\Omega\to\infty} \exp(iH_0'^\Omega t) \exp(-i(H'^\Omega - C^\Omega)(t - t_0))$$

$$\times \exp(-iH_0'^\Omega t_0)$$

(the limit should be understood in the sense explained in Section 9.3). Using the relation $H'^\Omega = W^\Omega H_\Omega (W^\Omega)^{-1}$, we can write that

$$S = \lim_{t\to\infty, t_0\to-\infty} \lim_{\Omega\to\infty} \exp(i(H_0'^\Omega + C^\Omega)t) W^\Omega \exp(-iH_\Omega(t - t_0))$$

$$\times (W^\Omega)^{-1} \exp(-i(H_0'^\Omega + C^\Omega)t_0). \tag{9.49}$$

The operator $(W^\Omega)^{-1}$ is called a dressing operator, since it transforms the bare vacuum θ and the bare single-particle state $a_{\mathbf{k}}^+\theta$ into stationary states of the Hamiltonian H_Ω that can be described as the dressed vacuum and dressed single-particle states (in fact, the operator $(W^\Omega)^{-1}$ transforms stationary states of the Hamiltonian H'^Ω into stationary states of the Hamiltonian H_Ω and θ and $a_{\mathbf{k}}^+\theta$ are clearly stationary states of H'^Ω).

The formula (9.49) suggests the following generalization of a dressing operator (more precisely, a family of dressing operators D^Ω depending on the parameter Ω).

The family of operators D^Ω acting on the space F_Ω is called a *family of dressing operators for the Hamiltonian H* if there exists a Hamiltonian H_{kb}^Ω of the form $\gamma^\Omega + \sum_{\mathbf{k}} \nu_{\mathbf{k}}^\Omega a_{\mathbf{k}}^+ a_{\mathbf{k}}$ such that the S-matrix for the Hamiltonian H can be written in the form

$$S = \lim_{t\to\infty, t_0\to-\infty} \lim_{\Omega\to\infty} \exp(iH_{kb}^\Omega t)(D^\Omega)^{-1}$$

$$\times \exp(-iH_\Omega(t - t-)) D^\Omega \exp(-iH_{kb}^\Omega t_0).$$

Formula (9.49) shows that the family of operators $(W_\Omega)^{-1}$ is a family of dressing operators in the sense of the above definition.

In Section 11.4, we will describe a broad class of families of dressing operators that includes the operators $(W^\Omega)^{-1}$.

The theorem establishing the equivalence of Faddeev's definition with other definitions of scattering matrices is a particular case of a theorem of invariance of scattering matrices under canonical transformations (theorems of this kind are called equivalence theorems).

Let us give a general definition of a canonical transformation.

Let us express the symbols $b^+(\mathbf{k}), b(\mathbf{k})$ in terms of the symbols $a^+(\mathbf{k}), a(\mathbf{k})$ by using the formulas

$$b(k) = \sum_{m,n} \int \sigma_{m,n}(\mathbf{k}_1, \ldots, \mathbf{k}_m | \mathbf{p}_1, \ldots, \mathbf{p}_n)$$

$$\times \, \delta(\mathbf{k} + \mathbf{k}_1 + \cdots + \mathbf{k}_m - \mathbf{p}_1 - \cdots - \mathbf{p}_n) a^+(\mathbf{k}_1)$$

$$\ldots a^+(\mathbf{k}_m) a(\mathbf{p}_1) \ldots a(\mathbf{p}_n) d^m \mathbf{k} d^n \mathbf{p}; \tag{9.50}$$

$$b^+(k) = \sum_{m,n} \int \overline{\sigma_{m,n}}(\mathbf{k}_1, \ldots, \mathbf{k}_m | \mathbf{p}_1, \ldots, \mathbf{p}_n)$$

$$\times \, \delta(\mathbf{k} + \mathbf{k}_1 + \cdots + \mathbf{k}_m - \mathbf{p}_1 - \cdots - \mathbf{p}_n) a^+(\mathbf{p}_1)$$

$$\ldots a^+(\mathbf{p}_1) a(\mathbf{k}_m) \ldots a(\mathbf{k}_1) d^m \mathbf{k} d^n \mathbf{p}, \tag{9.51}$$

where $\sigma_{m,n}$ are smooth functions whose derivatives do not grow faster than a power.

The transformation from the symbols $a^+(\mathbf{k}), a(\mathbf{k})$ into the symbols $b^+(\mathbf{k}), b(\mathbf{k})$ is called a *canonical transformation* if the symbols $b^+(\mathbf{k}), b(\mathbf{k})$ obey the same commutation (or anticommutation) relations as the symbols $a^+(\mathbf{k}), a(\mathbf{k})$ (i.e. obey CR). In the case of CCR, we can calculate the commutators for the symbols $b^+(\mathbf{k}), b(\mathbf{k})$ by using CCR for the operators $a^+(\mathbf{k}), a(\mathbf{k})$, distributivity, and the relation $[A, B, C] = [A, C]B + A[B, C]$ (we can calculate the anticommutator in the case of CAR analogously).

The above equivalence theorem can be reformulated (not quite rigorously) in the following way.

Theorem 9.1. *The canonical transformation (9.50) transforms the translation-invariant Hamitlonian H of the form (9.1) into the translation-invariant Hamiltonian $\tilde{H}$ having the same scattering matrix.*

(The Hamiltonian $\tilde{H}$ can be constructed in the following way: in the expression of the Hamiltonian H with respect to the symbols

$a^+(\mathbf{k}), a(\mathbf{k})$, we substitute against the symbols $b^+(\mathbf{k}), b(\mathbf{k})$, expressed in terms of $a^+(\mathbf{k}), a(\mathbf{k})$ using the formula (9.50); we then write the obtained expression into normal form using CR, discarding any infinite constants that arise.)

It is easy to see that the Hamiltonian H' obtained from the translation-invariant Hamiltonian using Faddeev's construction is connected with the Hamiltonian H through the canonical transformation (this canonical transformation is obtained from the canonical transformation (9.44) relating the operators $b_{\mathbf{k}}^+, b_{\mathbf{k}}$ with the operators $a_{\mathbf{k}}^+, a_{\mathbf{k}}$ in the limit $\Omega \to \infty$).

It is therefore clear from the equivalence theorem that the scattering matrices for the operators H and H' coincide; this demonstrates that Faddeev's construction leads to a conventional scattering matrix.

We will prove the equivalence theorem in the framework of perturbation theory in Section 11.3.

Note that for linear canonical transformations, the considerations in Section 11.3 prove the equivalence theorem outside the framework of perturbation theory.

9.5 Semiclassical approximation

We have already seen that the Hamiltonians of quantum field theory can be obtained through the quantization of classical systems with an infinite number of degrees of freedom. We will leverage this fact to obtain approximate solutions to problems in quantum field theory. As in Section 8.3, we will not make the assumption that $\hbar = 1$, as we do in the rest of this book.

Let us consider a quantum mechanical system with a finite number of degrees of freedom that is described by the Hamiltonian (8.22). In order to find weakly excited states (stationary states whose energy is close to the energy of the ground state), we can apply the following method. Let us find the minimum of the function $U(q_1, \ldots, q_n)$ (potential energy); suppose the minimum occurs at the point $(q_1^0, \ldots, q_n^0)$. Let us introduce the new variables $x_i = q_i - q_i^0$ (these should be understood as the deviation from the classical equilibrium state) and let us decompose the function U as a series in powers of x_i:

$$U(q_1, \ldots, q_n) = U(q_1^0, \ldots, q_n^0) + \frac{1}{2} \sum_{i,j} c_{ij} x_i x_j$$

$$+ \frac{1}{3!} \sum_{i,j,k} c_{ijk} x_i x_j x_k + \cdots .$$

If the matrix c_{ij} is positive definite, then the energy of a weakly excited state can be determined by the following approximate expression:

$$U(q_1^0, \ldots, q_n^0) + \frac{\hbar}{2} \sum \omega_i + \hbar \sum n_i \omega_i, \tag{9.52}$$

where $\omega_1^2, \ldots, \omega_n^2$ are eigenvalues of the matrix c_{ij} and $n_i = 0, 1, 2, \ldots$ are non-negative integers.

To check this, we will write the Hamiltonian (8.22) in terms of the operators $\hat{p}_i$ and $\hat{x}_i = \hat{q}_i - q_i^0$, satisfying the relations $[\hat{p}_i, \hat{p}_j] = [\hat{x}_i, \hat{x}_j] = 0$, $[\hat{p}_i, \hat{x}_j] = \frac{\hbar}{i} \delta_{ij}$; we obtain

$$H = H_0 + V;$$

$$H_0 = \frac{1}{2} \sum \hat{p}_i^2 + \frac{1}{2} \sum c_{ij} \hat{x}_i \hat{x}_j + U(q_1^0, \ldots, q_n^0);$$

$$V = \frac{1}{3!} \sum c_{ijk} \hat{x}_i \hat{x}_j \hat{x}_k + \cdots .$$

The operators $\hat{p}_i, \hat{x}_i$ satisfy the same commutation relations as $\hat{p}_i, \hat{q}_i$, i.e. the transformation between these operators can be considered to be a canonical transformation.

The eigenvalues of the operator H_0 are specified by the expression (9.52) (see Section 2.6). The eigenvalues of the operator $H = H_0 + V$ can be calculated in the framework of perturbation theory by considering the operator V as a perturbation. It is easy to see that the corrections to the energies of a weakly excited state are small. This becomes evident if we express H_0 and V in terms of the operators a_i^+, a_i, satisfying CCR (we rely on the considerations in

Section 2.6):

$$H_0 = U(q_1^0, \ldots, q_n^0) + \frac{1}{2}\hbar \sum \omega_i + \hbar \sum \omega_i a_i^+ a_i;$$

$$V = \sum_{m+n \geq 3} \hbar^{(\frac{1}{2})(m+n)} \sum_{i_1,\ldots,i_m,j_1,\ldots,j_n} \Gamma^{m,n}_{i_1,\ldots,i_m,j_1,\ldots,j_n} a_{i_1}^+ \ldots a_{j_n}$$

(the perturbation series here can be rewritten as a series in powers of $\hbar^{1/2}$, hence the correction to the energy has order at least $\hbar^{3/2}$).

We see that the analysis of weakly excited states of Hamiltonian H is based on the canonical transformation (transforming to the operators $\hat{p}_i, \hat{x}_i$) where we single out the quadratic part of H_0 and consider V as a perturbation.

Let us now consider a classical system specified by the Hamiltonian

$$\mathcal{H}(\pi, \phi) = \frac{1}{2}\sum_i \int \pi_i^2(\mathbf{x})d\mathbf{x} + \sum_m \sum_{i_1,\ldots,i_m} \int V_{i_1,\ldots,i_m}(\mathbf{x}_1,\ldots,\mathbf{x}_m)$$

$$\times \phi_{i_1}(\mathbf{x}_1)\ldots\phi_{i_m}(\mathbf{x}_m)d^m\mathbf{x}; \tag{9.53}$$

$V_{i_1,\ldots,i_m}(\mathbf{x}_1,\ldots,\mathbf{x}_m) = v_{i_1,\ldots,i_m}(\mathbf{x}_1 - \mathbf{x}_m,\ldots,\mathbf{x}_{m-1} - \mathbf{x}_m)$. Quantizing this system leads to the Hamiltonian

$$H = \frac{1}{2}\sum_i \int \hat{\pi}_i^2(\mathbf{x})d\mathbf{x}$$

$$+ \sum_m \sum_{i_1,\ldots,i_m} \int V_{i_1,\ldots,i_m}(\mathbf{x}_1,\ldots,\mathbf{x}_m)\hat{\phi}_{i_1}(\mathbf{x}_1)$$

$$\ldots \hat{\phi}_{i_m}(\mathbf{x}_m)d^m\mathbf{x}, \tag{9.54}$$

where

$$[\hat{\pi}_i(\mathbf{x}), \hat{\pi}_j(\mathbf{x}')] = [\hat{\phi}_i(\mathbf{x}), \hat{\phi}_j(\mathbf{x}')] = 0,$$

$$[\hat{\pi}_i(\mathbf{x}), \hat{\phi}_j(\mathbf{x}')] = \frac{\hbar}{i}\delta_{i,j}\delta(\mathbf{x} - \mathbf{x}')$$

(for a slightly less general Hamiltonian, see (8.23) discussed in Section 8.3; everything that is known for the Hamiltonian (8.23) can be easily proven for the Hamiltonians (9.54)).

As in the case of a finite number of degrees of freedom, the analysis of the Hamiltonian (9.54) can be performed with the canonical transformation: instead of the symbols $\hat{\phi}_i(\mathbf{x})$, we should introduce the symbols $\hat{\xi}_i(\mathbf{x}) = \hat{\phi}_i(\mathbf{x}) - \alpha_i$, where α_i are real numbers; the symbols $\hat{\pi}_i(\mathbf{x}), \hat{\xi}_i(\mathbf{x})$ satisfy the same commutation relations as $\hat{\pi}_i(\mathbf{x}), \hat{\phi}_j(\mathbf{x})$ (we could introduce the symbol $\hat{\xi}_i(\mathbf{x})$ by the more general formula $\hat{\xi}_i(\mathbf{x}) = \hat{\phi}_i(\mathbf{x}) - \alpha_i(\mathbf{x})$, but since our Hamiltonian is translation-invariant it is natural to assume that $\alpha_i(\mathbf{x})$ do not depend on $\mathbf{x}$). Let us consider a function of n variables

$$\nu(\phi_1, \ldots, \phi_n) = \sum_m \sum_{i_1,\ldots,i_m} \nu_{i_1,\ldots,i_m} \phi_{i_1,\ldots,i_m},$$

where

$$\nu_{i_1,\ldots,i_m} = \int \nu_{i_1,\ldots,i_m}(\mathbf{x}_1, \ldots, \mathbf{x}_{m-1}) d^{m-1}(\mathbf{x})$$

(the function ν has the physical meaning of energy density of the classical system in the state $\pi_1(\mathbf{x}) = \cdots = \pi_n(\mathbf{x}) \equiv 0, \phi_1(\mathbf{x}) \equiv \phi_1, \ldots, \phi_n(\mathbf{x}) \equiv \phi_n$). The numbers $\alpha_1, \ldots, \alpha_n$ specifying the canonical transformation will be found from the condition

$$\min_{\phi_1,\ldots,\phi_n} \nu(\phi_1, \ldots, \phi_n) = \nu(\alpha_1, \ldots, \alpha_n).$$

The point in n-dimensional space $\alpha = (\alpha_1, \ldots, \alpha_n)$, where the function $\nu(\phi_1, \ldots, \phi_n)$ achieves its minimum, will be called the classical vacuum.

It is easy to check, under certain conditions, that one can justify the choice of the numbers $\alpha_1, \ldots, \alpha_n$ by considering a limit of systems with finite number of degrees of freedom.

Replacing $\hat{\phi}_i(\mathbf{x})$ with $\hat{\xi}_i(\mathbf{x}) + \alpha_i$ in (9.54), we obtain the Hamiltonian H' with the same scattering matrix by the equivalence theorem. Extracting from H' the quadratic terms H_0, we can represent the

Hamiltonian H' in the form $H_0 + V$, where

$$H_0 = \frac{1}{2} \sum_i \int \hat{\pi}_i^2(\mathbf{x}) dx + \sum_{i,j} \frac{1}{2} \int c_{ij}(\mathbf{x} - \mathbf{y})$$

$$\times \hat{\xi}_i(\mathbf{x}) \hat{\xi}_j(\mathbf{y}) dxdy,$$

$$V = \sum_{m \geq 3} \sum_{i_1,\ldots,i_m} \int c_{i_1,\ldots,i_m}(\mathbf{x}_1 - \mathbf{x}_m, \ldots, \mathbf{x}_{m-1} - \mathbf{x}_m)$$

$$\times \hat{\xi}_{i_1}(\mathbf{x}_1) \ldots \hat{\xi}_{i_m}(\mathbf{x}_m) d^m \mathbf{x}.$$

Using the ideas of Section 8.3, we can construct an operator realization of the Hamiltonian H_0. We see that H_0 describes n types of particles with energy dependence on the momentum, defined by the formula $E_j(\mathbf{p}) = \hbar \omega_j(\frac{\mathbf{p}}{\hbar})$, where $\omega_1^2(\mathbf{k}), \ldots, \omega_n^2(\mathbf{k})$ are eigenvalues of the matrix $\tilde{c}_{ij}(\mathbf{k}) = (2\pi)^{-3} \int c_{ij}(\mathbf{x}) \exp(i\mathbf{k}\mathbf{x}) dx$. Understanding V as a perturbation, we obtain the corrections to the energy of single-particle states. By perturbation theory, we can also calculate the scattering matrix corresponding to the Hamiltonian H' and, therefore, to the Hamiltonian H.

Hence, the semiclassical approximation, as it is described in this section, prompts a reasonable choice of an initial approximation in the perturbation theory. However, one can show that semiclassical approximation leads to important qualitative conclusions. Let us consider, for example, the Hamiltonian

$$H = \frac{1}{2} \int \hat{\pi}^2(\mathbf{x}) dx + \frac{1}{2} \int (\nabla \hat{\phi})^2 dx + \int U(\hat{\phi}(\mathbf{x})) dx, \qquad (9.55)$$

where $U(\phi) = a_1 \phi + a_2 \phi^2 + a_3 \phi^3 + a_4 \phi^4$. (This Hamiltonian describes a Lorentz-invariant theory (see Section 12.3). The Hamiltonian H leads to ultraviolet divergences that should be treated by means of renormalization; we do not analyze these divergences here.) It is easy to see that for the Hamiltonian (9.55), $v(\phi) = U(\phi)$. If $U(\phi) = a_2 \phi^2 + a_4 \phi^4, a_2 > 0, a_4 > 0$, then the minimum of $v(\phi)$ is achieved at $\phi = 0$. This means that in the analysis of the Hamiltonian (9.55), we should use the perturbation theory where H_0 is specified in the usual way as follows:

$$H_0 = \frac{1}{2} \int \hat{\pi}^2(\mathbf{x}) d\mathbf{x} + \frac{1}{2} \int (\nabla \hat{\phi})^2 d\mathbf{x} + \int a_2 \hat{\phi}^2(\mathbf{x}) d\mathbf{x}. \qquad (9.56)$$

In general, it is unreasonable to specify H_0 by (9.56) (this is clear since the Hamiltonian (9.56), in the case $a_2 < 0$, describes particles with imaginary mass; it is impossible to construct its operator realization). In order to make the correct choice for an initial approximation, we should find

$$\min \nu(\phi) = \min U(\phi) = \min(a_1 \phi + a_2 \phi^2 + a_3 \phi^3 + a_4 \phi^4).$$

If this minimum is achieved for $\phi = \alpha$, then, using the canonical transformation, we get the Hamiltonian $H' = H_0 + V$, where

$$H_0 = \frac{1}{2} \int \hat{\pi}^2(\mathbf{x}) d\mathbf{x} + \frac{1}{2} \int (\nabla \hat{\xi})^2 d\mathbf{x} + \frac{1}{2} \mu^2 \int \hat{\xi}^2(\mathbf{x}) d\mathbf{x},$$

$$\mu^2 = U''(\alpha).$$

Now, we can apply the perturbation theory to the Hamiltonian H' considering V as a perturbation. In the case when $U(\phi) = a_2 \phi^2 + a_4 \phi^4$, $a_2 < 0, a_4 > 0$, the minimum of the function $U(\phi)$ is achieved at two points $\alpha = \pm \sqrt{\frac{a_2}{2a_4}}$ (there are two different classical vacua). Hence, the necessary canonical transformation and the Hamiltonian H' can be constructed in two ways: $H'_\pm = H_0 + V_\pm$, where

$$V_\pm = a_4 \int \hat{\xi}^4(\mathbf{x}) d\mathbf{x} \pm 2\sqrt{2a_2 a_4} \int \hat{\xi}^3(\mathbf{x}) d\mathbf{x}.$$

This allows us to conjecture that in the case at hand, one can construct different operator realizations of the Hamiltonian H. In fact, let us take an operator realization of the Hamiltonian H'_+ assuming that V_+ is a perturbation. This means that

$$\lim_{\hbar \to 0} \left\langle \hat{\xi}(\mathbf{x}, t)\Phi, \Phi \right\rangle = 0.$$

Starting with the operator realization of the Hamiltonian H'_+, we get an equivalent operator realization of the Hamiltonian H that gives

$$\lim_{\hbar \to 0} \left\langle \hat{\phi}(\mathbf{x}, t)\Phi, \Phi \right\rangle = \sqrt{\frac{a_2}{2a_4}} + \lim_{\hbar \to 0} \langle \hat{\xi}(\mathbf{x}, t)\Phi, \Phi \rangle$$

$$= \sqrt{\frac{a_2}{2a_4}} > 0.$$

For an operator realization of the Hamiltonian H, constructed in a similar way by means of H'_-, we get that

$$\lim_{\hbar \to 0} \langle \hat{\phi}(\mathbf{x}, t)\Phi, \Phi \rangle = -\sqrt{\frac{a_2}{2a_4}} < 0,$$

and hence, we have constructed two essentially different operator realizations.

Note that we have encountered an important phenomenon called symmetry breaking. The initial Hamiltonian function is invariant with respect to the transformation $\phi(\mathbf{x}) \to -\phi(\mathbf{x})$; however, the classical vacuum is not invariant with respect to this transformation (it transforms one classical vacuum into another). A similar situation arises in the quantum case: the Hamiltonian H is invariant with respect to the transformation $\hat{\phi}(\mathbf{x}) \to -\hat{\phi}(\mathbf{x})$ and therefore, from one operator realization of the Hamiltonian H, we can get another by replacing the operators $\hat{\phi}(\mathbf{x}, t)$ by the operators $-\hat{\phi}(\mathbf{x}, t)$. If $\langle \hat{\phi}(\mathbf{x}, t)\Phi, \Phi \rangle \neq 0$, then this replacement gives an operator realization that is not equivalent to the original.

Let us consider one more interesting example, the Hamiltonian

$$H = \frac{1}{2} \sum \int \hat{\pi}_i^2(\mathbf{x}) d\mathbf{x} + \int \left(\sum \hat{\phi}_i^2(\mathbf{x}) - a^2 \right)^2 d\mathbf{x}. \tag{9.57}$$

Here,

$$\nu(\phi_1, \ldots, \phi_n) = (\phi_1^2 + \cdots + \phi_n^2 - a^2)^2$$

and hence, the minimum of the function ν is achieved at the points $(\alpha_1, \ldots, \alpha_n)$ satisfying the condition $\sum \alpha_i^2 = a^2$ (the classical vacua fill an $(n-1)$-dimensional sphere). The Hamiltonian (9.57), as well as the corresponding classical Hamiltonian functional, is invariant with respect to the transformations

$$\hat{\phi}_i'(\mathbf{x}) = \sum a_{ij}\hat{\phi}_j(\mathbf{x}), \quad \hat{\pi}_i'(\mathbf{x}) = \sum a_{ij}\hat{\pi}_j(\mathbf{x}),$$

where $a_{ij} \in O(n)$ (here, $O(n)$ is the group of orthogonal matrices). However, in the operator realization of the Hamiltonian H, we have a smaller symmetry group — the group $O(n-1)$. To check this fact, we note that the operator realization of the Hamiltonian H corresponds to a classical vacuum and, for every classical vacuum, the transformations $g \in O(n)$ that sends this vacuum to itself (satisfying the condition $g\alpha = \alpha$) constitute a subgroup isomorphic to $O(n-1)$. Then, transformations that do not belong to this subgroup send an operator realization to another that is not equivalent to the original operator realization.

In this example, we obtain a symmetry breaking of a continuous symmetry group. In this situation, we necessarily have particles with energy tending to zero when the momentum tends to zero (Goldstone particles). It is easy to verify this statement when one can apply the semiclassical considerations of the present section. Indeed, in the initial approximation, the energy of the particles is determined by the eigenvalues of the matrix $\tilde{c}_{ij}(\mathbf{k})$. This matrix continuously depends on $\mathbf{k}$, and therefore, it is sufficient to analyze the matrix $\tilde{c}_{ij}(0)$ that coincides with the matrix of second derivatives $\frac{\partial^2 \nu}{\partial\phi_i\partial\phi_j}$ of the function ν at the minimum point. This matrix is necessarily degenerate, since in our assumption the minimum cannot be achieved at one isolated point. Hence, we conclude that in the zeroth approximation, the Goldstone particles exist. It is easy to check that taking into account the higher approximations of the perturbation theory with respect to V does not change this statement. In Lorentz-invariant theory, the energy of a particle with momentum $\mathbf{p}$ is equal to $\sqrt{\mathbf{p}^2 + m^2}$, where m is the mass of the particle, and hence Goldstone particles have zero mass.

We can conclude that the Hamiltonian (9.54) describes particles with energy having the form $\hbar\omega_j(\hbar^{-1}\mathbf{p})$ when $\hbar \to 0$. However, aside from these particles, that can be called elementary, the Hamiltonian (9.54) also describes the particles with energies having a finite limit as $\hbar \to 0$. These particles correspond to the so-called "particle-like" solutions of classical equations. The simplest particle-like solutions are *solitons*. A solution $\pi_i(\mathbf{x}, t), \phi_i(\mathbf{x}, t)$ of the Hamiltonian equations is called a soliton if the function $\phi_i(\mathbf{x}, t)$ can be represented in the form $s_i(\mathbf{x} - \mathbf{v}t)$ (we assume that the energy of this solution is finite). (Here, we assume that the energy is measured as a deviation from the classical vacuum, i.e. $\mathcal{H}(\pi, \phi) = 0$ if $\pi_i(\mathbf{x}) \equiv 0, \phi_i(\mathbf{x}) = \alpha_i$, where $(\alpha_1, \ldots, \alpha_n)$ is a classical vacuum. If the condition $\mathcal{H}(\pi, \phi) = 0$ is not satisfied, then we should replace the Hamiltonian functional (9.54) with the functional

$$\mathcal{H}(\pi, \phi) = \frac{1}{2} \sum_i \int \pi_i^2(\mathbf{x}) dx + \sum_m \sum_{i_1, \ldots, i_m} \int V_{i_1, \ldots, i_m}(\mathbf{x}_1, \ldots, \mathbf{x}_m)$$

$$\times (\phi_{i_1}(\mathbf{x}_1) \ldots \phi_{i_m}(\mathbf{x}_m) - \alpha_{i_1} \ldots \alpha_{i_m}) d^m \mathbf{x},$$

which leads to the same equations of motion.)

A soliton with a zero velocity $(\mathbf{v} = 0)$ is a solution of the equations of motion that does not depend on time. Note that it follows from translation-invariance of the Hamiltonian functional (9.53) that the momentum $\mathbf{P} = \sum_i \int \pi_i \frac{\partial \phi_i}{\partial \mathbf{x}} dx$ is an integral of motion of the Hamilton equations; the soliton with zero velocity has zero momentum. In the Lorentz-invariant case, we can obtain solitons with arbitrary momentum (and with aribtrary velocity that is less than the speed of light) from the solitons of zero velocity by means of Lorentz transformation.

One can check that there exists a quantum particle that corresponds to a stable soliton (the solution $\phi_i(\mathbf{x}, t), \pi_i(\mathbf{x}, t)$ is called stable if every solution $\phi_i'(\mathbf{x}, t), \pi_i'(\mathbf{x}, t)$ that is close to the original at $t = 0$ stays close to it at any other moment in time t). More precisely, if for every three-dimensional vector $\mathbf{p}$, we have a stable soliton of the Hamilton equations, having the momentum $\mathbf{p}$ and energy $\epsilon(\mathbf{p})$, then under certain conditions, we can prove that among the particles described by Hamiltonian (9.54) are particles with energy tending to $\epsilon(\mathbf{p})$ as $\hbar \to 0$ (in other words, in the space of operator realizations of the Hamiltonian (9.54), we have vector generalized functions $\Phi_\hbar(\mathbf{p})$ that are δ-normalized and obey the conditions $H\Phi_\hbar(\mathbf{p}) = E_\hbar(\mathbf{p})\Phi_\hbar(\mathbf{p}), \mathbf{P}\Phi_\hbar(\mathbf{p}) = \mathbf{p}\Phi_\hbar(\mathbf{p})$, where $\lim E_\hbar(\mathbf{p}) = \epsilon(\mathbf{p})$;

for more, see Takhtadzhyan and Faddeev (1974), Faddeev (1975), Tiupkin *et al.* (1975b), Polyakov (1974) and Tyupkin *et al.* (1975).

As an example, we can consider a classical system with the Hamiltonian

$$\mathcal{H}(\pi,\phi) = \frac{1}{2}\int \pi^2(\mathbf{x})d\mathbf{x} + \frac{1}{2}\int \left(\frac{d\phi}{d\mathbf{x}}\right)^2 d\mathbf{x}$$

$$+ \int (\phi^2(\mathbf{x}) - a^2)^2 d\mathbf{x},$$

where $x \in E^1$. The Hamiltonian equations take the form

$$\frac{\partial}{\partial t}\phi(\mathbf{x},t) = \pi(\mathbf{x},t),$$

$$\frac{\partial}{\partial t}\phi(\mathbf{x},t) = \frac{\partial^2}{\partial \mathbf{x}^2}\phi(\mathbf{x},t) - 4\phi(\mathbf{x},t)(\phi^2(\mathbf{x},t) - a^2).$$

If $\phi(\mathbf{x},t) = s(\mathbf{x})$, then the function $s(\mathbf{x})$ satisfies the equation

$$\frac{d^2 s}{d\mathbf{x}^2} = 4s(s^2 - a^2). \tag{9.58}$$

From the condition of finite energy, we can conclude that $\lim_{x\to\pm\infty}|s(\mathbf{x})| = a$; using this relation and (9.58), we obtain, in the case at hand, the solitons with zero velocity

$$s(\mathbf{x}) = \pm a \operatorname{th}\sqrt{2}ax,$$

they have energy $\mu = \frac{4\sqrt{2}}{3}a^3$. The soliton with momentum $\mathbf{p}$ can be obtained directly or by means of Lorentz transformation; the energy of this soliton is equal to $\sqrt{\mathbf{p}^2 + \mu^2}$. One can prove that there exist quantum particles corresponding to these solitons.

Let us note in conclusion that one can use the semiclassical approximation to analyze more general Hamiltonians (8.2). Namely, we should introduce the complex functions $\psi(\mathbf{x}) = \frac{1}{\sqrt{2}}(\phi(\mathbf{x}) + i\pi(\mathbf{x}))$, $\overline{\psi}(\mathbf{x}) = \frac{1}{\sqrt{2}}(\phi(\mathbf{x}) - i\pi(\mathbf{x}))$ in place of the generalized momenta $\pi(\mathbf{x})$ and coordinates $\phi(\mathbf{x})$. An arbitrary Hamiltonian $\mathcal{H}(\pi,\phi)$ can be expressed in terms of $\overline{\psi}(\mathbf{x}), \psi(\mathbf{x})$; let us assume that

$$\mathcal{H}(\overline{\psi},\psi) = \sum_{m,n}\int H_{m,n}(\mathbf{x}_1,\dots,\mathbf{x}_m|\mathbf{y}_1,\dots,\mathbf{y}_n)$$

$$\times \overline{\psi}(\mathbf{x}_1)\dots\overline{\psi}(\mathbf{x}_m)\psi(\mathbf{y}_1)\dots\psi(\mathbf{y}_n)d^m\mathbf{x}d^n\mathbf{y}.$$

By the quantization of the functions $\overline{\psi}(\mathbf{x}), \psi(\mathbf{x})$, we obtain the operator generalized functions $\hat{\psi}^{+}(\mathbf{x}), \hat{\psi}(\mathbf{x})$ obeying the commutation relations

$$[\hat{\psi}(\mathbf{x}), \hat{(\mathbf{x}')}] = [\hat{\psi}^{+}(\mathbf{x}), \hat{\psi}^{+}(\mathbf{x}')] = 0,$$

$$[\hat{\psi}(\mathbf{x}), \hat{\psi}^{+}(\mathbf{x}')] = \hbar\delta(\mathbf{x} - \mathbf{x}'),$$

and the Hamiltonian functional $\mathcal{H}$ specifies the quantum Hamiltonian

$$H = \sum_{m,n} \int H_{m,n}(\mathbf{x}_1, \ldots, \mathbf{x}_m | \mathbf{y}_1, \ldots, \mathbf{y}_n)$$

$$\times \hat{\psi}^{+}(\mathbf{x}_1) \ldots \hat{\psi}^{+}(\mathbf{x}_m)\hat{\psi}(\mathbf{y}_1) \ldots \hat{\psi}(\mathbf{y}_n) d^m\mathbf{x} d^n\mathbf{y}$$

(in the transition from the Hamiltonian functional $\mathcal{H}$ to the quantum Hamiltonian H, we should answer the question about the ordering of the operators $\hat{\psi}^{+}, \hat{\psi}$ in the expression of H; the choice we have made corresponds to the so-called "Wick quantization"). If the Hamiltonian functional $\mathcal{H}$ is translation-invariant, then using the formulas

$$\hat{\psi}^{+}(\mathbf{x}) = (2\pi)^{-3/2}\hbar^{1/2} \int \exp(i\mathbf{k}\mathbf{x})a^{+}(\mathbf{k})d\mathbf{k},$$

$$\hat{\psi}(\mathbf{x}) = (2\pi)^{-3/2}\hbar^{1/2} \int \exp(-i\mathbf{k}\mathbf{x})a(\mathbf{k})d\mathbf{k}$$

we can write down the Hamiltonian (9.35) in the form (8.2). This allows us to apply the considerations of this section to the analysis of the Hamiltonian (8.2).

Chapter 10

Axiomatic Scattering Theory

10.1 Main assumptions and the construction of the scattering matrix

Let us suppose that we have four commuting self-adjoint operators H and $\mathbf{P} = (P_1, P_2, P_3)$ defined on the Hilbert space $\mathcal{H}$. The operator H has the physical meaning of an energy operator and the (vector) operator $\mathbf{P}$ has the meaning of the momentum operator. The operator $\exp(i\mathbf{P}\mathbf{a})$ is called *the spatial shift operator* or spatial translation and the operator $\exp(-iHt)$ is *the time shift operator* (time translation). Let us assume that the energy operator has the unique ground state Φ and the state is invariant with respect to space shifts and time shifts: $H\Phi = \mathbf{P}\Phi = 0$. The vector Φ is called *the physical vacuum.*

In the situation at hand, we can apply the definition of a particle given in Section 5.3. We will formulate this definition in a slightly different way.

Let us suppose that for every function $f \in L^2(E^3)$ we have a corresponding vector $\Phi(f) \in \mathcal{H}$ that linearly depends on the function f. This relationship can be understood as a linear operator mapping from $L^2(E^3)$ to $\mathcal{H}$; however, we will prefer to view it as a vector generalized function $\Phi(\mathbf{k})$, where the vector $\Phi(f)$ can be written in the form $\Phi(f) = \int f(\mathbf{k})\Phi(\mathbf{k})d\mathbf{k}$. The vector generalized function $\Phi(\mathbf{k})$ is called a *particle* (a single-particle state), if for every

191

pair of functions $f, g \in L^2(E^3)$ we have the relations

$$H\Phi(f) = \Phi(\hat{h}f),$$

$$\mathbf{P}\Phi(f) = \Phi(\hat{\mathbf{p}}f),$$

$$\langle \Phi(f), \Phi(g) \rangle = \langle f, g \rangle,$$

where

$$(\hat{h}f)(\mathbf{k}) = \omega(\mathbf{k})f(\mathbf{k}),$$

$$(\hat{\mathbf{p}}f)(\mathbf{k}) = \mathbf{k}f(\mathbf{k}).$$

The vector $\Phi(f)$ represents the state of a particle with the wave function f and the function $\omega(\mathbf{k})$ has the meaning of the energy of the single-particle state (dispersion law). The relations above are equivalent to the relations (5.20)–(5.22). The last one implies that the operator taking the function f to the vector $\Phi(f)$ is an isometry; this is equivalent to the relation (5.22), which shows that $\Phi(\mathbf{k})$ is δ-function normalized. It is possible that there are several types of particles (several types of generalized functions $\Phi(\mathbf{k})$) that satisfy (5.20)–(5.22). A system of particles $\Phi_1(\mathbf{k}), \ldots, \Phi_s(\mathbf{k})$ is called *complete*, if every pair of particles is orthogonal (i.e. $\langle \Phi_i(\mathbf{k}), \Phi_j(\mathbf{k}') \rangle = 0$ for $i \neq j$) and there does not exist another particle that is orthogonal to all the particles in the system.

One can show that *every particle $\Phi(\mathbf{k})$ can be decomposed into the particles of a complete system* (i.e. we can find functions $f_i(\mathbf{k})$ so that $\Phi(\mathbf{k}) = \sum_{i=1}^{s} f_i(\mathbf{k})\Phi_i(\mathbf{k})$).

Let us fix a complete particle system $\Phi_1(\mathbf{k}), \ldots, \Phi_s(\mathbf{k})$ with the dispersion laws $\omega_1(\mathbf{k}), \ldots, \omega_s(\mathbf{k})$, assuming that the functions $\omega_i(\mathbf{k})$ are strongly convex. (We suppose for definiteness that we have a finite number s of particles, even though the case of an infinite number of particles is not much more difficult.) *The single-particle subspace $\mathcal{H}_1$* of the space $\mathcal{H}$ will be defined as the smallest subspace that contains all vectors of the form $\Phi_i(f) = \int f(\mathbf{k})\Phi_i(\mathbf{k})d\mathbf{k}$. *The multi-particle subspace $\mathcal{M}$* will be defined as the orthogonal complement of the direct sum $\mathcal{H}_0 + \mathcal{H}_1$ in $\mathcal{H}$, where $\mathcal{H}_0$ is the one-dimensional subspace generated by the physical vacuum Φ.

The spaces $\mathcal{H}_0, \mathcal{H}_1, \mathcal{M}$ are clearly invariant with respect to the operators $H, \mathbf{P}$. The joint spectrum of the system of commuting operators $H, \mathbf{P}$ in the space $\mathcal{H}_1$ (in the space $\mathcal{M}$) will be called the single-particle (correspondingly, the multi-particle) spectrum and we will denote it by $\sum_1$ (correspondingly, $\sum_{\mathcal{M}}$). The set $\sum_1$ clearly coincides with the union of the sets $\sum_1^{(i)}$, where $\sum_1^{(i)}$ denotes the set of points of the form $(\mathbf{k}, \omega_i(\mathbf{k}))$.

Let us assume that *the single-particle spectrum does not intersect the multi-particle spectrum and that there exists $\delta > 0$, such that the spectrum of the operator H, with the exception of the point 0, corresponding to the physical vacuum, is contained in the ray $[\delta, +\infty)$.* We will later show how this assumption can be relaxed.

Let us consider, along with the space $\mathcal{H}$, *the asymptotic state space* $\mathcal{H}_{as}$, which we will define as the Fock space $F(L^2(E^3 \times N))$, where N is a finite set whose elements are in one-to-one correspondence with the complete system of particles we have assumed previously. Let us suppose that the operator generalized functions $a_i^+(\mathbf{k})$, $a_i(\mathbf{k})$ obey CCR (see Section 3.2) and act on the space $\mathcal{H}_{as}$; these functions can be considered as the creation and annihilation operators of particles of the ith type with momentum $\mathbf{k} \in E^3$. We will define the asymptotic Hamiltonian H_{as} and the momentum operator $\mathbf{P}_{as}$ by the formulas

$$H_{as} = \sum_{i=1}^{s} \int \omega_i(\mathbf{k}) a_i^+(\mathbf{k}) a_i(\mathbf{k}) d\mathbf{k},$$

$$\mathbf{P}_{as} = \sum_{i=1}^{s} \int \mathbf{k} a_i^+(\mathbf{k}) a_i(\mathbf{k}) d\mathbf{k}.$$

If A is an operator on the space $\mathcal{H}$, then $A(\mathbf{x}, t)$ will denote the operator $\exp(i(Ht - \mathbf{P}\mathbf{x})) A \exp(-i(Ht - \mathbf{P}\mathbf{x}))$. Bounded operators A and B, acting on the space $\mathcal{H}$, will be said to be *asymptotically commuting* if for every natural number n we can find real numbers C and r such that

$$\|[A, B(\mathbf{x}, t)]\| \leq C \frac{1 + |t|^r}{1 + |\mathbf{x}|^n}.$$

A family of bounded operators $\mathcal{A}$ on the space $\mathcal{H}$ that contains the conjugate operator A^* for every operator A in $\mathcal{A}$ will be called *asymptotically Abelian* if two arbitrary operators in $\mathcal{A}$ are asymptotically commuting.

An asymptotically Abelian family that contains the identity operator will be called an *asymptotically Abelian algebra* (or asymptotically commutative algebra) if for every pair of operators A, B in $\mathcal{M}$, the operators

$$\lambda A + \mu B, \quad AB, \quad \int f(\mathbf{x}, t) A(\mathbf{x}, t) d\mathbf{x} dt$$

(here, λ, μ are complex numbers, $\mathbf{x} \in E^3$, $-\infty < t < \infty$, and $f(\mathbf{x}, t)$ is a function in the space $\mathcal{S}$ of smooth, fast-decaying functions) are in $\mathcal{M}$ as well. It is easy to show that every asymptotically Abelian family is contained in an asymptotically Abelian algebra. (The proof is based on the fact that adding any of the listed operators along with their adjoints to an asymptotically Abelian family preserves asymptotic commutativity.) Therefore, without loss of generality, we will consider asymptotically Abelian algebras instead of asymptotically Abelian families.

Let us fix an asymptotically Abelian algebra $\mathcal{A}$ of operators on the space $\mathcal{H}$; let us assume that the physical vacuum Φ is a cyclic vector of the algebra $\mathcal{A}$.

With the definitions and assumptions above, we will now show how we can construct a scattering matrix (S-matrix) corresponding to an asymptotically Abelian algebra $\mathcal{A}$ (we can then define a scattering matrix of an asymptotically Abelian family as the scattering matrix corresponding to the asymptotically Abelian algebra that contains this family). For the sake of simplicity of definition, we will assume that the full particle system consists of a single particle $\Phi(\mathbf{k})$.

Let us first make several preliminary definitions.

An operator B is called *smooth* if it can be written in the form $B = \int f(\mathbf{x}, t) A(\mathbf{x}, t) d\mathbf{x} dt$, where $A \in \mathcal{A}$ and the function f belongs to the space $\mathcal{S}$. (This name comes from the observation that the operator $B(\mathbf{x}, t)$, where B is a smooth operator, is infinitely differentiable in $\mathbf{x}$

and t in the sense of differentiation in norm.) The smooth operator B is called *good*, if (1) $B^*\Phi = 0$; (2) there exists a function ϕ such that $B\Phi = \int \phi(\mathbf{k})\Phi(\mathbf{k})d\mathbf{k}$.

Definition 10.1. An isometric operator S_- (S_+), transforming the space $\mathcal{H}_{\mathrm{as}}$ into the space $\mathcal{H}$, is called a Møller matrix if for every collection of good operators $B_1, \ldots, B_n$ and for every collection of smooth functions with compact support $f_1(\mathbf{p}), \ldots, f_n(\mathbf{p})$

$$\lim_{t \to \genfrac{}{}{0pt}{}{-\infty}{(+\infty)}} B_1(f_1, t) \cdots B_n(f_n, t)\Phi = S_{\genfrac{}{}{0pt}{}{-}{(+)}} a^+(\overline{f}_1\overline{\phi}_1) \cdots a^+(\overline{f}_n\overline{\phi}_n)\theta.$$

$$(10.1)$$

[Here, θ is a vacuum vector in the Fock space $\mathcal{H}_{\mathrm{as}}, \phi_j(\mathbf{k}) = \langle B_j\Phi, \Phi(\mathbf{k})\rangle$, the functions $\tilde{f}_j(\mathbf{x}|t)$ are determined by the relation

$$\tilde{f}_j(\mathbf{x}|t) = \int \exp(-i\omega(\mathbf{p})t + i\mathbf{p}\mathbf{x})f_j(\mathbf{p})\frac{d\mathbf{p}}{(2\pi)^3}, \qquad (10.2)$$

and the operators $B_j(f_j, t)$ are given by the formula

$$B_j(f_j, t) = \int \tilde{f}_j(\mathbf{x}|t)B_j(\mathbf{x}, t)dx.\Bigg] \qquad (10.3)$$

The vector $B_1(f_1, t) \cdots B_n(f_n, t)\Phi$, constructed with good operators $B_1, \ldots, B_n$ and the smooth functions with compact support $f_1, \ldots, f_n$, will be denoted by $\Psi(B_1, \ldots, B_n | f_1, \ldots, f_n | t)$. The limit of this vector as $t \to \pm\infty$ will be denoted by $\Psi_\pm(B_1, \ldots, B_n | f_1, \ldots, f_n)$.

Definition 10.2. The operator $S = S_+^* S_-$, acting on the space $\mathcal{H}_{\mathrm{as}}$, is called the scattering matrix, corresponding to the asymptotically Abelian algebra $\mathcal{A}$.

Let us prove the following statement.

With the previous assumptions, *the operators S_- and S_+ satisfying the conditions of Definition 10.1, exist, and are uniquely specified* *by these conditions. The following relations also hold:*

$$S_{\mp}\theta = \Phi, \quad S_{\pm}a^+(\mathbf{k})\theta = \Phi(\mathbf{k}), \tag{10.4}$$

$$HS_{\mp} = S_{\mp}H_{\text{as}}, \quad \mathbf{P}S_{\mp} = S_{\mp}\mathbf{P}_{\text{as}}. \tag{10.5}$$

It follows from relation (10.5) that *the scattering matrix $S = S_+^* S_-$ commutes with the Hamiltonian and the momentum operator*

$$SH_{\text{as}} = H_{\text{as}}S, \quad S\mathbf{P}_{\text{as}} = \mathbf{P}_{\text{as}}S,$$

and the relation (10.4) implies the *stability of the vacuum and single-particle states:*

$$S\theta = \theta,$$

$$Sa^+(\mathbf{k})\theta = a^+(\mathbf{k})\theta.$$

If the images of the operators S_- and S_+ coincide ($S_-\mathcal{H}_{\text{as}} = S_+\mathcal{H}_{\text{as}}$), then the scattering matrix is unitary and can be written in the form

$$S = S_+^{-1}S_-.$$

If, in addition to the algebra $\mathcal{A}$, we have a second asymptotically Abelian algebra $\mathcal{A}'$, for which the vector Φ is cyclic and the algebras $\mathcal{A}$ and $\mathcal{A}'$ asymptotically commute,[1] then the Møller matrices $S_{\mp}$ and the scattering matrix S, constructed with the algebra $\mathcal{A}$, coincide with the Møller matrix $S'_{\mp}$ and the scattering matrix S', constructed with the algebra $\mathcal{A}'$.

The proof of the listed statements is based on a sequence of lemmas. Let us now formulate the necessary lemmas and then prove the statements. The proofs of the lemmas are relegated to Section 10.2.

Lemma 10.1. *The set of vectors of the form $B\Phi$, where B runs over all good operators in the algebra $\mathcal{A}$, is dense in the single-particle space $\mathcal{H}_1$.*

---

[1] The algebras $\mathcal{A}$ and $\mathcal{A}'$ asymptotically commute if every operator in $\mathcal{A}$ asymptotically commutes with every operator in $\mathcal{A}'$.

Lemma 10.2. *Let $\omega(\mathbf{p})$ be a smooth, strongly convex function, $f(\mathbf{p})$ be a smooth function with compact support, and*

$$\tilde{f}(\mathbf{x}|t) = \int \exp(-i\omega(\mathbf{p})t + i\mathbf{p}\mathbf{x})f(\mathbf{p})\frac{d\mathbf{p}}{(2\pi)^3}, \qquad (10.6)$$

then

$$\sup_{\mathbf{x}} |\tilde{f}(\mathbf{x}|t)| \leq C_1|t|^{-3/2}, \qquad (10.7)$$

$$\int |\tilde{f}(\mathbf{x}|t)|d\mathbf{x} \leq C_2(|t|^{3/2} + 1). \qquad (10.8)$$

If U is a set of vectors of the form $\mathbf{v}(\mathbf{p}) = \nabla\omega(\mathbf{p})$, where $\mathbf{p} \in \operatorname{supp} f$ and U_ϵ is an ϵ-neighborhood of the set U, then for any n and ϵ, we can find a constant D such that

$$|\tilde{f}(\mathbf{x}|t)| \leq D(1 + x^2 + t^2)^{-n} \qquad (10.9)$$

whenever $\frac{\mathbf{x}}{t} \notin U_\epsilon$.

[The symbol $\operatorname{supp} f$, as usual, denotes the closure of the set of points where $f(\mathbf{p}) \neq 0$.]

Before we formulate Lemma 10.3, we need to introduce the concept of a truncated vacuum expectation value.

Let us define *the vacuum expectation value $\langle A \rangle$ of the operator A* as the number $\langle A\Phi, \Phi \rangle$. The truncated vacuum expectation value $\langle A_1 \ldots A_n \rangle^T$ of the product of n operators $A_1, \ldots, A_n$ is defined by the following formulas for $n = 1, 2, 3$:

$$\langle A_1 \rangle^T = \langle A_1 \rangle,$$

$$\langle A_1 A_2 \rangle^T = \langle A_1 A_2 \rangle - \langle A_1 \rangle^T \langle A_2 \rangle^T = \langle A_1 A_2 \rangle - \langle A_1 \rangle \langle A_2 \rangle,$$

$$\langle A_1 A_2 A_3 \rangle^T = \langle A_1 A_2 A_3 \rangle - \langle A_1 A_2 \rangle^T \langle A_3 \rangle^T - \langle A_1 \rangle^T \langle A_2 A_3 \rangle^T$$
$$- \langle A_2 \rangle^T \langle A_1 A_3 \rangle^T - \langle A_1 \rangle^T \langle A_2 \rangle^T \langle A_3 \rangle^T.$$

For arbitrary n, *the truncated vacuum expectation value is defined by the recurrence relation*

$$\langle A_1 \ldots A_n \rangle = \sum_{k=1}^{n} \sum_{\rho \in R_k} \langle A(\pi_1) \rangle^T \cdots \langle A(\pi_k) \rangle^T, \qquad (10.10)$$

where R_k is the collection of all partitions of the set $\{1, \ldots, n\}$ into k subsets; $\pi_1, \ldots, \pi_k$ are the subsets that constitute the partition $\rho \in R_k$; $\langle A(\pi) \rangle^T$ is a truncated vacuum expectation value of the product of the operators A_i with indices in the set π (the operators are ordered in ascending order by the index i).

Let us suppose that for every λ in the (finite or infinite) set Λ we have a corresponding operator A_λ from the algebra $\mathcal{A}$.

The expectation value

$$w_n(\lambda_1, \mathbf{x}_1, t_1, \ldots, \lambda_n, \mathbf{x}_n, t_n) = \langle A_{\lambda_1}(\mathbf{x}_1, t_1) \cdots A_{\lambda_n}(\mathbf{x}_n, t_n) \rangle$$

of the product $A_{\lambda_1}(\mathbf{x}_1, t_1) \cdots A_{\lambda_n}(\mathbf{x}_n, t_n)$ in the ground state Φ will be called the (n-point) Wightman function (see Section 7.1). Let us define the *truncated Wightman function* $w_n^T(\lambda_1, \mathbf{x}_1, t_1, \ldots, \lambda_n, \mathbf{x}_n, t_n)$ as the truncated vacuum expectation value of this product:

$$w_n^T(\lambda_1, \mathbf{x}_1, t_1, \ldots, \lambda_n, \mathbf{x}_n, t_n) = \langle A_{\lambda_1}(\mathbf{x}_1, t_1) \cdots A_{\lambda_n}(\mathbf{x}_n, t_n) \rangle^T .$$

In other words,

$$w_n(\lambda_1, \mathbf{x}_1, t_1, \ldots, \lambda_n, \mathbf{x}_n, t_n) = \sum_{k=1}^{n} \sum_{\rho \in R_k} w_{\alpha_1}^T(\pi_1) \cdots w_{\alpha_k}^T(\pi_k).$$

$$(10.11)$$

Here, α_r denotes the number of elements in the subset π_r; $w_{\alpha_r}^T(\pi_r)$ is a truncated Wightman function with the arguments $\lambda_{i_1}, \mathbf{x}_{i_1}, t_{i_1}, \ldots, \lambda_{i_{\alpha_r}}, \mathbf{x}_{i_{\alpha_r}}, t_{i_{\alpha_r}}$, where $i_1, \ldots, i_{\alpha_r} \in \pi_r$ (the order of the arguments is in order of increasing indices i). For $n = 1, 2, 3$, the relations connecting Wightman functions and truncated Wightman functions have the form

$$w_1(\lambda_1, \mathbf{x}_1, t_1) = w_1^T(\lambda_1, \mathbf{x}_1, t_1),$$

$$w_2(\lambda_1, \mathbf{x}_1, t_1, \lambda_2, \mathbf{x}_2, t_2) = w_2^T(\lambda_1, \mathbf{x}_1, t_1, \lambda_2, \mathbf{x}_2, t_2)$$

$$+ w_1^T(\lambda_1, \mathbf{x}_1, t_1) w_1^T(\lambda_2, \mathbf{x}_2, t_2),$$

$$w_3(\lambda_1, \mathbf{x}_1, t_1, \lambda_2, \mathbf{x}_2, t_2, \lambda_3, \mathbf{x}_3, t_3)$$

$$= w_3^T(\lambda_1, \mathbf{x}_1, t_1, \lambda_2, \mathbf{x}_2, t_2, \lambda_3, \mathbf{x}_3, t_3)$$

$$+ w_1^T(\lambda_1, \mathbf{x}_1, t_1) w_2^T(\lambda_2, \mathbf{x}_2, t_2, \lambda_3, \mathbf{x}_3, t_3)$$

$$+ w_1^T(\lambda_2, \mathbf{x}_2, t_2) w_2^T(\lambda_1, \mathbf{x}_1, t_1, \lambda_3, \mathbf{x}_3, t_3)$$

$$+ w_1^T(\lambda_3, \mathbf{x}_3, t_3) w_2^T(\lambda_1, \mathbf{x}_1, t_1, \lambda_2, \mathbf{x}_2, t_2)$$

$$+ w_1^T(\lambda_1, \mathbf{x}_1, t_1) w_1^T(\lambda_2, \mathbf{x}_2, t_2) w_1^T(\lambda_3, \mathbf{x}_3, t_3).$$

Lemma 10.3. *If all the operators A_{λ_i} are smooth, then the truncated Wightman function*

$$w_n^T(\lambda_1, \mathbf{x}_1, t_1, \ldots, \lambda_n, \mathbf{x}_n, t_n) = \langle A_{\lambda_1}(\mathbf{x}_1, t_1) \ldots A_{\lambda_n}(\mathbf{x}_n, t_n) \rangle^T,$$

for fixed $t_1, \ldots, t_n$, tends to zero faster than any power function of $D = \max_{1 \le i, j \le n} |\mathbf{x}_i - \mathbf{x}_j|$ as $D \to \infty$.

Lemma 10.4. *Let $f_0, \ldots, f_n$ be finite smooth functions and let $A_0, \ldots, A_n$ be smooth operators belonging to the algebra $\mathcal{A}$. Then for $n \ge 1$, we have the inequality*

$$|\langle A_0(f_0, t) \ldots A_n(f_n, t) \rangle^T| \le C(1 + |t|)^{-\frac{3}{2}(n-1)}, \qquad (10.12)$$

where $A_j(f_j, t) = \int \tilde{f}_j(\mathbf{x}|t) A_j(\mathbf{x}|t) d\mathbf{x}$. This inequality remains valid if in the expression for $\langle A_0(f_0, t) \ldots A_n(f_n, t) \rangle^T$ some of the operators $A_j(f_j, t)$ are replaced by their time derivatives $\dot{A}_j(f_j, t)$ and the operators $(A_j(f_j, t))^*$ are replaced by $(\dot{A}_j(f_j, t))^*$.

Let us now turn to the derivation of the statements made in the beginning of this chapter from the lemmas we have formulated. First, let us derive the relations (10.4). The first of these is self-evident; to prove the second relation, let us note that

$$\Psi(B|f|t) = B(f, t)\Phi$$

$$= \int \tilde{f}(\mathbf{x}|t) B(\mathbf{x}, t) \Phi d\mathbf{x}$$

$$= \int \tilde{f}(\mathbf{x}|t) \exp(itH - i\mathbf{Px}) B\Phi d\mathbf{x}$$

$$= \int \tilde{f}(\mathbf{x}|t) \exp(it\omega(\mathbf{p}) - i\mathbf{p}\mathbf{x})\phi(\mathbf{p})\Phi(\mathbf{p})d\mathbf{x}d\mathbf{p}$$

$$= \int f(\mathbf{p})\phi(\mathbf{p})\Phi(\mathbf{p})d\mathbf{p}, \tag{10.13}$$

i.e. $\Psi(B|f|t)$ does not depend on t:

$$\frac{d}{dt}\Psi(B|f|t) = \frac{d}{dt}B(f,t)\Phi = 0. \tag{10.14}$$

This implies that

$$\Psi_{\pm}(B,f) = \int f(\mathbf{p})\phi(\mathbf{p})\Phi(\mathbf{p})d\mathbf{p},$$

i.e.

$$S_{\mp}\left(\int f(\mathbf{p})\phi(\mathbf{p})d\mathbf{p}\theta\right) = \int f(\mathbf{p})\phi(\mathbf{p})\Phi(\mathbf{p})d\mathbf{p}. \tag{10.15}$$

It follows from Lemma 10.1 that vectors of the form

$$B\Phi = \int \phi_B(\mathbf{p})\Phi(\mathbf{p})d\mathbf{p},$$

where B runs over a set of good operators, constitute a dense set in the space of single-particle states. This implies that the functions $\phi_B(\mathbf{p})$, corresponding to good operators B, constitute a dense set in $L^2(E^3)$, since the correspondence between the function $\phi(\mathbf{p})$ and the vector $\int \phi(\mathbf{p})\Phi(\mathbf{p})d\mathbf{p}$ is an isometry between the space $L^2(E^3)$ and the space of single-particle states. It is clear that the functions $f(\mathbf{p})\phi_B(\mathbf{p})$, where $f(\mathbf{p})$ is a smooth function, also form a dense set in $L^2(E^3)$. The necessary relation then follows from (10.15).

Let us now show that the limit in the relation (10.1) exists. To show this, we will prove the inequality

$$\left\langle \frac{d\Psi}{dt}, \frac{d\Psi}{dt} \right\rangle \leq C|t|^{-3}, \tag{10.16}$$

where $\Psi(t) = \Psi(B_1, \ldots, B_n | f_1, \ldots, f_n | t)$. This inequality implies the convergence of the limit in (10.1) since

$$\|\Psi(t_2) - \Psi(t_1)\| \leq \int_{t_1}^{t_2} \left\| \frac{d\Psi}{dt} \right\| dt \leq \int_{t_1}^{t_2} C|t|^{-3/2} dt \to 0,$$

for $t_1 \leq t_2$ and $t_1 \to +\infty$ or $t_2 \to -\infty$.

In order to prove the inequality, let us write the expansion

$$\left\langle \frac{d\Psi}{dt}, \frac{d\Psi}{dt} \right\rangle = \sum_{i,j} \langle B_1(f_1, t) \ldots B_i(f_i, t)$$

$$\ldots B_n(f_n, t)\Phi, B_1(f_1, t) \ldots \dot{B}_j(f_j, t) \ldots B_n(f_n, t)\Phi \rangle$$

$$= \sum_{i,j} \langle (B_n(f_n, t))^* \ldots (\dot{B}_j(f_j, t))^* \ldots (B_1(f_1, t))^*$$

$$\times B_1(f_1, t) \ldots \dot{B}_i(f_i, t) \ldots B_n(f_n, t)\Phi, \Phi \rangle$$

as a sum of the products of truncated vacuum expectation values using the relation (10.10).

Every factor in an arbitrary term of this expansion has the form

$$I_{k,l}(t) = \langle (\dot{B}_{i_1}(f_{i_1}, t))^* \ldots (\dot{B}_{i_k}(f_{i_k}, t))^*$$

$$\times \dot{B}_{j_1}(f_{j_1}, t) \ldots \dot{B}_{j_l}(f_{j_l}, t) \rangle^T,$$

where the dots above the operators denote differentiation with respect to time, which can enter in one of the first k operators and in one of the last l operators. We can suppose without loss of generality that $k \geq 1$, $l \geq 1$ (otherwise, $I_{k,l} = 0$ by the relation $B^*\Phi = 0$). If each factor in the given term has $k = l = 1$, then the term is equal to zero because one of the factors either has the form

$$\langle (\dot{B}_i(f_i, t))^* \dot{B}_j(f_j, t) \rangle^T = \langle (\dot{B}_i(f_i, t))^* \dot{B}_j(f_j, t)\Phi, \Phi \rangle$$

or the form

$$\langle (\dot{B}_i(f_i, t))^* \dot{B}_j(f_j, t) \rangle^T = \langle (\dot{B}_i(f_i, t))^* \dot{B}_j(f_j, t)\Phi, \Phi \rangle$$

and both expressions are zero by the relation (10.14). Hence, all non-zero terms in the sum above contain either a factor $I_{k,l}$ with

$k + l \geq 4$ or at least two factors $I_{k,l}$ with $k + l = 3$ (a term where all factors except one have $k + l = 2$ and the remaining factor has $k + l = 3$ is not possible since the total number of operators is even).

The inequality

$$|I_{k,l}(t)| < C|t|^{-\frac{3}{2}(k+l-2)},$$

follows from Lemma 10.4. In particular, we have

$$|I_{1,1}(t)| < C,$$

$$|I_{k,l}(t)| < Ct^{-3/2},$$

for $k + l = 3$, and

$$|I_{k,l}(t)| < Ct^{-3}$$

for $k + l \geq 4$. From these inequalities and the statements shown above, we obtain the inequality (10.16) for $\left\langle \frac{d\Psi}{dt}, \frac{d\Psi}{dt} \right\rangle$ and therefore the existence of the limit in (10.1) follows.

Since the limit in (10.1) exists, the relation (10.1) specifies the operators $S_{\mp}$ on some subspace L of the space $\mathcal{H}_{\mathrm{as}}$. However, it is not clear if the action of the operators $S_{\mp}$ is uniquely defined, i.e. whether the vector $S_{\mp}\xi = \Psi_{\mp}(B_1, \ldots, B_n | f_1, \ldots, f_n)$ remains the same if we write ξ in two different forms

$$\xi = a^+(\overline{\phi}_1 \overline{f}_1) \ldots a^+(\overline{\phi}_n \overline{f}_n)\theta,$$

where the functions f_i are smooth and finite and the functions ϕ_i correspond to good operators.

Therefore, let us now prove that the relation (10.1) defines $S_{\mp}$ uniquely and, furthermore, this operator can be uniquely extended to a linear isometric operator defined on the whole space $\mathcal{H}_{\mathrm{as}}$. To show this, we will use the following general statement.

Let us suppose that on the total[2] subset L of the Hilbert space $\mathcal{H}$, we have a multi-valued isometric mapping α taking values in $\mathcal{H}'$ (i.e. to every point $\xi \in L$ corresponds a set of points $\alpha(\xi) \subset \mathcal{H}'$ in such a way that for any selection of points $\xi_1, \xi_2 \in L$, $x_1, x_2 \in \mathcal{H}'$ satisfying $x_1 \in \alpha(\xi_1)$, $x_2 \in \alpha(\xi_2)$ we have $\langle x_1, x_2 \rangle = \langle \xi_1, \xi_2 \rangle$). Then

---

[2]A subset is called total if the set of its linear combinations is dense in $\mathcal{H}$.

the mapping α is in fact single-valued (i.e. every set $\alpha(\xi)$ consists of a single point) and can be uniquely extended to a linear isometric operator defined on the whole space $\mathcal{H}$.

To prove this statement, let us first extend the mapping α by linearity to the multi-valued mapping $\tilde{\alpha}$, defined on the set $\tilde{L}$ of linear combinations of points in the set L. More precisely, we will assume that $x \in \tilde{\alpha}(\xi)$ if we can write ξ and x in the forms $\xi = \sum_{i=1}^{n} \lambda_i \xi_i$, $x = \sum_{i=1}^{n} \lambda_i x_i$ with $x_i \in \alpha(\xi_i)$. It is easy to see that the mapping $\tilde{\alpha}$ inherits linearity: if $\xi = \lambda_1 \xi_1 + \lambda_2 \xi_2$, $x_1 \in \tilde{\alpha}(\xi_1)$, $x_2 \in \tilde{\alpha}(\xi_2)$, then $\lambda_1 x_1 + \lambda_2 x_2 \in \tilde{\alpha}(\xi)$.

The mapping $\tilde{\alpha}$ is also a multi-valued isometric mapping (if $x^{(1)} \in \tilde{\alpha}(\xi^{(1)})$, $x^{(2)} \in \tilde{\alpha}(\xi^{(2)})$ then $\xi^{(1)} = \sum \lambda_i \xi_i^{(1)}$, $x^{(1)} = \sum \lambda_i x_i^{(1)}$, $\xi^{(2)} = \sum \mu_j \xi_j^{(2)}$, $x^{(2)} = \sum \mu_j x_j^{(2)}$, where $x_i^{(1)} = \alpha(\xi_i^{(1)})$, $x_j^{(2)} = \alpha(\xi_j^{(2)})$, and therefore

$$\left\langle x^{(1)}, x^{(2)} \right\rangle = \sum \lambda_i \bar{\mu}_j \left\langle x_i^{(1)}, x_j^{(2)} \right\rangle$$

$$= \sum \lambda_i \bar{\mu}_j \left\langle \xi_i^{(1)}, \xi_j^{(2)} \right\rangle = \left\langle \xi^{(1)}, \xi^{(2)} \right\rangle).$$

Let us now prove that the mapping $\tilde{\alpha}$ is in fact single-valued. Let us suppose that $x_1 \in \tilde{\alpha}(\xi)$, $x_2 \in \tilde{\alpha}(\xi)$. It follows from the isometry of the mapping that

$$\langle x_1, x_1 \rangle = \langle x_2, x_2 \rangle = \langle x_1, x_2 \rangle = \langle \xi, \xi \rangle .$$

From this equation, we obtain that $x_1 = x_2$.

Therefore, $\tilde{\alpha}$ is a well-defined isometric linear mapping defined on the set $\tilde{L}$. Since the set L is total and the set $\tilde{L}$ is dense in $\mathcal{H}$, we can continuously extend $\tilde{\alpha}$ to an isometric linear operator defined on all of $\mathcal{H}$.

To apply the statement above to the proof of the Møller matrix properties, we should check that the set L of vectors of the form $a^+(\overline{\phi}_1 \overline{f}_1) \ldots a^+(\overline{\phi}_n \overline{f}_n)\theta$, on which the operators $S_{\mp}$ are defined by (10.1), is total in the space $\mathcal{H}_{\mathrm{as}}$, and furthermore, we should check that for good operators $B_1, \ldots, B_n, B'_1, \ldots, B'_m$ and for smooth functions with compact support $f_1, \ldots, f_n, f'_1, \ldots, f'_m$ we have

$$\left\langle \Psi_{\mp}(B_1, \ldots, B_n | f_1, \ldots, f_n), \Psi_{\mp}(B'_1, \ldots, B'_m | f'_1, \ldots, f'_m) \right\rangle$$

$$= \left\langle a^+(\overline{\phi}_1 \overline{f}_1) \cdots a^+(\overline{\phi}_n \overline{f}_n)\theta, a^+(\overline{\phi}'_1 \overline{f}'_1) \cdots a^+(\overline{\phi}'_n \overline{f}'_n)\theta \right\rangle . \qquad (10.17)$$

The totality of the set L follows from the fact that functions of the form $\phi_B(\mathbf{k})f(\mathbf{k})$, where $f(\mathbf{k})$ is a smooth function with compact support and B is a good operator, are dense in $L^2(E^3)$ (as we have noted above, this fact follows from Lemma 10.1).

To prove relation (10.17), let us represent the left side of this relation in the form

$$\Big\langle \Psi_{\mp}(B_1,\ldots,B_n|f_1,\ldots,f_n),\, \Psi_{\mp}(B_1',\ldots,B_m'|f_1',\ldots,f_m')\Big\rangle$$

$$= \lim_{t\to\mp\infty} \Big\langle \Psi(B_1,\ldots,B_n|f_1,\ldots,f_n|t),\, \Psi(B_1',\ldots,B_m'|f_1',\ldots,f_m'|t)\Big\rangle$$

$$= \lim_{t\to\mp\infty} \Big\langle (B_m'(f_m',t))^*\cdots(B_1'(f_1',t))^* B_1(f_1,t)\cdots B_n(f_n,t)\Big\rangle$$

$$\tag{10.18}$$

and then expand the expression under the limit sign in terms of truncated Wightman functions. Every term of the resulting expansion consists of a product of factors of the form

$$R_{k,l}(t) = \Big\langle (B_{i_1}'(f_{i_1},t))^*\cdots(B_{i_k}'(f_{i_k},t))^* B_{j_1}(f_{j_1},t)\cdots B_{j_l}(f_{j_l},t)\Big\rangle^T .$$

It follows from Lemma 10.4 that for $k+l\geq 3$ the factors $R_{k,l}(t)$ converge to zero as $t\to\pm\infty$. On the other hand, the factors $R_{k,l}$ are equal to zero if $k=0$ or $l=0$, by the relation $B\Phi=0$. Therefore, the terms that differ from zero in the limit $t\to\mp\infty$ are the ones where all the factors have the form $R_{1,1}(t)$. It is easy to check that

$$R_{1,1}(t) = \Big\langle (B_i'(f_i',t))^* B_j(f_j,t)\Big\rangle^T = \Big\langle (B_i'(f_i',t)^* B_j(f_j,t))\Big\rangle$$

$$= \Big\langle B_j(f_j,t)\Phi,\, B_i'(f_i',t)\Phi\Big\rangle$$

$$= \left\langle \int f_j(\mathbf{k})\phi_j(\mathbf{k})\Phi(\mathbf{k})d\mathbf{k},\, \int f_i'(\mathbf{k})\phi_i'(\mathbf{k})\Phi(\mathbf{k})d\mathbf{k}\right\rangle$$

$$= \langle f_j\phi_j,\, f_i'\phi_i'\rangle = \Big\langle a^+(\overline{f}_j\overline{\phi}_j)\theta,\, a^+(\overline{f}_i'\overline{\phi}_i')\theta\Big\rangle$$

and that $R_{1,1}$ in fact does not depend on t [in the calculations we have used the formula (10.13)].

The statements proven above imply that the expression (10.18) is equal to $\delta_m^n \sum \prod_{s=1}^n \langle \phi_s f_s,\, \phi_{i_s}' f_{i_s}'\rangle$ [the sum is taken over all permutations $(i_1,\ldots,i_n)$]. This justifies equation (10.17).

Thus, we have proven that relation (10.1) specifies $S_\mp$ as an isometric mapping on a total set and we have shown that one can uniquely extend this mapping to an isometric operator defined on the space $\mathcal{H}_{\text{as}}$.

To prove relation (10.5), note that the equation

$$\exp(-iH\tau)B(f,t)\exp(iH\tau)$$
$$= \int \exp(-iH\tau)\tilde{f}(x|t)B(\mathbf{x},t)\exp(iH\tau)dx$$
$$= \int \tilde{f}(\mathbf{x}|t)B(\mathbf{x},t-\tau)dx$$
$$= \int \left(\int \exp(-i\omega(\mathbf{p})t+i\mathbf{px})f(\mathbf{p})\frac{d\mathbf{p}}{(2\pi)^3} \right) B(\mathbf{x},t-\tau)dx$$
$$= \int \left(\int \exp(-i\omega(\mathbf{p})(t-\tau)+i\mathbf{px})f(\mathbf{p})\exp(-i\omega(\mathbf{p})\tau)\frac{d\mathbf{p}}{(2\pi)^3} \right)$$
$$\times B(\mathbf{x},t-\tau)dx,$$

holds; this implies

$$\exp(-iH\tau)B(f,t)\exp(iH\tau) = B(f^\tau,t-\tau), \tag{10.19}$$

$[f^\tau$ denotes the function $f^\tau(\mathbf{p}) = f(\mathbf{p})\exp(-i\omega(\mathbf{p})\tau)]$. It follows from equation (10.19) that

$$\exp(-iH\tau)\Psi(B_1,\ldots,B_n|f_1,\ldots,f_n|t)$$
$$= \Psi(B_1,\ldots,B_n|f_1^\tau,\ldots,f_n^\tau|t-\tau). \tag{10.20}$$

Taking the limit $t \to \mp\infty$ in (10.20) and using the fact that

$$\exp(-iH_{\text{as}}\tau)a^+(\overline{\phi}_1\overline{f}_1)\ldots a^+(\overline{\phi}_n\overline{f}_n)\theta$$
$$= a^+(\overline{\phi}_1\overline{f}_1^\tau)\ldots a^+(\overline{\phi}_n\overline{f}_n^\tau)\theta,$$

we obtain

$$\exp(-iH\tau)S_\mp\xi = S_\mp \exp(-iH_{\text{as}}\tau)\xi$$

for any vector $\xi \in L$. By the totality of the set L, we have

$$\exp(-iH\tau)S_{\mp} = S_{\mp}\exp(-iH_{\mathrm{as}}\tau)$$

and, therefore, the first relation in (10.5) follows (the second can be proven by the same method, but is even easier).

To prove the final statement, let us note that Møller matrices do not change when we replace an asymptotically Abelian algebra by a larger one. Indeed, if the algebra $\mathcal{A}$, satisfying the necessary conditions, is contained in an asymptotically Abelian algebra $\tilde{\mathcal{A}}$, then in the construction of the operators $S_{\mp}$ corresponding to the algebra $\tilde{\mathcal{A}}$, we can use the good operators that belong to the algebra $\mathcal{A}$; hence, it is clear that Møller matrices corresponding to the algebras $\mathcal{A}$ and $\tilde{\mathcal{A}}$ are identical. If the algebras $\mathcal{A}$ and $\mathcal{A}'$ are asymptotically commuting, then the family $\mathcal{A} \cup \mathcal{A}'$ is asymptotically Abelian and can be included in an asymptotically Abelian algebra $\tilde{\mathcal{A}}$. Since $\mathcal{A} \subset \tilde{\mathcal{A}}$ and $\mathcal{A}' \subset \tilde{\mathcal{A}}$, the Møller matrices $S_{\mp}$ and $S'_{\mp}$, constructed with the algebras $\mathcal{A}$ and $\mathcal{A}'$, coincide with the Møller matrices constructed with the algebra $\tilde{\mathcal{A}}$, and therefore coincide with each other. The coincidence of Møller matrices clearly entails the coincidence of the scattering matrices constructed with the algebras $\mathcal{A}$ and $\mathcal{A}'$.

In conclusion, let us note that the Møller matrices and the scattering matrix do not depend on the choice of complete particle system (recall that there exists an isomorphism between any two choices of complete particle system for the space $\mathcal{H}_{\mathrm{as}}$; therefore, it makes sense to discuss the coincidence of scattering matrices constructed under different particle systems). To prove this, note that for any particle system with a single particle $\Phi(\mathbf{k})$, any other particle can be written in the form

$$\Phi'(\mathbf{k}) = \exp(i\alpha(\mathbf{k}))\Phi(\mathbf{k}),$$

where $\alpha(\mathbf{k})$ is a real-valued function.

The operators $a^+(\mathbf{k}), a(\mathbf{k})$ in the space $\mathcal{H}_{\mathrm{as}}$, corresponding to the particle $\Phi(\mathbf{k})$, are related to the operators $a'^+(\mathbf{k}), a'(\mathbf{k})$ on $\mathcal{H}_{\mathrm{as}}$, corresponding to $\Phi'(\mathbf{k})$, by the equations

$$a'(\mathbf{k}) = \exp(-i\alpha(\mathbf{k}))a(\mathbf{k}),$$
$$a'^+(\mathbf{k}) = \exp(i\alpha(\mathbf{k}))a^+(\mathbf{k}).$$

It is easy to check, using these relations, that for any operator B_j

$$a'^+(\overline{f}_j\overline{\phi}'_j) = a^+(\overline{f}_j\overline{\phi}_j), \qquad (10.21)$$

where

$$\phi'_j(\mathbf{k}) = \langle B_j\Phi, \Phi'(\mathbf{k})\rangle, \qquad \phi_j(\mathbf{k}) = \langle B_j\Phi, \Phi(\mathbf{k})\rangle.$$

Applying (10.21) and noting that the left part of (10.1) does not depend on the choice of the function $\Phi(\mathbf{k})$ representing the particle, we obtain that the Møller matrices $S_\pm$ do not depend on the choice of the function $\Phi(\mathbf{k})$.

10.2 Proof of lemmas

Proof of Lemma 10.1. Let $\lambda(\omega, \mathbf{p})$ be a function of four parameters with support in the set Δ. We will consider the operator $\lambda(H, \mathbf{P})$ as a function of the commuting self-adjoint operators $H, \mathbf{P}$. It is easy to check that if the set Δ does not intersect the spectrum of the operators $(H, \mathbf{P})$, then the operator $\lambda(H, \mathbf{P}) = 0$, and if the set Δ does not contain the origin and points belonging to the multi-particle spectrum $\sum_\mathcal{M}$, then the set of values of the operator $\lambda(H, \mathbf{P})$ belongs to the single-particle subspace.

Using this remark, we can construct a good operator by the following method.

Let us consider a function $\alpha(t, \mathbf{x}) \in \mathcal{S}(E^4)$ with the property that its Fourier transform

$$\hat{\alpha}(\omega, \mathbf{p}) = \int \alpha(t, \mathbf{x}) \exp(i(\omega t - \mathbf{p}\mathbf{x}))d\mathbf{x}dt$$

has support that does not intersect the multi-particle spectrum and the half-space $\omega \leq 0$. Then the *operator*

$$B = \int \alpha(t, \mathbf{x})A(\mathbf{x}, t)d\mathbf{x}dt,$$

where $A \in \mathcal{A}$, is good.

Indeed, the operator B is smooth. The vector

$$B\Phi = \int \alpha(t,\mathbf{x})A(\mathbf{x},t)\Phi dxdt$$

$$= \int \alpha(t,\mathbf{x})\exp(iHt - i\mathbf{P}\mathbf{x})A\exp(-iHt + i\mathbf{P}\mathbf{x})\Phi dxdt$$

$$= \int \alpha(t,\mathbf{x})\exp(i(Ht - \mathbf{P}\mathbf{x}))A\Phi dxdt = \hat{\alpha}(H,\mathbf{P})A\Phi$$

belongs to the single-particle subspace, by the remark above. The vector

$$B^*\Phi = \int \overline{\alpha(t,x)}\exp(i(Ht - \mathbf{P}\mathbf{x}))A^*\Phi dxdt$$

$$= \overline{\hat{\alpha}}(-H,-\mathbf{P})A^*\Phi = 0,$$

since the support of the function $\overline{\hat{\alpha}}(-\omega,-\mathbf{p})$ does not intersect the spectrum of the operators H, $\mathbf{P}$.

Let us now show that vectors $B\Phi$, where B is a good operator, constructed by the method above, are dense in the single-particle subspace.

To prove this, let us consider the vector $\Phi(\lambda) = \int \lambda(\mathbf{k})\Phi(\mathbf{k})d\mathbf{k}$, where $\lambda(\mathbf{k})$ is a smooth function with compact support, in $\mathcal{H}_1$. We will construct a sequence of good operators B_n such that $B_n\Phi \to \Phi(\lambda)$ (this is enough to prove the statement above since smooth functions with compact support $\lambda(\mathbf{k})$ are dense in $L^2(E^3)$ and therefore the corresponding vectors $\Phi(\lambda) = \int \lambda(\mathbf{k})\Phi(\mathbf{k})d\mathbf{k}$ are dense in $\mathcal{H}_1$).

By the cyclicity of the vacuum vector there exists a sequence of operators $A_n \in \mathcal{A}$ so that $A_n\Phi \to \Phi(\lambda)$. The necessary sequence of good operators can now be constructed by setting $B_n = \int \alpha(t,\mathbf{x}) A_n(\mathbf{x},t)dxdt$, where

$$\alpha(t,\mathbf{x}) = \int \hat{\alpha}(\omega,\mathbf{p})\exp[-i(\omega t - \mathbf{p}\mathbf{x})]\frac{d\omega d\mathbf{p}}{(2\pi)^4},$$

the properties of the supports of the functions $\hat{\alpha}$ ensure that the operators B_n are good, and the function $\hat{\alpha}$ satisfies the condition

$\hat{\alpha}(\omega(\mathbf{k}), \mathbf{k}) = 1$ for all $\mathbf{k}$, belonging to the support of the function $\lambda(\mathbf{k})$. Indeed, we have

$$\lim_{n\to\infty} B_n \Phi = \lim_{n\to\infty} \hat{\alpha}(H, \mathbf{P}) A_n \Phi$$

$$= \hat{\alpha}(H, \mathbf{P}) \Phi(\lambda)$$

$$= \int \lambda(\mathbf{k}) \hat{\alpha}(H, \mathbf{P}) \Phi(\mathbf{k}) d\mathbf{k}$$

$$= \int \lambda(\mathbf{k}) \hat{\alpha}(\omega(\mathbf{k}), \mathbf{k}) \Phi(\mathbf{k}) d\mathbf{k}$$

$$= \int \lambda(\mathbf{k}) \Phi(\mathbf{k}) d\mathbf{k} = \Phi(\lambda).$$

This completes the proof of Lemma 10.1. $\qquad\square$

Before we prove Lemma 10.2, we will conduct some non-rigorous but intuitive discussion. The integral (10.6) for large t can be approximated by the method of stationary phase. The stationary point $\mathbf{p}_0$ of the phase $\omega(\mathbf{p})t - \mathbf{p}\mathbf{x}$ is determined by the equation $\mathbf{v}(\mathbf{p}_0)t = \mathbf{x}$; denoting by $\mathbf{w}(\xi)$ the solution of the equation $\mathbf{v}(\mathbf{p}) = \xi$, we can write the stationary point in the form $\mathbf{p}_0 = \mathbf{w}(\frac{\mathbf{x}}{t})$. Applying the method of stationary phase to $\tilde{f}(\mathbf{x}|t)$ for large t, we obtain

$$\tilde{f}(\mathbf{x}|t) \approx C t^{-3/2} f\left(\mathbf{w}\left(\frac{\mathbf{x}}{t}\right)\right) \exp\left(-i\omega\left(\mathbf{w}\left(\frac{\mathbf{x}}{t}\right)\right)t + i\mathbf{w}\left(\frac{\mathbf{x}}{t}\right)\right),$$

where

$$C = (2\pi i)^{-3/2} |\det \gamma_{ik}|^{-1/2},$$

$$\gamma_{ik} = \frac{\partial^2}{\partial p_i \partial p_k} \omega(\mathbf{p})|_{\mathbf{p}=\mathbf{w}(\frac{\mathbf{x}}{t})}.$$

Hence, $\tilde{f}(\mathbf{x}|t)$, for large t, decays as $t^{-3/2}$. If $\frac{\mathbf{x}}{t} \notin U_\epsilon$, then $f(\mathbf{w}(\frac{\mathbf{x}}{t})) = 0$ and the stationary point is not contained in the support of $f(\mathbf{p})$; in this case, we should expect the function $\tilde{f}(\mathbf{x}|t)$ to be very small. Unfortunately, the simple considerations here cannot be applied to estimate the function $\hat{f}(\mathbf{x}|t)$ uniformly in $\mathbf{x}$; we will provide accurate but more cumbersome proofs for these inequalities.

Proof of Lemma 10.2. Let us first estimate $\tilde{f}(\mathbf{x}|t)$ when $\frac{\mathbf{x}}{t} \notin U_\epsilon$.

We will use the formula

$$\int \exp(-i\sigma(\mathbf{p}))\chi(\mathbf{p})d\mathbf{p} = \int \exp(-i\sigma(\mathbf{p}))(L^n\chi)(\mathbf{p})d\mathbf{p}, \qquad (10.22)$$

where $\chi(\mathbf{p})$ is a smooth function with compact support, $(L\chi)(\mathbf{p}) = -\mathrm{div}(\mathbf{u}(\mathbf{p})\chi(\mathbf{p}))$; $\mathbf{u}(\mathbf{p}) = \frac{\nabla\sigma(\mathbf{p})}{|\nabla\sigma(\mathbf{p})|^2}$ (see, for example, Fedoryuk (1971)).

In the case at hand, we have

$$\sigma(\mathbf{p}) = \omega(\mathbf{p})t - \mathbf{p}\mathbf{x}, \quad \chi(\mathbf{p}) = f(\mathbf{p}),$$

$$\mathbf{u}(\mathbf{p}) = \frac{\mathbf{v}(\mathbf{p})t - x}{|\mathbf{v}(\mathbf{p})t - \mathbf{x}|^2} = \frac{1}{t} \cdot \frac{\mathbf{v}(\mathbf{p}) - \frac{\mathbf{x}}{t}}{|\mathbf{v}(\mathbf{p}) - \frac{\mathbf{x}}{t}|^2}.$$

If $\frac{\mathbf{x}}{t} \notin U_\epsilon$, it is easy to check that

$$\sup_{\mathbf{p}\in\mathrm{supp}\,\chi} |D^{(\alpha)}\mathbf{u}(\mathbf{p})| \leq \frac{C}{|t|}$$

(here, $D^{(\alpha)} = \frac{\partial^{|\alpha|}}{\partial p_1^{\alpha_1} p_2^{\alpha_2} p_3^{\alpha_3}}$, $|\alpha| = \alpha_1 + \alpha_2 + \alpha_3$, and C, here and in the following, denotes a quantity independent of $\mathbf{x}$ and t, though it may depend on other parameters, for example, here it depends on α and ϵ). The inequality

$$\sup_{\mathbf{p}\in\mathrm{supp}\,\chi} |(L^n\chi)(\mathbf{p})| \leq C|t|^{-n} \qquad (10.23)$$

follows. The formula (10.22) and the inequality (10.23) lead to the inequality $|\tilde{f}(\mathbf{x}|t)| \leq \frac{C}{1+|t|^n}$, where $\frac{\mathbf{x}}{t} \notin U_\epsilon$ (n is an arbitrary number).[3] To prove the inequality (10.9), we should consider the cases $|\mathbf{x}| \leq at$ and $|\mathbf{x}| > at$, where $a = 2\sup_{\mathbf{p}\in\mathrm{supp}\,f} |\mathbf{v}(\mathbf{p})|$, separately. In the first case, the inequality (10.9) follows from (10.23); in the second case, we can use the fact that

$$\sup_{\substack{\mathbf{p}\in\mathrm{supp}\,f \\ |\mathbf{x}|>at}} |D^{(\alpha)}\mathbf{u}(\mathbf{p})| \leq \frac{C}{|\mathbf{x}|}.$$

---

[3]In the proof, we have assumed that the function $\chi(p)$ is smooth. One can relax this condition assuming that this function is m times continuously differentiable. Then we have the inequality (10.23) for $n = m$.

To prove inequality (10.7), let us take a smooth function $\mu(\mathbf{x})$ equal to zero for $|\mathbf{x}| \leq \nu_1$ and equal to one for $|\mathbf{x}| \geq \nu_2$.

Let us split the integral representing the function $\tilde{f}(\mathbf{x}|t)$ in two as follows:

$$I_1 = \int \exp(-i\omega(\mathbf{p})t + i\mathbf{p}\mathbf{x})\mu\left(t^\rho\left(\mathbf{v}(\mathbf{p}) - \frac{\mathbf{x}}{t}\right)\right) f(\mathbf{p})d\mathbf{p}; \quad (10.24)$$

$$I_2 = \int \exp(-i\omega(\mathbf{p})t + i\mathbf{p}\mathbf{x})\left(1 - \mu\left(t^\rho\left(\mathbf{v}(\mathbf{p}) - \frac{\mathbf{x}}{t}\right)\right)\right) f(\mathbf{p})d\mathbf{p},$$

$$(10.25)$$

where the number ρ is in the interval $\frac{5}{12} < \rho < \frac{1}{2}$. The integral I_1 can be estimated with the formula (10.22). Let us introduce the notation

$$\|\phi\|_n = \sup_{\substack{\mathbf{p} \in \Gamma \\ |\alpha| \leq n}} |D_\phi^{(\alpha)}(\mathbf{p})|,$$

where Γ denotes the set of points $\mathbf{p}$ such that $|\mathbf{v}(\mathbf{p}) - \frac{\mathbf{x}}{t}| \geq \nu_1|t|^{-\rho}$ (outside of Γ the integrand in the integral I_1 is zero). It is easy to check that $\|\mathbf{u}\|_n \leq C|t|^{-1+(n+1)\rho}$.

Furthermore, we have

$$\|L\chi\|_n \leq C\|\mathbf{u}\chi\|_{n+1} \leq C \sum_{\alpha+\beta=n+1} \|u\|_\alpha\|\chi\|_\beta,$$

which implies that

$$\|L^r\chi\|_n \leq C \sum_{\alpha_1+\cdots+\alpha_r+\beta=n+r} \|\mathbf{u}\|_{\alpha_1}\cdots\|\mathbf{u}\|_{\alpha_r}\|\chi\|_\beta$$

$$\leq C \sum_\beta |t|^{-r+(n+2r-\beta)\rho}\|\chi\|_\beta.$$

To estimate the integral I_1, we should introduce

$$\chi(\mathbf{p}) = \mu\left(t^\rho\left(\mathbf{v}(\mathbf{p}) - \frac{\mathbf{x}}{t}\right)\right)$$

into formula (10.22). Then, $\|\chi\|_\beta \leq C|t|^{\beta\rho}$ implies

$$|I_1| \leq C\|L^r\chi\|_0 \leq \sum_\beta C|t|^{-r+(2r-\beta)\rho}\|\chi\|_\beta \leq C|t|^{-r+2r\rho}.$$

When $r \geq \frac{1.5}{-2\rho+1}$, we can show that $|I_1| \leq C|t|^{-3/2}$.

In integral I_2, let us replace $\omega(\mathbf{p})$ with

$$\omega_1(\mathbf{p}) = \omega(\mathbf{p}_0) + \mathbf{v}(\mathbf{p}_0)(\mathbf{p} - \mathbf{p}_0) + \sum_{i=1}\sum_{j=1} \frac{1}{2}\frac{\partial^2\omega(\mathbf{p}_0)}{\partial p_i \partial p_j}$$
$$\times (p_i - p_{0i})(p_j - p_{0j}),$$

where $\mathbf{p}_0$ is any point with $|\mathbf{v}(\mathbf{p}_0) - \frac{\mathbf{x}}{t}| \leq \nu_2|t|^{-\rho}$. We will show that the error Δ resulting from this change does not exceed $C|t|^{-3/2}$. To show this, let us use the formula $\mathbf{w} = \mathbf{v}(\mathbf{p}) - \frac{\mathbf{x}}{t}$ to make a substitution in the integral

$$\Delta = \int (\exp(-i\omega(\mathbf{p})t + i\mathbf{p}\mathbf{x}) - \exp(-i\omega_1(\mathbf{p})t + i\mathbf{p}\mathbf{x}))$$
$$\times \left(1 - \mu\left(t^{-\rho}\left(\mathbf{v}(\mathbf{p}) - \frac{\mathbf{x}}{t}\right)\right)\right) f(\mathbf{p})d\mathbf{p}.$$

Since the function $\omega(\mathbf{p})$ is strongly convex and the function $f(\mathbf{p})$ has compact support, it follows that the Jacobian of this transformation is bounded above and below by positive constants.

By the properties of the functions μ, we can assume that the integral Δ is taken over the domain Γ_1, where $|\mathbf{w}| \leq \nu_2|t|^{-\rho}$; the volume of this domain equals $C|t|^{-3\rho}$. Over the domain Γ_1, we can use the estimate

$$|(\exp(-i\omega(\mathbf{p})t + i\mathbf{p}\mathbf{x}) - \exp(-i\omega_1(\mathbf{p})t + i\mathbf{p}\mathbf{x}))f(\mathbf{p})|$$
$$\leq C|t||\mathbf{p} - \mathbf{p}_0|^3 \leq C_1|t||\mathbf{w}|^3 \leq C_2|t|^{-3\rho+1},$$

and we then obtain that $|\delta| \leq C|t|^{1-6\rho} \leq C|t|^{1-6\cdot\frac{5}{12}} = C|t|^{-3/2}$.

In the remaining part of the proof, we need to estimate the expression

$$I_2 - \Delta = \int \exp(-i\omega_1(\mathbf{p})t + i\mathbf{p}\mathbf{x})f(\mathbf{p})d\mathbf{p}$$
$$- \int \exp(-i\omega_1(\mathbf{p}) + i\mathbf{p}\mathbf{x})\mu\left(t^{-\rho}\left(\mathbf{v}(\mathbf{p}) - \frac{\mathbf{x}}{t}\right)\right) d\mathbf{p}.$$

The second term in this expression can be estimated in the same way as the integral I_1. The first term can be estimated if we rewrite it using the Fourier transform, in the form

$$\int G(\mathbf{x} - \mathbf{y}, t) \tilde{f}(\mathbf{y}) d\mathbf{y},$$

where

$$\tilde{f}(\mathbf{x}) = \int f(\mathbf{p}) \exp(i\mathbf{p}\mathbf{x}) \frac{d\mathbf{p}}{(2\pi)^3},$$

$$G(\mathbf{x}, t) = \left(\frac{2\pi}{it}\right)^{3/2} |\det \gamma_{jk}|^{-1/2} \exp\left(\frac{i}{2} \sum_{j,k} \lambda_{jk} x_j x_k\right),$$

$$\gamma_{jk} = \frac{\partial^2 \omega(\mathbf{p}_0)}{\partial p_j \partial p_k},$$

λ_{jk} is the inverse matrix of the matrix γ_{jk}.

The inequality (10.8) follows from the inequality (10.7) in the case $\mathbf{x} \in U_\epsilon t$ and from the inequality (10.9) in the case $\mathbf{x} \notin U_\epsilon t$. $\qquad\square$

Remark 10.1. With some extra conditions on the behavior of the function $\omega(\mathbf{p})$ at infinity, we can prove Lemma 10.2 for any smooth, rapidly decaying function $f(\mathbf{p})$.

Proof of Lemma 10.3. Let us begin by proving the simple but important identity that connects the truncated vacuum expectation value

$$\langle AB(t) \rangle^T = \langle AB(t)\Phi, \Phi \rangle - \langle A\Phi, \Phi \rangle \langle B\Phi, \Phi \rangle$$

with the vacuum expectation value of the commutator $[A, B(t)]$. To prove this, let us consider the smooth function $h(\omega)$, satisfying the conditions $h(\omega) = 0$ for $\omega \le 0$ and $h(\omega) = 1$ for $\omega \ge \delta$ [the number $\delta > 0$ is chosen such that all points of the spectrum of the operator H, with the exception of 0, lie on the ray $(\delta, +\infty)$]. Let us note that

$$\begin{aligned} h(H) &= 1 - P_0, \\ h(-H) &= 0, \end{aligned} \tag{10.26}$$

where P_0 is a projection operator on the vacuum vector Φ (these equations can be checked by using the isomorphism between the spaces $\mathcal{H}$ and $L^2(M)$, which sends the operator H to an operator of multiplication by a function). Let us assign to a function $f \in \mathcal{S}$ the function

$$f_h(t) = \int h(\omega) \exp(i\omega(\tau - t)) f(\tau) \frac{d\omega d\tau}{2\pi}.$$

It is easy to check that the function $f_h(t)$ also belongs to the space $\mathcal{S}$ (the operator mapping f to the function f_h is in fact a composition of the Fourier transform over t, the operator of multiplication by $h(\omega)$, and the inverse Fourier transform; all these operators transform $\mathcal{S}$ into itself). If the function f only depends on t and not on the other variables, then the function f_h is constructed by means of the Fourier transform over t (other variables are considered as parameters). If the function $f(t, x_1, \ldots, x_n)$ belongs to the space $\mathcal{S}(E^{n+1})$, then the function f_h belongs to it as well.

Let us show that for any function $f \in \mathcal{S}$, we have

$$\int f(t) \langle AB(t) \rangle^T \, dt = \int f_h(t) \langle [A, B(t)] \rangle \, dt. \tag{10.27}$$

Indeed, we have

$$\int f_h(t) \langle AB(t) \rangle \, dt = \int f_h(t) \langle A \exp(iHt) B \rangle \, dt$$

$$= \langle A h(H) \tilde{f}(H) B \rangle = \langle A(1 - P_0) \tilde{f}(H) B \rangle$$

$$= \int f(t) \langle A(1 - P_0) \exp(iHt) B \rangle \, dt = \int f(t) \langle AB(t) \rangle^T \, dt, \tag{10.28}$$

$$\int f_h(t) \langle B(t) A \rangle \, dt = \int f_h(t) \langle B \exp(-iHt) A \rangle \, dt$$

$$= \langle B h(-H) \tilde{f}(-H) A \rangle = 0. \tag{10.29}$$

Here, we have used the relation (10.26) and the equations

$$\int f_h(t) \exp(i\omega t) dt = h(\omega) \tilde{f}(\omega),$$

$$\int f(t) \exp(i\omega t) dt = \tilde{f}(\omega).$$

Combining (10.28) and (10.29), we obtain (10.27).

Let us now prove the lemma for the case $n = 2$. We will suppose that A_1, A_2, $C \in \mathcal{A}$, $\alpha \in \mathcal{S}$, $A_2 = \int \alpha(\mathbf{x}, t) C(\mathbf{x}, t) d\mathbf{x} dt$ and introduce the notation $D = \int \alpha_h(\mathbf{x}, t) C(\mathbf{x}, t) d\mathbf{x} dt$. From the relation (10.27), it is easy to see that

$$\langle A_1 A_2(\mathbf{x}, t) \rangle^T = \langle [A_1, D(\mathbf{x}, t)] \rangle .$$

Since $D \in \mathcal{A}$ and all operators in $\mathcal{A}$ asymptotically commute, it follows that

$$|w_2^T(\mathbf{x}_1, t_1, \mathbf{x}_2, t_2)| = |\langle A_1(\mathbf{x}_1, t_1) A_2(\mathbf{x}_2, t_2) \rangle^T|$$

$$= |\langle [A_1, D(\mathbf{x}_2 - \mathbf{x}_1, t_2 - t_1)] \rangle| \leq \frac{C(1 + |t_2 - t_1|^r)}{1 + |\mathbf{x}_2 - \mathbf{x}_1|^n} .$$

Now that we have proved the lemma for the case $n = 2$, let us give the proof in the case of arbitrary Wightman functions. Let us consider the function

$$w_n(\mathbf{x}_1, t_1, \ldots, \mathbf{x}_n, t_n) = \langle A_1(\mathbf{x}, t_1) \ldots A_n(\mathbf{x}_n, t_n) \rangle$$

and the two functions

$$w_k^{(1)}(\mathbf{x}_{i_1}, t_{i_1}, \ldots, \mathbf{x}_{i_k}, t_{i_k}) = \langle A_{i_1}(\mathbf{x}_{i_1}, t_{i_1}) \ldots A_{i_k}(\mathbf{x}_{i_k}, t_{i_k}) \rangle ,$$

$$w_l^{(1)}(\mathbf{x}_{j_1}, t_{j_1}, \ldots, \mathbf{x}_{j_l}, t_{j_l}) = \langle A_{j_1}(\mathbf{x}_{j_1}, t_{j_1}) \ldots A_{j_l}(\mathbf{x}_{j_l}, t_{j_l}) \rangle ,$$

obtained by splitting the operators $A_1, \ldots, A_n$, entering the definition of the function w_n, into two groups $A_{i_1}, \ldots, A_{i_k}$ and $A_{j_1}, \ldots, A_{j_k}$ (here, $k + l = n$, the indices $i_1, \ldots, i_k$ and $j_1, \ldots, j_l$ are in ascending order, and the operators A_i are assumed to be smooth).

The set of indices $i_1, \ldots, i_k$, the indices of the operators in the first group, will be denoted by K; the set of indices $j_1, \ldots, j_l$ of the second group of operators will be denoted by L; clearly $K \cup L = \{1, 2, \ldots, n\}$.

It can be shown that when all the points $\mathbf{x}_i$, where $i \in K$, are far from the points $\mathbf{x}_j$, where $j \in L$, the following approximation holds:

$$w_n(\mathbf{x}_1, t_1, \ldots, \mathbf{x}_n, t_n)$$
$$\approx w_k^{(1)}(\mathbf{x}_{i_1}, t_{i_1}, \ldots, \mathbf{x}_{i_k}, t_{i_k}) w_l^{(2)}(\mathbf{x}_{j_1}, t_{j_1}, \ldots, \mathbf{x}_{j_l}, t_{j_l}).$$

More precisely, *for fixed $t_1, \ldots, t_n$, the absolute value of the function*

$$w_n(\mathbf{x}_1, t_1, \ldots, \mathbf{x}_n, t_n) - w_k^{(1)}(\mathbf{x}_{i_1}, t_{i_1}, \ldots, \mathbf{x}_{i_k}, t_{i_k}) w_l^{(2)}$$

$$\times (\mathbf{x}_{j_1}, t_{j_1}, \ldots, \mathbf{x}_{j_l}, t_{j_l})$$

does not exceed Cd^{-p}. Here, $d = \min_{i \in K, j \in L} |\mathbf{x}_i - \mathbf{x}_j|$; p is an arbitrary number; C is a constant depending on p, but not on $\mathbf{x}_1, \ldots, \mathbf{x}_n$. (This property of Wightman functions is often called *asymptotic factorization* or the cluster property.)

To prove the asymptotic factorization property, we will first consider the case when the set K consists of the points $(1, 2, \ldots, k)$ and the set L consists of the points $(k+1, \ldots, n)$. In this case, the difference we aim to estimate can be written in the form

$$\langle RU \rangle^T = \langle RU \rangle - \langle R \rangle \langle U \rangle,$$

where

$$R = A_1(\mathbf{x}_1, t_1) \ldots A_k(\mathbf{x}_k, t_k),$$

$$U = A_{k+1}(\mathbf{x}_{k+1}, t_{k+1}) \ldots A_n(\mathbf{x}_n, t_n).$$

Recall that the operators $A_1, \ldots, A_n$ are smooth and can therefore be written in the form

$$A_i = \int f_i(\mathbf{x}, t) B_i(\mathbf{x}, t) d\mathbf{x} dt,$$

where $f_i \in \mathcal{S}, B_i \in \mathcal{A}$. Noting that

$$U = \int f_{k+1}(\xi_1, \tau_1) \ldots f_n(\xi_{n-k}, \tau_{n-k}) B_{k+1}(\mathbf{x}_{k+1} + \xi_1, t_{k+1} + \tau_1)$$

$$\ldots B_n(\mathbf{x}_n + \xi_{n-k}, t_n + \tau_{n-k}) d^{n-k}\xi d^{n-k}\tau$$

$$= \int \alpha(t, \xi_1 - \mathbf{x}_{k+1}, \ldots, \xi_{n-k} - \mathbf{x}_n, \sigma_1, \ldots, \sigma_{n-k-1})$$

$$\times \exp(iHt) B_{k+1}(\xi_1, t_{k+1}) B_{k+2}(\xi_2, t_{k+2} + \sigma_1)$$

$$\ldots B_n(\xi_{n-k}, t_n + \sigma_{n-k-1}) \exp(-iHt) dt d^{n-k}\xi d^{n-k-1}\sigma,$$

where

$$\alpha(t, \xi_1, \ldots, \xi_{n-k}, \sigma_1, \ldots, \sigma_{n-k-1})$$

$$= f_{k+1}(\xi_1, t) f_{k+2}(\xi_2, t + \sigma_1) \cdots f_n(\xi_{n-k}, t + \sigma_{n-k-1}),$$

we can apply the relation (10.27). This leads to

$$\langle R, U \rangle^T = \int \alpha_h(t, \xi_1 - \mathbf{x}_{n+1}, \ldots, \xi_{n-k} - \mathbf{x}_n, \sigma_1, \ldots, \sigma_{n-k-1})$$
$$\times \langle [R, B_{k+1}(\xi_1, t + t_{k+1}) B_{k+2}(\xi_2, t + t_{k+2} + \sigma_1)$$
$$\ldots B_n(\xi_{n-k}, t + t_n + \sigma_{n-k-1})] \rangle \, dt d^{n-k} \xi d^{n-k-1} \sigma. \quad (10.30)$$

The commutator on the right-hand side of equation (10.30) can be written in the form

$$\sum_{\substack{1 \leq i \leq k \\ k+1 \leq j \leq n}} C_{ij}[A_i(\mathbf{x}_i, t_i), B_j(\xi_{j-k}, t + t_j + \sigma_{j-k-1})] D_{ij},$$

where C_{ij}, D_{ij} are operators whose norms are bounded by a constant not depending on $\mathbf{x}_i$, t_i, ξ_i, σ_i. Using this fact, asymptotic commutativity, and formula (10.30), we obtain the asymptotic factorization property for the case $K = \{1, \ldots, k\}$, $L = \{k+1, \ldots, n\}$. We can show that the general case can be reduced to this special case. Suppose the set K consists of the points $(i_1, \ldots, i_k)$ and the set L consists of the points $(j_1, \ldots, j_l)$, where $i_1 < \cdots < i_k$, $j_1 < \cdots < j_l$. If the quantity $d = \min_{i \in K, j \in L} |\mathbf{x}_i - \mathbf{x}_j|$ is large, we can use the approximation

$$\langle A_1(\mathbf{x}_1, t_1) \ldots A_n(\mathbf{x}_n, t_n) \rangle \approx \langle A_{i_1}(\mathbf{x}_{i_1}, t_{i_1}) \ldots A_{i_k}(\mathbf{x}_{i_k}, t_{i_k})$$
$$\times A_{j_1}(\mathbf{x}_{j_1}, t_{j_1}) \ldots A_{j_l}(\mathbf{x}_{j_l}, t_{j_l}) \rangle, \quad (10.31)$$

whose error does not exceed Cd^{-m}, where m is an arbitrary number; C is a constant depending on $m, n, t_1, \ldots, t_n$. We can justify this approximation by using the relation

$$\langle E A_i(\mathbf{x}_i, t_i) A_j(\mathbf{x}_j, t_j) F \rangle \approx \langle E A_j(\mathbf{x}_j, t_j) A_i(\mathbf{x}_i, t_i) F \rangle,$$

for $i \in K, j \in L$, that we have proved already (here, E, F are compositions of the operators $A_\alpha(\mathbf{x}_\alpha, t_\alpha)$, where $\alpha \in K \cup L$).

From the relation (10.31) and the asymptotic factorization property, we can now obtain the asymptotic factorization property in the general case by the same method as in the special case shown earlier.

Let us now show how to derive Lemma 10.3 from the asymptotic factorization property of Wightman functions. We will suppose that the lemma holds for functions w_{n-1}^T and then show it for functions w_n^T. However, we will first prove the following geometric lemma.

Let N be a set consisting of the points $\mathbf{x}_1, \ldots, \mathbf{x}_n$ in Euclidean space and let D be the diameter of N (i.e. $D = \max_{x_i, x_j \in N} |\mathbf{x}_i - \mathbf{x}_j|$). Then, we can partition N into two subsets P and Q in such a way so that $\rho(P, Q) = \min_{\mathbf{x}_i \in P, \mathbf{x}_j \in Q} |\mathbf{x}_i - \mathbf{x}_j| \geq \frac{D}{2n}$.

We will prove this fact by induction on n. Let $D = |\mathbf{x}_\alpha - \mathbf{x}_\beta|$. Let us delete from N the point $\mathbf{x}_\gamma$ which does not coincide with $\mathbf{x}_\alpha$ or with $\mathbf{x}_\beta$, obtaining as a result the set N', consisting of $n - 1$ points and having diameter D. By the inductive hypothesis, we can split the set N' into two sets P' and Q' so that $\rho(P', Q') = \min_{\mathbf{x}_i \in P, \mathbf{x}_j \in Q} |\mathbf{x}_i - \mathbf{x}_j| \geq \frac{D}{2^{n-1}}$. By the triangle inequality, the point $\mathbf{x}_\gamma$ cannot satisfy the inequalities $\rho(P', \mathbf{x}_\gamma) = \min_{\mathbf{x}_i \in P'} |\mathbf{x}_i - \mathbf{x}_\gamma| < \frac{D}{2n}$ and $\rho(Q', \mathbf{x}_\gamma) = \min_{\mathbf{x}_i \in Q'} |\mathbf{x}_i - \mathbf{x}_\gamma| < \frac{D}{2n}$ simultaneously; for concreteness, suppose that $\rho(P', \mathbf{x}_\gamma) \geq \frac{D}{2n}$. Then, it is clear that the necessary partition can be constructed by taking P to be P' and taking Q to be $Q' \cup \{\mathbf{x}_\gamma\}$.

Let us now suppose that Lemma 10.3 holds for functions w_k^T with $k < n$. To prove the lemma for functions w_n^T, we need to estimate $w_n^T(\mathbf{x}_1, t_1, \ldots, \mathbf{x}_n, t_n)$ under the condition $\max_{1 \leq i \leq j \leq n} |\mathbf{x}_i - \mathbf{x}_j| = D$ (with the times $t_1, \ldots, t_n$ fixed). Using the geometric lemma we just proved, let us partition the set $\{1, \ldots, n\}$ into the subsets K and L such that $\min_{i \in K, j \in L} |\mathbf{x}_i - \mathbf{x}_j| \geq \frac{D}{2n}$. To estimate the functions w_n^T, let us note that in the right-hand side of the relation (10.11) (the recurrence relation defining the functions w_n^T) all factors have the form $w_{\alpha_r}(\pi_r)$ with the set π_r containing both elements of K and L. By induction, we can obtain an estimate of the form CD^{-m}, where m is arbitrary and C depends on m. It therefore follows that every term on the right-hand side of (10.11) contains at least one factor of the form just described and therefore admits an estimate of the form CD^{-m}, since the factors $w_{\alpha_i}^T(\pi_i)$ do not exceed a constant, depending only on $\|A_1\|, \ldots, \|A_n\|$ and the number n (this statement becomes clear if we note that

$$|\langle A_1(\mathbf{x}_1, t_1) \ldots A_n(\mathbf{x}_n, t_n) \rangle| \leq \|A_1\| \ldots \|A_n\|).$$

Hence, with error not exceeding CD^{-m}, we can write the approximation

$$w_n(\mathbf{x}_1, t_1, \ldots, \mathbf{x}_n, t_n) \approx w_n^T(\mathbf{x}_1, t_1, \ldots, \mathbf{x}_n, t_n)$$

$$+ \sum_{k=2}^{n} \sum_{\rho} w_{\alpha_1}^T(\pi_1) \ldots w_{\alpha_k}^T(\pi_k), \qquad (10.32)$$

where the sum is taken over partitions ρ of the set $\{1, \ldots, n\}$ such that every set π_j is completely contained in either K or L.

It is easy to check that the sum in (10.32) equals the product $w_k(\mathbf{x}_{i_1}, t_{i_1}, \ldots, \mathbf{x}_{i_k}, t_{i_k}) w_l(\mathbf{x}_{j_1}, t_{j_1}, \ldots, \mathbf{x}_{j_l}, t_{j_l})$, where $(i_1, \ldots, i_k)$ are elements of K and $(j_1, \ldots, j_l)$ are elements of L.

Hence, we see that

$$w_n^T(\mathbf{x}_1, t_1, \ldots, \mathbf{x}_n, t_n)$$

$$\approx w_n(\mathbf{x}_1, t_1, \ldots, \mathbf{x}_n, t_n)$$

$$- w_k(\mathbf{x}_{i_1}, t_{i_1}, \ldots, \mathbf{x}_{i_k}, t_{i_k}) w_l(\mathbf{x}_{j_1}, t_{j_1}, \ldots, \mathbf{x}_{j_l}, t_{j_l}).$$

To finish the proof of Lemma 10.3, we apply the asymptotic factorization property of Wightman functions. $\qquad \square$

Remark 10.2. Because the operators A_i are smooth, the Wightman functions $w_n(\mathbf{x}_1, t_1, \ldots, \mathbf{x}_n, t_n)$ and the truncated Wightman functions $w_n^T(\mathbf{x}_1, t_1, \ldots, \mathbf{x}_n, t_n)$ are smooth, and hence the derivatives of these functions can be viewed as Wightman and truncated Wightman functions constructed with different smooth operators. (This is clear if we note that for a smooth operator $A = \int f(\tau, \mathbf{x}) B(\mathbf{x}, \tau) d\xi d\tau$, where $f \in \mathcal{S}(E^4)$, the following relation holds:

$$D^{(\alpha)} A(\mathbf{x}, t) = D^{(\alpha)} \int f(\tau - t, \xi - \mathbf{x}) B(\xi, \tau) d\xi d\tau$$

$$= \int g(\tau - t, \xi - \mathbf{x}) B(\xi, \tau) d\xi d\tau = C(\mathbf{x}, t),$$

where $D^{(\alpha)} = \dfrac{\partial^{|\alpha|}}{\partial t^{\alpha_0} \partial x_1^{\alpha_1} \partial x_2^{\alpha_2} \partial x_3^{\alpha_3}}$ is a differential operator; $g(t, \mathbf{x}) = (-1)^{|\alpha|} D^{(\alpha)} f(t, \mathbf{x})$; C is a smooth operator defined by $C = \int g(\tau, \xi) B(\xi, \tau) d\xi d\tau$; operator derivatives are understood as norm derivatives.) Using this fact and the statement of Lemma 10.3, we can say that for fixed $t_1, \ldots, t_n$ and fixed $\mathbf{x}_n$, the function $w_n^T(\mathbf{x}_1, t_1, \ldots, \mathbf{x}_n, t_n)$, constructed with smooth operators, belongs to the space $\mathcal{S}(E^{3(n-1)})$.

The requirement that the operators $A_1, \ldots, A_n$ be smooth in Lemma 10.3 can be relaxed. However, if we completely remove this requirement, then we have to weaken the statement of the lemma. Namely, one can prove that without the smoothness condition on the operators $A_1, \ldots, A_n$, the function

$$\nu(\mathbf{x}_1, t_1, \ldots, \mathbf{x}_n, t_n)$$

$$= \int \gamma(\mathbf{x}_1 - \xi_1, t_1 - \tau_1, \ldots, \mathbf{x}_n - \xi_n, t_n - \tau_n)$$

$$\times w_n^T(\xi_1, \tau_1, \ldots, \xi_n, \tau_n) d^n \xi d^n \tau,$$

where $\gamma \in \mathcal{S}(E^{4n})$, for fixed $t_1, \ldots, t_n$ and fixed $\mathbf{x}_n$, belongs to the space $\mathcal{S}(E^{3(n-1)})$.

Proof of Lemma 10.4. To prove this lemma, we need to estimate the quantity

$$I = \langle A_0(f_0, t) \ldots A_n(f_n, t) \rangle^T$$

$$= \int \tilde{f}_0(\mathbf{x}_0|t) \ldots \tilde{f}_n(\mathbf{x}_n|t) \langle A_0(\mathbf{x}_0, t) \ldots A_n(\mathbf{x}_n, t) \rangle^T d\mathbf{x}_0 \ldots d\mathbf{x}_n.$$

$$(10.33)$$

Let us change variables in integral (10.33) to the new variables $\mathbf{x}_0, \xi_1, \ldots, \xi_n$, where $\xi_j = \mathbf{x}_j - \mathbf{x}_0$, and apply the estimate

$$|\langle A_0(\mathbf{x}_0, t) \ldots A_n(\mathbf{x}_n, t) \rangle^T| \leq \frac{C}{1 + \|\xi\|^m}$$

$$= \frac{C}{1 + (\xi_1^2 + \cdots + \xi_n^2)^{m/2}},$$

following from Lemma 10.3 (here, m is arbitrary and C depends on m). Applying this estimate, we obtain

$$|I| \leq \int |\tilde{f}_0(\mathbf{x}_0|t)| \left(\frac{C}{1 + |t|^{3/2}} \right)^n \frac{C}{1 + \|\xi\|^m} d\mathbf{x}_0 d^n \xi$$

$$\leq \text{const}(1 + |t|)^{-\frac{3}{2}n} \int |\tilde{f}_0(\mathbf{x}_0|t)| d\mathbf{x}_0$$

$$\leq \text{const}(1 + |t|)^{-\frac{3}{2}(n-1)}.$$

$$(10.34)$$

In the derivation above, we have used the inequalities $|f_j(\mathbf{x}|t)| \leq C(1 + |t|^{3/2})^{-1}$, for $j \geq 1$, and $\int |\tilde{f}_0(\mathbf{x}|t)|d\mathbf{x} \leq C(1 + |t|)^{3/2}$, which follow from Lemma 10.2; the number m is chosen to be large enough so that the integral $\int (1 + \|\xi\|)^{-m}d^n\xi$ converges.

The inequality (10.34) gives the necessary estimate of the quantity I. To finish the proof of Lemma 10.4, we need an analogous estimate in the case when in the expression $\langle A_1(f_1, t) \ldots A_n(f_n, t) \rangle^T$ some of the operators $A_j(f_j, t)$ are replaced by the operators $\dot{A}_j(f_j, t) = \frac{d}{dt}A_j(f_j, t)$.

Let us note that

$$\dot{A}_j(f_j, t) = \frac{d}{dt} \int \tilde{f}_j(\mathbf{x}|t) A_j(\mathbf{x}, t)d\mathbf{x}$$

$$= \int \left(\frac{d}{dt}\tilde{f}_j(\mathbf{x}|t) \right) A_j(\mathbf{x}, t)d\mathbf{x} + \int \tilde{f}_j(\mathbf{x}|t)\frac{d}{dt}A_j(\mathbf{x}, t)d\mathbf{x}$$

$$= \int \tilde{g}_j(\mathbf{x}|t) A_j(\mathbf{x}, t)d\mathbf{x} + \int \tilde{f}_j(\mathbf{x}|t)K_j(\mathbf{x}, t)d\mathbf{x},$$

where we have used the notation

$$K = i[H, A],$$

$$\tilde{g}_j(\mathbf{x}|t) = \frac{d}{dt}\tilde{f}_j(\mathbf{x}|t)$$

$$= \int \exp(-i\omega(\mathbf{p})t + i\mathbf{p}\mathbf{x})g_j(\mathbf{p})\frac{d\mathbf{p}}{(2\pi)^3},$$

$$g_j(\mathbf{p}) = -i\omega(\mathbf{p})f_j(\mathbf{p}).$$

Hence, the operator $\dot{A}_j(f_j, t)$ can be written in the form

$$\dot{A}_j(f_j, t) = A_j(g_j, t) + K_j(f_j, t), \tag{10.35}$$

where g_j is a smooth function with compact support; K_j is a smooth operator. [The smoothness of K_j follows if we note that if the operator A_j is represented in the form $A_j = \int \phi_j(\tau, \xi)B_j(\xi, \tau)d\xi d\tau$, the operator K_j is equal to $-\int (\frac{d}{d\tau}\phi_j(\tau, \xi))B_j(\xi, \tau)d\xi d\tau.$]

By formula (10.35), the proof of the lemma in the case when some of the operators $A_j(f_j, t)$ are replaced by the operators $\dot{A}_j(f_j, t)$ clearly reduces to the case already considered. The proof of the lemma in the case when some of the operators $A_j(f_j, t)$ are replaced with the operators $(A_j(f_j,t))^*$ or $(\dot{A}_j(f_j,t))^*$ requires no new ideas. $\square$

10.3 Asymptotic fields (in- and out-operators)

In Section 10.1, we defined the Møller matrices $S_\pm$ corresponding to the asymptotically Abelian algebra $\mathcal{A}$. With the Møller matrices, one can define the *in-* and *out-operators* (*asymptotic fields*) $a^+_{\substack{\text{in}\\\text{out}}}(\mathbf{k})$ and $a_{\substack{\text{in}\\\text{out}}}(\mathbf{k})$ by the relations

$$S_- a^+(\mathbf{k}) = a^+_{\text{in}}(\mathbf{k}) S_-,$$

$$S_- a(\mathbf{k}) = a_{\text{in}}(\mathbf{k}) S_-,$$

$$S_+ a^+(\mathbf{k}) = a^+_{\text{out}}(\mathbf{k}) S_+,$$

$$S_+ a^+(\mathbf{k}) = a_{\text{out}}(\mathbf{k}) S_+.$$

These equations clearly define $a^+_{\text{in}}(\mathbf{k})$, $a_{\text{in}}(\mathbf{k})$ as operator generalized functions on the space $\mathcal{H}_{\text{in}} = S_-\mathcal{H}_{\text{as}}$ and $a^+_{\text{out}}(\mathbf{k})$, $a_{\text{out}}(\mathbf{k})$ as operator generalized functions on the space $\mathcal{H}_{\text{out}} = S_+\mathcal{H}_{\text{as}}$.

In this section, we will show that in- and out-operators describe the asymptotic behavior of the operators $A(\mathbf{x},t)$, where $A \in \mathcal{A}$, as $t \to \pm\infty$ (this justifies the name asymptotic fields).

Before we begin precise formulations, let us introduce the following definition.

Let us assign to every function $f \in \mathcal{S}(E^3)$ the set $U(f) \in E^3$ consisting of points of the form $\frac{\partial\omega(\mathbf{k})}{\partial\mathbf{k}}$, where $\mathbf{k} \in \text{supp}\,f$ (i.e. $\mathbf{k}$ belongs to the support of f).

A family of functions $f_1,\ldots,f_n$ will be called *non-overlapping* if the sets $U(f_j)$ do not intersect pairwise (i.e. the sets $U(f_i)$ and $U(f_j)$ do not have common points if $i \neq j$).[4]

---

[4]Under the assumption of strong convexity on the functions $\omega(\mathbf{k})$, the family $f_1,\ldots,f_n$ will be non-overlapping if and only if the supports of these functions $\text{supp}\,f_j$ are pairwise non-intersecting. Therefore, in the case at hand, the definition of non-overlapping family can be simplified. However, in the case when we have multiple particles, the definition in the main text is necessary.

The following remark elucidates the physical meaning of the new notion.

If $f_1, \ldots, f_n$ is a non-overlapping family of functions, then the dynamics on the space $\mathcal{H}_{\text{as}}$ as $t \to \pm\infty$ transform the vector $a^+(\overline{f}_1) \ldots a^+(\overline{f}_n)\theta$ into a set of far apart particles. More precisely, we have

$$\exp(-iH_{\text{as}}t)a^+(\overline{f}_1) \ldots a^+(\overline{f}_n)\theta = a^+(\overline{f_1^t}) \ldots a^+(\overline{f_n^t})\theta$$

$$= (2\pi)^{\frac{3}{2}n} \int \tilde{f}_1(\mathbf{x}_1|t) \ldots \tilde{f}_n(\mathbf{x}_n|t)a^+(\mathbf{x}_1) \ldots a^+(\mathbf{x}_n)d\mathbf{x}_1 \ldots d\mathbf{x}_n\theta,$$

where

$$f_j^t(\mathbf{k}) = f_j(\mathbf{k})\exp(-i\omega(\mathbf{k})t),$$

$$\tilde{f}_j(\mathbf{x}|t) = \int f_j^t(\mathbf{k})\exp(i\mathbf{k}\mathbf{x})\frac{d\mathbf{k}}{(2\pi)^3}$$

$$= \int f_j(\mathbf{k})\exp(-i\omega(\mathbf{k})t + i\mathbf{k}\mathbf{x})\frac{d\mathbf{k}}{(2\pi)^3}.$$

It follows from Lemma 10.2 that the function $\tilde{f}_j(\mathbf{x}|t)$, for large t, is small outside the set $tU_\epsilon(f_j)$, where the symbol $U_\epsilon(f_j)$ denotes an ϵ-neighborhood of the set $U(f_j)$. Under the assumed conditions, for small enough ϵ and large enough t, the sets $tU_\epsilon(f_j)$ are far apart.

Let us consider vectors of the form $\Psi_-(B_1, \ldots, B_n | f_1, \ldots, f_n)$ [correspondingly, of the form $\Psi_+(B_1, \ldots, B_n | f_1, \ldots, f_n)$], where $B_1, \ldots, B_n$ are good operators and $f_1, \ldots, f_n$ are functions with compact support. We will call them *non-overlapping* in-*vectors* (out-*vectors*) if $f_1, \ldots, f_n$ is a non-overlapping family of functions. The set of linear combinations of non-overlapping in-vectors (out-vectors) will be denoted by $\mathcal{D}_{\text{in}}$ (correspondingly, $\mathcal{D}_{\text{out}}$); $\mathcal{D}_{\text{in}}$ and $\mathcal{D}_{\text{out}}$ are linear subspaces that are dense in $\mathcal{H}_{\text{in}}$ and $\mathcal{H}_{\text{out}}$, respectively.

Let us now prove a theorem that describes the asymptotic behavior of operators $A(\mathbf{x}, t)$, where $A \in \mathcal{A}$, in terms of in- and out-operators. For concreteness, we will formulate the theorem in the case $t \to -\infty$.

If $A_1, \ldots, A_n$ are good operators, $\Psi_1 \in \mathcal{D}_{\text{in}}$, then

$$\lim_{t \to -\infty} A_1(f_1, t, \sigma_1) \ldots A_n(f_n, t, \sigma_n)\Psi_1$$

$$= a_{\text{in}}(\overline{\phi}_1 \overline{f}_1, \sigma_1) \ldots a_{\text{in}}(\overline{\phi}_n \overline{f}_n, \sigma_n)\Psi_1. \tag{10.36}$$

If $A_1, \ldots, A_n$ are smooth operators in the algebra $\mathcal{A}$, satisfying the condition $\langle A_j \Phi, \Phi \rangle = 0$, $\Psi_1 \in \mathcal{D}_{\text{in}}$, $\Psi_2 \in \mathcal{D}_{\text{in}}$, then

$$\lim_{t \to -\infty} \langle A_1(f_1, t, \sigma_1) \ldots A_n(f_n, t, \sigma_n)\Psi_1, \Psi_2 \rangle$$

$$= \langle a_{\text{in}}(\overline{\phi}_1 \overline{f}_1, \sigma_1) \ldots a_{\text{in}}(\overline{\phi}_n \overline{f}_n, \sigma_n)\Psi_1, \Psi_2 \rangle \tag{10.37}$$

under the condition that the σ_j, corresponding to smooth operators A_j that are not good, coincide.

In (10.36) and (10.37), the functions f_j are smooth and have compact support. The functions ϕ_j are given by the relation $\phi_j(\mathbf{k}) = \langle A_j \Phi, \Phi(\mathbf{k}) \rangle$ and $A(f, t, \sigma)$ are defined by the formulas

$$A(f, t, 1) = A(f, t) = \int \tilde{f}(\mathbf{x}|t) A(\mathbf{x}, t) d\mathbf{x},$$

$$A(f, t, -1) = A^*(f, t) = \int \overline{\tilde{f}(\mathbf{x}|t)} A^*(\mathbf{x}, t) d\mathbf{x}.$$

(In other words, for arbitrary smooth operators $A_1, \ldots, A_n$, we have weak convergence of the operators $A_1(f_1, t, \sigma_1) \ldots A_n(f_n, t, \sigma_n)$ to the operators $a_{\text{in}}(\overline{\phi}_1 \overline{f}_1, \sigma_1) \ldots a_{\text{in}}(\overline{\phi}_n \overline{f}_n, \sigma_n)$ in the limit as $t \to -\infty$ on the set $\mathcal{D}_{\text{in}}$ under the condition that $\sigma_1 = \cdots = \sigma_n$; for good operators $A_1, \ldots, A_n$, one can prove that the convergence is strong on $\mathcal{D}_{\text{in}}$ and the condition $\sigma_i = \sigma_j$ is not necessary.)

Let us begin the proof of the theorem.

Lemma 10.5. *Let $A_1, \ldots, A_n \in \mathcal{A}, f_1, \ldots, f_n$ be smooth functions with compact support and $\sigma_1, \ldots, \sigma_n = \pm 1$. If for at least one pair of indices i, j, where $1 \leq i \leq j \leq n$, the sets $U(f_i)$ and $U(f_j)$ do not intersect and the $\sigma_i = \sigma_j$, then for any m*

$$|\langle A_1(f_1, t, \sigma_1) \ldots A_n(f_n, t, \sigma_n) \rangle^T| \leq C(1 + |t|^m)^{-1}.$$

This statement holds if we replace some of the operators $A_i(f_i, t, \sigma_i)$ with their time derivatives $\frac{d}{dt} A_i(f_i, t, \sigma_i)$ in the expression $\langle A_1(f_1, t, \sigma_1) \dots A_n(f_n, t, \sigma_n) \rangle^T$.

Proof. Let us write the expression

$$\langle A_1(f_1, t, \sigma_1) \dots A_n(f_n, t, \sigma_n) \rangle^T \qquad (10.38)$$

in the form

$$\int f_1(\mathbf{p}_1, \sigma_1) \exp(-i\sigma_1 \omega(\mathbf{p}_1)t) \dots f_n(\mathbf{p}_n, \sigma_n)$$

$$\times \exp(-i\sigma_n \omega(\mathbf{p}_n)t) \tilde{w}_n^T(\mathbf{p}_1, \sigma_1, \dots, \mathbf{p}_n, \sigma_n) d^n \mathbf{p}.$$

[Here, we have introduced the notation

$$\tilde{w}_n^T(\mathbf{p}_1, \sigma_1, \dots, \mathbf{p}_n, \sigma_n)$$

$$= \int \exp\left(i \sum \mathbf{p}_j \mathbf{x}_j\right) \langle A_1(\mathbf{x}_1, t, \sigma_1) \dots A_n(\mathbf{x}_n, t, \sigma_n) \rangle^T d^n \mathbf{x},$$

$$A(\mathbf{x}, t, 1) = A(\mathbf{x}, t), \quad A(\mathbf{x}, t, -1) = A^*(\mathbf{x}, t).]$$

It follows from Lemma 10.3 in Section 10.1 that the function $\tilde{w}_n^T$ has the form $\tilde{w}_n^T(\mathbf{p}_1, \sigma_1, \dots, \mathbf{p}_n, \sigma_n) = \nu_n(\mathbf{p}_2, \dots, \mathbf{p}_n, \sigma_1, \dots, \sigma_n)\delta(\mathbf{p}_1 + \dots + \mathbf{p}_n)$, where ν_n is a smooth function [one can check that $\nu_n \in \mathcal{S}(E^{3(n-1)})$].

Let us assume for concreteness that $\sigma_1 = \sigma_2$ and $U(f_1)$ does not intersect $U(f_2)$. It then becomes convenient to write (10.38) in the form

$$\int \exp(-i\Omega(\mathbf{p}_2, \dots, \mathbf{p}_n)t)\chi(\mathbf{p}_2, \dots, \mathbf{p}_n)d\mathbf{p}_2 \dots d\mathbf{p}_n,$$

where $\chi \in \mathcal{S}(E^{3(n-1)})$;

$$\Omega(\mathbf{p}_2, \dots, \mathbf{p}_n) = \sigma_1 \omega(-\mathbf{p}_2 - \dots - \mathbf{p}_n)$$

$$+ \sigma_2 \omega(\mathbf{p}_2) + \sigma_3 \omega(\mathbf{p}_3) + \dots + \sigma_n \omega(\mathbf{p}_n).$$

Let us note that

$$\frac{\partial \Omega}{\partial \mathbf{p}_2} = -\sigma_1 \left.\frac{\partial \omega}{\partial \mathbf{p}}\right|_{\mathbf{p}=-\mathbf{p}_2-\cdots-\mathbf{p}_n} + \sigma_2 \left.\frac{\partial \omega}{\partial \mathbf{p}}\right|_{\mathbf{p}=\mathbf{p}_2}$$

$$= \sigma_2 \left(\left.\frac{\partial \omega}{\partial \mathbf{p}}\right|_{\mathbf{p}=\mathbf{p}_2} - \left.\frac{\partial \omega}{\partial \mathbf{p}}\right|_{\mathbf{p}=\mathbf{p}_1} \right),$$

from which it is clear that

$$\frac{\partial \Omega}{\partial \mathbf{p}_2} \neq 0$$

on the support of the function χ.

We can therefore use formula (10.22), assuming that all variables, except for $\mathbf{p}_2$, are fixed. In the case at hand, we have

$$\mathbf{u}(\mathbf{p}) = \frac{1}{t} \left(\frac{\partial \Omega}{\partial \mathbf{p}_2} \right) \left| \frac{\partial \Omega}{\partial \mathbf{p}_2} \right|^{-2},$$

and therefore

$$|(L^r \chi)(\mathbf{p})| \leq C |t|^{-r}. \tag{10.39}$$

The inequality (10.39) gives us the inequality we seek. $\qquad\square$

The proof in the case when the operators $A_i(f_i, t, \sigma_i)$ are replaced by their derivatives is similar to the proof of Lemma 10.4.

Lemma 10.6. *If $f_1, \ldots, f_n$ are non-overlapping smooth functions with compact support and $B_1, \ldots, B_n$ are good operators, then for all m*

$$\lim_{t \to -\infty} t^m \| \Psi_-(B_1, \ldots, B_n | f_1, \ldots, f_n)$$

$$- \Psi(B_1, \ldots, B_n | f_1, \ldots, f_n | t) \| = 0.$$

Proof. It is enough to show that for any m, we have

$$\left| \left\langle \frac{d}{dt} \Psi(B_1, \ldots, B_n | f_1, \ldots, f_n | t), \frac{d}{dt} \Psi(B_1, \ldots, B_n | f_1, \ldots, f_n | t) \right\rangle \right|$$

$$\leq C (1 + |t|^m)^{-1}.$$

This inequality can be obtained in the same way as the inequality with $m = 3$ in Section 10.1, by using the just-proved Lemma 10.5 instead of Lemma 10.4. $\qquad\square$

Lemma 10.7. *The relations* (10.36) *and* (10.37) *hold in the case of* $\Psi_1 = \Phi.$

Proof. To prove this lemma, we first note that the norm of the vector $\xi(t)$, defined by the formula

$$\xi(t) = A_1(f_1, t, \sigma_1) \ldots A_n(f_n, t, \sigma_n)\Phi,$$

is bounded above by a constant not depending on t. Indeed, $\langle \xi(t), \xi(t) \rangle$ can be expanded in terms of truncated Wightman functions, and by Lemma 10.4 from Section 10.1, all the resulting truncated Wightman functions are bounded. To prove the lemma, it is enough to consider the case when the vector Ψ_2 has the form

$$\Psi_2 = \Psi_-(B_1, \ldots, B_m | g_1, \ldots, g_m) = \lim_{t \to -\infty} \Psi_2(t),$$

where

$$\Psi_2(t) = \Psi(B_1, \ldots, B_m | g_1, \ldots, g_m | t).$$

It follows from the boundedness of $\|\xi(t)\|$ and $\|\Psi_2(t)\|$ that

$$\lim_{t \to -\infty} \langle \xi(t), \Psi_2 \rangle = \lim_{t \to -\infty} \langle \xi(t), \Psi_2(t) \rangle.$$

The inner product

$$\langle \xi(t), \Psi_2(t) \rangle$$

$$= \langle A_1(f_1, t, \sigma_1) \ldots A_n(f_n, t, \sigma_n)\Phi, B_1(g_1, t) \ldots B_m(g_m, t)\Phi \rangle$$

can be expanded in terms of truncated Wightman functions. By Lemma 10.4 of Section 10.1, as $t \to -\infty$, the products that differ from zero contain two-point truncated Wightman functions. It therefore follows that

$$\lim_{t \to +\infty} \langle \xi(t), \Psi_2(t) \rangle$$

$$= \langle a_{\text{in}}(\overline{\phi}_1 \overline{f}_1, \sigma_1) \ldots a_{\text{in}}(\overline{\phi}_n \overline{f}_n, \sigma_n)\Phi, a_{\text{in}}^+(\overline{\psi}_1, \overline{g}_1) \ldots a_{\text{in}}^+(\overline{\psi}_m \overline{g}_m)\Phi \rangle$$

$$= \langle a_{\text{in}}(\overline{\phi}_1 \overline{f}_1, \sigma_1) \ldots a_{\text{in}}(\overline{\phi}_n, \overline{f}_n, \sigma_n)\Phi, \Psi_-(B_1, \ldots, B_m | g_1, \ldots, g_m) \rangle$$

$$\tag{10.40}$$

(here, the symbol ψ_i denotes a generalized function defined by the formula $B_i\Phi = \int \psi_i(\mathbf{k})\Phi(\mathbf{k})d\mathbf{k}$).

In the derivation of (10.40), we should use the formula

$$\lim_{t\to\infty} \langle A_i(f_i, t, \sigma_i) A_j(f_j, t, \sigma_j)\rangle^T$$
$$= \langle a_{\mathrm{in}}(\overline{\phi}_i\overline{f}_i, \sigma_i)a_{\mathrm{in}}(\overline{\phi}_j\overline{f}_j, \sigma_j)\rangle,$$

which holds if at least one of the operators A_i, A_j is good or is adjoint to a good operator or, without this assumption, if we have $\sigma_i = \sigma_j$ (in particular, this formula can be applied in the case $A_j = B_j$, $f_j = g_j$; then $\phi_j = \psi_j$).

The relation (10.40) implies equation (10.37) for $\Psi_1 = \Phi$.

In the case when $A_1, \ldots, A_n$ are good operators, we can, as in Section 10.1, prove that the vector $\xi(t)$ has a limit as $t \to -\infty$.

It follows from the statements we have already proved that the vector $\xi - \eta$, where $\xi = \lim_{t\to-\infty}\xi(t), \eta = a_{\mathrm{in}}(\overline{\phi}_1\overline{f}_1, \sigma_1)$ $\ldots a_{\mathrm{in}}(\overline{\phi}_n\overline{f}_n, \sigma_n)\Phi$, is orthogonal to the space $\mathcal{H}_{\mathrm{in}}$. On the other hand, a simple calculation, based on expansion in terms of truncated Wightman functions, shows that $\langle \xi, \xi\rangle = \langle \eta, \eta, \rangle$. Since $\eta \in \mathcal{H}_{\mathrm{in}}$, this implies that $\xi = \eta$. Therefore, relation (10.36) holds for $\Psi_1 = \Phi$, which finishes the proof of Lemma 10.6 in this case. $\square$

Let us now prove the relations (10.36) and (10.37) in the general case.

Let $\Psi_1 = \Psi_-(E_1, \ldots, E_r | h_1, \ldots, h_r)$, where $E_1, \ldots, E_r$ are good operators, $h_1, \ldots, h_r$ are non-overlapping functions, and $\Psi_1(t) = \Psi(E_1, \ldots, E_r | h_1, \ldots, h_r | t)$.

It follows from (10.8) that $\|A_i(f, t, \sigma)\| \leq C(1 + |t|^{3/2})$ and therefore from Lemma 10.3 it follows that for $t \to \infty$ we have

$$\|A_1(f_1, t, \sigma_1) \ldots A_n(f_n, t, \sigma_n)(\Psi_1 - \Psi_1(t))\| \to 0.$$

Therefore, we have

$$\lim_{t\to-\infty} A_1(f_1, t, \sigma_1) \ldots A_n(f_n, t, \sigma_n)\Psi_1$$

$$= \lim_{t\to-\infty} A_1(f_1, t, \sigma_1) \ldots A_n(f_n, t, \sigma_n)\Psi_1(t)$$

$$= \lim_{t\to-\infty} A_1(f_1, t, \sigma_1) \ldots A_n(f_n, t, \sigma_n)E_1(h_1, t, 1) \ldots E_r(h_r, t, 1)\Phi.$$

To finish the proof, it remains to apply Lemma 10.3.

Remark 10.3. Relation (10.37) holds in the case when Ψ_2 is an arbitrary vector in $\mathcal{H}_{\text{in}}$.

10.4 Dressing operators

In this section, we will prove that the scattering matrix construction provided in Section 10.1 is equivalent to a different construction which is more closely connected to the physical picture of scattering.

Let us begin by introducing the notion of a dressing operator.

An operator D, acting from space $\mathcal{H}_{\text{as}}$ to the space $\mathcal{H}$, will be called in-*dressing* (out-*dressing*), if

$$S_- = \operatorname*{slim}_{t \to -\infty} \exp(iHt) D \exp(-iH_{\text{as}}t), \tag{10.41}$$

$$S_+ = \operatorname*{slim}_{t \to +\infty} \exp(iHt) D \exp(-iH_{\text{as}}t). \tag{10.42}$$

If both relations are satisfied, then the operator D will be called a *dressing operator*.

It is useful to note that it follows from relation (10.4) that both in-dressing and out-dressing operators satisfy the conditions

$$D\theta = \Phi, \tag{10.43}$$

$$Da^+(\mathbf{k})\theta = \Phi(\mathbf{k}). \tag{10.44}$$

However, these conditions are not sufficient for an operator to be a dressing operator. Note that, in particular, the S_- operator is an in-dressing operator and the S_+ operator is an out-dressing operator.

Let us now give sufficient conditions for an operator D to be in-dressing. Namely, let us prove the following theorem.

Theorem 10.1. *Let us suppose that* (a) *the operator D has norm 1,* (b) *for all non-overlapping families of functions with compact support* $f_1(\mathbf{k}), \ldots, f_n(\mathbf{k})$ *and for all smooth operators* $A_1, \ldots, A_n \in \mathcal{A}$, *we have, for $t < 0$,*

$$\langle Da^+(\overline{f}_1^t) \ldots a^+(\overline{f}_n^t)\theta, A_1(g_1) \ldots A_n(g_n)\Phi\rangle$$

$$\approx \sum_{\pi \in P} \langle \Phi(f_1^t), A_{i_1}(g_{i_1})\Phi\rangle \ldots \langle \Phi(f_n^t), A_{i_n}(g_{i_n})\Phi\rangle \tag{10.45}$$

with error not exceeding $c\nu^n(g)|t|^{-N}$ (here, $f_i^t(\mathbf{k}) = f_i(\mathbf{k})\exp(-i\omega(\mathbf{k})t)$, P is the set of permutations $\pi = (i_1,\ldots,i_n)$ of indices $(1,\ldots,n)$, $\nu(g) = \max_i(\sup|g_i(\mathbf{x})| + \int |g_i(\mathbf{x})|d\mathbf{x})$; $g_i \in \mathcal{S}(E^3)$; N is an arbitrary number; C is a constant depending on N but not on the functions g_i). Then the operator D is in-dressing.

The condition for an operator to be out-dressing can be obtained from the theorem above by replacing $t < 0$ with $t > 0$.

To prove the above theorem, let us consider the vector $x = S_- y$, where $x = \Psi_-(B_1,\ldots,B_n|f_1,\ldots,f_n)$; $y = a^+(\overline{\phi}_1\overline{f}_1)\ldots a^+(\overline{\phi}_n\overline{f}_n)\theta$; $B_1,\ldots,B_n$ are good operators and $f_1,\ldots,f_n$ are non-overlapping functions, and

$$x(t) = \Psi_-(B_1,\ldots,B_n|f_1^t,\ldots,f_n^t)$$
$$= S_- a^+(\overline{\phi}_1\overline{f}_1^t)\ldots a^+(\overline{\phi}_n\overline{f}_n^t)\theta.$$

It is clear that

$$\exp(-iH_{\mathrm{as}}t)y = a^+(\overline{\phi}_1\overline{f}_1^t)\ldots a^+(\overline{\phi}_n\overline{f}_n^t)\theta \qquad (10.46)$$

and therefore, by the relation (10.5), we have

$$x(t) = \exp(-iHt)x. \qquad (10.47)$$

For large t, the vector $x(t)$ negligibly differs from the vector $\xi(t) = \Psi(B_1,\ldots,B_n|f_1^t,\ldots,f_n^t|0) = B_1(f_1^t,0)\ldots B_n(f_n^t,0)\Phi$ (to prove this statement, note that for $t \to \pm\infty$ we have

$$\|x(t) - \xi(t)\| = \|\exp(iHt)(x(t) - \xi(t))\|$$
$$= \|\Psi_-(B_1,\ldots,B_n|f_1,\ldots,f_n)$$
$$- \Psi(B_1,\ldots,B_n|f_1,\ldots,f_n|t)\| \to 0). \qquad (10.48)$$

On the other hand, by relation (10.46), we have that

$$\lim_{t\to-\infty} \langle D\exp(-iH_{\mathrm{as}}t)y, \xi(t)\rangle$$

$$= \lim_{t\to-\infty} \langle Da^+(\overline{\phi}_1\overline{f}_1^t)\ldots a^+(\overline{\phi}_n\overline{f}_n^t)\theta, B_1(f_1^t,0)\ldots B_n(f_n^t,0)\Phi\rangle$$

$$= \lim_{t\to-\infty} \sum_{\pi\in P} \langle \Phi(\phi_1 f_1^t), B_{i_1}(f_{i_1}^t,0)\Phi\rangle \ldots \langle \Phi(\phi_n f_n^t), B_{i_n}(f_{i_n}^t,0)\Phi\rangle$$

$$= \langle y, y\rangle. \qquad (10.49)$$

Recalling that $\|D\exp(-iH_{\mathrm{as}}t)y\| \leq \|y\|$ and that $\lim_{t\to-\infty}\|\xi(t) - x(t)\| = 0$, we obtain from (10.49) the relation

$$\lim_{t\to-\infty} \langle D\exp(-iH_{\mathrm{as}}t)y, x(t)\rangle = \langle y, y\rangle, \tag{10.50}$$

from which it follows by (10.47) that

$$\lim_{t\to-\infty} \langle \exp(iHt)D\exp(-iH_{\mathrm{as}}t)y, S_-y\rangle$$

$$= \lim_{t\to-\infty} \langle \exp(iHt)D\exp(-iH_{\mathrm{as}}t)y, x\rangle$$

$$= \lim_{t\to-\infty} \langle D\exp(-iH_{\mathrm{as}}t)y, x(t)\rangle = \langle y, y\rangle = \langle S_-y, S_-y\rangle.$$

Here, we apply the following simple lemma.

If $\lim \langle\beta_i, \eta\rangle = \langle\eta, \eta\rangle$ and $\|\beta_t\| \leq \|\eta\|$, then $\lim \beta_t$ exists and equals η (to prove this statement, we write $\langle\beta_t, \eta\rangle$ in the form

$$\langle\beta_t, \eta\rangle = \|\beta_t\| \cdot \|\eta\| \cos\phi_t,$$

where ϕ_t is the angle between β_t and η, and note that it follows from the inequalities $\|\beta_t\| \leq \|\eta\|, |\cos\phi_t| \leq 1$ that $\|\beta_t\| \cos\phi_t$ converges to $\|\eta\|$ only in the case when $\lim\|\beta_t\| = \|\eta\|, \lim\cos\phi_t = 1$).

Applying the lemma to the vectors $\beta_t = \exp(iHt)D\exp(-iH_{\mathrm{as}}t)y$, $\eta = S_-y$, we see that $\lim_{t\to-\infty}\exp(iHt)D\exp(-iH_{\mathrm{as}}t)y$ exists and equals S_-y.

Since the vectors $y = a^+(\bar\phi_1\bar f_1)\ldots a^+(\bar\phi_n\bar f_n)\theta$, which we have shown to satisfy the relation

$$\lim_{t\to-\infty} \exp(iHt)D\exp(-iH_{\mathrm{as}}t)y = S_-y,$$

constitute a total set in the space $\mathcal{H}_{\mathrm{as}}$, it is clear that the relation holds for a dense set of vectors and hence for any vector $y \in \mathcal{H}_{\mathrm{as}}$ (see Appendix A.5).

We have thus obtained sufficient conditions for an operator D to be in-dressing and out-dressing. These conditions can be relaxed by changing the requirement that $\|D\| = 1$ with

$$\lim_{\substack{t\to-\infty \\ (t\to+\infty)}} \|Da^+(\bar f_1^t)\ldots a^+(\bar f_n^t)\theta\| = \|a^+(\bar f_1^t)\ldots a^+(\bar f_n^t)\theta\| \tag{10.51}$$

(the proof remains the same).

The above statements also give conditions guaranteeing that the operator D is dressing (since the operator is dressing if it is simultaneously in- and out-dressing). These conditions can be easily formulated in a slightly different form. Namely, we have the following theorem.

Theorem 10.2. *Let us suppose the operator D has norm 1 and for all functions $\tilde{f}_1(\mathbf{x}), \ldots, \tilde{f}_n(\mathbf{x}) \in \mathcal{S}(E^3)$, whose supports are far from each other, and for all smooth operators $A_1, \ldots, A_n \in \mathcal{A}$, the following approximation holds:*

$$\langle Da^+(\overline{f}_1)\ldots a^+(\overline{f}_n)\theta, A_1(g_1)\ldots A_n(g_n)\Phi\rangle$$

$$\approx \sum_{\pi \in P} \langle \Phi(f_1), A_{i_1}(g_{i_1})\Phi\rangle \ldots \langle \Phi(f_n), A_{i_n}(g_{i_n})\Phi\rangle \quad (10.52)$$

with error not exceeding $C\nu^n(f)\nu^n(g)d^{-N}$ (here, $\tilde{f}_i(\mathbf{x}) = \int f_i(\mathbf{k})$ $\exp(i\mathbf{k}\mathbf{x})\frac{d\mathbf{k}}{(2\pi)^3}$; the functions $g_i \in \mathcal{S}(E^3)$; d is the minimal distance between the supports of the functions $\tilde{f}_i(\mathbf{x})$; N is an arbitrary number; C is a constant depending on N but not depending on the functions $f_1, \ldots, f_n, g_1, \ldots, g_n$; the sum is taken over all permutations $\pi = (i_1, \ldots, i_n)$). Then the operator D is a dressing operator.

Indeed, relation (10.52) implies relation (10.45) for $t \leq 0$ as well as for $t \geq 0$. To show this, note that by Lemma 10.2 from Section 10.1 the "essential supports" of the functions $\tilde{f}_i(\mathbf{x}|t) = \int \exp(i\mathbf{k}\mathbf{x})f_i^t(\mathbf{k})\frac{d\mathbf{k}}{(2\pi)^3}$ are far from each other in the limit $t \to \pm\infty$. (By this lemma, the function $\tilde{f}_i(\mathbf{x}|t)$ is small outside the set $tU_\epsilon^{(i)}$, where $U_\epsilon^{(i)}$ is an ϵ-neighborhood of the set $U^{(i)}$ consisting of points of the form $\frac{\partial\omega(\mathbf{k})}{\partial\mathbf{k}}$, where $\mathbf{k}$ runs over $\mathrm{supp} f_i$. Hence, we can say that the set $tU_\epsilon^{(i)}$ is the "essential support" of the function $f_i(\mathbf{x}|t)$, recognizing that the values of the function outside of this set are very small.)

To provide more rigorous reasoning, let us introduce smooth functions $h_i(\mathbf{x}) \in \mathcal{S}$, equal to 1 on the set $U_\epsilon^{(i)}$ and 0 outside the set $U_{2\epsilon}^{(i)}$. Then, the functions ${}^t\tilde{r}_i = h_i(t\mathbf{x})\tilde{f}_i(\mathbf{x}|t)$ will have far away supports as $t \to \pm\infty$ and the quantity $\langle Da^+({}^t\overline{r}_1)\ldots$

$a^+(^t\overline{r}_n)\theta, A_1(g_1)\dots A_n(g_n)\Phi\rangle$ can be approximated with the relation (10.52). From the other side, we have

$$a^+(\overline{f}_1^t)\dots a^+(\overline{f}_n^t)\theta \approx a^+(^t\overline{r}_1)\dots a^+(^t\overline{r}_n)\theta$$

by Lemma 10.2 of Section 10.1. Using these remarks, we obtain that relation (10.52) implies relation (10.45).

Let us note that the condition $\|D\| = 1$ can be relaxed here as well, by assuming that either the condition (10.51) or the condition

$$\|Da^+(\overline{f}_1)\dots a^+(\overline{f}_n)\theta\| \approx \|a^+(\overline{f}_1)\dots a^+(\overline{f}_n)\overline{\theta}\| \tag{10.53}$$

holds. (In (10.53), we assume that the functions $\tilde{f}_1(\mathbf{x}),\dots,\tilde{f}_n(\mathbf{x}) \in \mathcal{S}$ have distant supports; the error should not exceed

$$C\nu^n(f)d^{-N},$$

where N is an arbitrary number, d is the minimal distance between the supports, and C depends on N).

The following statement is used in Section 11.5.

Theorem 10.3. *Let $\tilde{S}$ and S be scattering matrices constructed for energy operators $\tilde{H}$ and H, the momentum operator $\mathbf{P}$, and a family of operators $\mathcal{A}$ (this family must be asymptotically commutative relative to the operators $\tilde{H}, \mathbf{P}$, as well as relative to the operators $H, \mathbf{P}$). Let us suppose that T is a unitary operator satisfying the following conditions:*

(1) *$T\Phi = \tilde{\Phi}$,*
(2) *$T\Phi(\mathbf{k}) = \tilde{\Phi}(\mathbf{k})$,*
(3) *if $A \in \mathcal{A}$, then $T^{-1}AT \in \mathcal{A}$ (here, Φ and $\tilde{\Phi}$ are ground states and $\Phi(\mathbf{k})$ and $\tilde{\Phi}(\mathbf{k})$ are single-particle states corresponding to the energy operators H and $\tilde{H}$).*

If D is a dressing operator for the operator H, satisfying the conditions of Theorem 10.2, then the operator $\tilde{D} = TD$ is a dressing operator for the energy operator $\tilde{H}$.

The proof of this theorem consists of checking that the conditions of Theorem 10.2 apply to the operator $\tilde{D}$. This verification can be done with the following transformations:

$$\langle \tilde{D} a^+(\overline{f}_1)\ldots a^+(\overline{f}_n)\theta, A_1(g_1)\ldots A_n(g_n)\tilde{\Phi}\rangle$$

$$= \langle D a^+(\overline{f}_1)\ldots a^+(\overline{f}_n)\theta, T^{-1}A_1(g_1)\ldots A_n(g_n)TT^{-1}\tilde{\Phi}\rangle$$

$$= \langle D a^+(\overline{f}_1)\ldots a^+(\overline{f}_n)\theta, A_1^T(g_1)\ldots A_n^T(g_n)\tilde{\Phi}\rangle$$

$$\approx \sum_\pi \langle \Phi(f_1), A_{i_1}^T(g_{i_1})\Phi\rangle \ldots \langle \Phi(f_n), A_{i_n}^T(g_{i_n})\Phi\rangle$$

$$= \sum_\pi \langle T\Phi(f_1), A_{i_1}(g_{i_1})T\Phi\rangle \ldots \langle T\Phi(f_n), A_{i_n}(g_{i_n})T\Phi\rangle$$

$$= \sum_\pi \langle \tilde{\Phi}(f_1), A_{i_1}(g_{i_1})\tilde{\Phi}\rangle \ldots \langle \tilde{\Phi}(f_n), A_{i_n}(g_{i_n})\tilde{\Phi}\rangle$$

(here, we have used the notation $T^{-1}A_iT = A_i^T$).

The statements we have proved in this section allow us to give a new definition of Møller matrices and the scattering matrix. Namely, if the operator D satisfies conditions (10.52) and (10.53), then we can define Møller matrices using equations (10.41) and (10.42). The scattering matrix, as usual, can be defined by the relation $S = S_+^* S_-$; in the case when the scattering matrix is unitary, this definition is equivalent to the formula

$$S = S_+^{-1} S_-$$

$$= \operatorname*{slim}_{\substack{t\to+\infty,\\ t_0\to-\infty}} \exp(iH_{\mathrm{as}}t)D^{-1}\exp(-iH(t-t_0))D\exp(-iH_{\mathrm{as}}t_0).$$

$$(10.54)$$

Let us now consider how the the formulas (10.41), (10.42) and (10.54) are connected to the physical picture of particle scattering. For this discussion, we will fix an operator D, satisfying the conditions (10.52) and (10.53).

Let us first consider the collision of two particles. A single particle with the wave function f will be written as the vector

$$\Phi(f) = \int f(\mathbf{k})\Phi(\mathbf{k})d\mathbf{k} = Da^+(\overline{f})\theta.$$

Let us ask the following question: which vector in the space $\mathcal{H}$ should be used to describe a state with two particles having wave functions f_1 and f_2?

Before we answer this question, note that its formulation in such a general form is physically meaningless: we should assume that the particles are far from each other. More formally, if two particles have wave functions $f_1(\mathbf{k}), f_2(\mathbf{k})$, we should assume that the supports of the functions $\tilde{f}_1(\mathbf{x}), \tilde{f}_2(\mathbf{x})$ (the wave functions of these particles in coordinate representation) are far from each other. With this narrowed framing of the question, we can now give a definite answer. Namely, we should represent such a state with the vector $Da^+(\overline{f}_1)a^+(\overline{f}_2)\theta$.

Indeed, condition (10.52), imposed on the operator D, corresponds to the physical representation of the state with two distant particles $\Phi(f_1)$ and $\Phi(f_2)$.

Analogously, the vector $Da^+(\overline{f}_1)\ldots a^+(\overline{f}_n)\theta$, in the case when the supports of the functions $\tilde{f}_1(\mathbf{x}),\ldots,\tilde{f}_n(\mathbf{x})$ are far from each other, describes the state of n spatially separated particles.

It should now be easy to see that the collision of two particles can be described by the vector

$$x(t) = \exp(-iHt)S_- a^+(\overline{f}_1)a^+(\overline{f}_2)\theta,$$

where f_1, f_2 are non-overlapping functions. Indeed, from the relation (10.41), it follows that for $t \to -\infty$, we have

$$x(t) \approx D\exp(-iH_{\mathrm{as}}t)a^+(\overline{f}_1)a^+(\overline{f}_2)\theta = Da^+(\overline{f}_1^t)a^+(\overline{f}_2^t)\theta,$$

and hence as $t \to -\infty$, $x(t)$ describes the state of two spatially separated particles $\Phi(f_1^t)$ and $\Phi(f_2^t)$ (technically, it is wrong to say that the supports of the functions $\tilde{f}_1^t$, $\tilde{f}_2^t$ are far from each other; as discussed above, it is in fact their essential supports that can be shown to be far from each other).

In particular, the vector $x(0) = S_- a^+(\overline{f}_1)a^+(\overline{f}_2)\theta$ depicts the state of a system of two particles at $t = 0$, whose wave functions at $t \to -\infty$ are f_1^t, f_2^t (if each particle moves freely, then they can be described at $t = 0$ by the wave functions f_1 and f_2). Analogously, we can interpret other vectors of the form $S_-\xi$ and $S_+\eta$.

It should now be clear that the operator S defines the transition from the initial state to the final state in a particle collision. Indeed, if $\eta = S\xi$, then the vector

$$x(t) = \exp(-iHt)S_-\xi = \exp(-iHt)S_+\eta$$

depicts the collision process where we begin (at $t \to -\infty$) in the state

$$x(t) \approx D\exp(-iH_{\mathrm{as}}t)\xi,$$

and end (at $t \to +\infty$) in the state

$$x(t) \approx D\exp(-iH_{\mathrm{as}}t)\eta.$$

Let us now investigate the relationship between the matrix entries of the scattering matrix (scattering amplitudes) and the experimentally observed quantities.

For simplicity, let us consider the collision of two particles. We assume that a beam of particles with momentum $\mathbf{q}$ scatters on a resting particle. Let us assume that the beam has unit flux (the meaning of this, in both the classical and quantum situations, is explained in the Chapter 5, Section 5.2). In the simplest case, two particles remain at the end of the scattering process, however we will consider the more general situation where new particles can be born and we can thus have more than two particles at the end. We should predict the probability (or, more precisely, the probability density) that at the end we observe n particles with momenta $(\mathbf{p}_1, \dots, \mathbf{p}_n)$. It will be more convenient to consider the probability σ_G that the vector of momenta lies in the domain G in $3n$-dimensional space E^{3n} (if the particles are identical, we should assume that the domain G is invariant with respect to permutations). This number describes how many collisions per second end up with n particles, whose momentum vector $\mathbf{p}_1, \dots, \mathbf{p}_n$ belongs to the domain G. This number is called the effective collision cross-section (and called the effective differential cross-section in the case of an infinitesimally small domain G).

Let us now show that the collision cross-section σ_G can be expressed using the scattering amplitudes by the following formula:

$$\sigma_G = \frac{1}{n!}\frac{1}{|\mathbf{v}(\mathbf{q})|}\int_G |s_{n,2}(\mathbf{p}_1, \dots, \mathbf{p}_n|\mathbf{q}, 0)|^2 \delta(\mathbf{p}_1 + \cdots + \mathbf{p}_n - \mathbf{q})$$

$$\times \delta(\omega(\mathbf{p}_1) + \cdots + \omega(\mathbf{p}_n) - \omega(\mathbf{q}) - \omega(0))d\mathbf{p}_1 \dots d\mathbf{p}_n. \tag{10.55}$$

Here, $s_{n,k}$ are functions that are related to the matrix elements $S_{n,k}$ of the scattering matrix by the relation

$$S_{n,k}(\mathbf{p}_1, \ldots, \mathbf{p}_n | \mathbf{q}_1, \ldots, \mathbf{q}_k)$$
$$= \langle S a^+(\mathbf{q}_1) \ldots a^+(\mathbf{q}_k)\theta, a^+(\mathbf{p}_1) a^+(\mathbf{p}_n)\theta \rangle$$
$$= s_{n,k}(\mathbf{p}_1, \ldots, \mathbf{p}_n | \mathbf{q}_1, \ldots, \mathbf{q}_k)\delta(\mathbf{p}_1 + \cdots + \mathbf{p}_n - \mathbf{q}_1 - \cdots - \mathbf{q}_k)$$
$$\times \delta(\omega(\mathbf{p}_1) + \cdots + \omega(\mathbf{p}_n) - \omega(\mathbf{q}_1) - \cdots - \omega(\mathbf{q}_k)) \qquad (10.56)$$

(sometimes the term scattering amplitudes refers to the functions $s_{n,k}$ as well). The symbol $\mathbf{v}(\mathbf{k})$, as always, denotes the function $\frac{\partial \omega(\mathbf{k})}{\partial \mathbf{k}}$; we assume that $\mathbf{v}(0) = 0$. Formula (10.55) is proved below under the assumption that $\mathbf{q} \neq 0$ and the function $s_{n,2}$ is continuous at points $(\mathbf{p}_1, \ldots, \mathbf{p}_n | \mathbf{q}, 0)$, where $(\mathbf{p}_1, \ldots, \mathbf{p}_n) \in G$.

To understand the collision cross-sections in the case at hand, we will use the construction described in Section 5.2. In general, the following discussion is largely similar to the discussion in Section 5.2.

Namely, let us assume that the process of collision is described by the vector $D \exp(-iH_0 t)\xi_\alpha$ for $t \to -\infty$, where

$$\xi_\alpha = \left(\int \exp(i\mathbf{k}\alpha) f(\mathbf{k}) a^+(\mathbf{k}) d\mathbf{k} \right) \left(\int g(\mathbf{k}) a^+(\mathbf{k}) d\mathbf{k} \right) \theta, \qquad (10.57)$$

$f(\mathbf{k})$ and $g(\mathbf{k})$ are normalized wave functions differing from zero only in a small neighborhood of the points $\mathbf{q}$ and 0, respectively; α is a vector orthogonal to the vector $\mathbf{v}(\mathbf{q})$ (as explained in Section 5.2, this vector is analogous to the impact parameter in classical mechanics.)

The collision process is described by the vector

$$x_\alpha(t) = \exp(-iHt)S_- \xi_\alpha,$$

and at the end of the process (with $t \to +\infty$) we obtain the state

$$D \exp(-iH_0 t)(S\xi_\alpha).$$

The probability that for initial state (10.57) we obtain the final state $S\xi_\alpha$ of n particles with momenta $(\mathbf{p}_1, \ldots, \mathbf{p}_n)$, belonging to the domain G, can be written in the form

$$w_{\boldsymbol{\alpha}}(G) = \frac{1}{n!} \int_G \left| \int d\mathbf{k}_1 d\mathbf{k}_2 S_{n,2}(\mathbf{p}_1, \ldots, \mathbf{p}_n | \mathbf{k}_1, \mathbf{k}_2) \right.$$

$$\left. \times\, f(\mathbf{k}_1) \exp(i\mathbf{k}_1\boldsymbol{\alpha}) g(\mathbf{k}_2) \right|^2 d\mathbf{p}_1 \ldots d\mathbf{p}_n \qquad (10.58)$$

$\left[\vphantom{\sum}\right.$ we have used the relation

$$S\xi_{\boldsymbol{\alpha}} = \sum_m \frac{1}{m!} \int S_{m,2}(\mathbf{p}_1, \ldots, \mathbf{p}_m | \mathbf{k}_1, \mathbf{k}_2) f(\mathbf{k}_1)$$

$$\times\, \exp(i\mathbf{k}_1\boldsymbol{\alpha}) g(\mathbf{k}_2) a^+(\mathbf{p}_1) \ldots a^+(\mathbf{p}_m)\theta d\mathbf{p}_1 \ldots d\mathbf{p}_m d\mathbf{k}_1 d\mathbf{k}_2 \left.\vphantom{\sum}\right].$$

The collision cross-section σ_G is equal to[5]

$$\sigma_G = \int_{\boldsymbol{\alpha}\perp\mathbf{v(q)}} d\boldsymbol{\alpha}\, w_{\boldsymbol{\alpha}}(G). \qquad (10.59)$$

Substituting into (10.59) the expression (10.58) and integrating over $\boldsymbol{\alpha}$, we see that

$$\sigma_G = \frac{1}{n!} \int_G d\mathbf{p}_1 \ldots d\mathbf{p}_n \int d\mathbf{k}_1 d\mathbf{k}_1' d\mathbf{k}_2 d\mathbf{k}_2' (s_{n,2}(\mathbf{p}_1,$$

$$\ldots, \mathbf{p}_n | \mathbf{k}_1, \mathbf{k}_2) \bar{s}_{n,2}(\mathbf{p}_1, \ldots, \mathbf{p}_n | \mathbf{k}_1', \mathbf{k}_2') f(\mathbf{k}_1) \bar{f}(\mathbf{k}_1')$$

$$\times\, g(\mathbf{k}_2) \bar{g}(\mathbf{k}_2') \delta(\mathbf{k}_1^T - \mathbf{k}_1'^T))$$

$$= \frac{1}{n!} \int_G d\mathbf{p}_1 \ldots d\mathbf{p}_n \int d\mathbf{k}_1 d\mathbf{k}_1' d\mathbf{k}_2 d\mathbf{k}_2' (s_{n,2}(\mathbf{p}_1, \ldots, \mathbf{p}_n | \mathbf{k}_1, \mathbf{k}_2)$$

$$\times\, \bar{s}_{n,2}(\mathbf{p}_1, \ldots, \mathbf{p}_n | \mathbf{k}_1', \mathbf{k}_2') f(\mathbf{k}_1) \bar{f}(\mathbf{k}_1') g(\mathbf{k}_2) \bar{g}(\mathbf{k}_2')$$

$$\times\, \delta(\mathbf{k}_1^T - \mathbf{k}_1'^T) \delta(\mathbf{p}_1 + \cdots + \mathbf{p}_n - \mathbf{k}_1 - \mathbf{k}_2) \delta(\mathbf{p}_1 + \cdots + \mathbf{p}_n$$

$$-\, \mathbf{k}_1' - \mathbf{k}_2') \delta(\omega(\mathbf{p}_1) + \cdots + \omega(\mathbf{p}_n) - \omega(\mathbf{k}_1) - \omega(\mathbf{k}_2))$$

$$\times\, \delta(\omega(\mathbf{p}_1) + \cdots + \omega(\mathbf{p}_n) - \omega(\mathbf{k}_1') - \omega(\mathbf{k}_2')) \qquad (10.60)$$

---

[5]Strictly speaking, this formula contains a limit: the probabilities $w_{\boldsymbol{\alpha}}(G)$ should be defined with the functions $f_\nu(\mathbf{k})$, $g_\nu(\mathbf{k})$, whose supports contract to the points $\mathbf{q}, 0$, respectively, as $\nu \to \infty$.

[here the symbols $\mathbf{k}_1^T$, $\mathbf{k}_1'^T$ denote the projections of the vectors $\mathbf{k}_1$, $\mathbf{k}_1'$ on the plane orthogonal to the vector $\mathbf{v}(\mathbf{q})$].

In further calculations, we will use the following relations:

$$\delta(\mathbf{p}_1 + \cdots + \mathbf{p}_n - \mathbf{k}_1 - \mathbf{k}_2)\delta(\mathbf{p}_1 + \cdots + \mathbf{p}_n - \mathbf{k}_1' - \mathbf{k}_2')$$
$$= \delta(\mathbf{p}_1 + \cdots + \mathbf{p}_n - \mathbf{k}_1 - \mathbf{k}_2)\delta(\mathbf{k}_1 + \mathbf{k}_2 - \mathbf{k}_1' - \mathbf{k}_2'),$$
$$\delta(\omega(\mathbf{p}_1) + \cdots + \omega(\mathbf{p}_n) - \omega(\mathbf{k}_1) - \omega(\mathbf{k}_2))\delta(\omega(\mathbf{p}_1) + \cdots$$
$$+ \omega(\mathbf{p}_n) - \omega(\mathbf{k}_1') - \omega(\mathbf{k}_2'))$$
$$= \delta(\omega(\mathbf{p}_1) + \cdots + \omega(\mathbf{p}_n)$$
$$- \omega(\mathbf{k}_1) - \omega(\mathbf{k}_2))\delta(\omega(\mathbf{k}_1) + \omega(\mathbf{k}_2) - \omega(\mathbf{k}_1') - \omega(\mathbf{k}_2')),$$
$$\delta(\mathbf{k}_1^T - \mathbf{k}_1'^T)\delta(\mathbf{k}_1 + \mathbf{k}_2 - \mathbf{k}_1' - \mathbf{k}_2')\delta(\omega(\mathbf{k}_1) + \omega(\mathbf{k}_2) - \omega(\mathbf{k}_1') - \omega(\mathbf{k}_2'))$$
$$= \frac{|\mathbf{v}(\mathbf{q})|}{(\mathbf{v}(\mathbf{k}_1) - \mathbf{v}(\mathbf{k}_2))\mathbf{v}(\mathbf{q})}\delta(\mathbf{k}_1 - \mathbf{k}_1')\delta(\mathbf{k}_2 - \mathbf{k}_2')$$
$$+ \boldsymbol{\alpha}(\mathbf{k}_1, \mathbf{k}_2)\delta(\mathbf{k}_1' - \boldsymbol{\beta}(\mathbf{k}_1, \mathbf{k}_2))\delta(\mathbf{k}_2' - \boldsymbol{\gamma}(\mathbf{k}_1, \mathbf{k}_2)),$$

where $\boldsymbol{\beta}(\mathbf{k}_1, \mathbf{k}_2)$, $\boldsymbol{\gamma}(\mathbf{k}_1, \mathbf{k}_2)$ are non-trivial solutions of the following system:

$$\omega(\mathbf{k}_1) + \omega(\mathbf{k}_2) - \omega(\boldsymbol{\beta}) - \omega(\boldsymbol{\gamma}) = 0,$$
$$\mathbf{k}_1 + \mathbf{k}_2 - \boldsymbol{\beta} - \boldsymbol{\gamma} = 0,$$
$$\mathbf{k}_1^T - \boldsymbol{\beta}^T = 0$$

(i.e. solutions other than $\boldsymbol{\beta} = \mathbf{k}_1, \boldsymbol{\gamma} = \mathbf{k}_2$).

Using these formulas, we can transform expression (10.60) into the form

$$\sigma_G = \frac{1}{n!}\int_G d\mathbf{p}_1 \ldots d\mathbf{p}_n \int d\mathbf{k}_1 d\mathbf{k}_2 \left(\frac{|\mathbf{v}(\mathbf{q})|}{(\mathbf{v}(\mathbf{k}_1) - \mathbf{v}(\mathbf{k}_2)) \cdot \mathbf{v}(\mathbf{q})}\right.$$
$$\times |s_{n,2}(\mathbf{p}_1, \ldots, \mathbf{p}_n|\mathbf{k}_1, \mathbf{k}_2)|^2 |f(\mathbf{k}_1)|^2 |g(\mathbf{k}_2)|^2$$
$$\times \delta(\mathbf{p}_1 + \cdots + \mathbf{p}_n - \mathbf{k}_1 - \mathbf{k}_2)\delta(\omega(\mathbf{p}_1) + \cdots + \omega(\mathbf{p}_n)$$
$$\left. - \omega(\mathbf{k}_1) - \omega(\mathbf{k}_2))\right)$$

$$+\frac{1}{n!}\int_G d\mathbf{p}_1\ldots d\mathbf{p}_n\int d\mathbf{k}_1 d\mathbf{k}_2(s_{n,2}(\mathbf{p}_1,$$

$$\ldots,\mathbf{p}_n|\mathbf{k}_1,\mathbf{k}_2)\bar{s}_{n,2}(\mathbf{p}_1,\ldots,\mathbf{p}_n|\boldsymbol{\beta}(\mathbf{k}_1,\mathbf{k}_2),\boldsymbol{\gamma}(\mathbf{k}_1,\mathbf{k}_2))$$

$$\times f(\mathbf{k}_1)\bar{f}(\boldsymbol{\beta}(\mathbf{k}_1,\mathbf{k}_2)g(\mathbf{k}_2)\bar{g}(\boldsymbol{\gamma}(\mathbf{k}_1,\mathbf{k}_2))\boldsymbol{\alpha}(\mathbf{k}_1,\mathbf{k}_2)$$

$$\times \delta(\mathbf{p}_1+\cdots+\mathbf{p}_n-\mathbf{k}_1-\mathbf{k}_2)\delta(\omega(\mathbf{p}_1)$$

$$+\cdots+\omega(\mathbf{p}_n)-\omega(\mathbf{k}_1)-\omega(\mathbf{k}_2))).$$

To obtain the final expression for σ_G, we should use the fact that for normalized functions $f(\mathbf{k})$ with support in a small neighborhood of the point $\mathbf{q}$ and for functions $\lambda(\mathbf{k})$, continuous at the point $\mathbf{q}$, we have the relations

$$\int \lambda(\mathbf{k})|f(\mathbf{k})|^2 d\mathbf{k}=\lambda(\mathbf{q}),$$

$$\int \lambda(\mathbf{k})f(\mathbf{k})\bar{f}(\rho(\mathbf{k}))d\mathbf{k}=0 \tag{10.61}$$

(in formulas (10.61) there is a limit in which the support of the function $f(\mathbf{k})$ contracts to the point $\mathbf{q}$; $\rho(\mathbf{k})$ is a function that is continuous at the point $\mathbf{q}$ and satisfying the condition $\rho(\mathbf{q})\neq \mathbf{q}$).

10.5 Generalizations

In this section, we describe some generalizations of the results of previous sections of this chapter.

1. The condition of strong convexity imposed on the dispersion law $\omega(\mathbf{p})$ can be substantially relaxed. Namely, it is enough to require that for any open set $G\subset E^3$ the function $\omega(\mathbf{p})$ is not linear (i.e. for any neighborhood of a point $\mathbf{p}_0\in E^3$ there is another point $\mathbf{p}_1\in E^3$ such that $\frac{\partial\omega(\mathbf{p})}{\partial\mathbf{p}}|_{\mathbf{p}=\mathbf{p}_1}\neq\frac{\partial\omega(\mathbf{p})}{\partial\mathbf{p}}|_{\mathbf{p}=\mathbf{p}_0}$).

In this case, the definition of the Møller matrix should be modified: in the formula (10.1), we should consider only non-overlapping families of functions $f_1,\ldots,f_n$. The main modification required in the proofs is to use Lemma 10.5 of Section 10.3 in place of Lemma 10.2 of Section 10.1.

These modifications in the definitions and proofs allow us to construct a scattering theory in the one- and two-dimensional cases, i.e. in the case when the momentum operator $\mathbf{P}$ has only one or two components instead of three. (In higher dimensions, there are no essential modifications.)

2. Up to this point, we have only considered situations with a single one-particle state. The generalization to an arbitrary number of one-particle states is straightforward.

A good operator B, in this case, can be defined as a smooth operator with (1) $B^*\Phi = 0$; (2) we can find a particle $\Phi_i(\mathbf{k})$ and a function $\phi_B(\mathbf{k})$ for which $B\Phi = \Phi_i(\phi_B) = \int \phi_B(\mathbf{k})\Phi_i(\mathbf{k})d\mathbf{k}$. [Recall that in Section 10.1 we have agreed to fix a complete orthogonal system of particles $\Phi_1(\mathbf{k}), \dots, \Phi_s(\mathbf{k})$.]

The modifications of the statements and their proofs consist mostly of adding indices; perhaps the most important change comes from the need to use the word "total" instead of "dense" in the formulation of Lemma 10.1.

It is easy to check that the Møller matrices and the scattering matrix do not depend on the choice of complete system of particles (i.e. they depend only on the operators $H, \mathbf{P}$, and the algebra $\mathcal{A}$).

3. All the constructions in this chapter can be adapted, with the appropriate modifications, to the case when we replace the condition of asymptotic commutatitivty by the condition of asymptotic anticommutatitivity of the family of operators $\mathcal{A}$ (asymptotic anticommutativity means that for any two operators $A, B \in \mathcal{A}$ and any n, we can find C and r such that

$$\| [A, B(\mathbf{x}, t)]_+ \| \leq C \frac{1 + |t|^r}{1 + |\mathbf{x}|^n} \Big).$$

If we have an asymptotically anticommutative family $\mathcal{A}$, we impose an additional condition that any vector $A_1 \dots A_{2k+1}\Phi$, where $A_i \in \mathcal{A}$, is orthogonal to the vacuum Φ (this condition becomes necessary to construct the scattering matrix). The space $\mathcal{H}$ can then be expanded into a direct sum of two subspaces $\mathcal{H}_g$ and $\mathcal{H}_u$ in such a way that the vector $A_1 \dots A_n\Phi$ belongs to the space $\mathcal{H}_g$, if n is even, and to the space $\mathcal{H}_u$, if n is odd. We will call vectors in the spaces $\mathcal{H}_g$ and $\mathcal{H}_u$ even and odd, respectively.

Let us use the symbol $\tilde{\mathcal{A}}$ to denote the smallest algebra of operators that contains the family $\mathcal{A}$. The algebra $\tilde{\mathcal{A}}$, like the linear space above, can be represented as a direct sum $\tilde{\mathcal{A}}_g + \tilde{\mathcal{A}}_u$, where every operator $A \in \tilde{\mathcal{A}}_g$ asymptotically commutes with every operator $B \in \tilde{\mathcal{A}}$ and every operator $A \in \tilde{\mathcal{A}}_u$ asymptotically anticommutes with every operator $B \in \tilde{\mathcal{A}}_u$ (this can be easily checked if we note that the operator $A_1 \ldots A_m$ and the operator $B_1 \ldots B_n$, where $A_i \in \mathcal{A}, B_j \in \mathcal{A}$, asymptotically commute in the case when one of the numbers m, n is even and asymptotically anticommute in the case when both numbers m and n are odd). Operators from the family $\tilde{\mathcal{A}}_g$ will be called even and operators from the family $\tilde{\mathcal{A}}_u$ will be called odd. It is easy to check that an even operator preserves the parity of vectors (i.e. transforms the spaces $\mathcal{H}_g$ and $\mathcal{H}_u$ into themselves), while an odd operator transforms an even vector into an odd vector and an odd vector into an even vector.

Let us suppose that the particle system $\Phi_1(\mathbf{k}), \ldots, \Phi_s(\mathbf{k})$ is chosen in such a way that each particle has a definite parity. In other words, the particles can be split into even, whose vectors $\Phi_i(f) \in \mathcal{H}_g$, and odd, whose vectors $\Phi_i(f) \in \mathcal{H}_u$. Even particles should be considered as bosons, while odd particles as fermions. This implies that the space $\mathcal{H}_{\text{as}}$ can be constructed as the Fock space representation of the relations

$$[a_i(\mathbf{k}), a_j(\mathbf{k}')]_{\mp} = [a_i^+(\mathbf{k}), a_j^+(\mathbf{k}')]_{\mp} = 0,$$

$$[a_i(\mathbf{k}), a_j^+(\mathbf{k}')]_{\mp} = \delta_j^i \delta(\mathbf{k} - \mathbf{k}'),$$

where the anticommutator is used in the case when both particles $\Phi_i(\mathbf{k}), \Phi_j(\mathbf{k})$ are odd and the commutator is used in the other cases. One can say that

$$\mathcal{H}_{\text{as}} = F^s(L^2(E^3 \times N_1)) \otimes F^a(L^2(E^3 \times N_2)),$$

where N_1 is the set of even particles and N_2 is the set of odd particles.

In the case considered in this subsection, the construction of the scattering matrices is performed in the same way as when the family $\mathcal{A}$ is asymptotically commutative.

In the situation at hand, the elementary particles are fermions while bosons arise only as composite particles. If the elementary particles can be either fermions or bosons, then the family $\mathcal{A}$ must be a union of an asymptotically commutative family $\mathcal{A}_1$ and an asymptotically anticommutative family $\mathcal{A}_2$, where each operator from $\mathcal{A}_1$ must asymptotically commute with any operator from $\mathcal{A}_2$.

All the results of Chapter 10, as well as the results of the following chapters, can be easily extended to the case when we have both fermions and bosons as elementary particles.

4. We have previously assumed that the single-particle spectrum does not intersect the multi-particle spectrum (in other words, that the laws of conservation of energy and momentum prohibit the decay of particles). This condition is not always satisfied. However, in the theory of elementary particles, if a particle is stable, then its decay in all known cases is prohibited either by the conservation of energy and momentum or by some other conservation law. In this case, one can prove all the results established above by the same methods. More precisely, in axiomatic scattering theory, it is enough to assume that the space $\mathcal{H}$ expands into a direct sum of the spaces $\mathcal{N}_i$, that are invariant with respect to the operators $H, \mathbf{P}$, in such a way that for every i, the spectrum of the operators $H, \mathbf{P}$ in the subspace $\mathcal{H}_1 \cap \mathcal{N}_i$ does not intersect the spectrum of these operators in the subspace $M \cap \mathcal{N}_i$ (here, M is the multi-particle subspace). In this case, however, it is necessary to tighten the cyclicity requirement by assuming that vectors of the form $A\Phi$, where $A \in \mathcal{A}$, are dense in each space $\mathcal{N}_i$.

5. Let us describe a generalization of the concept of an asymptotically commutative algebra that allows us to study asymptotic commutative algebras containing unbounded operators.

Let us suppose that the energy operator H and the momentum operator $\mathbf{P} = (P_1, P_2, P_3)$ act on the Hilbert space $\mathcal{H}$. We will assume, as always, that H, P_1, P_2, P_3 are commuting self-adjoint operators, whose spectrum satisfies the conditions outlined in Section 10.1.

Let us fix a linear subspace D that is dense in $\mathcal{H}$, contains the ground state Φ of the energy operator H, and is invariant with respect to operators of the form $\exp(iHt - i\mathbf{P}\mathbf{x})$. (In this subsection, we will assume that these conditions always apply to the operators $H, \mathbf{P}$ and the set D.) A family $\mathcal{A}$, consisting of operators defined on the set D that transform the set into itself, will be called *asymptotically Abelian algebra*, if

C1. For every operator $A \in \mathcal{A}$, the operators A^+ and $A(\mathbf{x}, t)$ also belong to $\mathcal{A}$; for every pair of operators $A, B \in \mathcal{A}$, their linear combination $\lambda A + \mu B$ and product AB also belong to $\mathcal{A}$.

C2. For any operators $A_1, \ldots, A_r \in \mathcal{A}$ and arbitrary function $f \in \mathcal{S}(E^{4r})$, the operator

$$\int f(\mathbf{x}_1, t_1, \ldots, \mathbf{x}_r, t_r) A_1(\mathbf{x}_1, t_1) \ldots A_r(\mathbf{x}_r, t_r) d^r \mathbf{x} d^r t \quad (10.62)$$

belongs to the algebra $\mathcal{A}$ and the number

$$\left\langle \int f(\mathbf{x}_1, t_1, \ldots, \mathbf{x}_r, t_r) A_1(\mathbf{x}_1, t_1) \ldots A_r(\mathbf{x}_r, t_r) d^r \mathbf{x} d^r t \Phi, \Phi \right\rangle$$
$$(10.63)$$

continuously depends on the function $f \in \mathcal{S}$ (the integral in (10.62) is understood in a weak sense).[6]

C3. For any operators $A, B, A_1, \ldots, A_r \in \mathcal{A}$ and any number n, we can find numbers C and s such that

$$| \langle [A(\mathbf{x}, t), B] A_1(\xi_1, \tau_1) \ldots A_j(\xi_j, \tau_j) \Phi,$$
$$A_{j+1}(\xi_{j+1}, \tau_{j+1}) \ldots A_r(\xi_r, \tau_r) \Phi \rangle |$$
$$\leq \frac{C(1 + |t|^s)}{1 + |\mathbf{x}|^n} \left(1 + \sum_{i=1}^{r} |\xi_i| + \sum_{i=1}^{r} |\tau_i| \right)^k, \quad (10.64)$$

where k is a number not depending on n.

We will assume that the vector Φ is a cyclic vector of the asymptotic commutative algebra $\mathcal{A}$. Then, using the earlier considerations of this chapter, we can define Møller matrices and scattering matrices, corresponding to the operators $H, \mathbf{P}$, and the algebra $\mathcal{A}$, prove their existence, and transfer to this case the results of earlier considerations. (The definitions, theorems, and proofs do not require any significant modifications.)

---

[6]The second part of the condition C2 implies that the function $\langle A_1(\mathbf{x}_1, t_1) \ldots A_r(\mathbf{x}_r, t_r) \Phi, \Phi \rangle$ is a locally summable function of moderate growth.

Let us add to these results the following statement which will be used in Section 11.4.

Suppose that $B(\mathbf{k}|t)$ is an operator function, which is a generalized function in the variable $\mathbf{k}$, that satisfies the following conditions:

(1) $\exp(iH\tau - i\mathbf{P}\mathbf{a})B(\mathbf{k}|t)\exp(-iH\tau + i\mathbf{P}\mathbf{a}) = \exp(-i\mathbf{k}\mathbf{a})B(\mathbf{k}|t+\tau)$;
(2) $B^+(\mathbf{k}|t)\Phi = 0$;
(3) vectors of the form $\int f(\mathbf{k})B(\mathbf{k}|t)d\mathbf{k}\Phi$ belong to the single-particle subspace;
(4) the functions $\rho_{m,n}$ defined by the equation

$$\rho_{m,n}(\mathbf{k}_1, t_1, \ldots, \mathbf{k}_m, t_m | \mathbf{k}'_1, t'_1, \ldots, \mathbf{k}'_{n-1}, t'_{n-1})$$

$$\times \delta(\mathbf{k}_1 + \cdots + \mathbf{k}_m - \epsilon_1 \mathbf{k}'_1 - \cdots - \epsilon_n \mathbf{k}'_n)$$

$$= \left\langle A_1(\mathbf{k}_1|t_1)\ldots A_m(\mathbf{k}_m|t_m)B(\mathbf{k}'_1, \epsilon_1|t'_1)\ldots B(\mathbf{k}'_n, \epsilon_n|0)\right\rangle^T,$$

where $A_i(\mathbf{k}|t) = (2\pi)^{-3}\int A_i(\mathbf{x}, t)\exp(i\mathbf{k}\mathbf{x})d\mathbf{x}$; $A_i \in \mathcal{A}$; $B(\mathbf{k}, 1|t) = B^+(\mathbf{k}|t)$; $B(\mathbf{k}, -1|t) = B(\mathbf{k}|t)$ belongs to the space $\mathcal{S}(E^{3(m+n-1)})$ over the variables $\mathbf{k}_1, \ldots, \mathbf{k}_m, \mathbf{k}'_1, \ldots, \mathbf{k}'_{n-1}$ for fixed $t_1, \ldots, t_m, t'_1, \ldots, t'_{n-1}$. [We will call such generalized functions "good"; this name is motivated by the fact that for every good operator B, we can construct a good operator generalized function $B(\mathbf{k}|t) = (2\pi)^{-3}\int B(\mathbf{x}, t)\exp(i\mathbf{k}\mathbf{x})d\mathbf{x}$.] Then, for any system of smooth functions with compact support $f_1(\mathbf{k}), \ldots, f_m(\mathbf{k})$, we have

$$\lim_{t\to\pm\infty} B(f_1^t|t)\ldots B(f_n^t|t)\Phi = S_\pm a^+(\overline{\phi f_1})\ldots a^+(\overline{\phi f_n})\theta,$$

where

$$f_i^t(\mathbf{k}) = f_i(\mathbf{k})\exp(-i\omega(\mathbf{k})t),$$

$$B(f_i^t|t) = \int f_i^t(\mathbf{k})B(\mathbf{k}|t)d\mathbf{k},$$

$$B(\mathbf{k}|0)\Phi = \phi(\mathbf{k})\Phi(\mathbf{k}).$$

The proof of this statement follows the reasoning in Section 10.1.

Let us now define the notion of asymptotically commutative families of operator generalized functions.

A family $\mathcal{B}$ of operator generalized functions $A(\mathbf{x}, t)$ is called asymptotically commutative if there exists an asymptotically commutative algebra $\mathcal{A}$ that contains all operators of the form $A(f) = \int f(\mathbf{x}, t) A(\mathbf{x}, t) d\mathbf{x} dt$, where $A(\mathbf{x}, t)$ belongs to the family $\mathcal{B}, f \in \mathcal{S}(E^4)$. (We will assume that the vector Φ is a cyclic vector of the family of operators $A(f)$. The operators $A(f)$ should have a common domain of definition D and transform it into itself; the domain of definition D' of operators belonging to the algebra $\mathcal{A}$ may coincide with the set D, but in general must at least contain it. When we say that the algebra $\mathcal{A}$ contains the operator $A(f)$, we mean that in this algebra, we can find an operator coinciding with $A(f)$ on the set D.)

The scattering matrix corresponding to the operators $H, \mathbf{P}$ and the asymptotically commutative family of operator generalized functions $\mathcal{B}$ can be defined as the scattering matrix constructed for $H, \mathbf{P}$, and the asymptotic algebra $\mathcal{A}$, containing all operators $A(f)$. (It is easy to see that this scattering matrix does not depend on the choice of algebra $\mathcal{A}$.) We can define Møller matrices analogously.

The following theorem gives sufficient conditions for the asymptotic commutativity of a family of operator generalized functions.

Let us suppose that $\mathcal{B}$ is a family of operator generalized functions satisfying the following conditions:

D1. For any operator generalized function $A(\mathbf{x}, t) \in \mathcal{B}$, the operators $A(f) = \int f(\mathbf{x}, t) A(\mathbf{x}, t) d\mathbf{x} dt$, where $f \in \mathcal{S}(E^4)$, are defined on a set D and transform it into itself.

D2. The functional $\langle A(f)\psi_1, \psi_2 \rangle$, where $A \in \mathcal{B}, \psi_1, \psi_2 \in D, f \in \mathcal{S}(E^4)$, continuously depends on the function $f \in \mathcal{S}(E^4)$ (i.e. the real-valued generalized function $\langle A(\mathbf{x}, t)\psi_1, \psi_2 \rangle$ is a generalized function of moderate growth).

D3. If $A \in \mathcal{B}$, then

$$\exp(iH\tau - i\mathbf{P}\mathbf{a}) A(\mathbf{x}, t) \exp(-iH\tau + i\mathbf{P}\mathbf{a}) = A(\mathbf{x} + \mathbf{a}, t + \tau).$$

D4. For every function $A(\mathbf{x}, t)$ in family $\mathcal{B}$, the family also contains the adjoint operator generalized function $A^+(\mathbf{x}, t)$.

D5. For any operator generalized functions $A_1(\mathbf{x}, t), \ldots, A_r(\mathbf{x}, t)$, $A(\mathbf{x}, t), B(\mathbf{x}, t)$ in $\mathcal{B}$, we can find a $\delta > 0$ such that the real-valued generalized function

$$T(\mathbf{x}, t, \xi_1, \tau_1, \ldots, \xi_r, \tau_r)$$

$$= \langle [A(\mathbf{x} + \mathbf{a}, t + \alpha), B(\mathbf{a}, \alpha)] A_1(\xi_1, \tau_1)$$

$$\ldots A_j(\xi_j, \tau_j)\Phi, A_{j+1}(\xi_{j+1}, \tau_{j+1}) \ldots A_r(\xi_r, \tau_r)\Phi \rangle$$

in the domain satisfying $|t| < \delta|\mathbf{x}|$ can be represented in the form

$$T(\mathbf{x}, t, \xi_1, \tau_1, \ldots, \xi_r, \tau_r)$$

$$= \sum_{|\alpha| \leq N(n)} D^{(\alpha)} \left[\left(1 + \sum_i |\xi_i|^2 + \sum_i |\tau_i|^2 \right)^k \right.$$

$$\left. \times (1 + |\mathbf{x}|^2)^{-n} \, \sigma_{\alpha,n}(\mathbf{x}, t, \xi_1, \tau_1, \ldots, \xi_n, \tau_n) \right]$$

(here, n is an arbitrary natural number; $\sigma_{\alpha,n}$ is a bounded continuous function; $D^{(\alpha)}$ is a differential operator of order $|\alpha|$ with constant coefficients; k is a number not depending on n).

D6. The vector Φ is a cyclic vector of the family of operators $A(f)$, where $A(\mathbf{x}, t) \in \mathcal{B}$, $f \in \mathcal{S}(E^4)$.

Then, the family $\mathcal{B}$ is asymptotically Abelian.

To prove this theorem, let us consider the family $\mathcal{A}$ consisting of linear combinations of operators of the form

$$\int f(\mathbf{x}_1, t_1, \ldots, \mathbf{x}_n, t_n) A_1(\mathbf{x}_1, t_1) \ldots A_n(\mathbf{x}_n, t_n) d^n\mathbf{x} d^n t, \qquad (10.65)$$

where $f \in \mathcal{S}(E^{4n})$, $A_i(\mathbf{x}, t) \in \mathcal{B}$. As it is shown in Appendix A.7, expressions of the form (10.65) can be seen as operators defined on some linear subspace D' and transform D' into itself (it follows that we can take D' to be the set of all vectors of the form $A\Psi$, where $A \in \mathcal{A}$; $\Psi \in D$; expression of the form (10.65) defines an operator on the set D by the operator analog of the kernel theorem). It is easy to check that the set D' contains the set D and is invariant with respect to the operators $\exp(iHt - i\mathbf{Pa})$.

It is easy to check that the family $\mathcal{A}$, when viewed as a family of operators defined on the set D', is asymptotically commutative. Conditions C1 and C2 can be easily checked with the help of the statements proven in Appendix A.7. For example, if $A \in \mathcal{A}$ is an operator of the form (10.65), then the operators

$$A^+ = \int \bar{f}(\mathbf{x}_1, t_1, \dots, \mathbf{x}_n, t_n) A_n^+(\mathbf{x}_1, t_1) \dots A_1^+(\mathbf{x}_n, t_n) d^n\mathbf{x} d^n t,$$

$$A(\mathbf{x}, t) = \int f(\mathbf{x}_1 - \mathbf{x}, t_1 - t, \dots, \mathbf{x}_n - \mathbf{x}, t_n - t)$$

$$\times A_1(\mathbf{x}_1, t_1) \dots A_n(\mathbf{x}_n, t_n) d^n\mathbf{x} d^n t$$

are also contained in $\mathcal{A}$. Inequality (10.64) follows from condition D5.

Let us now give a simple example of objects satisfying the conditions of axiomatic scattering theory.

Let us consider the Fock space $\mathcal{H} = F(L^2(E^3))$ with the free Hamiltonian $H_0 = \int \epsilon(\mathbf{k}) a^+(\mathbf{k}) a(\mathbf{k}) d\mathbf{k}$ and the momentum operator $\mathbf{P} = \int \mathbf{k} a^+(\mathbf{k}) a(\mathbf{k}) d\mathbf{k}$. If $\epsilon(\mathbf{k})$ is a smooth, strongly convex function, satisfying the condition $\epsilon(\mathbf{k}_1) + \epsilon(\mathbf{k}_2) > \epsilon(\mathbf{k}_1 + \mathbf{k}_2)$, then the spectrum of these operators satisfies the conditions outlined in Section 10.1. [The ground state of this system is the Fock vacuum θ and the single-particle state is defined by the formula $\Phi(\mathbf{k}) = a^+(\mathbf{k})\theta$.] The space $\mathcal{H}_{\mathrm{as}}$ can be identified with the space $\mathcal{H}$; the Hamiltonian H_{as} becomes the Hamiltonian H_0 in this identification.

The operator generalized functions

$$a^+(\mathbf{x}, t) = (2\pi)^{-3/2} \int \exp(-i\mathbf{k}\mathbf{x}) a^+(\mathbf{k}, t) d\mathbf{k},$$

$$a(\mathbf{x}, t) = (2\pi)^{-3/2} \int \exp(i\mathbf{k}\mathbf{x}) a(\mathbf{k}, t) d\mathbf{k},$$

where

$$a^+(\mathbf{k}, t) = \exp(iH_0 t) a^+(\mathbf{k}) \exp(-iH_0 t)$$

$$= \exp(i\omega(\mathbf{k}) t) a^+(\mathbf{k}),$$

$$a(\mathbf{k}, t) = \exp(iH_0 t) a(\mathbf{k}) \exp(-iH_0 t) \exp(-i\omega(\mathbf{k}) t) a(\mathbf{k}),$$

constitute an asymptotically commutative family $\mathcal{B}_0$ of generalized functions (as always, we assume that these operator generalized functions act on the linear manifold $\mathcal{S}_\infty$ consisting of vectors of the form

$$\sum_n \int \phi_n(\mathbf{k}_1,\ldots,\mathbf{k}_n)a^+(\mathbf{k}_1)\ldots a^+(\mathbf{k}_n)\theta d^n\mathbf{k},$$

where the functions ϕ_n belong to the space $\mathcal{S}$ and only a finite number of these functions are non-zero). The family $\mathcal{B}_0$ can be shown to be asymptotically commutative with the aid of the theorem proven above. A different proof of the same fact can be obtained if we note that operators of the form

$$\int f(\mathbf{x},t)a^+(\mathbf{x},t)d\mathbf{x}dt \text{ and } \int f(\mathbf{x},t)a(\mathbf{x},t)d\mathbf{x}dt,$$

where $f \in \mathcal{S}(E^4)$ are contained in the asymptotically commutative algebra $\mathcal{A}_0$ consisting of operators of the form

$$\sum_{m,n} \int f_{m,n}(\mathbf{k}_1,\ldots,\mathbf{k}_m|\mathbf{p}_1,\ldots,\mathbf{p}_n)a^+(\mathbf{k}_1)$$

$$\ldots a^+(\mathbf{k}_m)a(\mathbf{p}_1)\ldots a(\mathbf{p}_n)d^m\mathbf{k}d^n\mathbf{p}, \tag{10.66}$$

where $f_{m,n} \in \mathcal{S}(E^{3(m+n)})$, the sum in (10.66) is finite, and operators of the form (10.66) are viewed as operators, defined on the set $\mathcal{S}_\infty$. It is easy to check directly that $\mathcal{A}_0$ is an asymptotically commutative algebra; this can also be shown by results we formulate below. It is easy to verify that the Møller matrix $S_\pm$ and the scattering matrix S, constructed for the operators $H_0,\mathbf{P}$ and the family $\mathcal{A}_0$, are trivial (i.e. they are all the identity operator).

It is sometimes convenient to assume that an asymptotically commutative algebra is endowed with a topology (this will be useful in Sections 10.6 and 11.5). We will now give some conditions that when applied to a topological operator algebra guarantee that it will be asymptotically commutative.

Let us assume that on the Hilbert space $\mathcal{H}$ we have, as before, an energy operator H, a momentum operator $\mathbf{P}$, and a linear subspace D, satisfying the conditions outlined above, and let us consider a set of operators $\mathcal{A}$ obeying the following conditions:

T1. The set $\mathcal{A}$ consists of operators that are defined on the set D and transform D into itself.

T2. For any A, B in $\mathcal{A}$, the following operators are also in $\mathcal{A}$: the linear combination $\lambda A + \mu B$ (i.e. the set can be viewed as a linear space), the product AB (this implies that $\mathcal{A}$ is an algebra of operators), and the adjoints A^+, B^+.

T3. The linear space $\mathcal{A}$ is equipped with a locally convex topology; the space $\mathcal{A}$ must be complete, and the product and conjugation operations must be continuous in the topology $\mathcal{A}$.[7]

T4. The vector $A\Phi \in \mathcal{H}$ continuously depends on $A \in \mathcal{A}$ in the topology of the algebra $\mathcal{A}$ [i.e. we can find a seminorm $p(A)$ on $\mathcal{A}$ such that

$$\|A\Phi\| \leq p(A)].$$

T5. If $A \in \mathcal{A}$, then the operator $A(\mathbf{x}, t)$ is in $\mathcal{A}$ and it continuously depends on $\mathbf{x}, t$ in the topology on $\mathcal{A}$; for any seminorm $p(A)$ on $\mathcal{A}$ and any compact set $F \subset \mathcal{A}$, we can find another seminorm $q(A)$ and a number k, such that for $A \in F, \mathbf{x} \in E^3, -\infty < t < \infty$ we have

$$p(A(\mathbf{x}, t)) \leq (1 + |\mathbf{x}|^k + |t|^k)q(A).$$

T6. For any seminorm $p(A)$ on $\mathcal{A}$, any compact set $F \subset \mathcal{A}$, and any natural number n, there exists a seminorm $q(A)$ such that

$$p([A(\mathbf{x}), B]) \leq \frac{q(A)q(B)}{1 + |\mathbf{x}|^n}$$

for all $A, B \in F, \mathbf{x} \in E^3$.

(If the locally convex topology on $\mathcal{A}$ is specified by a system of seminorms $\|A\|_\alpha$, then conditions T5 and T6 should be reformulated in the following way: for all seminorms $\|A\|_\alpha$, compact sets $F \subset \mathcal{A}$, and natural numbers n, there must exist a seminorm $\|A\|_\lambda$ in the system and numbers k, C such that

$$\|A(\mathbf{x}, t)\|_\alpha \leq C(1 + |\mathbf{x}|^k + |t|^k)\|A\|_\lambda,$$

$$\|[A(\mathbf{x}), B]\|_\alpha \leq \frac{C\|A\|_\lambda\|B\|_\lambda}{1 + |\mathbf{x}|^n}$$

for any $A, B \in F$.)

---

[7]Conditions T1–T3 mean that the set $\mathcal{A}$ is a complete locally convex topological algebra with involution.

T7. The vector Φ is a cyclic vector with respect to the family of operators $\mathcal{A}$.

We can now state the following proposition:

A family of operators $\mathcal{A}$ satisfying conditions T1–T6 is an asymptotically commutative algebra.

Indeed, condition C1 follows from conditions T1 and T2. To check C2, note that for any $A_1, \ldots, A_n \in \mathcal{A}$ the operators $A_1(\mathbf{x}_1, t_1) \ldots A_n(\mathbf{x}_n, t_n)$ continuously depend on $\mathbf{x}_1, t_1, \ldots, \mathbf{x}_n, t_n$ in the topology on $\mathcal{A}$ and for any seminorm $p(A)$ on $\mathcal{A}$ we can find a number k such that

$$p(A_1(\mathbf{x}_1, t_1) \ldots A_n(\mathbf{x}_n, t_n)) \leq 1 + |\mathbf{x}|^k + |t|^k \tag{10.67}$$

(this follows from conditions T3 and T5). These remarks and the completeness of $\mathcal{A}$ imply that the integral (10.62) converges in the topology $\mathcal{A}$ and defines an element of the algebra $\mathcal{A}$. It follows from condition T4 and inequality (10.67) that

$$\langle A_1(\mathbf{x}_1, t_1) \ldots A_n(\mathbf{x}_n, t_n)\Phi, \Phi \rangle$$

is a continuous function of moderate growth in the variables $\mathbf{x}_1, t_1, \ldots, \mathbf{x}_n, t_n$. This proves the second part of condition C2. Finally, we can easily check C3 by using T3–T6.

The statement we have just proved allows us to say that conditions T1–T7 are sufficient to construct a scattering theory starting with the operators $H, \mathbf{P}$ and the algebra $\mathcal{A}$.

Let us now return to the operators $H_0 = \int \epsilon(\mathbf{k})a^+(\mathbf{k})a(\mathbf{k})d\mathbf{k}, \mathbf{P} = \int \mathbf{k}a^+(\mathbf{k})a(\mathbf{k})d\mathbf{k}$ and the algebra $\mathcal{A}_0$, consisting of operators of the form (10.66). We will show that the algebra can be endowed with a topology in a way that satisfies conditions T1–T7. The algebra $\mathcal{A}_0$ can be represented as the union of linear spaces $\mathcal{A}_{(m,n)}$ consisting of operators of the form

$$A = \int f_{m,n}(\mathbf{k}_1, \ldots, \mathbf{k}_m | \mathbf{p}_1, \ldots, \mathbf{p}_n)a^+(\mathbf{k}_1) \ldots a^+(\mathbf{k}_m)a(\mathbf{p}_1)$$

$$\ldots a(\mathbf{p}_n)d^m\mathbf{k}d^n\mathbf{p}.$$

Elements of the space $\mathcal{A}_{(m+n)}$ are in one-to-one correspondence with functions in the space $\mathcal{S}(E^{3(m+n)})$; this allows us to transfer the topology of the space $\mathcal{S}(E^{3(m+n)})$ to the space $\mathcal{A}_{(m+n)}$. Hence, we can consider $\mathcal{A}_{(m+n)}$ to be a complete locally convex space.

Let us consider a system of seminorms on algebra $\mathcal{A}_0$ that are continuous functions on each subspace $\mathcal{A}_{(m,n)} \subset \mathcal{A}_0$. This system of seminorms defines a topology on $\mathcal{A}_0$ which is called the *inductive limit topology*; the algebra $\mathcal{A}_0$ is complete in this topology (see, for example, Robertson *et al.* (1980)).

It is easy to check that the algebra $\mathcal{A}_0$, with the topology introduced above, satisfies conditions T1–T7. To verify conditions T5 and T6, it helps to use the following lemmas.

Lemma 10.8. *For any norm* $\|f\|_{\alpha,\beta}$ *on the space* $\mathcal{S}(E^{3(m+n)})$ *we can find a norm* $\|f\|_{\gamma,\delta}$ *on* $\mathcal{S}$ *and numbers* C, k *such that for any function* $f \in \mathcal{S}(E^{3(m+n)})$ *we have*

$$\|U_t V_{\mathbf{x}} f\|_{\alpha,\beta} \leq C(1 + |\mathbf{x}|^k + |t|^k)\|f\|_{\gamma,\delta}$$

[here, $V_{\mathbf{x}}$ *and* U_t *denote operators in* $\mathcal{S}(E^{3(m+n)})$ *transforming the function* $f(\mathbf{k}_1, \ldots, \mathbf{k}_m | \mathbf{p}_1, \ldots, \mathbf{p}_n)$ *to the functions*

$$\exp\left(\left(i\sum_{j=1}^{m}\mathbf{k}_j - t\sum_{j=1}^{n}\mathbf{p}_j\right)\mathbf{x}\right) f(\mathbf{k}_1, \ldots, \mathbf{k}_m | \mathbf{p}_1, \ldots, \mathbf{p}_n),$$

$$\exp\left(-i\left(\sum_{j=1}^{m}\epsilon(\mathbf{k}_j) - \sum_{j=1}^{n}\epsilon(\mathbf{p}_j)\right)t\right) f(\mathbf{k}_1, \ldots, \mathbf{k}_m | \mathbf{p}_1, \ldots, \mathbf{p}_n),$$

respectively].

Lemma 10.9. *For any norm* $\|f\|_{\alpha,\beta}$ *on the space* $\mathcal{S}(E^{3(m+m'+n+n'-2r)})$ *and any number* N, *we can find norms* $\|f\|_{\gamma,\delta}$ *and* $\|f\|_{\gamma',\delta'}$ *on the spaces* $\mathcal{S}(E^{3(m+n)})$ *and* $\mathcal{S}(E^{3(m'+n')})$ *and a number* C *so that*

$$\|\lambda_r(V_{\mathbf{x}} f, g)\|_{\alpha,\beta} \leq \frac{C}{1 + |\mathbf{x}|^N}\|f\|_{\gamma,\delta}\|g\|_{\gamma',\delta'}.$$

[Here, $r > 0$ *and* $\lambda_r(f, g)$ *denotes the function*

$$\int f(\mathbf{k}_1, \ldots, \mathbf{k}_m | \mathbf{p}_1, \ldots, \mathbf{p}_{n-r}, \mathbf{q}_1, \ldots, \mathbf{q}_r)$$

$$\times g(\mathbf{q}_1, \ldots, \mathbf{q}_r, \mathbf{k}'_1, \ldots, \mathbf{k}'_{m'-r} | \mathbf{p}'_1, \ldots, \mathbf{p}'_n) d\mathbf{q}_1 \ldots d\mathbf{q}_r.]$$

It follows from Lemma 10.8 that for any seminorm $p(A)$ on the space $\mathcal{A}_{(m+n)}$ we can find a seminorm $q(A)$ on $\mathcal{A}_{(m+n)}$ and a number k so that for any operator $A \in \mathcal{A}_{m,n}$ we have

$$p(A(\mathbf{x}, t)) \leq (1 + |\mathbf{x}|^k + |t|^k)q(A).$$

It follows from Lemma 10.9 that for any seminorm $p(A)$ on the space $\mathcal{A}_0$ and for any number N we can find seminorms $q(A)$ and $q'(A)$ on the spaces $A_{(m,n)}$ and $A_{(m',n')}$ so that for any $A \in \mathcal{A}_{(m,n)}, B \in A_{(m',n')}$, we have

$$p([A(\mathbf{x}), B]) \leq \frac{q(A)q'(B)}{1 + |\mathbf{x}|^N}.$$

To finish checking conditions T5 and T6, we should note that every compact subset of the algebra $\mathcal{A}_0$ is contained in a direct sum of a finite number of subspaces $\mathcal{A}_{(m,n)}$ (see Robertson *et al.* (1980)).

Therefore, the algebra $\mathcal{A}_0$ satisfies conditions T1–T7; as we have remarked above, this implies that $\mathcal{A}_0$ is asymptotically commutative.

10.6 Adiabatic theorem in axiomatic scattering theory

We will now show how in axiomatic scattering theory one can express Møller matrices and scattering matrices in terms of their adiabatic analogs.

First, we will define the adiabatic Møller matrix and the adiabatic scattering matrix in a slightly more general situation than was considered in Section 4.1.

Let the self-adjoint operator $H(g)$ act on the Hilbert space $\mathcal{H}$, with the parameter g taking values in the interval $[0, 1]$. We will fix a continuous function $h(\tau)$ of a real variable τ which rapidly decays at infinity and equals 1 at $\tau = 0$. The symbol $U_\alpha(t, t_0)$ will denote the evolution operator, constructed with the time-dependent Hamiltonian $H(h(\alpha\tau))$, and the symbol $S_\alpha(t, t_0)$ will denote the operator $\exp(iH(0)t)U_\alpha(t, t_0)\exp(-iH(0)t_0)$. Otherwise, we can define the operator $S_\alpha(t, t_0)$ as the solution to the equation

$$i\frac{dS_\alpha(t, t_0)}{dt} = \exp(iH(0)t)(H(h(\alpha t)) - H(0))\exp(-iH(0)t)S_\alpha(t, t_0)$$

with the initial condition $S_\alpha(t_0, t_0) = 1$.

We will similarly use $U(t, t_0|g(\tau))$ to denote the evolution operator corresponding to the time-dependent Hamiltonian $H(g(\tau))$, where $g(\tau)$ is a function taking values in $[0, 1]$; using this notation, we can write $U_\alpha(t, t_0) = U(t, t_0|h(\alpha\tau))$.

We will call the operators $S_\alpha(0, \pm\infty) = \text{slim}_{t\to\pm\infty} S_\alpha(0, t)$ the adiabatic Møller matrices and the the operator $S_\alpha = S_\alpha(\infty, -\infty) = \text{slim}_{\substack{t\to\infty \\ t_0\to-\infty}} S_\alpha(t, t_0)$ will be called the adiabatic S-matrix. The construction of the adiabatic S-matrix and the adiabatic Møller matrices provided in Section 4.1 for the pair of operators H, H_0 corresponds to $H(g) = H_0 + g(H - H_0)$ and $h(\tau) = \exp(-|\tau|)$.

In what follows, for the sake of simplifying the proofs, we will assume that the function $h(\tau)$ satisfies a few somewhat stronger conditions than we have assumed above. Namely, we will assume that this function is smooth, even, has compact support, and satisfies $h(0) = 1$ in a neighborhood of the point $\tau = 0$. The radius of this neighborhood will be denoted by the symbol δ and the radius of the support of the function $h(\tau)$ will be denoted by Δ (i.e. $h(\tau) = 1$ for $|\tau| < \delta$ and $h(\tau) = 0$ for $|\tau| > \Delta$).

Let us now consider the situation when the energy operator in an axiomatic scattering theory depends on a parameter g. More precisely, suppose that we have an energy operator $H(g)$ that acts on the Hilbert space $\mathcal{H}$, where the parameter g is in the interval $[0, 1]$, a momentum operator $\mathbf{P} = (P_1, P_2, P_3)$, and a family of operators $\mathcal{A}$, satisfying the following conditions:

(a) The self-adjoint operators $H(g), P_1, P_2, P_3$ commute for every g. In the space $\mathcal{H}$, there exists a vector Φ (the ground state of the energy operator $H(g)$, not depending on the parameter g) and a vector function $\Phi(\mathbf{k}|g)$, generalized with respect to $\mathbf{k}$ and continuous in the parameter g [single-particle state of the operator $H(g)$], for which we have

(1) $H(g)\Phi = \mathbf{P}\Phi = 0$;
(2) $H(g)\Phi(\mathbf{k}|g) = \omega(\mathbf{k}|g)\Phi(\mathbf{k}|g)$,

where $\omega(\mathbf{k}|g)$ is a positive function, infinitely differentiable in $\mathbf{k}, g$ and strongly convex in $\mathbf{k}$;

(3) $\mathbf{P}\Phi(\mathbf{k}|g) = \mathbf{k}\Phi(\mathbf{k}|g)$;
(4) $\langle \Phi(\mathbf{k}|g), \Phi(\mathbf{k}'|g) \rangle = \delta(\mathbf{k} - \mathbf{k}')$;
(5) for any $\mathbf{k}_0 \in E^3, 0 \leq g_0 \leq 1$, we can find an operator $A \in \mathcal{A}$, such that $\langle \Phi(\mathbf{k}_0|g_0), A\Phi \rangle \neq 0$;
(6) for any $\mathbf{k}_0 \in E^3, 0 \leq g_0 \leq 1$, there exist $\epsilon > 0, \delta > 0$, so that for any point $(\omega, \mathbf{k})$ belonging to the multi-particle spectrum of the operators $(H(g), \mathbf{P})$ and satisfying $|\mathbf{k} - \mathbf{k}_0| < \delta, |g - g_0| < \delta$, we have the inequality $\omega > \omega(\mathbf{k}_0|g_0) + \epsilon$.

(b) Let us denote by D the set of vectors of the form $A\Phi$ where $A \in \mathcal{A}$. We assume that D is dense in $\mathcal{H}$ and invariant with respect to the operators $\exp(i\mathbf{P}\mathbf{x})$ and $U(t, t_0|g(\tau))$, where $g(\tau)$ is a smooth function taking values in $[0, 1]$. For any two operators $A, B \in \mathcal{A}$, the linear combination $\lambda A + \mu B$, the product AB, and the adjoint operators A^+, B^+ all belong to $\mathcal{A}$ (in other words, the family $\mathcal{A}$ is an operator algebra with an involution). We will assume that $\mathcal{A}$ is equipped with a locally convex topology, in which $\mathcal{A}$ is complete and the product and involution operations are continuous.

The topology on $\mathcal{A}$ should satisfy the following conditions:

(1) $\langle A\Phi, \Phi \rangle$ continuously depends on $A \in \mathcal{A}$;
(2) if $A \in \mathcal{A}$, $g(\tau)$ is a smooth function with values in $[0, 1]$, $\mathbf{x} \in E^2$, and $-\infty < t_0, t_1 < \infty$, then the operator

$$U(t_1, t_0|g(\tau))A(\mathbf{x})U(t_0, t_1|g(\tau)) \in \mathcal{A};$$

for every seminorm $p(A)$ on $\mathcal{A}$ and every compact $F \subset \mathcal{A}$ we can find a seminorm $q(A)$ on $\mathcal{A}$ and a number k such that for $A \in F$, we have

$$p(U(t_1, t_0|g(\tau))A(\mathbf{x})U(t_0, t_1|g(\tau))$$

$$\leq (1 + |t_1 - t_0|^k + |\mathbf{x}|^k)q(A). \tag{10.68}$$

In particular, $A(\mathbf{x}, t|g) \in \mathcal{A}$; we will suppose that this operator is a smooth function in g with respect to the topology $\mathcal{A}$, the operator $\frac{\partial^m}{\partial g^m} A(\mathbf{x}, t|g)$ continuously depends on $\mathbf{x}, t, g$ and

$$p\left(\frac{\partial^m}{\partial g^m} A(\mathbf{x}, t|g)\right) \leq (1 + |\mathbf{x}|^k + |t|^k) q(A), \qquad (10.69)$$

where $p(A)$ is any seminorm on $\mathcal{A}$; m is an arbitrary whole number; the seminorm $q(A)$ on $\mathcal{A}$ and the number k depend on the seminorm p, the number m, and the compact set F. [Here and further, $A(\mathbf{x})$ denotes the operator $\exp(-i\mathbf{Px})A\exp(i\mathbf{Px})$ and $A(\mathbf{x}, t|g)$ denotes $\exp(iH(g)t)A(\mathbf{x})\exp(-iH(g)t)$];

(3) for any seminorm $p(A)$ on $\mathcal{A}$, any number n, and any compact $F \subset \mathcal{A}$, we can find a seminorm $q(A)$ on $\mathcal{A}$ such that for all $A, B \in \mathcal{A}, \mathbf{x} \in E^3, A \in F$ we have

$$p([A(\mathbf{x}), B]) \leq \frac{q(A)q(B)}{1 + |\mathbf{x}|^n}. \qquad (10.70)$$

These conditions are sufficient to guarantee that starting from the operators $H(g), \mathbf{P}$, and the algebra $\mathcal{A}$, we can construct the Møller matrices $S_\pm(g)$ and the scattering matrix $S(g) = S_+^*(g)S_-(g)$. (Not very precisely, one can say that we require the operators $H(g), \mathbf{P}$, and the algebra $\mathcal{A}$ to satisfy the conditions T1–T7 of the Section 10.5, which guarantee the existence of Møller matrices, but we additionally require these conditions to be satisfied uniformly in g).

We will now show that we can derive the following relations from the above conditions:

$$S_-(1) = \operatorname*{slim}_{\alpha \to 0} S_\alpha(0, -\infty) \exp\left(\frac{i}{\alpha} \int \rho(\mathbf{k}) a_{\mathrm{in}}^+(\mathbf{k}) a_{\mathrm{in}}(\mathbf{k}) dk\right),$$

$$\tag{10.71}$$

$$S_+(1) = \operatorname*{slim}_{\alpha \to 0} S_\alpha(0, +\infty) \exp\left(\frac{i}{\alpha} \int \rho(\mathbf{k}) a_{\mathrm{out}}^+(\mathbf{k}) a_{\mathrm{out}}(\mathbf{k}) dk\right),$$

$$\tag{10.72}$$

where

$$\rho(\mathbf{k}) = \int_{-\infty}^{0} (\omega(\mathbf{k}|h(\sigma)) - \omega(\mathbf{k}|0))d\sigma,$$

$$a_{\mathrm{in}}(\mathbf{k}) = S_{-}(0)a(\mathbf{k})S_{-}^{-1}(0),$$

$$a_{\mathrm{out}}(\mathbf{k}) = S_{+}(0)a(\mathbf{k})S_{+}^{-1}(0)$$

are in- and out-operators, constructed for the energy operator $H(0)$. The proof of these relations (*the adiabatic theorem*) is the main result of this section.

Of course, one can easily state the analog of relations (10.71) and (10.72) for the operators $S_{\pm}(g)$, where $0 \le g \le 1$; we will not stop to do this.

The relations (10.71) and (10.72) are clearly equivalent to the relations

$$S_{-}(1) = \operatorname*{slim}_{\alpha \to 0} \operatorname*{slim}_{t \to -\infty} U_{\alpha}(0, t)S_{-}(0)W_{\alpha}(t), \qquad (10.73)$$

$$S_{+}(1) = \operatorname*{slim}_{\alpha \to 0} \operatorname*{slim}_{t \to +\infty} U_{\alpha}(0, t)S_{+}(0)W_{\alpha}(t), \qquad (10.74)$$

where $W_{\alpha}(t)$ is an operator in the space $\mathcal{H}_{\mathrm{as}}$, defined by the formula

$$W_{\alpha}(t) = \exp\left(i \int r_{\alpha}(\mathbf{k}|t)a^{+}(\mathbf{k})a(\mathbf{k})d\mathbf{k}\right),$$

$$r_{\alpha}(\mathbf{k}|t) = \int_{t}^{0} \omega(\mathbf{k}|h(\alpha\tau))d\tau$$

(to verify this equivalence, it is enough to note that

$$\exp(-iH(0)t)S_{-}(0) = S_{-}(0)\exp\left(-i \int \omega(\mathbf{k}|0)ta^{+}(\mathbf{k})a(\mathbf{k})d\mathbf{k}\right);$$

$$\frac{1}{\alpha}\rho(\mathbf{k}) = \lim_{t \to -\infty} (r_{\alpha}(\mathbf{k}|t) - t\omega(\mathbf{k}|0))).$$

We will prove the adiabatic theorem in the form (10.73) and (10.74). The proof will use the following series of lemmas.

Lemma 10.10. *For any seminorm $p(A)$ on $\mathcal{A}$, any compact $F \subset \mathcal{A}$, and any number n, we can find a seminorm $q(A)$ on $\mathcal{A}$ and a number* *k, such that*

$$p([U(t_1, t_0|g(\tau))A(\mathbf{x})U(t_0, t_1|g(\tau)), B])$$

$$\leq \frac{1 + |t_1 - t_0|^k}{1 + |\mathbf{x}|^n} q(A)q(B)$$

(here, $A, B \in F$, $g(\tau)$ is a smooth function with values in $[0, 1]$, $-\infty < t_1, t_0 < \infty, \mathbf{x} \in E^3$).

The proof of this lemma quickly follows from inequalities (10.68) and (10.70) and the remark that the operator $U(t_1, t_0|g(\tau))$ commutes with $\exp(i\mathbf{P}\mathbf{x})$.

Lemma 10.11. *If $\phi(\mathbf{x}, t)$ is a piecewise-continuous function, decaying faster than a power function, then the operator*

$$\int \phi(\mathbf{x}, t)A(\mathbf{x}, t|g)d\mathbf{x}dt,$$

where $A \in \mathcal{A}, 0 \leq g \leq 1$, belongs to the algebra $\mathcal{A}$ and is infinitely differentiable with respect to g in the topology of $\mathcal{A}$, and furthermore the operator

$$\frac{d^k}{dg^k} \int \phi(\mathbf{x}, t)A(\mathbf{x}, t|g)d\mathbf{x}dt$$

continuously depends on $\phi \in T$, $A \in \mathcal{A}$, $g \in [0, 1]$ in the topology on $\mathcal{A}$.

[We will assume that the space T of piecewise-continuous functions with decay faster than a power function is endowed with a topology by means of the norm family

$$\|\phi\|_\lambda = \sup_{\mathbf{x} \in E^3, -\infty < t < \infty} (1 + |\mathbf{x}|^\lambda + |t|^\lambda)\phi(\mathbf{x}, t).\Big]$$

The proof of this lemma is based on the completeness of the algebra $\mathcal{A}$ and condition b(2).

Lemma 10.12. *Suppose that A_g, $B_g \in \mathcal{A}$ are two families of operators, infinitely differentiable in g in the topology on $\mathcal{A}$. Let us* *assume that $B_g^*\Phi = 0$ for all g (this condition is satisfied if B_g is a good operator). Then the function*

$$\rho(\mathbf{k}|g) = \int \langle A_g(\mathbf{x})\Phi, B_g\Phi \rangle \exp(i\mathbf{kx})d\mathbf{x}$$

is infinitely differentiable with respect to $\mathbf{k}, g$.

Indeed, let us consider the function

$$\tilde{\rho}(\mathbf{x}|g) = \langle A_g(\mathbf{x})\Phi, B_g\Phi \rangle = \langle B_g^* A_g(\mathbf{x})\Phi, \Phi \rangle.$$

By Lemma 10.10, the function

$$\frac{\partial^m}{\partial g^m}\tilde{\rho}(\mathbf{x}|g) = \sum_{k=0}^{m} C_m^k \left\langle \frac{\partial^k B_g^*}{\partial g^k} \frac{\partial^{m-k} A_g(\mathbf{x})}{\partial g^{m-k}}\Phi, \Phi \right\rangle$$

$$= \sum_{k=0}^{m} C_m^k \left\langle \left[\frac{\partial^k B_g^*}{\partial g^k}, \frac{\partial^{m-k} A_g(\mathbf{x})}{\partial g^{m-k}}\right]\Phi, \Phi \right\rangle$$

obeys the inequality

$$\left|\frac{\partial^m}{\partial g^m}\tilde{\rho}(\mathbf{x}|g)\right| \leq \frac{C}{1 + |\mathbf{x}|^n}$$

(here, m and n are arbitrary natural numbers). It follows from this inequality that the function $\rho(\mathbf{k}|g) = \int \tilde{\rho}(\mathbf{x}|g) \exp(i\mathbf{kx})d\mathbf{x}$ is infinitely differentiable.

Before we move on to the next lemma, let us note that the vector generalized function $\Phi(\mathbf{k}|g)$ defined by conditions a(2), a(3), a(4) is not uniquely specified: a vector generalized function $\Phi'(\mathbf{k}|g) = \exp(i\lambda(\mathbf{k}|g))\Phi(\mathbf{k}|g)$, where $\lambda(\mathbf{k}|g)$ is a real measurable function, will satisfy the same conditions.

Lemma 10.13. *For any operator $A \in \mathcal{A}$, we can find a real measurable function $\lambda(\mathbf{k}|g)$ so that*

$$\langle A\Phi, \Phi'(\mathbf{k}|g) \rangle = \langle A\Phi, \exp(i\lambda(\mathbf{k}|g))\Phi(\mathbf{k}|g) \rangle$$

is infinitely differentiable in $\mathbf{k}$ and g.

Let us suppose that B_g^1, B_g^2 are two families of good operators that are infinitely differentiable in g in the topology of $\mathcal{A}$; the functions $r_1(\mathbf{k}|g)$ and $r_2(\mathbf{k}|g)$ will be defined by the relation

$$B_g^i \Phi = \int r_i(\mathbf{k}|g)\Phi(\mathbf{k}|g)d\mathbf{k}.$$

Let us consider the function

$$\nu(\mathbf{x}|g) = \left\langle B_g^1(\mathbf{x})\Phi, B_g^2\Phi \right\rangle$$

and note that

$$\nu(\mathbf{x}|g) = \left\langle \exp(i\mathbf{Px}) \int r_1(\mathbf{k}|g)\Phi(\mathbf{k}|g)d\mathbf{k}, \right.$$

$$\left. \times \int r_2(\mathbf{k}|g)\Phi(\mathbf{k}|g)d\mathbf{k} \right\rangle$$

$$= \int \exp(i\mathbf{kx})r_1(\mathbf{k}|g)\overline{r_2(\mathbf{k}|g)}d\mathbf{k}. \tag{10.75}$$

It follows from relation (10.75) and Lemma 10.12 that the function $r_1(\mathbf{k}|g)\overline{r_2(\mathbf{k}|g)}$ is infinitely differentiable in $\mathbf{k}$ and g.

Let us now show that for all points $(\mathbf{k}_0, g_0)$ we can find an infinitely differentiable family of good operators B_g, such that

$$\langle B_{g_0}\Phi, \Phi(\mathbf{k}_0|g_0)\rangle \neq 0.$$

We will use the fact that for every operator $A \in \mathcal{A}$ and every smooth function with compact support $\sigma(\mathbf{k}, \omega|g)$, which is equal to 0 for $\omega \leq 0$ and for any point $(\omega, \mathbf{k})$ belonging to the multi-particle spectrum of the operators $H(g), \mathbf{P}$, we can construct a family of good operators $B_g^{\sigma,A}$ by means of the formula

$$B_g^{\sigma,A} = \int \tilde{\sigma}(\mathbf{x}, t|g)A(\mathbf{x}, t|g)d\mathbf{x}dt,$$

where

$$\tilde{\sigma}(\mathbf{x}, t|g) = \frac{1}{(2\pi)^4} \int \exp(i\mathbf{kx} - i\omega t)\sigma(\mathbf{k}, \omega|g)d\mathbf{k}d\omega$$

(see Lemma 10.1 in Section 10.1). By Lemma 10.11, this family is infinitely differentiable in g. It is easy to check that

$$\langle B_g^{\sigma,A}\Phi, \Phi(\mathbf{k}|g)\rangle = \sigma(\mathbf{k}, \omega(\mathbf{k}|g)|g) \langle A\Phi, \Phi(\mathbf{k}|g)\rangle. \tag{10.76}$$

For every point $\mathbf{k}_0, g_0$, we can find a function σ such that $\sigma(\mathbf{k}_0, \omega(\mathbf{k}_0|g_0)|g_0) \neq 0$. This observation, combined with condition a(5) shows that for every point $\mathbf{k}_0$, g_0 we can find an operator $A \in \mathcal{A}$ such that

$$\langle B_{g_0}^{\sigma, A}\Phi, \Phi(\mathbf{k}_0|g_0)\rangle \neq 0.$$

We can now use the following statement.

Let $\mathcal{F} = \{\phi_\gamma(k)\}$ be a family of measurable complex functions on the convex subset M of the Euclidean space E^n, satisfying (a) for any two functions $\phi_\gamma \in \mathcal{F}$, $\phi_\delta \in \mathcal{F}$, the function $\bar{\phi}_\gamma(k)\phi_\delta(k)$ is infinitely differentiable; (b) for any point $k \in M$, we can find a function $\phi_\gamma \in \mathcal{F}$, not equalling zero at the point k. Then we can find a measurable real function $\lambda(k)$ with the property that the product of any function $\phi_\gamma \in \mathcal{F}$ with $\exp(i\lambda(k))$ is a smooth function.

(The proof of this statement, based on several theorems from the topology of fiber bundles, is given in Fateev and Shvarts (1973).)

Applying this statement, we can find a real function $\lambda(\mathbf{k}|g)$ such that for any family of good operators $B_g \in \mathcal{A}$, which are infinitely differentiable in g in the topology of $\mathcal{A}$, the function

$$\exp(i\lambda(\mathbf{k}|g))r(\mathbf{k}|g) = \exp(i\lambda(\mathbf{k}|g))\langle B_g\Phi, \Phi(\mathbf{k}|g)\rangle$$

is infinitely differentiable.

It follows from relation (10.76) that the same choice of the function $\lambda(\mathbf{k}|g)$ implies the differentiability of all functions of the form $\exp(i\lambda(\mathbf{k}|g))\langle A\Phi, \Phi(\mathbf{k}|g)\rangle$, where $A \in \mathcal{A}$; this finishes the proof of the lemma.

Remark 10.4. Using the same arguments, one can prove, without requiring condition a(5), that for any operator $A \in \mathcal{A}$, the function $|\langle \Phi(\mathbf{k}|g), A\Phi\rangle|^2$ is infinitely differentiable. This allows us to consider the value of the function $|\langle \Phi(\mathbf{k}|g), A\Phi\rangle|$ at individual points. Therefore, condition a(5) has a precise meaning.

In what follows, we will assume that the single-particle state $\Phi(\mathbf{k}|g)$ is chosen in such a way that for any operator $A \in \mathcal{A}$, the function $\langle A\Phi, \Phi(\mathbf{k}|g)\rangle$ is infinitely differentiable in $\mathbf{k}, g$ [Lemma 10.13

implies that this can always be done by replacing $\Phi(\mathbf{k}|g)$ with $\Phi'(\mathbf{k}|g) = \exp(i\lambda(\mathbf{k}|g))\Phi(\mathbf{k}|g)$ if necessary].

Lemma 10.14. *For any smooth function with compact support $f(\mathbf{k}|g)$, we can find a family of good operators B_g^f, infinitely differentiable in g in the topology of $\mathcal{A}$, such that*

$$B_g^f \Phi = \int f(\mathbf{k}|g)\Phi(\mathbf{k}|g)d\mathbf{k}.$$

To prove this lemma, it is convenient to use the relation

$$B_g^{\psi\phi} = \int \tilde{\phi}(\mathbf{x}|g)B_g^\psi(\mathbf{x})d\mathbf{x} \tag{10.77}$$

$\Big[$more precisely, if a family of good operators $B_g^\psi \in \mathcal{A}$ is infinitely differentiable in g and

$$B_g^\psi \Phi = \int \psi(\mathbf{k}|g)\Phi(\mathbf{k}|g)d\mathbf{k},$$

$\phi(\mathbf{k}|g)$ is smooth function with compact support, and

$$\tilde{\phi}(\mathbf{x}|g) = \frac{1}{(2\pi)^3}\int \phi(\mathbf{k}|g)\exp(i\mathbf{k}\mathbf{x})d\mathbf{k},$$

then formula (10.77) specifies a family of good operators, infinitely differentiable in g and satisfying the condition

$$B_g^{\psi\phi}\Phi = \int \psi(\mathbf{k}|g)\phi(\mathbf{k}|g)\Phi(\mathbf{k}|g)d\mathbf{k}\Big].$$

Let us note that for every point $(\mathbf{k}_0, g_0)$ there exists a neighborhood U, an operator $A \in \mathcal{A}$, and a function σ, such that the function $\langle B_g^{\sigma,A}, \Phi, \Phi(\mathbf{k}|g)\rangle$ is non-zero in the neighborhood U (here, $B_g^{\sigma,A}$ is a family of good operators, constructed in the proof of Lemma 10.12). If the support of the function $f(\mathbf{k}|g)$ is contained in the neighborhood U, then the necessary family B_g^f can be obtained with the help of relation (10.77), using $B_g^\psi = B_g^{\sigma,A}$, $B_g^{\psi\phi} = B_g^f$ in the relation. In other words, we should choose function (10.76) as the function $\psi(\mathbf{k}|g)$ and the function

$$\frac{f(\mathbf{k}|g)}{\langle B_g^{\sigma,A}\Phi, \Phi(\mathbf{k}|g)\rangle}$$

as the function $\phi(\mathbf{k}|g)$ (clearly, the latter function is smooth and has compact support if we define it to be zero at the values where both the numerator and the denominator are zero).

An arbitrary function f can be represented as a finite sum of functions f_i, for which the family $B_g^{f_i}$ can be obtained by the construction above. It is clear that the operators

$$B_g^f = \sum B_g^{f_i}$$

are infinitely differentiable in g and satisfy the conditions

$$B_g^f \Phi = \int f(\mathbf{k}|g)\Phi(\mathbf{k}|g)d\mathbf{k},$$

$$(B_g^f)^*\Phi = 0.$$

However, these functions do not, *a priori*, satisfy the condition of smoothness entering the definition of a good operator. To satisfy this condition, we can replace the operators B_g^f by the operators

$$\tilde{B}_g^f = \int \tilde{\mu}(\mathbf{x},t|g)B(\mathbf{x},t|g)d\mathbf{x}dt,$$

where

$$\tilde{\mu}(\mathbf{x},t|g) - \frac{1}{(2\pi)^4}\int \mu(\mathbf{k},\omega|g)\exp(i\mathbf{k}\mathbf{x} - i\omega t)d\mathbf{k}d\omega,$$

$\mu(\mathbf{k},\omega|g)$ is a smooth function with compact support, equal to 1 if $(\mathbf{k},g) \in \operatorname{supp} f$, $\omega = \omega(\mathbf{k}|g)$.

Lemma 10.15. *The function $\nu(\mathbf{k}|g)$ defined by the formula*

$$\left\langle \frac{d\Phi(f|g)}{dg}, \Phi(f'|g)\right\rangle = \int \nu(\mathbf{k}|g)f(\mathbf{k})\overline{f'(\mathbf{k})}d\mathbf{k}, \qquad (10.78)$$

where $f(\mathbf{k}), f'(\mathbf{k})$ are smooth functions with compact support, is smooth in $\mathbf{k}$ and g.

Let us first note that, by Lemma 10.14, the vector $\Phi(f|g)$ can be represented in the form $B_g^f \Phi$, where the operator B_g^f is infinitely differentiable in g; hence, the derivative $\frac{d\Phi(f|g)}{dg}$ exists. Furthermore, the bilinear form

$$A(f, f') = \left\langle \frac{d\Phi(f|g)}{dg}, \Phi(f'|g) \right\rangle$$

is translation-invariant [i.e. $A(f, f') = A(f_{\mathbf{x}}, f'_{\mathbf{x}})$, where $f_{\mathbf{x}}(\mathbf{k}) = \exp(-i\mathbf{k}\mathbf{x})f(\mathbf{k})$] and, therefore, can be written in the form (10.78).

Let us consider the function

$$\zeta(\mathbf{x}|g) = \left\langle \frac{d\Phi(f|g)}{dg}, \Phi(f'_{\mathbf{x}}|g) \right\rangle$$

$$= \left\langle \frac{d\Phi(f|g)}{dg}, \exp(-i\mathbf{P}\mathbf{x})\Phi(f'|g) \right\rangle$$

$$= \left\langle \frac{dB_g^f}{dg}\Phi, B_g^{f'}(\mathbf{x})\Phi \right\rangle.$$

It is clear that

$$\zeta(\mathbf{x}|g) = \int \exp(i\mathbf{k}\mathbf{x})\nu(\mathbf{k}|g)f(\mathbf{k})\overline{f'(\mathbf{k})}d\mathbf{k}. \tag{10.79}$$

It follows from relation (10.79) and Lemma 10.12 that the function $\nu(\mathbf{k}|g)f(\mathbf{k})\overline{f'(\mathbf{k})}$ is smooth. Since $f(\mathbf{k}), f'(\mathbf{k})$ are arbitrary smooth functions, the proof of the lemma follows.

It follows from Lemma 10.15 that the single-particle state $\Phi(\mathbf{k}|g)$ can be chosen in such a way that for every f, f' we have

$$\left\langle \frac{d\Phi(f|g)}{dg}, \Phi(f'|g) \right\rangle = 0. \tag{10.80}$$

[If equality (10.80) is not satisfied, then we can satisfy it if we replace $\Phi(\mathbf{k}|g)$ with the generalized vector function

$$\tilde{\Phi}(\mathbf{k}|g) = \exp(i\tau(\mathbf{k}|g))\Phi(\mathbf{k}|g),$$

where $\tau(\mathbf{k}|g)$ is a solution of the equation

$$i\frac{\partial\tau(\mathbf{k}|g)}{\partial g} = -\nu(\mathbf{k}|g).\Big]$$

Lemma 10.15 implies that $\tau(\mathbf{k}|g)$ is a smooth function and hence the function $\langle A\Phi, \tilde{\Phi}(\mathbf{k}|g)\rangle$ is smooth for all $A \in \mathcal{A}$.

In what follows, we will assume that the single-particle state $\Phi(\mathbf{k}|g)$ is chosen such that it satisfies condition (10.80).

Let us denote by t_0 a moment in time, satisfying the relation $t_0 \leq -\frac{\Delta}{\alpha}$ (recall that the symbol Δ denotes the support radius of the fixed function $h(\tau)$, so that $h(\tau) = 0$ for $\tau \leq \alpha t_0$).

Lemma 10.16. *For any smooth function with compact support $f(\mathbf{k})$, we have*

$$U_\alpha(t, t_0)\Phi(f|0)$$
$$= \exp(ir_\alpha(\mathbf{P}|t) - ir_\alpha(\mathbf{P}|t_0))(\Phi(f|h(\alpha t)) + \alpha\xi_1(f|\alpha t)$$
$$+ \cdots + \alpha^n\xi_n(f|\alpha t) + \alpha^{n+1}\eta_{n+1}(f, \alpha, t, t_0)),$$

where $\xi_i(f|\sigma)$ is defined by the equations

$$i\frac{d\xi_{i-1}(f|\sigma)}{d\sigma} = (H(h(\sigma)) - \omega(\mathbf{P}|h(\sigma)))\xi_i(f|\sigma), \qquad (10.81)$$

$$\left\langle \frac{d\xi_i(f|\sigma)}{d\sigma}, \Phi(f'|h(\sigma)) \right\rangle = 0, \qquad (10.82)$$

$$\xi_0(f|\sigma) = \Phi(f|h(\sigma)), \qquad (10.83)$$

$$\xi_0(f|\sigma_0) = 0 \ for \ j > 0 \qquad (10.84)$$

and $\|\eta_{n+1}(f, \alpha, t, t_0)\| \leq C$ (where C does not depend on α, t, t_0, but may depend on f and n).

The proof of this lemma is based on arguments similar to the arguments in Section 4.3. The vector $\Psi(t) = U_\alpha(t, t_0)\Phi(f|0)$ satisfies the equation

$$i\frac{d\Psi}{dt} = H(h(\alpha t))\Psi(t).$$

By using the change of variables $\sigma = \alpha t$, this equation can be written in the form

$$i\alpha\frac{d\Psi}{d\sigma} = H(h(\sigma))\Psi(\sigma). \qquad (10.85)$$

We will look for solutions of (10.85) of the form

$$\Psi(\sigma) = \exp\left(-\frac{i}{\alpha}\int_{\sigma_0}^{\sigma}\omega(\mathbf{P}|h(\sigma'))d\sigma'\right)\left(\xi_0(f|\sigma) + \alpha\xi_1(f|\sigma)\right.$$

$$\left. + \cdots + \alpha^n\xi_n(f|\sigma) + \alpha^{n+1}\eta_{n+1}(f,\alpha,\sigma)\right).$$

Let us note that equation (10.81) specifies the vector $\xi_i(f|\sigma)$ up to a constant belonging to the kernel of the operator $H(h(\sigma)) - \omega(\mathbf{P}|h(\sigma))$ (i.e. the vector is specified up to a vector from the single-particle subspace). Equations (10.82)–(10.84) fix this constant vector uniquely. Further on (see Lemma 10.17), the vector $\xi_i(f|\sigma)$ is infinitely differentiable in σ.

The function $\eta_{n+1}(f,\alpha|\sigma)$ satisfies the equation

$$i\alpha\frac{d\eta_{n+1}(f,\alpha|\sigma)}{d\sigma} = (H(h(\sigma)) - \omega(\mathbf{P}|h(\sigma)))\eta_{n+1}(f,\alpha|\sigma)$$

$$- i\frac{d\xi_n(f|\sigma)}{d\sigma} \tag{10.86}$$

with initial condition $\eta_{n+1}(f,\dot\alpha|\sigma_0) = 0$. From this equation and the remark that $\left\|\frac{d\xi_n(f|\sigma)}{d\sigma}\right\|$ is bounded above by a constant, not depending on σ, it is easy to obtain the inequality

$$\|\eta_{n+1}(f,\alpha|\sigma)\| \leq \frac{C}{\alpha}$$

(the constant C depends on f). Noting the fact that

$$\eta_n(f,\alpha|\sigma) = \xi_n(f|\sigma) + \alpha\eta_{n+1}(f,\alpha|\sigma),$$

we see that

$$\|\eta_n(f,\alpha|\sigma)\| \leq \text{const.}$$

This finishes the proof of the lemma.

Lemma 10.17. *The vector $\xi_i(f|\sigma)$, entering the formulation of Lemma 10.16, can be written in the form*

$$\xi_i(f|\sigma) = D_i^{f,\sigma}\Phi,$$

where $D_i^{f,\sigma}$ is a family of operators, infinitely differentiable in σ in the topology on $\mathcal{A}$, hence, this vector is infinitely differentiable in σ.

We will prove this lemma by induction. Suppose the operator $D_{i-1}^{f,\sigma}$ is given. Let us consider the vector $\zeta_i(f|\sigma)$, defined by the relations

$$i\frac{d\xi_{i-1}(f|\sigma)}{d\sigma} = (H(h(\sigma)) - \omega(\mathbf{P}|h(\sigma)))\zeta_i(f|\sigma),$$

$$\langle \zeta_i(f|\sigma), \Phi(f'|h(\sigma)) \rangle = 0,$$

where f, f' are arbitrary smooth functions with compact support. This vector can be written in the form

$$\zeta_i(f|\sigma) = \gamma(H(h(\sigma)), \mathbf{P}|h(\sigma))\frac{d\xi_{i-1}(f|\sigma)}{d\sigma}. \tag{10.87}$$

Here,

$$\gamma(\omega, \mathbf{p}|g) = i(\omega - \omega(\mathbf{p}|g))^{-1},$$

if $\mathbf{p} \in \operatorname{supp} f$, and $(\omega, \mathbf{p})$ belongs to the spectrum of the operators $H(g), \mathbf{P}$, but $\omega \neq \omega(\mathbf{p}|g)$; if $\omega = \omega(\mathbf{p}|g)$, then $\gamma(\omega, \mathbf{p}|g)$ should be equal to zero. The function $\gamma(\omega, \mathbf{p}|g)$ can be chosen to be smooth and equal to zero for large $|\mathbf{p}|$.

Using relation (10.87), we can write

$$\zeta_i(f|\sigma) = \int \tilde{\gamma}(t, \mathbf{x}|h(\sigma))E_{i-1}^{f,\sigma}(\mathbf{x}, t|\sigma)\Phi d\mathbf{x}dt, \tag{10.88}$$

where $\tilde{\gamma}(t, \mathbf{x}|g)$ is the Fourier transform of $\gamma(\omega, \mathbf{p}|g)$ in the variables $\omega, \mathbf{p}$ and $E_{i-1}^{f,\sigma} = (d/d\sigma)D_{i-1}^{f,\sigma}$. Equation (10.88) implies that $\zeta_i(f|\sigma) = F_i^{f,\sigma}\Phi$, where $F_i^{f,\sigma} = \int \tilde{\gamma}(t, \mathbf{x}|h(\sigma))E_{i-1}^{f,\sigma}(\mathbf{x}, t|\sigma)d\mathbf{x}dt$ (this operator is infinitely differentiable in σ, by Lemma 10.11). The vector $\xi_i(f|g)$ can be expressed in terms of the vector $\zeta_i(f|\sigma)$ in the following way:

$$\xi_i(f|\sigma) = \zeta_i(f|\sigma) + \Phi(\lambda_i^\sigma f|h(\sigma)), \tag{10.89}$$

where $\lambda_i^\sigma(\mathbf{k})$ is a function satisfying the equation

$$\int \frac{\partial \lambda_i^\sigma(\mathbf{k})}{\partial\sigma} f(\mathbf{k})\overline{f'(\mathbf{k})}d\mathbf{k} = -\left\langle \frac{d\zeta_i(f|\sigma)}{d\sigma}, \Phi(f'|h(\sigma)) \right\rangle \tag{10.90}$$

with the initial condition $\lambda_i^{\sigma_0}(\mathbf{k}) = 0$ for $\sigma_0 < -\Delta$. By introducing the vector generalized functions $\xi_i(\mathbf{k}|\sigma)$ and $\zeta_i(\mathbf{k}|\sigma)$ with the relations $\xi_i(f|\sigma) = \int f(\mathbf{k})\xi_i(\mathbf{k}|\sigma)d\mathbf{k}$, $\zeta_i(f|\sigma) = \int f(\mathbf{k})\zeta_i(\mathbf{k}|\sigma)d\mathbf{k}$, the formulas (10.89) and (10.90) can be written in the form

$$\xi_i(\mathbf{k}|\sigma) = \zeta_i(\mathbf{k}|\sigma) + \lambda_i^\sigma(\mathbf{k})\Phi(\mathbf{k}|h(\sigma)),$$

$$\frac{\partial\lambda_i^\sigma(\mathbf{k})}{\partial\sigma}\delta(\mathbf{k} - \mathbf{k}') = -\left\langle \frac{\partial\zeta_i(\mathbf{k}|\sigma)}{\partial\sigma}, \Phi(\mathbf{k}'|h(\sigma)) \right\rangle.$$

The function $\lambda_i^\sigma(\mathbf{k})$ is infinitely differentiable in σ; this can be shown using the ideas in the proof of Lemma 10.15. We can therefore select an operator $D_i^{f,\sigma}$ of the form

$$D_i^{f,\sigma} = F_i^{f,\sigma} + B_{h(\sigma)}^{\lambda_i^\sigma f}.$$

It is easy to see that the operator $D_i^{f,\sigma}$ is infinitely differentiable in σ and, therefore, satisfies the conditions of the lemma (the differentiability of the operator $F_i^{f,\sigma}$ has already been shown, while operator $B_{h(\sigma)}^{\lambda_i^\sigma f}$ can be written in the form

$$B_{h(\sigma)}^{\lambda_i^\sigma f} = B_{h(\sigma)}^{\mu_i^\sigma f} = \int \tilde{\mu}_i^\sigma(\mathbf{x})B_{h(\sigma)}^f(\mathbf{x})d\mathbf{x},$$

where $\mu_i^\sigma(\mathbf{k}) = \beta(\mathbf{k})\lambda_i^\sigma(\mathbf{k})$ and the symbol $\beta(\mathbf{k})$ denotes a smooth function with compact support, equal to 1 when $\mathbf{k} \in \mathrm{supp}f$).

Remark 10.5. For all σ in the interval $[-\delta, \delta]$, where the function h equals 1, we can show the following equations by induction:

$$\frac{d\xi_i(f|\sigma)}{d\sigma} = 0; \quad \zeta_i(f|\sigma) = 0; \quad F_i^{f,\sigma} = 0;$$

$$\frac{\partial\lambda_i^\sigma(\mathbf{k})}{\partial\sigma} = 0; \quad D_i^{f,\sigma} = B_1^{\lambda_i^\sigma f}.$$

Let us define the functions $\tilde{\phi}^{t,\alpha}(\mathbf{x})$, $\phi^{t,\alpha}(\mathbf{k})$ by the formulas

$$\tilde{\phi}^{t,\alpha}(\mathbf{x}) = (2\pi)^{-3}\int \phi^{t,\alpha}(\mathbf{k})\exp(i\mathbf{k}\mathbf{x})d\mathbf{k},$$

$$\phi^{t,\alpha}(\mathbf{k}) = \phi(\mathbf{k})\exp(ir_\alpha(\mathbf{k}|t)).$$

Lemma 10.18. *Let $\phi(\mathbf{k})$ be a smooth function with compact support. For any whole number n and any $\epsilon > 0$ we can find a number $C_{n,\epsilon}$,* *not depending on* $\mathbf{x}, t,$ *and* α, *so that*

$$|\tilde{\phi}^{t,\alpha}(\mathbf{x})| \leq \frac{C_{n,\epsilon}}{1 + (\rho(\mathbf{x}, V_\alpha^t(\phi)) - \epsilon|t|)^n}$$

(here, $V_\alpha^t(\phi)$ *is a set of points which can be written in the form* $\nabla r_\alpha(\mathbf{k}|t)$, *where* $\mathbf{k} \in \operatorname{supp}\phi$, *and* $\rho(\mathbf{x}, A)$ *is the distance between the point* $\mathbf{x}$ *and the set* A*). There exists a constant* C, *not depending on* $\mathbf{x}, t, \alpha$, *so that*

$$|\tilde{\phi}^{t,\alpha}(\mathbf{x})| \leq C|t|^{-3/2},$$

$$\int |\tilde{\phi}^{t,\alpha}(\mathbf{x})|d\mathbf{x} \leq C(1 + |t|^{3/2}). \tag{10.91}$$

The proof is analogous to the proof of Lemma 10.2 in Section 10.1.

Lemma 10.19. *If* $f(\mathbf{k})$, $\phi(\mathbf{k})$ *are smooth functions with compact support, then the vector*

$$\Gamma(t) = U_\alpha(t, t_0)S_-(0)W_\alpha(t_0)a^+(\bar{f}\bar{\phi})\theta,$$

where $t_0 \leq -\frac{\Delta}{\alpha}$, *can be written in the form*

$$\Gamma(t) = Q_s^t \Phi + R_s(t),$$

where

$$Q_s^t = Q_s^t(f, \phi, \alpha) = \int \tilde{\phi}^{t,\alpha}(\mathbf{x})N_s^{t,\alpha}(\mathbf{x})d\mathbf{x}, \tag{10.92}$$

$$N_s^{t,\alpha} = \sum_{i=0}^{s} D_i^{f,\alpha t},$$

$$\|R_s(t)\| \leq C_s(f, \phi)\alpha^{s+1}(1 + |t|^{3/2}). \tag{10.93}$$

Indeed,

$$W_\alpha(i_0)a^+(\bar{f}\bar{\phi})\theta = a^+(\bar{f}\,\overline{\phi^{t_0,\alpha}})\theta.$$

Therefore,

$$\Gamma(t) = U_\alpha(t, t_0)\Phi(f\phi^{t_0,\alpha}|0)$$

$$= U_\alpha(t, t_0)\phi^{t_0,\alpha}(\mathbf{P})\Phi(f|0)$$

$$= \phi^{t_0,\alpha}(\mathbf{P})U_\alpha(t, t_0)\Phi(f|0).$$

Hence, we can conclude from Lemmas 10.16–10.18 that

$$\Gamma(t) = \phi^{t_0,\alpha}(\mathbf{P}) \exp(ir_\alpha(\mathbf{P}|t) - ir_\alpha(\mathbf{P}|t_0))$$

$$\times \left(\sum_{i=0}^{s} \alpha^i \xi_i(f|\alpha t) + \alpha^{s+1} \eta_{s+1}(f,\alpha,t,t_0) \right)$$

$$= \phi^{t,\alpha}(\mathbf{P}) \left(\sum_{i=0}^{s} D_i^{f,\alpha t} \Phi \right) + R_s(t)$$

$$= \phi^{t,\alpha}(\mathbf{P}) N_s^{t,\alpha} \Phi + R_s(t)$$

$$= \int \tilde{\phi}^{t,\alpha}(\mathbf{x}) N_s^{t,\alpha}(\mathbf{x}) \Phi d\mathbf{x} + R_s(t),$$

which proves the necessary statement [to obtain an estimate for R_s, we should use an estimate for $\|\eta_{s+1}\|$, used in Lemma 10.16, and inequality (10.91)].

Remark 10.6. The relation

$$U_\alpha(t',t) U_\alpha(t,t_0) = U_\alpha(t',t_0)$$

implies that

$$U_\alpha(t',t)\Gamma(t) = \Gamma(t'). \tag{10.94}$$

Using equation (10.94) and Lemma 10.18, we can conclude that

$$U_\alpha(t',t) Q_s^t \Phi = Q_s^{t'} \Phi + R_s(t,t'), \tag{10.95}$$

where $\|R_s(t,t')\| \leq C\alpha^{s+1}$ and the constant C depends on f and ϕ (but does not depend t,t',α). We will use relation (10.95) later.

Lemma 10.20. *Let $\phi_1(\mathbf{k})$, $\phi_2(\mathbf{k})$ be non-overlapping smooth functions with compact support. Then there exists a number $c > 0$ such that the distance between $V_\alpha^t(\phi_1)$ and $V_\alpha^t(\phi_2)$ is larger than $c|t|$.*

It follows from the strong convexity of $\omega(\mathbf{k}|g)$ that

$$d^2\omega(\mathbf{k}|g) \geq \lambda d\mathbf{k}^2 \tag{10.96}$$

(here, λ is positive and does not depend on $\mathbf{k}, g$, if $\mathbf{k}$ runs over a bounded set; the second differential is taken over only the variable $\mathbf{k}$).

Inequality (10.96) implies that $d^2 r_\alpha(\mathbf{k}|t) \geq \lambda|t|d\mathbf{k}^2$, from which we can conclude that the second derivative of $r_\alpha(\mathbf{k}|t)$, in any direction, cannot be less than $\lambda|t|$. Indeed, we can set

$$\mathbf{e} = \frac{\mathbf{k}_1 - \mathbf{k}_2}{|\mathbf{k}_1 - \mathbf{k}_2|}, \quad \nu(\zeta) = r_\alpha(\mathbf{k}_2 + \zeta\mathbf{e}|t)$$

and conclude that $\nu''(\zeta) \geq \lambda|t|$. Hence, it is clear that

$$|\nabla r_\alpha(\mathbf{k}_1|t) - \nabla r_\alpha(\mathbf{k}_2|t)|$$

$$\geq |v'(|\mathbf{k}_1 - \mathbf{k}_2|) - v'(0)| \geq \lambda|t||\mathbf{k}_1 - \mathbf{k}_2|.$$

Therefore, if the distance between the sets $\operatorname{supp}\phi_1$ and $\operatorname{supp}\phi_2$ is larger than ϵ, then the distance between the sets $V_\alpha^t(\phi_1)$ and $V_\alpha^t(\phi_2)$ is larger than $\lambda|t|\epsilon$. This finishes the proof of the lemma.

Lemma 10.21. *Let $f_1(\mathbf{k})$, $f_2(\mathbf{k})$, $\phi_1(\mathbf{k})$, $\phi_2(\mathbf{k})$ be smooth functions with compact support and suppose that the supports of the functions ϕ_1, ϕ_2 do not intersect. Then*

$$p([U_\alpha(t + \tau, t)Q_s^{1,t}U_\alpha(t, t + \tau), Q_s^{2,t}]) \leq \frac{C|\tau|^r}{1 + |t|^n}$$

[here, $p(A)$ is an arbitrary seminorm on the algebra $\mathcal{A}$; n is an arbitrary number; $-\infty < \tau, t < \infty$, $|\tau| \leq |t|$; C and r do not depend on t, τ, and α; the symbol $Q_s^{i,t}$ denotes the operator

$$Q_s^{i,t} = Q_s^t(f_i, \phi_i, \alpha),$$

defined by the relation (10.92)].

The proof of this lemma follows directly from Lemmas 10.10, 10.18, and 10.20.

Lemma 10.22. *Let $\phi_1(\mathbf{k}), \ldots, \phi_n(\mathbf{k})$ be smooth functions with compact and non-intersecting supports, let $f_1(\mathbf{k}), \ldots, f_n(\mathbf{k})$ be smooth functions with compact support, and let $t \leq -a\alpha^{-\epsilon}$, $t_0 \leq -\frac{\Delta}{\alpha}$, $a > 0$,*

$\epsilon > 0$, $s \geq \frac{3}{2}n + 1$. *Then the vector*

$$\Psi_n(t) = U_\alpha(t, t_0) S_-(0) W_\alpha(t_0) b^+(\bar{f}_1 \bar{\phi}_1) \ldots b^+(\bar{f}_n \bar{\phi}_n) \theta$$

can be written in the form

$$\Psi_n(t) = Q_s^{1,t} \ldots Q_s^{n,t} \Phi + \pi(t), \tag{10.97}$$

where

$$Q_s^{i,t} = Q_s^t(f_i, \phi_i, \alpha),$$

$$\|\pi(t)\| \leq C|t - t_0|\alpha^2,$$

and the number C does not depend on t, t_0, and α.

It is easy to see that the vector $\Psi_n(t)$ does not depend on t_0, if t_0 is less than $-\Delta/\alpha$, hence we may assume that t_0 is the largest integer obeying $t_0 < -\Delta/\alpha$. We will assume that t is an integer and we will prove the relation (10.97) by induction on t; the generalization to real-valued t does not present any obstacles.

Let us suppose that

$$\Psi_n(t) \approx Q_s^{1,t} \ldots Q_s^{n,t} \Phi. \tag{10.98}$$

Using equation (10.95), we can conclude that

$$\begin{aligned}
\Psi_n(t+1) &= U_\alpha(t+1, t) \Psi_n(t) \\
&\approx U_\alpha(t+1, t) Q_s^{1,t} \ldots Q_s^{n,t} \Phi \\
&= U_\alpha(t+1, t) Q_s^{1,t} \ldots Q_s^{n-1,t} U_\alpha(t, t+1) U_\alpha(t+1, t) Q_s^{n,t} \Phi \\
&\approx U_\alpha(t+1, t) Q_s^{1,t} \ldots Q_s^{n-2,t} U_\alpha(t, t+1) U_\alpha(t+1, t) \\
&\quad \times Q_s^{n-1,t} U_\alpha(t, t+1) Q_s^{n,t+1} \Phi
\end{aligned} \tag{10.99}$$

An additional error, arising in the transition from approximation (10.98) to approximation (10.99) by (10.95), can be written in the form

$$U_\alpha(t+1, t) Q_s^{1,t} \ldots Q_s^{n-1,t} U_\alpha(t, t+1) R_s(t, t+1),$$

where $\|R_s(t, t+1)\| \leq C\alpha^{s+1}(1 + |t|^{3/2})$ and any seminorm of the operator $Q_s^{i,t}$ on the $\mathcal{A}$ does not exceed $C(1 + |t|^{3/2})$ [the last statement follows from (10.91)]. Using these remarks and the inequalities $|t| \leq |t_0|, s \geq \frac{3}{2}n + 1$, we can show that as we go from (10.98) to (10.99) the error does not increase by more than $C\alpha^2$.

By Lemma 10.21, we can see that

$$\Psi_n(t+1) \approx U_\alpha(t+1, t)Q_s^{1,t} \ldots Q_s^{n-2,t}$$

$$\times U_\alpha(t, t+1)Q_s^{n,t+1}U_\alpha(t+1, t)Q_s^{n-1,t}\Phi. \qquad (10.100)$$

Similarly, the additional error obtained from transferring to (10.100) from (10.99) does not exceed $C\alpha^2$ (moreover, this error does not exceed $C\alpha^r$, where r is arbitrary). To prove this, we should use, in addition to Lemma 10.21, the inequality $t < -a\alpha^{-\epsilon}$ and the estimate for $Q_s^{i,t}$ shown above.

Using equation (10.95) again, we obtain

$$\Psi_n(t+1) \approx U_\alpha(t+1, t)Q_s^{1,t}$$

$$\ldots Q_s^{n-2,t}U_\alpha(t, t+1)Q_s^{n,t+1}Q_s^{n-1,t+1}\Phi.$$

Applying equation (10.95) several times and then applying Lemma 10.21, we obtain

$$\Psi_n(t+1) \approx Q_s^{n,t+1} \ldots Q_s^{1,t+1}\Phi,$$

$$\approx Q_s^{1,t+1} \ldots Q_s^{n,t+1}\Phi. \qquad (10.101)$$

The error arising from each transformation does not exceed $C\alpha^2$. Hence, we have derived (10.101) from (10.98) and have proven that the error in (10.101) also does not exceed $C\alpha^2$. The inductive step is shown. Since the case $t = t_0$ for equation (10.97) is obvious, Lemma 10.22 is proven.

We can now finally easily show the main result of this section.

In the formulation of Lemma 10.22, let us take $t = -\frac{\delta}{\alpha}$, where δ is such that $h(\tau) \equiv 1$ for $-\delta \leq \tau \leq \delta$. Then we can show that

$$\tilde{\phi}_i^{t,\alpha}(\mathbf{x}) = \tilde{\phi}_i(\mathbf{x}, t)$$

$$= (2\pi)^{-3} \int \phi_i(\mathbf{k}) \exp(i\mathbf{k}\mathbf{x} - i\omega(\mathbf{k}|1)t)d\mathbf{k}, \qquad (10.102)$$

$$\Psi_n(t) \approx \prod_{i=1}^{n} \left(\int \tilde{\phi}_i(\mathbf{x}, t) \left(\sum_{j=0}^{s} \alpha^i D_j^{f_i, -\delta}(\mathbf{x}) \right) d\mathbf{x} \right) \Phi,$$

$$D_0^{f_i, -\delta} = B_1^{f_i},$$

$$D_j^{f_i, -\delta} = B_1^{\lambda_j^{-\delta} f_i}. \tag{10.103}$$

The error in (10.102) does not exceed $C\alpha^2(-\frac{\delta}{\alpha} + \frac{\Delta}{\alpha}) = \text{const } \alpha$ and inequality (10.103) follows from the remark to Lemma 10.17.

By the definition of the Møller matrix S_- and the relations (10.102) and (10.103) we obtain

$$S_-(1)b^+(\bar{f}_1\bar{\phi}_1)\dots b^+(\bar{f}_n\bar{\phi}_n)\theta$$

$$= \lim_{t\to-\infty} \prod_{i=1}^{n} \left(\int \tilde{\phi}_i(\mathbf{x}, t) B_1^{f_i}(\mathbf{x}, t|1) d\mathbf{x} \right) \Phi$$

$$= \lim_{t\to-\infty} \exp(iH(1)t) \prod_{i=1}^{n} \left(\int \tilde{\phi}_i(\mathbf{x}, t) B_1^{f_i}(\mathbf{x}) d\mathbf{x} \right) \Phi$$

$$= \lim_{t\to-\infty} \exp(-iH(1)t) \prod_{i=1}^{n} \left(\int \tilde{\phi}_i(\mathbf{x}, t) D_0^{f_i, -\delta}(\mathbf{x}) d\mathbf{x} \right) \Phi$$

$$= \lim_{\alpha\to 0} \exp\left(-iH(1)\frac{\delta}{\alpha}\right) \prod_{i=1}^{n} \left(\int \tilde{\phi}_i\left(\mathbf{x}, -\frac{\delta}{\alpha}\right) \right.$$

$$\times \left. \left(\sum_{j=0}^{s} \alpha^j D_j^{f_i, -\delta}(\mathbf{x}) \right) d\mathbf{x} \right) \Phi$$

$$= \lim_{\alpha\to 0} \exp\left(-iH(1)\frac{\delta}{\alpha}\right) \Psi_n\left(-\frac{\delta}{\alpha}\right)$$

$$= \lim_{\alpha\to 0} U_\alpha(0, t_0) S_-(0) W_\alpha(t_0) b^+(\bar{f}_1\bar{\phi}_1)\dots b^+(\bar{f}_n\bar{\phi}_n)\theta. \tag{10.104}$$

To derive equality (10.73) from relation (10.104), it is enough to recall that if $\lim A_n x = Ax$ for all x in a dense set and the norms of the operators A_n are uniformly bounded, then $\text{slim } A_n = A$. Relation (10.74) follows analogously.

Chapter 11

Translation-Invariant Hamiltonians
(Further Investigations)

11.1 Connections between the axiomatic theory and the Hamiltonian formalism

Let H be a formal translation-invariant Hamiltonian of the form (8.3). Let us suppose that there exists an operator realization $(\mathcal{H}, \hat{H}, \hat{\mathbf{P}}, a(\mathbf{k}, \epsilon, t), \Phi)$ of the Hamiltonian H (recall that $\mathcal{H}$ is a Hilbert space; $\hat{H}, \hat{\mathbf{P}}$ are the operators of energy and momentum; $a(\mathbf{k}, \epsilon, t)$ are operator generalized functions; Φ is a ground state).

Let us consider a family $\mathcal{A}$ consisting of two operator generalized functions

$$a(\mathbf{x}, 1, t) = (2\pi)^{-3/2} \int \exp(i\mathbf{k}\mathbf{x}) a(\mathbf{k}, 1, t) d\mathbf{k},$$

$$a(\mathbf{x}, -1, t) = (2\pi)^{-3/2} \int \exp(-i\mathbf{k}\mathbf{x}) a(\mathbf{k}, -1, t) d\mathbf{k}.$$

We will use the term *scattering matrix of the formal Hamiltonian* H to refer to the scattering matrix, constructed with the operators $\hat{H}, \hat{\mathbf{P}}$, and the family $\mathcal{A}$ (here, we assume that the family $\mathcal{A}$ and the operators $\hat{H}, \hat{\mathbf{P}}$ satisfy the conditions that guarantee the existence of the scattering matrix; see Section 10.5).

The relation between this definition and the definition of the scattering matrix provided in Chapter 9 will be addressed later. Currently, we will note that in the case when the Hamiltonian H satisfies the conditions in Section 9.2, then it follows from the results of Section 10.3 that the scattering matrix constructed in Section 9.2 coincides with the above definition.

Indeed, let $\lambda(\mathbf{k}) \in \mathcal{S}(E^3)$, $f(\mathbf{k})$, and $\mu(\tau)$ be smooth functions with compact support and $B = \int \lambda(\mathbf{k})\mu(\tau)a^+(\mathbf{k},\tau)d\mathbf{k}d\tau$. It is clear that

$$\int \tilde{f}(\mathbf{x}|t)B(\mathbf{x},t)d\mathbf{x}$$

$$= \int \tilde{f}(\mathbf{x}|t)\exp(-i\mathbf{kx})\lambda(\mathbf{k})\mu(\tau)a^+(\mathbf{k},t+\tau)d\mathbf{k}d\mathbf{x}d\tau$$

$$= \int f(\mathbf{k})\lambda(\mathbf{k})\mu(\tau)a^+(\mathbf{k},t+\tau)d\mathbf{k}d\tau,$$

and, therefore, in the conditions of Section 9.2, we have

$$\operatorname*{wlim}_{t\to-\infty} \int \tilde{f}(\mathbf{x}|t)B(\mathbf{x},t)d\mathbf{x} = \int f(\mathbf{k})\lambda(\mathbf{k})\mu(\tau)\overline{\Lambda}^{-1}(\mathbf{k})a_{\mathrm{in}}^+(\mathbf{k},\tau)d\mathbf{k}d\tau$$

$$= \int f(\mathbf{k})\lambda(\mathbf{k})\tilde{\mu}(\omega(\mathbf{k}))\overline{\Lambda}^{-1}(\mathbf{k})a_{\mathrm{in}}^+(\mathbf{k})d\mathbf{k},$$

$$(11.1)$$

where $a_{\mathrm{in}}^+(\mathbf{k},\tau)$, $a_{\mathrm{in}}(\mathbf{k},\tau)$ are in-operators, defined in Section 9.2; $\tilde{\mu}(\omega(\mathbf{k})) = \int \mu(\tau)\exp(i\omega(\mathbf{k})\tau)d\tau$.

Relations (9.18) and (9.21) imply that

$$\langle B\Phi, a_{\mathrm{in}}^+(\mathbf{p})\Phi \rangle = \int \lambda(\mathbf{k})\mu(\tau)\left\langle a^+(\mathbf{k},\tau)\Phi, a_{\mathrm{in}}^+(\mathbf{p})\Phi \right\rangle d\mathbf{k}d\tau$$

$$= \int \lambda(\mathbf{k})\tilde{\mu}(\omega(\mathbf{k}))\left\langle a^+(\mathbf{k})\Phi, a_{\mathrm{in}}^+(\mathbf{p})\Phi \right\rangle d\mathbf{k}$$

$$= \lambda(\mathbf{p})\tilde{\mu}(\omega(\mathbf{p}))\rho(1,\mathbf{p}) = \lambda(\mathbf{p})\tilde{\mu}(\omega(\mathbf{p}))\overline{\Lambda}^{-1}(\mathbf{p}).$$

$$(11.2)$$

Comparing equations (11.1) and (11.2) with relation (10.37), we see that the in-operators, defined in Section 9.2, coincide with the in-operators defined in Section 10.3; an analogous proof can be carried out for out-operators and for scattering matrices.

We should note, however, that the present definition of scattering matrix can also be used in the presence of bound states. (We define bound state as a single-particle state $\Phi(\mathbf{k})$ satisfying the condition

$\langle \Phi(\mathbf{k}), a^{+}(\mathbf{k}', t)\Phi \rangle = 0$. Other single-particle states are called elementary. The construction of Section 9.2 works only for elementary particles.)

The questions of the existence of an operator realization of the formal Hamiltonian H and whether the operators $\hat{H}, \hat{\mathbf{P}}$, and the operator $\mathcal{A}$ satisfy the requirements in Section 10.5 turn out to be non-trivial.

In this section, we will show that these questions can be answered in the framework of perturbation theory; more precisely, we will construct the relevant objects as formal power series in the coupling constant.[1]

Let us consider the Hermitian and translation-invariant formal expression

$$H = H_0 + V$$

$$= \int \epsilon(\mathbf{k}) a^{+}(\mathbf{k}) a(\mathbf{k}) d\mathbf{k} + \sum_{r \geq 1} g^r \sum_{m,n} \int v^{(r)}_{m,n}(\mathbf{k}_1, \ldots, \mathbf{k}_m | \mathbf{p}_1, \ldots, \mathbf{p}_n)$$

$$\times \delta(\mathbf{k}_1 + \cdots + \mathbf{k}_m - \mathbf{p}_1 - \cdots - \mathbf{p}_n) a^{+}(\mathbf{k}_1) \cdots a^{+}(\mathbf{k}_m) a(\mathbf{p}_1)$$

$$\cdots a(\mathbf{p}_n) d^m \mathbf{k} d^n \mathbf{p}. \tag{11.3}$$

$a^{+}(\mathbf{k}), a(\mathbf{k})$ are, as usual, symbols satisfying CCR. We will assume that $\epsilon(\mathbf{k})$ is a smooth function whose derivatives are of moderate growth, the functions $v^{(r)}_{m,n}$ belong to the space $\mathcal{S}$ and for every r, only a finite number of the functions $v^{(r)}_{m,n}$ are non-zero. The number g will be called the coupling constant. If the sum in (11.3) is infinite, then we will consider it simply as a formal power series without assuming

---

[1] In other words, we will consider formal expressions $\sum_n A_n g^n$, where A_n are numbers, vectors, or operators; if A_n are operators, we will assume that they are all defined on the same domain. The sums and products of two (number or operator) formal series $\sum_n A_n g^n$ and $\sum_n B_n g^n$ are understood as the series $\sum_n (A_n + B_n) g^n$ and $\sum_n (\sum_{k+l=n} A_k B_l) g^n$. The rest of this section will be considered in the framework of perturbation theory (without always stating so). In order to give precise mathematical meaning to the results in this section and their proofs, all the relations should be understood in terms of formal power series.

anything about its convergence. *The set of all formal expressions having these properties will be denoted by $\mathcal{M}$.*

An expression $H \in \mathcal{M}$ can be seen as a formal translation-invariant Hamiltonian (the most interesting are the Hamiltonians in the set $\mathcal{M}$ that can be written in the form $H = H_0 + gW$, where W does not already contain the coupling constant g; however, it will be convenient to consider the general case).

An operator realization of a Hamiltonian $H \in \mathcal{M}$ will be defined to be a Hilbert space $\mathcal{H}$ with the action of the energy operator $\hat{H} = \sum H_r g^r$ and the momentum operator $\mathbf{P}$, the operator generalized functions $a(\mathbf{k}, \epsilon, t) = \sum g^r a_r(\mathbf{k}, \epsilon, t)$ and the vector Φ, satisfying the operator realization conditions in every order with respect to g (the series $\sum H_r g^r$ and $\sum g^r a_r(\mathbf{k}, \epsilon, t)$ are to be understood as formal power series in g without convergence assumptions). For the operator $\hat{H} = \sum g^r H_r$, we can construct the operators $\exp(i\hat{H}t) = \sum g^r U_r(t)$ which constitute a group of unitary operators.[2] The operators $U_r(t)$ and the operators $\int f(\mathbf{k}) a_r(\mathbf{k}, \epsilon, t) d\mathbf{k}$ must be defined on the same domain D and must transform it to itself; the set D must contain the vector Φ.

In Section 11.3, we will show that *for every Hamiltonian $H \in \mathcal{M}$, we can construct an operator realization in the above sense* (for Hamiltonians that generate vacuum polarization, we will assume that the function $\epsilon(\mathbf{k})$, entering in the definition of the Hamiltonian H_0, satisfies the condition $\epsilon(\mathbf{k}_1 + \mathbf{k}_2) < \epsilon(\mathbf{k}_1) + \epsilon(\mathbf{k}_2)$, however this requirement is not necessary).

Any translation-invariant Hamiltonian $H = H_0 + W$ generates a family of Hamiltonians $H_0 + gW$, depending on the parameter g. If the Hamiltonian $H_0 + W$ satisfies the conditions which guarantee that the Heisenberg equations (8.5) make sense (these conditions are described in Section 8.1), then $H_0 + gW \in \mathcal{M}$; therefore, for

---

[2]The operators $\sum g^r U_r(t)$ constitute a group of unitary operators if

$$\sum_{k+l=r} U_k(t) U_l(t') = U_r(t + t'),$$

$$U_k(0) = 0 \quad \text{for } k > 0,$$
$$U_0(t) = 1.$$

these Hamiltonians, an operator realization can be constructed in the framework of perturbation theory.

In Section 11.3, we will show that the operators $\hat{H} = \sum g^r H_r, \hat{\mathbf{P}}$ and the family of operator generalized functions consisting of two functions $a(\mathbf{x}, \epsilon, t) = \sum g^r a_r(\mathbf{x}, \epsilon, t) = \sum g^r \int \exp(i\epsilon\mathbf{k}\mathbf{x})a_r(\mathbf{k}, \epsilon, t)d\mathbf{k}$, $\epsilon = \pm 1$, satisfy the conditions of Section 10.5 in every order. The proof uses the additional assumption that the Hamiltonian in consideration belongs to the subset $\tilde{\mathcal{M}}$ of $\mathcal{M}$, whose Hamiltonians obey the condition that their functions $\epsilon(\mathbf{k})$ satisfy $\epsilon(\mathbf{k}_1 + \mathbf{k}_2) < \epsilon(\mathbf{k}_1) + \epsilon(\mathbf{k}_2)$ and are not linear on any open subset.

It follows that the Møller matrices and the scattering matrices exist in the framework of perturbation theory (more precisely, the Møller matrices $S_\pm$ and the scattering matrix S can be written as formal power series $S_\pm = \sum g^r S_\pm^{(r)}, S = \sum g^r S^{(r)}$). To derive this, it is necessary to analyze the construction of Møller matrices, given in Chapter 10, and check that the statements proven in this chapter hold for formal power series in g as well. To see this, note that we must modify the Møller matrix construction by assuming that the functions $f_1(\mathbf{k}), \ldots, f_n(\mathbf{k})$, entering the formula (10.1), form a non-overlapping family of functions[3] (for details, see Section 11.5).

11.2 Heisenberg equations and canonical transformations

Let us consider a formal translation-invariant Hamiltonian H, belonging to the class $\mathcal{M}$ described in Section 11.1. As noted in Section 8.1, if $v_{m,0}^{(r)} \neq 0$ (the Hamiltonian generates vacuum polarization), then the expression H does not define a self-adjoint operator on Fock space. Furthermore, the operator $\exp(iHt)$ cannot be constructed in this case even in the framework of perturbation theory (in terms of a formal power series in g).

This can be easily shown by calculating the operator $S(t, t_0) = \exp(iH_0t)\exp(-iH(t - t_0))\exp(-iH_0t_0)$ by means of the diagram techniques developed in Section 6.4. For example, in the expression

---

[3]The same modification of the Møller matrix definition is used for a different purpose in Section 10.5, part 5.

Fig. 11.1   Divergent diagram.

corresponding to the diagram depicted in Fig. 11.1, we run into the meaningless expression

$$\delta(\mathbf{k}_1 + \cdots + \mathbf{k}_m)\delta(\mathbf{k}_1 - \mathbf{k}_{m+1})\cdots\delta(\mathbf{k}_m - \mathbf{k}_{2m})\delta(\mathbf{k}_{m+1} + \cdots + \mathbf{k}_{2m})$$
$$= \delta^2(\mathbf{k}_1 + \cdots + \mathbf{k}_m)\delta(\mathbf{k}_1 - \mathbf{k}_{m+1})\cdots\delta(\mathbf{k}_m - \mathbf{k}_{2m}).$$

A similar situation arises for all diagrams without outer vertices (for vacuum loops): in the expression corresponding to the diagram, the integrand contains a product of δ-functions whose arguments are linearly dependent, rendering the expression meaningless.

Therefore, we cannot write a solution to the Schrödinger equation $i\frac{d\psi}{dt} = H\psi$ even in perturbation theory.

However, using the expression (11.3), we can formally write the Heisenberg equations which remain meaningful in perturbation theory.

Indeed, the Heisenberg equations,

$$\frac{\partial a(\mathbf{k}, t)}{\partial t} = i[H, a(\mathbf{k}, t)], \tag{11.4}$$

$$\frac{\partial a^+(\mathbf{k}, t)}{\partial t} = i[H, a^+(\mathbf{k}, t)], \tag{11.5}$$

can be written in the form

$$i\frac{\partial a(\mathbf{k}, t)}{\partial t} - \epsilon(\mathbf{k})a(\mathbf{k}, t) = I(\mathbf{k}, t), \tag{11.6}$$

$$-i\frac{\partial a^+(\mathbf{k}, t)}{\partial t} - \epsilon(\mathbf{k})a^+(\mathbf{k}, t) = I^+(\mathbf{k}, t), \tag{11.7}$$

where

$$I(\mathbf{k}, t) = \sum_{r \geq 1} g^r \sum_{m,n} m \int v^{(r)}_{m,n}(\mathbf{k}, \mathbf{k}_1, \ldots, \mathbf{k}_{m-1} | \mathbf{p}_1, \ldots, \mathbf{p}_n) \quad (11.8)$$

$$\times \delta \left(\mathbf{k} + \sum_{i=1}^{m-1} \mathbf{k}_i - \sum_{j=1}^{n} \mathbf{p}_j \right) a^+(\mathbf{k}_1, t) \quad (11.9)$$

$$\cdots a^+(\mathbf{k}_{m-1}, t) a(\mathbf{p}_1, t) \cdots a(\mathbf{p}_n, t) d^{m-1}\mathbf{k} d^n \mathbf{p}. \quad (11.10)$$

(These equations can be obtained by calculating the commutator on the right-hand side of the equations (11.4) and (11.5) by using CCR, distributivity, and the relation $[AB, C] = [A, C]B + A[B, C]$.) By using the initial conditions $a^+(\mathbf{k}, 0) = a^+(\mathbf{k}), a(\mathbf{k}, 0) = a(\mathbf{k})$, equations (11.6) and (11.7) can be written in the form

$$a(\mathbf{k}, t) = \exp(-i\epsilon(\mathbf{k})t)a(\mathbf{k})$$

$$+ \int_0^t \exp(-i\epsilon(\mathbf{k})(t - \tau))I(\mathbf{k}, \tau)d\tau, \quad (11.11)$$

$$a^+(\mathbf{k}, t) = \exp(i\epsilon(\mathbf{k})t)a^+(\mathbf{k})$$

$$+ \int_0^t \exp(i\epsilon(\mathbf{k})(t - \tau))I^+(\mathbf{k}, \tau)d\tau. \quad (11.12)$$

We then look for solutions to equations (11.11) and (11.12) in the form

$$a(\mathbf{k}, t) = \exp(-i\epsilon(\mathbf{k})t)a(\mathbf{k})$$

$$+ \sum_{r \geq 1} g^r \sum_{m,n} \int f^{r,t}_{m,n}(\mathbf{k}_1, \ldots, \mathbf{k}_m | \mathbf{p}_1, \ldots, \mathbf{p}_n)$$

$$\times \delta \left(\mathbf{k} + \sum_{i=1}^{m} \mathbf{k}_i - \sum_{j=1}^{n} \mathbf{p}_j \right) a^+(\mathbf{k}_1) \cdots a^+(\mathbf{k}_m)a(\mathbf{p}_1)$$

$$\cdots a(\mathbf{p}_n) d^m \mathbf{k} d^n \mathbf{p}, \quad (11.13)$$

$$a^+(\mathbf{k}, t) = \exp(i\epsilon(\mathbf{k})t)a^+(\mathbf{k})$$

$$+ \sum_{r \geq 1} g^r \sum_{m,n} \int \overline{f_{m,n}^{r,t}}(\mathbf{k}_1, \ldots, \mathbf{k}_m | \mathbf{p}_1, \ldots, \mathbf{p}_n)$$

$$\times \delta \left(\mathbf{k} + \sum_{i=1}^{m} \mathbf{k}_i - \sum_{j=1}^{n} \mathbf{p}_j \right) a^+(\mathbf{p}_1) \cdots a^+(\mathbf{p}_n)a(\mathbf{k}_1)$$

$$\cdots a(\mathbf{k}_n)d^m \mathbf{k} d^n \mathbf{p}. \tag{11.14}$$

Substituting these expressions for $a(\mathbf{k}, t), a^+(\mathbf{k}, t)$ into (11.11) and (11.12) and equating terms having the same order in g, we obtain a recurrence relation for $f_{m,n}^{r,t}$, from which it is clear that the functions $f_{m,n}^{r,s}$ belong to the space $\mathcal{S}$ and for every order of g (i.e. for every r), only a finite number of functions $f_{m,n}^{r,t}$ are non-zero.

(Here, we view $a(\mathbf{k}, t)$ as formal power series consisting of the symbols $a(\mathbf{k}), a^+(\mathbf{k})$, obeying CCR; all expressions arising in the solution can be converted to normal form using CCR. The series in g are viewed formally.)

It is easy to check that $\|f_{m,n}^{r,t}\|_{\alpha,\beta}$ does not grow faster than a power of t, i.e.

$$\|f_{m,n}^{r,t}\|_{\alpha,\beta} \leq C(1 + |t|^k) \tag{11.15}$$

(here, $\| \cdot \|_{\alpha,\beta}$ is a norm on the space $\mathcal{S}$; C and k depend on r, α, β). Inequality (11.15) follows from the following statement.

Theorem 11.1. *Let $\sigma(k_1, \ldots, k_n)$ be a smooth real function whose derivatives do not grow faster than a power. Then for any norm $\| \cdot \|_{\alpha,\beta}$ on the space $\mathcal{S}$, we can find a norm $\| \cdot \|_{\gamma,\delta}$ on the space $\mathcal{S}$ and numbers C, r such that for any function $\phi \in \mathcal{S}$*

$$\| \exp(i\sigma(k_1, \ldots, k_n)t)\phi(k_1, \ldots, k_n)\|_{\alpha,\beta}$$

$$\leq C(1 + |t|^r)\|\phi(k_1, \ldots, k_n)\|_{\gamma,\delta}. \tag{11.16}$$

Indeed, using inequality (11.16), we see that the norm of the function

$$g_t(k_1, \ldots, k_n) = \int_0^t \exp(i\sigma(k_1, \ldots, k_n)(t - \tau))h_\tau(k_1, \ldots, k_n)d\tau$$

satisfies the inequality

$$\|g_t\|_{\alpha,\beta} \le C|t|(1+|t|^r) \sup_{0\le\tau\le t} \|h_\tau\|_{\gamma,\delta},$$

which implies inequality (11.15) by induction. It is easy to prove that for fixed t, the expression $a^+(\mathbf{k},t), a(\mathbf{k},t)$ satisfies CCR:

$$[a(\mathbf{k},t), a(\mathbf{k}',t)] = [a^+(\mathbf{k},t), a^+(\mathbf{k}',t)] = 0,$$

$$[a(\mathbf{k},t), a^+(\mathbf{k}',t)] = \delta(\mathbf{k}-\mathbf{k}') \tag{11.17}$$

(it is important to note that the calculations of the commutators in (11.17) do not require limits: all the involved series depend on a finite number of convergent integrals). In order to check relation (11.17), we can differentiate them in t, for example,

$$\frac{\partial}{\partial t}[a(\mathbf{k},t), a^+(\mathbf{k}',t)] = i[[H, a(\mathbf{k},t)], a^+(\mathbf{k}',t)]$$

$$+ i[a(\mathbf{k},t), [H, a^+(\mathbf{k}',t)]]$$

$$= i[H, [a(\mathbf{k},t), a^+(\mathbf{k}',t)]],$$

which, with the initial condition $[a(\mathbf{k},0), a^+(\mathbf{k}',0)] = [a(\mathbf{k}), a^+(\mathbf{k}')] = \delta(\mathbf{k}-\mathbf{k}')$, allows us to conclude that $[a(\mathbf{k},t), a^+(\mathbf{k}',t)] = \delta(\mathbf{k}-\mathbf{k}')$.

We will say that the formulas

$$b(\mathbf{k}) = \alpha(\mathbf{k})a(\mathbf{k}) + \sum_{r\ge 1} g^r \sum_{m,n} \int h^{(r)}_{m,n}(\mathbf{k}_1,\ldots,\mathbf{k}_m|\mathbf{p}_1,\ldots,\mathbf{p}_n)$$

$$\times \delta\left(\mathbf{k} + \sum_{i-1}^{m}\mathbf{k}_i - \sum_{j=1}^{n}\mathbf{p}_j\right) a^+(\mathbf{k}_1)$$

$$\cdots a^+(\mathbf{k}_m)a(\mathbf{p}_1)\cdots a(\mathbf{p}_n)d^m\mathbf{k}d^n\mathbf{p}, \tag{11.18}$$

$$b^+(\mathbf{k}) = \overline{\alpha}(\mathbf{k})a^+(\mathbf{k}) \tag{11.19}$$

$$+ \sum_{r\ge 1} g^r \sum_{m,n} \int \overline{h}^{(r)}_{m,n}(\mathbf{k}_1,\ldots,\mathbf{k}_m|\mathbf{p}_1,\ldots,\mathbf{p}_n)$$

$$\times \delta(\mathbf{k} + \sum_{i=1}^{m}\mathbf{k}_i - \sum_{j=1}^{n}\mathbf{p}_j)a^+(\mathbf{p}_1)\cdots a^+(\mathbf{p}_n)$$

$$\times a(\mathbf{k}_1)\cdots a(\mathbf{k}_m)d^m\mathbf{k}d^n\mathbf{p}, \tag{11.20}$$

where $\alpha(\mathbf{k})$ is a smooth function whose derivatives do not grow faster than a power of $|\mathbf{k}|$, $h_{m,n}^{(r)} \in \mathcal{S}$, and for every r, all but a finite number of the functions $h_{m,n}^{(r)}$ are zero, specify a *canonical transformation* if the following relations are satisfied:

$$[b(\mathbf{k}), b(\mathbf{k}')] = [b^+(\mathbf{k}), b^+(\mathbf{k}')] = 0,$$

$$[b(\mathbf{k}), b^+(\mathbf{k}')] = \delta(\mathbf{k} - \mathbf{k}').$$

Roughly speaking, the canonical transformations turn the symbols $a^+(\mathbf{k}), a(\mathbf{k})$, satisfying CCR, into new symbols $b^+(\mathbf{k}), b(\mathbf{k})$, also satisfying CCR.

It is easy to prove that the *set of canonical transformations constitutes a group*. This means that the composition of two canonical transformations λ and μ produces a new canonical transformation $\lambda\mu$ and that the inverse of a canonical transformation is also a canonical transformation.

It was shown above that the transformation from $a^+(\mathbf{k}), a(\mathbf{k})$ to the Heisenberg operators $a^+(\mathbf{k}, t), a(\mathbf{k}, t)$ is a canonical transformation; it therefore follows that any expression, $H \in \mathcal{M}$ defines a one-parameter family of canonical transformations; we will denote these transformations by σ_H^t.

It is useful to note that the canonical transformation σ_H^t can be obtained as a limit of canonical transformations in finite volume. Namely, using the Hamiltonian H_Ω, obtained as a finite-volume cutoff of H, we can introduce the operators $a_{\mathbf{k}}(t), a_{\mathbf{k}}^+(t)$, where $\mathbf{k} \in T_\Omega$, by the formulas

$$a_{\mathbf{k}}(t) = \exp(iH_\Omega t)a_{\mathbf{k}} \exp(-iH_\Omega t),$$

$$a_{\mathbf{k}}^+(t) = \exp(iH_\Omega t)a_{\mathbf{k}}^+ \exp(-iH_\Omega t).$$

The operators $a_{\mathbf{k}}^+(t), a_{\mathbf{k}}(t)$, for fixed t, clearly satisfy CCR; the transformation from the operators $a_{\mathbf{k}}^+, a_{\mathbf{k}}$ to the operators $a_{\mathbf{k}}^+(t), a_{\mathbf{k}}(t)$ specifies a canonical transformation in the volume Ω. Let us write the operators $a_{\mathbf{k}}^+(t), a_{\mathbf{k}}(t)$ in normal form:

$$a_{\mathbf{k}}(t) = \exp(-i\epsilon(\mathbf{k})t)a_{\mathbf{k}}$$
$$+ \sum_{r \geq 1} g^r \sum_{\substack{\mathbf{k}_i \in T_\Omega \\ \mathbf{p}_j \in T_\Omega}} {}^\Omega f_{m,n}^{r,t}(\mathbf{k}_1, \ldots, \mathbf{k}_m | \mathbf{p}_1, \ldots, \mathbf{p}_n)$$
$$\times \delta_{\mathbf{p}_1+\cdots+\mathbf{p}_n}^{\mathbf{k}_1+\cdots+\mathbf{k}_m} \left(\frac{2\pi}{L}\right)^{\frac{3}{2}(m+n-2)} a_{\mathbf{k}_1}^+ \cdots a_{\mathbf{k}_m}^+ a_{\mathbf{p}_1} \cdots a_{\mathbf{p}_n},$$

$$a_{\mathbf{k}}^+(t) = \exp(i\epsilon(\mathbf{k})t)a_{\mathbf{k}}^+$$
$$+ \sum_{r \geq 1} g^r \sum_{\substack{\mathbf{k}_i \in T_\Omega \\ \mathbf{p}_j \in T_\Omega}} {}^\Omega \overline{f}_{m,n}^{r,t}(\mathbf{k}_1, \ldots, \mathbf{k}_m | \mathbf{p}_1, \ldots, \mathbf{p}_n)$$
$$\times \delta_{\mathbf{p}_1+\cdots+\mathbf{p}_n}^{\mathbf{k}_1+\cdots+\mathbf{k}_m} \left(\frac{2\pi}{L}\right)^{\frac{3}{2}(m+n-2)} a_{\mathbf{p}_1}^+ \cdots a_{\mathbf{p}_n}^+ a_{\mathbf{k}_1} \cdots a_{\mathbf{k}_m}.$$

For the functions ${}^\Omega f_{m,n}^{r,t}$, just as for the functions $f_{m,n}^{r,t}$, we can derive recurrent formulas from the Heisenberg equations. It is easy to check that the recurrent formulas admit the limit $\Omega \to \infty$, and therefore, it follows that

$$\lim_{\Omega \to \infty} {}^\Omega f_{m,n}^{r,t} = f_{m,n}^{r,t}.$$

This equality is what we meant by the statement that the canonical transformation σ_H^t can be obtained as a limit of canonical transformations in finite volume.

Let the Hamiltonian $H \in \mathcal{M}$ be given by the formula (11.3) and the canonical transformation λ by the relations (11.18) and (11.20).

Let us further replace in H the symbols $a^+(\mathbf{k}), a(\mathbf{k})$ by the expressions $b^+(\mathbf{k}), b(\mathbf{k})$, defined by the formulas (11.18) and (11.20), and bring the resulting formal expression to normal form using CCR, while discarding infinite constants.

It is easy to check that we will obtain an expression $\tilde{H} \in \mathcal{M}$; we will say that the Hamiltonian $\tilde{H}$ is obtained from the Hamiltonian H by the canonical transformation λ, and write $\tilde{H} = \lambda(H)$.

Let us note that if the canonical transformation λ can be written in the form $\lambda = \sigma_A^t$, where $A \in \mathcal{M}$, then

$$\tilde{H} = \lambda(H) = \lim_{\Omega \to \infty} \lambda_\Omega(H),$$

where

$$\lambda_\Omega(H) = \exp(itA_\Omega) H_\Omega \exp(-itA_\Omega) - c_\Omega$$

(the limit should be understood in the sense of the limits of the coefficients of the functions in normal form of the operator $\lambda_\Omega(H)$ to the coefficients of the functions in the expression $\lambda(H)$). The proof of this follows from the fact that, as stated above, the canonical transformation σ_A^t can be obtained as a limit of canonical transformations in finite volume; the constant c_Ω arises from discarding infinite constants after the canonical transformation; we can assume that

$$c_\Omega = \langle \exp(itA_\Omega) H_\Omega \exp(-itA_\Omega)\theta, \theta \rangle .$$

If we formally calculate the commutator $[A, B]$ of two expressions $A, B \in \mathcal{M}$ and multiply it by i, then we obtain a new expression belonging to $\mathcal{M}$. It is easy to check that the set $\mathcal{M}$, equipped with the operation $i[A, B]$ and the obvious structure of vector space, is a Lie algebra. The group of canonical transformations $\mathcal{G}$ can be seen as an infinite-dimensional Lie group. It is possible to show that $\mathcal{M}$ is a Lie algebra of the group $\mathcal{G}$. The canonical transformations σ_H^t clearly constitute a one-parameter subgroup of $\mathcal{G}$, corresponding to the element H of the Lie algebra $\mathcal{M}$; the map taking the element $H \in \mathcal{M}$ to the element σ_H^1 is the exponential map from $\mathcal{M}$ to $\mathcal{G}$. Each element of the Lie group defines an automorphism of the corresponding Lie algebra; in our case of interest, the canonical transformation λ corresponds to the automorphism of the Lie algebra $\mathcal{M}$ that maps the formal expression H to the formal expression $\lambda(H)$.

11.3   Construction of an operator realization

Let us consider a Hamiltonian $H = \sum H_r g^r \in \mathcal{M}$ that does not generate vacuum polarization. The formal expressions H_r therefore define Hermitian operators H_r on Fock space with domain $\mathcal{S}_\infty$ (see Section 3.2, where it is shown that in the case of vacuum polarization, this statement ceases to be true). In the framework of perturbation theory, we can construct the operator $\exp(iHt) = \sum g^r U_r(t)$, where the operators $U_r(t)$ are defined on $\mathcal{S}_\infty$ and transform $\mathcal{S}_\infty$ into itself. This statement can be proved by means of diagram techniques described in Section 6.4; roughly speaking, the proof depends on the remark that in the case at hand, all the diagrams represent mathematically sensible expressions. (Before constructing a perturbation theory series for $\exp(iHt)$, it is necessary to express the Hamiltonian H in the form $H_0' + V'$, where

$$H_0' = H_0 + \sum_{r \geq 1} g^r \int v_{1,1}^{(r)}(\mathbf{k}) a^+(\mathbf{k}) a(\mathbf{k}) d\mathbf{k},$$

and use the diagram techniques for the operator $S(0,t) = \exp(-iH_0't)\exp(iHt)$.) Another proof is laid out in Section 11.5.

The operator generalized functions $a(\mathbf{k}, \epsilon, t) = \sum g^r a_r(\mathbf{k}, \epsilon, t)$ are defined by the formula

$$a(\mathbf{k}, \epsilon, t) = \exp(iHt) a(\mathbf{k}, \epsilon) \exp(-iHt),$$

where $a(\mathbf{k}, 1) = a^+(\mathbf{k}), a(\mathbf{k}, -1) = a(\mathbf{k})$ are the operators of annihilation and creation in Fock space. Since the operators $U_r(t)$ and $\int f(\mathbf{k}) a(\mathbf{k}, \epsilon) d\mathbf{k}$ transform the set $\mathcal{S}_\infty$ into itself, the operators

$$\int f(\mathbf{k}) a_r(\mathbf{k}, \epsilon, t) d\mathbf{k} = \sum_{s-0}^{r} U_s(t) \left(\int f(\mathbf{k}) a(\mathbf{k}, \epsilon) d\mathbf{k} \right) U_{r-s}(-t)$$

inherit the property for any function $f \in \mathcal{S}$.

It is clear that the energy operator $H = \sum H_r g^r$, the momentum operator $\mathbf{P} = \int \mathbf{k} a^+(\mathbf{k}) a(\mathbf{k}) d\mathbf{k}$, the operator generalized functions $a(\mathbf{k}, \epsilon, t) = \sum g^r a_r(\mathbf{k}, \epsilon, t)$, and the vector $\Phi = \theta$ constitute an *operator realization* of the Hamiltonian $H \in \mathcal{M}$.

To construct an operator realization for a Hamiltonian generating vacuum polarization, we will use the following statement.

Theorem 11.2. *Let λ be a canonical transformation and let $\tilde{H} = \lambda(H) \in \mathcal{M}$ be the Hamiltonian obtained from the Hamiltonian H*

by the canonical transformation λ. Then the operator realization of the Hamiltonian H can be obtained from the operator realization $(\mathcal{H}, \hat{H}, \hat{\mathbf{P}}, \tilde{a}(\mathbf{k}, \epsilon, t), \Phi)$ of the Hamiltonian $\tilde{H}$, by keeping the space $\mathcal{H}$, the operators of energy and momentum $\hat{H}, \hat{\mathbf{P}}$, and the vector Φ the same and replacing the operator generalized functions $\tilde{a}(\mathbf{k}, \epsilon, t)$ by the operator generalized functions

$$a(\mathbf{k}, -1, t) = a(\mathbf{k}, t) = \alpha(\mathbf{k})\tilde{a}(\mathbf{k}, t)$$

$$+ \sum_{r \geq 1} g^r \sum_{m,n} \int h_{m,n}^{(r)}(\mathbf{k}_1, \ldots, \mathbf{k}_m | \mathbf{p}_1, \ldots, \mathbf{p}_n)$$

$$\times \delta \left(\sum_{i=1}^{m} \mathbf{k}_i - \sum_{j=1}^{n} \mathbf{p}_j - \mathbf{k} \right) \tilde{a}^+(\mathbf{k}_1, t)$$

$$\cdots \tilde{a}^+(\mathbf{k}_m, t)\tilde{a}(\mathbf{p}_1, t) \cdots \tilde{a}(\mathbf{p}_n, t) d^m\mathbf{k} d^n\mathbf{p}, \qquad (11.21)$$

$$a(\mathbf{k}, 1, t) = a^+(\mathbf{k}, t) = \overline{\alpha}(\mathbf{k})\tilde{a}(\mathbf{k}, t)$$

$$+ \sum_{r \geq 1} g^r \sum_{m,n} \int \overline{h}_{m,n}^{(r)}(\mathbf{k}_1, \ldots, \mathbf{k}_m | \mathbf{p}_1, \ldots, \mathbf{p}_n)$$

$$\times \delta \left(\sum_{i=1}^{m} \mathbf{k}_i - \sum_{j=1}^{n} \mathbf{p}_j - \mathbf{k} \right) \tilde{a}^+(\mathbf{p}_1, t)$$

$$\cdots \tilde{a}^+(\mathbf{p}_n, t)\tilde{a}(\mathbf{k}_1, t) \cdots \tilde{a}(\mathbf{k}_n, t) d^m\mathbf{k} d^n\mathbf{p} \qquad (11.22)$$

[we assume that the canonical transformation λ is given by the formulas (11.18) and (11.20)].

To prove the formulated statement, let us note that the only dependence on the concrete form of the translation-invariant Hamiltonian in the definition of an operator realization is in the condition that the operator generalized function $a(\mathbf{k}, \epsilon, t)$ satisfies the Heisenberg equations.

In order to check that this condition is satisfied, it is necessary to differentiate equation (11.21) with respect to t, using the fact that

$$\frac{\partial \tilde{a}(\mathbf{k}, \epsilon, t)}{\partial t} = i[\tilde{H}, \tilde{a}(\mathbf{k}, \epsilon, t)]$$

by the definition of the operator realization of the Hamiltonian $\tilde{H}$. Expressing $\tilde{a}$ in terms of a by the canonical transformation λ^{-1}, we obtain an equation for $a(\mathbf{k}, \epsilon, t)$; simple but cumbersome calculations allow us to bring the resulting equations to the form

$$\frac{\partial a(\mathbf{k}, \epsilon, t)}{\partial t} = i[H, a(\mathbf{k}, \epsilon, t)].$$

We will further consider the Hamiltonian $H = H_0 + V \in \mathcal{M}$, assuming that the function $\epsilon(\mathbf{k})$, entering the definition of the Hamiltonian H_0, satisfies the inequality $\epsilon(\mathbf{k}_1 + \mathbf{k}_2) < \epsilon(\mathbf{k}_1) + \epsilon(\mathbf{k}_2)$. With the help of the statement proven above, we can easily convert the construction of the operator realization of this Hamiltonian to the already-solved problem of construction of the operator realization of a Hamiltonian that does not generate vacuum polarization. We should use the following statement proven (in different terms) in Section 9.4.

Theorem 11.3. *Let the Hamiltonian $H = H_0 + V \in \mathcal{M}$ be given and suppose the function $\epsilon(\mathbf{k})$ satisfies the inequality $\epsilon(\mathbf{k}_1 + \mathbf{k}_2) < \epsilon(\mathbf{k}_1) + \epsilon(\mathbf{k}_2)$. Then there exists a canonical transformation λ such that the Hamiltonian $H = \lambda(H) \in \mathcal{M}$ can be represented in the form*

$$\tilde{H} = H_0 + \tilde{V}, \tag{11.23}$$

where

$$\tilde{V} = \sum_r g^r \sum_{m,n} \int \tilde{v}_{m,n}^{(r)}(\mathbf{k}_1, \ldots, \mathbf{k}_m | \mathbf{p}_1, \ldots, \mathbf{p}_n)$$

$$\times \delta\left(\sum_{i=1}^m \mathbf{k}_i - \sum_{j=1}^n \mathbf{p}_j\right) a^+(\mathbf{k}_1) \cdots a^+(\mathbf{k}_m) a(\mathbf{p}_1)$$

$$\cdots a(\mathbf{p}_n) d^m \mathbf{k} d^n \mathbf{p},$$

$\tilde{V}$ does not contain summands of type $(m, 0)$ or summands of type $(m', 1)$ with $m' \neq 1$ $\left(\text{i.e. } \tilde{v}_{m,0}^{(r)} = 0, \tilde{v}_{m',1}^{(r)} = 0 \text{ for } m' \neq 1\right)$.

The transformation λ was previously called a Faddeev transformation; let us recall that it was constructed in the form σ_A^1, where $A \in \mathcal{M}$.

One can say that *the Faddeev canonical transformation transforms the Hamiltonian H into the Hamiltonian $\tilde{H}$, for which the bare vacuum coincides with the dressed vacuum and the bare single-particle state coincides with the dressed single-particle state.*

Let us now show that *any two operator realizations of a Hamiltonian $H \in \mathcal{M}$ are unitarily equivalent.* With the help of Faddeev transformation, we can reduce the problem to the case when the Hamiltonian does not generate vacuum polarization [recall that above we also assumed the condition $\epsilon(\mathbf{k}_1 + \mathbf{k}_2) < \epsilon(\mathbf{k}_1) + \epsilon(\mathbf{k}_2)$]. Indeed, from two unitarily equivalent operator realizations of the Hamiltonian H by means of the Faddeev transformation, we can obtain two unitarily equivalent operator realizations of the Hamiltonian not generating vacuum polarization.

In order to prove the uniqueness of the operator realization in the case when the Hamiltonian $H \in \mathcal{M}$ does not generate vacuum polarization, it is enough to check that for any operator realization $(\mathcal{H}', H', \mathbf{P}', a'(\mathbf{k}, \epsilon, t), \Phi')$ of the Hamiltonian H, the relation $a'(\mathbf{k}, -1, t)\Phi' = 0$ is satisfied (then the unitary operator U establishing the unitary equivalence of this operator realization and the operator realization constructed above can be defined as the operator satisfying the condition $Ua(\mathbf{k}, \epsilon, 0) = a'(\mathbf{k}, \epsilon, 0)U, U\Phi = \Phi'$).[4] Let us now check the relation $a'(\mathbf{k}, -1, t)\Phi' = (\sum g^r a_r(\mathbf{k}, -1, t))\Phi' = 0$. Consider the equations for $a'_r(\mathbf{k}, -1, t)$ following from the Heisenberg equations. It is easy to check that they have the form

$$i\frac{\partial a'_r(\mathbf{k}, -1, t)}{\partial t} - \epsilon(\mathbf{k})a'_r(\mathbf{k}, -1, t) = F_r, \qquad (11.24)$$

where F_r is an expression composed of $a'_s(\mathbf{k}, 1, t)$ and $a'_s(\mathbf{k}, -1, t)$ with $s < r$. In order to prove the necessary relation, we should show that $a'_r(\mathbf{k}, -1, t)\Phi' = 0$; we will perform the proof by induction in r.

---

[4]The existence of the operator U satisfying these conditions follows from the results of Section 6.1 (more precisely, from their analogs concerning the unitary equivalence of the Fock representations of CCR, written in the form of a formal expression in powers of g). Using the Heisenberg equations, we can show that $Ua(\mathbf{k}, \epsilon, t) = a'(\mathbf{k}, \epsilon, t)U$. From the condition (b) in the definition of operator realization, we obtain that $UH = H'U, U\mathbf{P} = \mathbf{P}'U$.

Suppose that $a'_s(\mathbf{k}, -1, t)\Phi' = 0$ for $s < r$. By the absence of vacuum polarization, every summand in the expression F_r will contain at least one operator of the form $a_s(\mathbf{k}, -1, t)$ and therefore $F_r\Phi' = 0$. Using equation (11.24) on the vector Φ, we can see that

$$i\frac{\partial}{\partial t}a'_r(\mathbf{k}, -1, t)\Phi' = \epsilon(\mathbf{k})a'_r(\mathbf{k}, -1, t)\Phi',$$

from which it follows that

$$a'_r(\mathbf{k}, -1, t)\Phi' = \exp(-i\epsilon(\mathbf{k})t)a'_r(\mathbf{k}, -1, 0)\Phi'.$$

Let us now calculate the expectation value of the energy in the state $\int f(\mathbf{k})a'(\mathbf{k}, -1, t)d\mathbf{k}\Phi'$:

$$\left\langle H' \int f(\mathbf{k})a'(\mathbf{k}, -1, t)d\mathbf{k}\Phi', \int f(\mathbf{k})a'(\mathbf{k}, -1, t)d\mathbf{k}\Phi' \right\rangle$$

$$= \left\langle \frac{1}{i} \int f(\mathbf{k})\frac{\partial}{\partial t}a'(\mathbf{k}, -1, t)d\mathbf{k}\Phi', \int f(\mathbf{k})a'(\mathbf{k}, -1, t)d\mathbf{k}\Phi' \right\rangle$$

$$= \sum_{s \geq r} \frac{1}{i} \int f(\mathbf{k})\overline{f(\mathbf{k}')}\left\langle \frac{\partial}{\partial t}a'_s(\mathbf{k}, -1, t)\Phi', a'_{s'}(\mathbf{k}', -1, t)\Phi' \right\rangle d\mathbf{k}d\mathbf{k}'$$

$$= C_{2r}g^{2r} + C_{2r+1}g^{2r+1} + \cdots,$$

where

$$C_{2r} = \frac{1}{i} \int f(\mathbf{k})\overline{f(\mathbf{k}')}\left\langle \frac{\partial}{\partial t}a'_r(\mathbf{k}, 1, t)\Phi', a'_r(\mathbf{k}', -1, t)\Phi' \right\rangle d\mathbf{k}d\mathbf{k}'$$

$$= -\int \epsilon(\mathbf{k})f(\mathbf{k})\overline{f(\mathbf{k}')}\exp(-i(\epsilon(\mathbf{k}) - \epsilon(\mathbf{k}'))t)$$

$$\times \left\langle a'_r(\mathbf{k}, -1, 0)\Phi', a'_r(\mathbf{k}', -1, 0)\Phi' \right\rangle d\mathbf{k}d\mathbf{k}'.$$

Since

$$\left\langle a'_r(\mathbf{k}, -1, 0)\Phi', a'_r(\mathbf{k}', -1, 0)\Phi' \right\rangle = \sigma(\mathbf{k})\delta(\mathbf{k} - \mathbf{k}'),$$

where $\sigma(\mathbf{k}) \geq 0$, we see that

$$C_{2r} = -\int \epsilon(\mathbf{k})|f(\mathbf{k})|^2\sigma(\mathbf{k})d\mathbf{k} \leq 0.$$

If $a'_r(\mathbf{k},-1,t)\Phi' \not\equiv 0$, then $\sigma(\mathbf{k}) \not\equiv 0$, and therefore, we can find a function $f(\mathbf{k})$ such that $C_{2n} < 0$. This contradicts the non-negativity assumption of the operator H'.

Let us now check the applicability conditions of axiomatic field theory in each power of g. Let us first consider the spectrum of the operators $\hat{H}, \hat{\mathbf{P}}$ in the operator realization $(\mathcal{H}, \hat{H}, \hat{\mathbf{P}}, \tilde{a}(\mathbf{k},\epsilon,t), \Phi)$ of a Hamiltonian of the form (11.23). For this Hamiltonian, the (bare) Fock vacuum θ and the bare single-particle state $a^+(\mathbf{k})\theta$ satisfy the conditions

$$\hat{H}\theta = \hat{\mathbf{P}}\theta = 0,$$

$$\hat{H}a^+(\mathbf{k})\theta = \left(\epsilon(\mathbf{k}) + \sum g^r \tilde{v}_{11}^{(r)}(\mathbf{k})\right)a^+(\mathbf{k})\theta,$$

$$\mathbf{P}a^+(\mathbf{k})\theta = \mathbf{k}a^+(\mathbf{k})\theta.$$

Therefore, θ can be viewed as the ground state of the operator $\hat{H}$ (dressed vacuum) and $a^+(\mathbf{k})\theta$ can be viewed as a single-particle state of operators $\hat{H}, \hat{\mathbf{P}}$ (dressed single-particle state). The joint spectrum of the operators $\hat{H}, \hat{\mathbf{P}}$ consists of the point 0, corresponding to the ground state, the set of points $(\omega(\mathbf{k}), \mathbf{k}) = (\epsilon(\mathbf{k}) + \sum g^r \tilde{v}_{1,1}^r(\mathbf{k}), \mathbf{k})$ (single-particle spectrum), and the multi-particle spectrum, whose points have the form $(\omega + \sum_{r\geq 1} g^r \omega_r, \mathbf{k} + \sum_{r\geq 1} g^r \mathbf{k}_r)$, where the point $(\omega, \mathbf{k})$ belongs to the multi-particle spectrum of the operators $\hat{H}_0, \hat{\mathbf{P}}$ (i.e. can be presented in the form $(\epsilon(\mathbf{k}_1) + \cdots + \epsilon(\mathbf{k}_m), \mathbf{k}_1 + \cdots + \mathbf{k}_m)$).

If the Hamiltonian H belongs to the class $\tilde{\mathcal{M}}$ [i.e. the function $\epsilon(\mathbf{k})$ is not linear on any open set and satisfies the condition $\epsilon(\mathbf{k}_1 + \mathbf{k}_2) < \epsilon(\mathbf{k}_1) + \epsilon(\mathbf{k}_2)$], *then the spectrum of the operators $\hat{H}, \hat{\mathbf{P}}$ satisfies the conditions formulated in Section 10.5.*

In order to finish the exposition of the case at hand, it remains to be checked that the family $\hat{\mathcal{B}}$ of operator generalized functions, consisting of two functions

$$\tilde{a}(\mathbf{x},\epsilon,t) = (2\pi)^{-3/2} \int \exp(i\epsilon\mathbf{k}\mathbf{x})\tilde{a}(\mathbf{k},\epsilon,t)d\mathbf{k}, \epsilon = \pm 1,$$

is asymptotically commutative, in the sense of Section 10.5, in every order of perturbation theory with respect to g. In order to show that $\mathcal{B}$ is asymptotically commutative, let us note that the operators of the form $\int f(\mathbf{x})\tilde{a}(\mathbf{x}, \epsilon, t)d\mathbf{x}$ that we have considered belong to the algebra $\mathcal{A}$, consisting of operators of the form

$$\sum g^r \sum_{m,n} \int f^{(r)}_{m,n}(\mathbf{k}_1, \ldots, \mathbf{k}_m | \mathbf{p}_1, \ldots, \mathbf{p}_n)a^+(\mathbf{k}_1)$$

$$\cdots a^+(\mathbf{k}_m)a(\mathbf{p}_1)\cdots a(\mathbf{p}_n)d^m\mathbf{k}d^n\mathbf{p},$$

where the functions $f^{(r)}_{m,n} \in \mathcal{S}$ and for every r, only finitely many functions $f^{(r)}_{m,n}$ are non-zero. It is easy to check (using the results of Section 11.2, specifically inequality (11.15)) that the algebra $\mathcal{A}$ is asymptotically commutative in every order in g. (This statement also follows from the results of Section 11.5.)

Therefore, the family $\tilde{\mathcal{B}}$ is asymptotically commutative.

Let us return to considering an arbitrary Hamiltonian $H = H_0 + V \in \tilde{\mathcal{M}}$. Using the operator realization of this Hamiltonian constructed by means of Faddeev transformation, we can obtain an operator realization of the Hamiltonian $\tilde{H}$ of the form (11.23). In these constructions, the energy and momentum operators in the operator realization of the Hamiltonian H coincide with the energy and momentum operators in the operator realization of the Hamiltonian $\tilde{H}$; this remark implies that they satisfy the necessary spectrum conditions. Operator generalized functions $a(\mathbf{k}, \epsilon, t)$, entering in the operator realization of the Hamiltonian H, are related to the operator generalized functions $a(\mathbf{k}, \epsilon, t)$ in the operator realization of the Hamiltonian $\tilde{H}$ through the canonical transformation (see the relation (11.21)). It therefore follows that the family $\mathcal{B}$ of two operator generalized functions $a(\mathbf{x}, \epsilon, t), \epsilon - \pm 1$, with which we may construct the scattering matrix for the Hamiltonian H, is an asymptotically commutative algebra in every order of g, and therefore the operators $\int f(\mathbf{x})a(\mathbf{x}, \epsilon, t)d\mathbf{x}$ belong to the asymptotically commutative family $\mathcal{A}$.

These statements imply that *the scattering matrix of the Hamiltonian $H \in \tilde{\mathcal{M}}$ can be constructed via axiomatic scattering theory in terms of formal series in g.*

It is now possible to apply axiomatic scattering theory to study (in the framework of perturbation theory) scattering matrices in the Hamiltonian formalism.

First, let us note that the discussions imply that *the scattering matrix of the Hamiltonian H coincides with the scattering matrix of the Hamiltonian $\tilde{H}$, obtained from H by means of Faddeev canonical transformation.*[5] Indeed, these two scattering matrices are built on the asymptotically commutative families $\mathcal{B}$ and $\tilde{\mathcal{B}}$, which are both contained in the same asymptotically commutative algebra $\mathcal{A}$; by the results of Chapter 10, the Møller and scattering matrices based on these families are identical.

The statement we have just proved implies the following equivalence theorem.

If the Hamiltonians $H_1 \in \tilde{\mathcal{M}}$ and $H_2 \in \tilde{\mathcal{M}}$ are related by a canonical transformation, then their Møller and scattering matrices coincide.

Indeed, let α be a canonical transformation transforming the Hamiltonian H_1 to the Hamiltonian H_2; let β be the Faddeev canonical transformation transforming the Hamiltonian H_2 to the Hamiltonian $\tilde{H}$ of the form (11.23). Then the canonical transformation $\alpha\beta$ transforms the Hamiltonian H_1 to the Hamiltonian $\tilde{H}$.

The scattering matrices of the Hamiltonians H_1 and H_2 above coincide with the scattering matrices of the Hamiltonian $\tilde{H}$ and, therefore, coincide with each other. The same reasoning applies to Møller matrices.

11.4 Dressing operators for translation-invariant Hamiltonians

In this section, we will consider translation-invariant Hamiltonians from class $\mathcal{M}$.

Let us first examine the case where the Hamiltonian $H \in \mathcal{M}$ does not generate vacuum polarization. The dressing operator for

---

[5]It is important to note that in this statement in place of the Faddeev transformation, we can select any canonical transformation transforming the Hamiltonian H into a Hamiltonian of the form (11.23), not necessarily the concrete transformation constructed in Section 9.4.

this Hamiltonian is defined as an operator satisfying the condition

$$S_\pm = \operatorname*{slim}_{t\to\pm\infty} \exp(iHt)D\exp(-iH_{\mathrm{as}}t). \tag{11.25}$$

(The Hamiltonian in consideration defines the operator H on Fock space $\mathcal{H} = F(L^2(E^3)))$, the asymptotic space $\mathcal{H}_{\mathrm{as}} = F(L^2(E^3))$ is naturally identified with the space $\mathcal{H}$, therefore, the Møller matrices $S_\pm$ act on the space $\mathcal{H}$. The symbol H_{as} denotes the asymptotic Hamiltonian which we also view as an operator on $\mathcal{H}$:

$$H_{\mathrm{as}} = \int \omega(\mathbf{k})a^+(\mathbf{k})a(\mathbf{k})d\mathbf{k},$$

where $\omega(\mathbf{k})$ is the energy of the single-particle state of the Hamiltonian H. All operators in the formula (11.25), just as for the rest of this section, are viewed as formal series in g. The limit in (11.25) is understood as a strong limit in each perturbation order on the set of linear combinations of vectors of the form $a^+(f_1)\cdots a^+(f_n)\theta$, where $f_1(\mathbf{k}),\ldots,f_n(\mathbf{k})$ are non-overlapping smooth functions with compact support.)

The following theorem describes a broad class of dressing operators for Hamiltonians $H \in \tilde{\mathcal{M}}$ that do not generate vacuum polarization.

Theorem 11.4. *Let D be an operator on Fock space $\mathcal{H} = F(L^2(E^3))$, satisfying the following conditions:*

(1) $D\theta = \theta;$ $\hspace{7.5cm}$ (11.26)

(2) *the vector generalized function $Da^+(\mathbf{k})\theta$ is δ-normalized and is an eigenvector for the operator H;*

(3) *the operator D can be represented as a product of operators of the form $\exp A$ or $N(\exp A)$, where $A = A_1 + iA_2$, $A_1 \in \mathcal{M}$, $A_2 \in \mathcal{M}$.*

Then the operator D is a dressing operator for the Hamiltonian H. (*The symbol $N(\exp A)$ denotes the normal exponent; see Section 6.3.*)

This theorem can be proved with the help of description of dressing operators in axiomatic scattering theory (see Section 10.4). In the following section, we present a proof of this theorem in the case when the operator D is represented in the form of a product of operators of the form $\exp A$, where $A \in \mathcal{M}$. In the section at hand, however, we will provide a different proof of the formulated theorem.

To begin, let us consider the case where $D = N(\exp A)$, where

$$A = \sum_m \int A_{m,1}(\mathbf{k}_1, \ldots, \mathbf{k}_m)\delta(\mathbf{k}_1 + \cdots + \mathbf{k}_m - \mathbf{p})$$

$$\times a^+(\mathbf{k}_1) \cdots a^+(\mathbf{k}_m)a(\mathbf{p})d\mathbf{k}_1 \cdots d\mathbf{k}_m d\mathbf{p}.$$

Let us first note that in this case, we have the relation

$$[D, a^+(\mathbf{k})] = B(\mathbf{k})D, \tag{11.27}$$

where $B(\mathbf{k})$ is a generalized operator function defined by the formula

$$B(\mathbf{k}) = \sum_m \int A_{m,1}(\mathbf{k}_1, \ldots, \mathbf{k}_m)\delta(\mathbf{k} - \mathbf{k}_1 - \cdots - \mathbf{k}_m)$$

$$\times a^+(\mathbf{k}_1) \cdots a^+(\mathbf{k}_m)d^m\mathbf{k}$$

(the relation (11.27) follows immediately from the equality

$$[N(A^n), a^+(\mathbf{k})] = B(\mathbf{k}) \cdot nN(A^{n-1}),$$

which we obtained using the formula

$$[R_1 \cdots R_m, R] = \sum_{1 \leq i \leq m} R_1 \cdots R_{i-1}[R_i, R]R_{i+1} \cdots R_m).$$

Let us rewrite relation (11.27) in the form

$$Da^+(\mathbf{k}) = E(\mathbf{k})D, \tag{11.28}$$

where $E(\mathbf{k}) = a^+(\mathbf{k}) + B(\mathbf{k})$. Using formulas (11.28) and (11.26), we see that

$$Da^+(\mathbf{k}_1) \cdots a^+(\mathbf{k}_n)\theta = E(\mathbf{k}_1) \cdots E(\mathbf{k}_n)\theta,$$

in particular

$$Da^+(\mathbf{k})\theta = E(\mathbf{k})\theta$$

$$= a^+(\mathbf{k}) + \sum_m \int A_{m,1}(\mathbf{k}_1, \ldots, \mathbf{k}_m)$$

$$\times \delta(\mathbf{k} - \mathbf{k}_1 - \cdots - \mathbf{k}_m)a^+(\mathbf{k}_1) \cdots a^+(\mathbf{k}_m)d^m\mathbf{k}. \tag{11.29}$$

It follows from relation (11.29) that

$$\exp(iHt)D\exp(-iH_0t)a^+(\bar{f}_1)\cdots a^+(\bar{f}_n)\theta$$

$$= \exp(iHt)Da^+(\overline{f_1^t})\cdots a^+(\overline{f_n^t})\theta$$

$$= \exp(iHt)\left(\int f_1^t(\mathbf{k}_1)E(\mathbf{k}_1)d\mathbf{k}_1\right)\cdots\left(\int f_n^t(\mathbf{k}_n)E(\mathbf{k}_n)d\mathbf{k}_n\right)\theta$$

$$= \left(\int f_1^t(\mathbf{k}_1)E(\mathbf{k}_1,t)d\mathbf{k}_1\right)\cdots\left(\int f_n^t(\mathbf{k}_n)E(\mathbf{k}_n,t)d\mathbf{k}_n\right)\theta$$

(here, $f^t(\mathbf{k}) = \exp(-it\omega(\mathbf{k}))f(\mathbf{k})$).

To check that D is a dressing operator, it is sufficient to establish that E is a good operator generalized function in the sense of Section 10.5; this can be easily done following the analysis of Heisenberg operators $a(\mathbf{k}, \epsilon, t)$ provided in Section 11.2.

Let us remark that knowing the single-particle state $\Phi(\mathbf{k}) = Da^+(\mathbf{k})\theta$ and using the formula (11.29), we can find the function $A_{m,1}$ and the rest of the dressing operator of the considered type. From this remark, it is easy to derive that the condition that the operator A belongs to the class $\mathcal{M} + i\mathcal{M}$, for the case at hand, is unnecessary (it is necessary to use the following expression of the single-particle state in terms of perturbation theory

$$\Phi(\mathbf{k}) = \Phi(\mathbf{k}|g)$$

$$= \exp(i\sigma(\mathbf{k}|g))\left(a^+(\mathbf{k}) + \sum_m\int \phi_m(\mathbf{k}_1,\ldots,\mathbf{k}_m)\right.$$

$$\left.\times \delta(\mathbf{k} - \mathbf{k}_1 - \cdots - \mathbf{k}_m)a^+(\mathbf{k}_1)\cdots a^+(\mathbf{k}_m)d^m\mathbf{k}\right)\theta,$$

where $\exp(i\sigma(\mathbf{k}|g))$ is an arbitrary phase factor; $\phi_m(\mathbf{k}_1,\ldots,\mathbf{k}_m) = \sum_m \phi_m^{(r)}(\mathbf{k}_1,\ldots,\mathbf{k}_m)g^r$; the functions $\phi_m^{(r)}$ belong to the space $\mathcal{S}$, and for every r only finitely many of these functions are non-zero).

The proven statement allows us to conclude that the definition of scattering matrix given in Section 9.1 (formula (9.10)) is equivalent to the other definitions.

Let us now return to the slightly more general case where $D = N(\exp A)$, where $A \in \mathcal{M}+i\mathcal{M}$ (i.e. $A = A_1+iA_2$, $A_1 \in \mathcal{M}$, $A_2 \in \mathcal{M}$).

We will write A in the form $A = \sum A_{m,n}$, where $A_{m,n} = \sum_r A^{(r)}_{m,n} g^r$,

$$A^{(r)}_{m,n} = \int A^{(r)}_{m,n}(\mathbf{k}_1, \ldots, \mathbf{k}_m | \mathbf{p}_1, \ldots, \mathbf{p}_n)$$

$$\times \, \delta(\mathbf{k}_1 + \cdots + \mathbf{k}_m - \mathbf{p}_1 - \cdots - \mathbf{p}_n) a^+(\mathbf{k}_1)$$

$$\cdots a^+(\mathbf{k}_m) a(\mathbf{p}_1) \cdots a(\mathbf{p}_n) d^m \mathbf{k} d^n \mathbf{p},$$

and define the operator A' in the same way as the operator A except only keeping terms of type $(m, 1)$ (i.e. we set $A' = \sum_m A_{m,1}$).

The operator $D' = N(\exp A')$ satisfies the conditions

$$D'\theta = \theta,$$

$$D'a^+(\mathbf{k})\theta = Da^+(\mathbf{k})\theta$$

(terms containing no less than two annihilation operators go to zero when acting on $a^+(\mathbf{k})\theta$).

As was shown above, the operator $D' = N(\exp A')$ is a dressing operator, i.e.

$$\operatorname*{slim}_{t \to \pm\infty} \exp(iHt) D' \exp(-iH_{\mathrm{as}}t) = S_{\pm}. \tag{11.30}$$

In order to derive from formula (11.30), the necessary relation

$$\operatorname*{slim}_{t \to \pm\infty} \exp(iHt) D \exp(-iH_{\mathrm{as}}t) = S_{\pm},$$

it is enough to show that

$$\operatorname*{slim}_{t \to \pm\infty} \exp(iHt)(D - D') \exp(-iH_{\mathrm{as}}t) = 0. \tag{11.31}$$

The proof of equation (11.31) in each order of perturbation theory immediately follows from the relation

$$\lim_{t \to \pm\infty} N\big(A^{(r_1)}_{m_1 n_1} \cdots A^{(r_s)}_{m_s n_s} \big) a^+\big(\overline{f^t_1}\big) \cdots a^+\big(\overline{f^t_s}\big)\theta = 0,$$

where at least one of the operators $A^{(r_i)}_{m_i n_i}$ satisfies the condition $n_i \geq 2$ (i.e. contains at least two annihilation operators), and the functions $f_1(\mathbf{k}), \ldots, f_s(\mathbf{k})$ constitute a non-overlapping family of smooth functions with compact support. The above completes the proof of the theorem in the case when the operator D takes the form $N(\exp A)$. The proof of the theorem in the general case can be reduced to the cases already considered using the following two lemmas.

Lemma 11.1. *Any operator of the form* $\exp(A_1 + iA_2)$, *where* $A_1, A_2 \in \mathcal{M}$, *can be written in the form* $N(\exp(A'_1 + iA'_2))$, *where* $A'_1, A'_2 \in \mathcal{M}$.

Lemma 11.2. *Any operator of the form* $\exp(A_1 + iA_2) \times \exp(A'_1 + iA'_2)$, *where* $A_1, A_2, A'_1, A'_2 \in \mathcal{M}$, *can be written in the form* $N(\exp(A''_1 + iA''_2))$, *where* $A''_1, A''_2 \in \mathcal{M}$.

We will not provide proofs of these lemmas.

Let us now consider the case of an arbitrary Hamiltonian $H \in \mathcal{M}$. The symbol H_Ω, as usual, denotes the operator H with finite volume cutoff (see Section 8.1); the operator H_Ω acts on the Fock space $F_\Omega = F(L^2(\Omega))$. In perturbation theory, we can calculate the operator $\exp(iH_\Omega t)$ (in other words, H_Ω can be viewed as a self-adjoint operator in the framework of perturbation theory).

The family of operators D^Ω will be called a *family of dressing operators* for the Hamiltonian H if the scattering matrix S of the Hamiltonian H can be written in the form

$$S = \lim_{\substack{t \to \infty \\ t_0 \to -\infty}} \lim_{\Omega \to \infty} \exp(i(c_\Omega + H_{\mathrm{as}\Omega})t)(D^\Omega)^{-1}$$

$$\times \exp(-iH_\Omega(t - t_0))D^\Omega \exp(-i(c_\Omega + H_{\mathrm{as}\Omega})t_0). \quad (11.32)$$

(The operators D^Ω act on the space F_Ω, the symbols $H_{\mathrm{as}\Omega}$ denote the asymptotic Hamiltonian H_{as} with volume cutoff, and c_Ω is a constant depending on the volume. The limit in (11.32) can be understood in the sense of the convergence of matrix elements (see Section 9.3) or via the relation

$$S = \operatorname*{slim}_{\substack{t \to \infty \\ t_0 \to -\infty}} \operatorname*{slim}_{\Omega \to \infty} i_\Omega \exp(i(c_\Omega + H_{\mathrm{as}\Omega})t)(D^\Omega)^{-1}$$

$$\times \exp(iH_\Omega(t - t_0))D^\Omega \exp(-i(c_\Omega + H_{\mathrm{as}\Omega})t_0)i^*_\Omega,$$

where i_Ω is a natural embedding of the Fock space $F_\Omega = F(L^2(\Omega))$ into the Fock space $F(L^2(E^3))$.)

Let us sketch the proof of the following theorem.

Theorem 11.5. *Let the family of operators D^Ω satisfy the following conditions*:

(1) *The operator D^Ω transforms the Fock vector θ and the bare single-particle state $a_k^+\theta$ into normalized eigenvectors of the operator H_Ω.*

(2) *The operator D^Ω can be represented as the product of operators of the form $N(\exp A_\Omega)$ and $\exp A_\Omega$, where A_Ω are operators obtained from a volume cutoff of operators of the form $A = A_1 + iA_2$ and A_1, A_2 are formal expressions from the class $\mathcal{M}$, not depending on Ω.*

Then D^Ω is a family of dressing operators. (More precisely, it can be shown that formula (11.32) holds, where c_Ω is defined by the formula $H_\Omega D^\Omega \theta = c_\Omega D^\Omega \theta$.)

Let us begin with the remark that with the help of analogs of Lemmas 11.1 and 11.2 for operators on the space F_Ω, we can reduce the proof to the case when the operator D^Ω has the form $N(\exp A_\Omega)$, where $A = A_1 + iA_2, A_1 \in \mathcal{M}, A_2 \in \mathcal{M}$. Let us first suppose that the Hamiltonian H does not generate vacuum polarization. For such a Hamiltonian, the operators $\exp(iH_\Omega t), D^\Omega = N(\exp A_\Omega), \exp(iH_{\mathrm{as}\Omega}t)$ have limits as $\Omega \to \infty$; they converge, respectively, to the operators $\exp(iHt), N(\exp A) = D, \exp(iH_{\mathrm{as}}t)$. This can be derived from the relations $H\theta = 0, H_{\mathrm{as}}\theta = 0, A\theta = 0$ (the last of these relations follows from the equation $D_\Omega \theta = \theta$). It is easy to check that on the right-hand side of equation (11.32), we can transform the limit of products into a product of limits. Noting that the constant c_Ω in the case at hand is equal to 0, we obtain that equation (11.32) is equivalent to the relation

$$S = \operatorname*{slim}_{\substack{t \to \infty \\ t_0 \to -\infty}} \exp(iH_{\mathrm{as}}t)D^{-1}\exp(-iH(t - t_0))$$

$$\times D \exp(-iH_{\mathrm{as}}t_0),$$

which follows from the already-proven relations (11.25) and the formula $S = S_+^* S_-$. Thus, the necessary statement is proven for

Hamiltonians not generating vacuum polarization. The case of an arbitrary Hamiltonian $H \in \tilde{\mathcal{M}}$ reduces to the considered case by Faddeev transformation. Indeed, by the results of Section 9.4, there exists a formal expression $W \in \mathcal{M}$, such that as $\Omega \to \infty$, we have

$$\exp(iW_\Omega) H_\Omega \exp(-iW_\Omega) \approx H'_\Omega + c_\Omega,$$

and, therefore,

$$\exp(iW_\Omega) e^{iH_\Omega t} \exp(-iW_\Omega) \approx e^{i(H'_\Omega + c_\Omega)t} \tag{11.33}$$

where H'_Ω is obtained from a Hamiltonian H' that does not generate vacuum polarization by means of volume cutoff; c_Ω is the energy of the ground state of the operator H_Ω. Using formula (11.33), we conclude that

$$\lim_{\Omega \to \infty} \exp(i(c_\Omega + H_{\mathrm{as}\Omega})t)(D^\Omega)^{-1} \exp(-iH_\Omega(t - t_0))$$

$$\times D^\Omega \exp(-i(c_\Omega + H_{\mathrm{as}\Omega})t_0)$$

$$= \lim_{\Omega \to \infty} \exp(iH'_{\mathrm{as}\Omega}t')(D'^\Omega)^{-1} \exp(-iH'_\Omega(t - t_0))$$

$$\times D'^\Omega \exp(-iH'_{\mathrm{as}\Omega}t_0), \tag{11.34}$$

where

$$D'^\Omega = \exp(iW_\Omega) D^\Omega$$

(to derive equation (11.34), one needs to use the coincidence of the asymptotic Hamiltonians H_{as} and H'_{as}, corresponding to the Hamiltonians H and H'). Applying the statement of the theorem to a Hamiltonian H' not generating vacuum polarization, we see that the operators D'^Ω constitute a family of dressing operators for the Hamiltonian H'. This implies that the right-hand side of equation (11.34), in the limit as $t \to \infty$, $t_0 \to -\infty$, equals the scattering matrix of the Hamiltonian H'. Since the scattering matrices of the Hamiltonians H and H' coincide, equation (11.34) proves the theorem in the general case.

11.5 Perturbation theory via the axiomatic approach

Let us suppose that on the Hilbert space $\mathcal{H}$, we have defined an energy operator H_0, a momentum operator $\mathbf{P}$, a linear dense subspace D of $\mathcal{H}$, and a complete locally convex topological algebra with involution $\mathcal{A}$, consisting of operators acting on D. Let us further assume that these objects satisfy conditions T1–T7 from Section 10.5.

Suppose that W is an operator from the algebra $\mathcal{A}$, satisfying the condition $W\Phi = 0$. Let us now prove that *the family of operators*

$$H(g) = H_0 + gV = H_0 + g \int W(\mathbf{x})d\mathbf{x}$$

together with momentum operator $\mathbf{P}$ *and algebra* $\mathcal{A}$, *satisfy conditions T1–T7 from Section 10.5 in each power of the parameter* g (the integral is understood in the strong sense). Imposing some conditions on the joint spectrum of the operators $H_0, \mathbf{P}$, we will prove that starting with the operators $H(g), \mathbf{P}$, and the algebra $\mathcal{A}$, one can construct the Møller matrices $S_\pm(g)$ as well as the scattering matrix $S(g)$, at least as formal series in powers of g.[6]

Let us first note that the operator $V \in \int W(\mathbf{x})d\mathbf{x}$ is defined on the set $\mathcal{A}\Phi$ (i.e. on the set of vectors of the form $A\Phi$, where $A \in \mathcal{A}$) and transforms this set into itself. Indeed,

$$\left(\int W(\mathbf{x})d\mathbf{x} \right) A\Phi = \int [W(\mathbf{x}), A]\Phi d\mathbf{x}.$$

By condition T6, it follows that for any seminorm $p(A)$ on $\mathcal{A}$ and any n, we have

$$p([W(\mathbf{x}), A]) \leq \frac{C}{1 + |\mathbf{x}|^n}. \tag{11.35}$$

By the completeness of the algebra $\mathcal{A}$ and equation (11.35), the integral $\int [W(\mathbf{x}), A]d\mathbf{x}$ converges in the topology of $\mathcal{A}$ and defines an element of the algebra $\mathcal{A}$. This completes the proof.

Later in the chapter, we will prove that when calculating the operator $\exp(-iH(g)t) = \exp(-i(H_0+gV)t)$ in terms of perturbation

---

[6]This statement, just as the rest of the statements in this section, applies to the more general class of Hamiltonians of the form $H_0 + \sum_{r \geq 1} g^r \int W_r(\mathbf{x})d\mathbf{x}$, and $W_r \in \mathcal{A}, W_r\Phi = 0$, and the series in g is understood as a formal series.

theory in each order of g, we obtain an operator, defined on the set $\mathcal{A}\Phi$, that transforms this set into itself (this implies that the operator $H(g)$ can be seen as a self-adjoint operator, at least in perturbation theory). Indeed, from the results of Section 4.1, it follows that

$$\exp(-iH(g)t) = \sum \left(\frac{g}{i}\right)^r U_r(t),$$

where

$$U_r(t) = \frac{1}{r!} \exp(-iH_0 t) \int_0^t d\tau_1 \int_0^{\tau_1} d\tau_2 \cdots \int_0^{\tau_{r-1}} d\tau_r V(\tau_1) \cdots V(\tau_r)$$

$$= \frac{1}{r!} \exp(-iH_0 t) \int_0^t d\tau_1 \int_0^{\tau_1} d\tau_2 \cdots \int_0^{\tau_{r-1}} d\tau_r \int d^r\mathbf{x} W(\mathbf{x}_1, \tau_1)$$

$$\cdots W(\mathbf{x}_r, \tau_r),$$

$$V(\tau) = \exp(iH_0\tau)V \exp(-iH_0\tau) = \int W(\mathbf{x}, \tau)d\mathbf{x},$$

$$W(\mathbf{x}, \tau) = \exp(i(H_0\tau - \mathbf{Px}))W \exp(-i(H_0\tau - \mathbf{Px})).$$

Let us define the operator $R_r(\mathbf{x}_1, t_1, \ldots, \mathbf{x}_r, t_r | A)$, where $A \in \mathcal{A}$, by induction and the relations

$$R_0(A) = A,$$

$$R_r(\mathbf{x}_1, t_1, \ldots, \mathbf{x}_r, t_r | A)$$

$$= [W(\mathbf{x}_1, t_1), R_{r-1}(\mathbf{x}_2, t_2, \ldots, \mathbf{x}_r, t_r | A)].$$

It is easy to check that the operator R_r continuously depends on $\mathbf{x}_1, t_1, \ldots, \mathbf{x}_r, t_r$ in the topology of the algebra $\mathcal{A}$ and

$$p(R_r(\mathbf{x}_1, t_1, \ldots, \mathbf{x}_r, t_r | A)) \leq \frac{C(1 + |t_1|^s + \cdots + |t_r|^s)}{1 + |\mathbf{x}_1|^n + \cdots + |\mathbf{x}_r|^n} \tag{11.36}$$

(here, $p(A)$ is any seminorm on $\mathcal{A}$, n is an arbitrary number and the numbers s and C depend on the seminorm p and the number n). Inequality (11.36) can be easily derived by induction and by means of conditions T5 and T6.

Let us now note that

$$W(\mathbf{x}_1, t_1) \cdots W(\mathbf{x}_r, t_r) A \Phi = R_r(\mathbf{x}_1, t_1, \ldots, \mathbf{x}_r, t_r | A) \Phi$$

(this can also be easily checked by induction). Thus,

$$U_r(t) A \Phi = B_r(t) \Phi,$$

where

$$B_r(t) = \frac{1}{r!} \int_0^t d\tau_1 \int_0^{\tau_1} d\tau_2 \cdots \int_0^{\tau_{r-1}} d\tau_r \int d^r \mathbf{x} \exp(-iH_0 t)$$
$$\times R_r(\mathbf{x}_1, \tau_1, \ldots, \mathbf{x}_r, \tau_r | A) \exp(iH_0 t).$$

From the inequality (11.36) and the completeness of algebra $\mathcal{A}$, it follows that for any $A \in \mathcal{A}$, the integral defining the operator $B_r(t)$ converges in the topology of $\mathcal{A}$ and, therefore, $B_r(t) \in \mathcal{A}$. Thus, the operator $U_r(t)$ is defined on the set $\mathcal{A}\Phi$ and transforms the set into itself. It therefore follows that the operator $A(\mathbf{x}, t | g) = \exp(iH(g)t - i\mathbf{P}\mathbf{x}) A \exp(-iH(g)t + i\mathbf{P}\mathbf{x})$, in each order of the perturbation series, is defined on the set $\mathcal{A}\Phi$ and then transforms the set into itself:

$$A(\mathbf{x}, t | g) = \sum g^r A_r(\mathbf{x}, t),$$

where the operator

$$A_r(\mathbf{x}, t) = \sum_{\alpha=0}^r U_\alpha(-t) A(\mathbf{x}) U_{r-\alpha}(t) \left(\frac{1}{i}\right)^r$$

acts on the set $\mathcal{A}\Phi$).

Let us now show that the operator $A_r(\mathbf{x}, t) \in \mathcal{A}$ is continuous in $\mathbf{x}, t$ in the topology on $\mathcal{A}$ and that for any seminorm $p(A)$ on $\mathcal{A}$, we have

$$p(A_r(\mathbf{x}, t)) \le (1 + |\mathbf{x}|^s + |t|^s) q(A), \tag{11.37}$$

where the number s and the seminorm $q(A)$ on $\mathcal{A}$ depend on the seminorm p and the number r; A runs over a compact set $F \subset \mathcal{A}$ (in other words, we will prove that condition T5 from Section 10.5 holds for operators $H(g), \mathbf{P}$, and the algebra $\mathcal{A}$ in each order of the perturbation theory in g). To do this, let us note that from the Heisenberg equation,

$$\frac{1}{i}\frac{\partial A(\mathbf{x}, t|g)}{\partial t} = [H_0 + gV, A(\mathbf{x}, t|g)],$$

follows the relation

$$\frac{1}{i}\frac{\partial A_r(\mathbf{x}, t)}{\partial t} = [H_0, A_r(\mathbf{x}, t)] + [V, A_{r-1}(\mathbf{x}, t)]. \tag{11.38}$$

With the relation (11.38) and the initial condition $A_r(\mathbf{x}, 0) = 0$ for $r \geq 1$, we can express $A_r(\mathbf{x}, t)$ through $A_{r-1}(\mathbf{x}, t)$:

$$A_r(\mathbf{x}, t) = \exp(iH_0 t)\tilde{A}_r(\mathbf{x}, t)\exp(-iH_0 t),$$

$$\tilde{A}_r(\mathbf{x}, t) = i\int_0^t \exp(-iH\tau)[V, A_{r-1}(\mathbf{x}, \tau)]\exp(iH\tau)d\tau$$

$$= i\int_0^t d\tau \int d\xi \exp(-iH\tau)[W(\xi), A_{r-1}(\mathbf{x}, \tau)]\exp(iH\tau). \tag{11.39}$$

Using this expression, the necessary statement follows by induction. We obtain

$$p(\exp(-iH\tau)[W(\xi), A_{r-1}(\mathbf{x}, \tau)]\exp(iH\tau)) \leq \frac{p_1(A)(1 + |\mathbf{x}|^2 + |t|^s)}{1 + |\xi|^n}, \tag{11.40}$$

where $A \in F$, p is an arbitrary seminorm, n is an arbitrary number, and the seminorm p_1 and the number s depend on p and n.

Inequality (11.40) implies that the integral in (11.39) converges uniformly in $\mathbf{x}, t$ in the topology on $\mathcal{A}$ if $\mathbf{x}, t$ belongs to a bounded set. From this remark, it follows that $A_r(\mathbf{x}, t) \in \mathcal{A}$ and that it continuously depends on $\mathbf{x}, t$; the inequality (11.37) follows immediately from the inequality (11.40).

It is now easy to prove that the operators $H(g)$, $\mathbf{P}$, and the algebra $\mathcal{A}$, whose elements take the form $D' = \mathcal{A}\Phi$, satisfy conditions T1–T7 in Section 10.5 in each order of the perturbation series in g. Indeed, we have already shown that the set D' is invariant in each order of perturbation expansion of the operators $\exp(iH(g)t)$. Its invariance with respect to the operators $\exp(i\mathbf{Px})$ follows from the relation $\exp(-i\mathbf{Px})A\Phi = A(\mathbf{x})\Phi$. Conditions T1–T4, T6, and T7 do not depend on the choice of the energy operator $H(g)$ and, therefore, are satisfied in the case at hand. We have already checked condition T5 by perturbation theory.

Let us now show that *the joint spectrum of the operators* $H(g), \mathbf{P}$ *satisfies the conditions on the spectrum of energy and momentum operators specified in Section 9.5* (at least in perturbation theory).

Let us suppose that the necessary conditions[7] are satisfied for $g = 0$ and let us further assume that for any point $\mathbf{k} \in E^3$, we can find an operator $A \in \mathcal{A}$, such that $|\langle \Phi_0(\mathbf{k}), A\Phi \rangle| \neq 0$ (the symbol $\Phi_0(\mathbf{k})$ denotes a single-particle state of the operator H_0). The arguments applied in the slightly more complicated situation in the proof of Lemmas 10.12–10.14 in Section 10.6, allow us to show that the single-particle state $\Phi(\mathbf{k})$ can be chosen such that the functions $\langle \Phi(\mathbf{k}), A\Phi \rangle$ for all $A \in \mathcal{A}$ will be infinitely differentiable. If the single-particle state $\Phi(\mathbf{k})$ is chosen this way, then for any smooth function with compact support, we can find a good operator B_f such that $B_f\Phi = \int f(\mathbf{k})\Phi(\mathbf{k})d\mathbf{k}$.

It follows from $W\Phi = 0$ that the ground state Φ of the operator H_0 is a stationary state of the operator $H(g)$ for any g; this implies that when we calculate the ground state of the operator $H(g)$ in perturbation theory, we obtain the vector Φ. We can find single-particle states of the operator $H(g)$ in perturbation theory in the form

$$\Phi(\mathbf{k}|g) = \sum_{r=0}^{\infty} g^r \Phi_r(\mathbf{k}),$$

using the relations

$$(H_0 + gV)\left(\sum g^r \Phi_r(\mathbf{k})\right) = \left(\sum g^r \omega_r(\mathbf{k})\right)\left(\sum g^r \Phi_r(\mathbf{k})\right), \quad (11.41)$$

---

[7]In order to prove the existence of Møller matrices and scattering matrices constructed with the operators $H(g), \mathbf{P}$, and the algebra $\mathcal{A}$, it is enough to assume that the joint spectrum of the operators $H_0, \mathbf{P}$ satisfies conditions 4 and 5 from Section 10.5.

$$\mathbf{P}\left(\sum g^r \Phi_r(\mathbf{k})\right) = \mathbf{k}\left(\sum g^r \Phi_r(\mathbf{k})\right), \qquad (11.42)$$

$$\left\langle \frac{\partial}{\partial g} \sum g^r \Phi_r(\mathbf{k}), \sum g^r \Phi_r(\mathbf{k}') \right\rangle = 0. \qquad (11.43)$$

By the relations (11.41) and (11.42), we obtain that

$$\int \omega_r(\mathbf{k}) f(\mathbf{k}) \overline{f'(\mathbf{k})} d\mathbf{k}$$

$$= \left\langle V\Phi_{r-1}(f), \Phi_0(f') \right\rangle - \left\langle \Phi_{r-1}(\omega_1 f), \Phi_0(f') \right\rangle - \cdots$$

$$- \left\langle \Phi_1(\omega_{r-1} f), \Phi_0(f') \right\rangle, \qquad (11.44)$$

$$\Phi_r(f) = \gamma(H_0, \mathbf{P})(-V\Phi_{r-1}(f) + \Phi_{r-1}(\omega_1 f) + \cdots + \Phi_0(\omega_r f))$$

$$+ \Phi_0(\lambda_r f), \qquad (11.45)$$

where $f(\mathbf{k})$ is a smooth function with compact support; $\Phi_r(f) = \int f(\mathbf{k})\Phi_r(\mathbf{k})$; $\gamma(\omega, \mathbf{k})$ is a smooth function with compact support equal to $(\omega - \omega_0(\mathbf{k}))^{-1}$, if $\mathbf{k} \in \operatorname{supp} f$ and $(\omega, \mathbf{k})$ belongs to the multi-particle spectrum of the operators $H, \mathbf{P}$ and also if $\mathbf{k} = 0, \omega = 0$. It should be equal to zero if $\omega = \omega_0(\mathbf{k})$. Using relations (11.43)–(11.45), we can find functions $\lambda_r(\mathbf{k})$, $\omega_r(\mathbf{k})$, and vectors $\Phi_r(\mathbf{k})$; simultaneously by induction on r, we can prove that $\omega_r(\mathbf{k})$ and $\lambda_r(\mathbf{k})$ are smooth functions and the vector $\Phi_r(f)$ can be written in the form

$$\Phi_r(f) = B_f^{(r)}\Phi, \qquad (11.46)$$

where $B_f^{(r)} \in \mathcal{A}$. (The representation (11.46) follows from equation (11.45) with the help of the reasoning employed in the proof of Lemma 10.17 in Section 10.6; the infinite differentiability of $\omega_r(\mathbf{k})$ and $\lambda_r(\mathbf{k})$ can be shown by a slight modification of the proof of Lemma 10.15 in Section 10.6.)

Let us now show that *the Møller matrices $S_\pm(g)$, corresponding to the operators $H(g), \mathbf{P}$, and the algebra $\mathcal{A}$, can be constructed at least in the framework of perturbation theory.* More precisely, we will construct the Møller matrix as a formal series

$$S_\pm(g) = \sum g^r S_\pm^{(r)},$$

where $S_\pm^{(r)}$ are operators defined on the set $D \subset \mathcal{H}_{as}$, consisting of linear combinations of vectors of the form $a^+(f_1) \cdots a^+(f_n)\theta$ (here, $f_1(\mathbf{k}), \ldots, f_n(\mathbf{k})$ is a non-overlapping family of smooth functions with compact support. (Recall that this condition on a family of functions means that for any $\mathbf{k} \in \operatorname{supp} f_i, \mathbf{k}' \in \operatorname{supp} f_j, i \neq j$, the gradients of the functions $\omega_0(\mathbf{k})$ at the points $\mathbf{k}$ and $\mathbf{k}'$ do not coincide.)

To construct Møller matrices, we repeat the construction in Section 10.1, considering all objects, entering into the formula (10.1), as formal series in powers of g. The proof of the correctness of this definition is nearly identical to the proof in Chapter 10. It is important to note that the estimate (10.7) cannot be proven in every order in g, in other words, if $f(\mathbf{k})$ is a smooth function with compact support with

$$\tilde{f}(\mathbf{x}|t) = (2\pi)^{-3} \int \exp\left(-it \sum_{r \geq 0} g^r \omega_r(\mathbf{k}) + i\mathbf{k}\mathbf{x}\right) f(\mathbf{k})d\mathbf{k}$$

$$= \sum_{r \geq 0} g^r \tilde{f}_r(\mathbf{x}|t),$$

then we cannot prove the inequality

$$|\tilde{f}_r(\mathbf{x}|t) \leq C|t|^{-3/2}].$$

Therefore, we also cannot prove that the Lemma 10.4 of Section 10.1 holds in each order of perturbation theory. However, in the proof of the correctness of the definition of the Møller matrix relying on formula (10.1), we can substitute this lemma with Lemma 10.5 from Section 10.3, which holds in each order in g (the family of functions $f_1, \ldots, f_n$ in formula (10.1) should be assumed to be a non-overlapping family of smooth functions with compact support).

The scattering matrix S is defined, as always, by the relation $S = S_+^* S_-$. (From what we have proved, it does not follow that S has the form $S = \sum g^r S^{(r)}$, where $S^{(r)}$ are operators with a common domain, however, the matrix elements

$$\langle Sa^+(\mathbf{k}_1) \cdots a^+(\mathbf{k}_m)\theta, a^+(\mathbf{p}_1) \cdots a^+(\mathbf{p}_n)\theta\rangle$$

$$= \langle S_- a^+(\mathbf{k}_1) \cdots a^+(\mathbf{k}_m)\theta, S_+ a^+(\mathbf{p}_1) \cdots a^+(\mathbf{p}_n)\theta\rangle \quad (11.47)$$

for $\mathbf{k}_i \neq \mathbf{k}_j, \mathbf{p}_i \neq \mathbf{p}_j$ can be viewed as formal series in g. More precisely, these matrix elements can be written in the form $\sum_{r \geq 0} g^r S^{(r)}(\mathbf{k}_1, \ldots, \mathbf{k}_m \,|\, \mathbf{p}_1, \ldots, \mathbf{p}_n)$, where $S^{(r)}$ are generalized functions, defined on the class of test functions $\phi(\mathbf{k}_1, \ldots, \mathbf{k}_m \,|\, \mathbf{p}_1, \ldots, \mathbf{p}_n) \in \mathcal{S}$, having support belonging to the set where $\mathbf{k}_i \neq \mathbf{k}_j, \mathbf{p}_i \neq \mathbf{p}_j$. Hence, by scattering matrix, we will mean the set of matrix elements (11.47).)

We can now use the results of Chapter 10 to investigate the Møller matrices $S_{\pm}(g)$ and the scattering matrix $S(g)$, corresponding to the operators $H(g), \mathbf{P}$, and the algebra $\mathcal{A}$. Let us note that, in particular, with the help of the statements proven above, we can easily prove the assumptions in the adiabatic theorem from Section 10.6 in each order of g. Unfortunately, this is not enough to prove, in each order of perturbation theory, the theorem that gives an expression of Møller matrix $S_{\pm}(g)$ in terms of the adiabatic Møller matrix $S_{\alpha}(0, \pm\infty | g)$. Inspection of the proof of the adiabatic theorem shows that the proof depends on relations of the form

$$\left| \exp\left(\frac{i}{\alpha} \nu(\mathbf{k}|g) \right) \right| \leq \text{const},$$

where $\nu(\mathbf{k}|g) = \sum_{r \geq 0} g^r \nu_r(\mathbf{k})$ is a real function; these relations clearly do not hold in each order of g. In particular, to prove Lemma 10.25, from Section 10.6, it is essential that

$$p(Q^t(f_i, \phi_i, \alpha)) = p(Q_i^t) \leq C(1 + |t|^n),$$

where C is a constant not depending on α; this inequality with $n = \frac{3}{2}$ follows from the inequality

$$\int |\phi_i^{\tilde{t},\alpha}(\mathbf{x})| d\mathbf{x} \leq C(1 + |t|^{3/2}),$$

which does not hold in the framework of perturbation theory. However, it is easy to verify that all the proofs in Section 10.6 hold in the framework of perturbation theory if we allow the use of partial summation of perturbation series. This allows us to say that the adiabatic theorem can be proven in the framework of perturbation theory with partial summation.

Let us now come back to Hamiltonians that do not generate vacuum polarization using the results we have proved.

Let us take as our Hilbert space $\mathcal{H}$ the space $F(L^2(E^3))$ and let the energy operator H_0 and the momentum operator $\mathbf{P}$ be defined by the formulas

$$H_0 = \int \epsilon(\mathbf{k}) a^+(\mathbf{k}) d\mathbf{k},$$

$$\mathbf{P} = \int \mathbf{k} a^+(\mathbf{k}) a(\mathbf{k}) d\mathbf{k}.$$

The function $\epsilon(\mathbf{k})$ is presumed to be a smooth function whose derivatives grow not faster than a power. We also assume that the function $\epsilon(\mathbf{k})$ satisfies the condition $\epsilon(\mathbf{k}_1 + \mathbf{k}_2) < \epsilon(\mathbf{k}_1) + \epsilon(\mathbf{k}_2)$ and does not coincide with a linear function on any open set.) Let us suppose that the algebra $\mathcal{A}$ is the algebra of operators of the form

$$\sum_{m,n}^{\infty} \int f_{m,n}(\mathbf{k}_1, \ldots, \mathbf{k}_m | \mathbf{p}_1, \ldots, \mathbf{p}_n)$$

$$\times a^+(\mathbf{k}_1) \cdots a^+(\mathbf{k}_m) a(\mathbf{p}_1) \cdots a(\mathbf{p}_n) d^m \mathbf{k} \, d^n \mathbf{p}, \qquad (11.48)$$

where $f_{m,n} \in \mathcal{S}$ and the summation in (11.48) is presumed to be finite (operators of the form (11.48) can be viewed as operators defined on the set $\mathcal{S}_\infty$). In Section 10.5, it was shown that the operators $H_0, \mathbf{P}$, and the algebra $\mathcal{A}$ satisfy conditions T1–T7. This allows us to apply to them all the statements proven in this chapter.

Let us now note that the operator

$$H_0 + \sum_{r \geq 1} \int W_r(\mathbf{x}) d\mathbf{x}, \qquad (11.49)$$

where $W_r \in \mathcal{A}, W_r \theta = 0$, can be written in the form

$$H_0 + \sum_{r \geq 1} g^r \sum_{m,n} \int w_{m,n}^{(r)}(\mathbf{k}_1, \ldots, \mathbf{k}_m | \mathbf{p}_1, \ldots, \mathbf{p}_n)$$

$$\times \delta(\mathbf{k}_1 + \cdots + \mathbf{k}_m - \mathbf{p}_1 - \cdots - \mathbf{p}_n)$$

$$\times a^+(\mathbf{k}_1) \cdots a^+(\mathbf{k}_m) a(\mathbf{p}_1) \cdots a(\mathbf{p}_n) d^m \mathbf{k} \, d^n \mathbf{p},$$

i.e. it is a translation-invariant Hamiltonian not generating vacuum polarization. (It is easy to check that the Hamiltonian (11.49) belongs to the class $\tilde{\mathcal{M}}$ described in Section 11.1.)

As was shown in this chapter, we can associate Møller matrices and scattering matrix written as formal series in powers of g to the Hamiltonian (11.49) (we have already discussed this in Section 11.3, where we have provided a different proof of this statement).

We can now apply the results of Chapter 10 to gain insight about the scattering matrix of a translation-invariant Hamiltonian not generating vacuum polarization.

For example, the following statement follows immediately from Theorem 10.3, Section 10.4 if we note that for any free Hamiltonian H_0, the identity operator is a dressing operator and satisfies the conditions of Theorem 10.2 in Section 10.4.

Let $H \in \tilde{\mathcal{M}}, B_1, \ldots, B_n \in \mathcal{M}$ be Hamiltonians not generating vacuum polarization. Suppose that the operator $D = \exp(iB_1) \cdots \exp(iB_n)$ transforms the bare single-particle state $a^+(\mathbf{k})\theta$ into the single-particle state of the Hamiltonian H. Then the operator D is a dressing operator for the Hamiltonian H.

Furthermore, let $H = H_0 + V \in \tilde{\mathcal{M}}$ be a Hamiltonian not generating vacuum polarization and let the function $\epsilon(\mathbf{k})$ be strictly convex. Then from the results of Section 10.6, we can conclude that *the definition of the Møller matrix in terms of the adiabatic Møller matrix given in Section 9.1 is equivalent to the other definitions.* It therefore follows that in the case at hand, *Definition 9.2 of the scattering matrix in Section 9.3 is equivalent to the other definitions* (both of these statements, as noted above, hold in the framework of perturbation theory with partial summation).

Chapter 12

Axiomatic Lorentz-Invariant Quantum Field Theory

12.1 Axioms describing Lorentz-invariant scattering matrices

Let $\mathcal{H}$ be a Hilbert space of states. To discuss the Lorentz invariance of a theory, we should first of all assume that we have a unitary representation of an inhomogeneous Lorentz group $\mathcal{P}$ (Poincaré group) on the space $\mathcal{H}$. We will denote the unitary operator corresponding to the element $g \in \mathcal{P}$ by $U(g)$. The group $\mathcal{P}$ contains the group of space–time translations T as a subgroup; we will define the energy operator H and the momentum operator $\mathbf{P}$ by the relation

$$\exp(-i(Ha_0 - \mathbf{Pa})) = U(1, a), \tag{12.1}$$

where the symbol $(1, a)$ denotes the translation $x' = x + a$ (in general, (Λ, a) denotes the element of $\mathcal{P}$ defined by the transformation $x' = \Lambda x + a$).

However, if we only have a representation of the Poincaré group, still the corresponding scattering matrix cannot be defined. We will assume, as in the axiomatic approach to scattering theory described in Chapter 10, that we have an asymptotically commutative algebra $\mathcal{A}$ on the space $\mathcal{H}$; we impose conditions guaranteeing the Lorentz invariance of the scattering matrix.

To begin, we will assume that the presentation $U(g)$ and the algebra $\mathcal{A}$ satisfy the following axioms:

A1. *Spectrality*: The energy operator H has a single ground state Φ; the vector Φ is Lorentz-invariant (i.e. $U(g)\Phi = \Phi$ for any transformation $g \in \mathcal{P}$).

The vector Φ is called the *physical vacuum*.

A2. *Cyclicity*: The vector Φ is a cyclic vector of the algebra $\mathcal{A}$.

A3. *Lorentz invariance of the algebra $\mathcal{A}$*: If the operator $A \in \mathcal{A}$, then for any transformation $g \in P$, the operator $U(g)AU^{-1}(g)$ also belongs to the algebra $\mathcal{A}$.

In what follows, axiom A3 can be replaced with the following weaker condition.

A3′. For any transformation $g \in P$, the algebra $\mathcal{A}$ asymptotically commutes with the algebra $U(g)\mathcal{A}U^{-1}(g)$ (i.e. with the algebra consisting of operators of the form $U(g)AU^{-1}(g)$, where $A \in \mathcal{A}$).

Let us analyze the consequences of axiom A1 on the joint spectrum of the operators $H, \mathbf{P}$.

Lorentz invariance of the vector Φ implies, in particular, that $\exp(-iHt)\Phi = \Phi$ and $\exp(i\mathbf{P}\mathbf{a})\Phi = \Phi$; hence,

$$H\Phi = \mathbf{P}\Phi = 0. \tag{12.2}$$

It is easy to check that the converse holds as well (i.e. from equation (12.2) and the uniqueness of the ground state, the Lorentz invariance of Φ follows).

Since the ground state Φ of the energy operator H has zero energy, by equation (12.2), the operator H is non-negative. This implies that the joint spectrum of the operators $H, \mathbf{P}$ is contained in the half-space consisting of points $(\omega, \mathbf{p})$ where $\omega \geq 0$. However, the joint spectrum of the operators $H, \mathbf{P}$ must be invariant with respect to homogenous Lorentz transformations (see Appendix A.9). Therefore, the spectrum of the operators $H, \mathbf{P}$ lies in the cone W, consisting of points $(\omega, \mathbf{p})$, that satisfy the inequality $\omega^2 \geq \mathbf{p}^2$ (any point with $\omega^2 < \mathbf{p}^2$ can be transformed to a point $(\omega', \mathbf{p}')$ with $\omega' < 0$ by a homogenous Lorentz transformation). The physical vacuum Φ corresponds to the point $0 = (0, 0, 0, 0) \in W$.

Let us consider an irreducible invariant subspace K of the space $\mathcal{H}$ [i.e. a subspace invariant with respect to the operators $U(g)$ and not containing non-trivial subspaces that are also invariant with respect to the operators $U(g)$]. For example, the subspace A, consisting of vectors of the form $\lambda\Phi$ (the vacuum subspace), is irreducible and invariant. All other irreducible invariant subspaces cannot contain the vector Φ (otherwise, they would contain the subspace A and not be irreducible). We will therefore assume that the subspace K does not contain the vector Φ.

A representation of the Poincaré group $\mathcal{P}$ by the operators $U(g)$, viewed as operators on the subspace K, is irreducible. The energy operator H on the space K cannot vanish, otherwise every vector $x \in K$ is a ground state of the operator H, which contradicts the uniqueness of the ground state (axiom A1). On the other hand, the same axiom implies that the operator H on the subspace K (and on the whole space) is non-negative.

As follows from known facts about irreducible representations of the group $\mathcal{P}$ (see Appendix A.9), the joint spectrum of the operators H and $\mathbf{P}$ on the space K constitutes the set U_μ, consisting of points $(\omega, \mathbf{p}) \in E^4$ with $\omega^2 - \mathbf{p}^2 = \mu^2, \omega \geq 0$. Here, μ is an arbitrary non-negative number, the set U_μ with $\mu > 0$ forms the top-half of a hyperboloid. Furthermore, one can prove that there exists a unitary map (an isomorphism) Ψ between the space $L^2(E^3 \times N)$, where N is a finite set, and the space K, having the property

$$U(g)\Psi(f) = \Psi(V(g)f), \tag{12.3}$$

where $V(g)$ is an irreducible presentation of type (μ, n), described in Appendix A.9. It follows from the relation (12.3) that

$$\begin{aligned} H\Psi(f) &= \Psi(\hat{h}f), \\ \mathbf{P}\Psi(f) &= \Psi(\hat{\mathbf{p}}f), \end{aligned} \tag{12.4}$$

where $\hat{h}$ and $\hat{\mathbf{p}}$ are operators in the space $L^2(E^3 \times N)$ defined by the formulas

$$\begin{aligned} (\hat{\mathbf{p}}f)(\mathbf{k}, \alpha) &= \mathbf{k}f(\mathbf{k}, \alpha), \\ (\hat{h}f)(\mathbf{k}, \alpha) &= \sqrt{\mathbf{k}^2 + \mu^2}\,f(\mathbf{k}, \alpha). \end{aligned} \tag{12.5}$$

The mapping Ψ can be viewed as n generalized vector functions $\Psi_\alpha(\mathbf{k})$ related to Ψ by the formula

$$\Psi(f) = \sum_{\alpha \in N} \int f(\mathbf{k}, \alpha)\Psi_\alpha(\mathbf{k})d\mathbf{k}$$

(here, n is the number of elements in the set N and α runs over the set N). From the unitarity of the mapping Ψ, it follows that the generalized vector functions $\Psi_\alpha(\mathbf{k})$ satisfy the relations

$$\langle \Psi_\alpha(\mathbf{k}), \Psi_{\alpha'}(\mathbf{k}') \rangle = \delta_\alpha^{\alpha'} \delta(\mathbf{k} - \mathbf{k}')$$

(orthogonality and normalization by a δ-function); from equations (12.4) and (12.5), it follows that

$$\mathbf{P}\Psi_\alpha(\mathbf{k}) = \mathbf{k}\Psi_\alpha(\mathbf{k}),$$

$$H\Psi_\alpha(\mathbf{k}) = \sqrt{\mathbf{k}^2 + \mu^2}\,\Psi_\alpha(\mathbf{k}).$$

This implies that the functions $\Psi_\alpha(\mathbf{k})$ can be viewed as single-particle states in the sense of the definitions previously used in this book. In ordinary terminology, the collection of functions $\Psi_1(\mathbf{k}), \ldots, \Psi_n(\mathbf{k})$ (i.e. the mapping Ψ) is identified with a particle having mass μ and spin $s = \frac{n-1}{2}$. A particle having spin 0 is called *scalar*.

The smallest subspace, containing all irreducible invariant subspaces differing from A, will be called the *single-particle subspace* and will be denoted by B.

It is easy to check that the space B can be represented as a direct sum of mutually orthogonal irreducible invariant subspaces K_i; using the isomorphisms Ψ_i, described above, between the spaces $L^2(E^3 \times N_i)$ and K_i, we can construct an isomorphism $\Psi = \sum_i \Psi_i$ between the space $L^2(E^3 \times N) = \sum_i L^2(E^3 \times N_i)$ and the space B (N is the union of the sets N_i).

It is clear from these remarks that the single-particle subspace B defined above coincides with the single-particle subspace in the sense of the definition in Section 10.1.

The multi-particle subspace M will be defined as the orthogonal completion of the direct sum $A + B$ (the space $A + B$ can be written as the sum of all irreducible invariant subspaces).

The space $\mathcal{H}_{\mathrm{as}}$ will be defined as the Fock space $F(B)$.

Using the isomorphism between the space B and the space $L^2(E^3 \times N)$, constructed above, we can introduce the operator generalized functions $a^+(\mathbf{k}, s), a(\mathbf{k}, s)$ on $\mathcal{H}_{\mathrm{as}}$, where $\mathbf{k} \in L^3, s \in N$, satisfying the relations

$$[a(\mathbf{k}, s), a(\mathbf{k}', s')] = [a^+(\mathbf{k}, s), a^+(\mathbf{k}', s')] = 0,$$

$$[a(\mathbf{k}, s), a^+(\mathbf{k}', s')] = \delta^s_{s'}\delta(\mathbf{k} - \mathbf{k}').$$

The space B is invariant with respect to the operators $U(g)$. The representation of $\mathcal{P}$ by the operators $U(g)$ on the space B clearly induces a representation on $\mathcal{H}_{as}$; we will denote the operators of this representation by $U_{as}(g)$. *The asymptotic Hamiltonian $\mathcal{H}_{as}$ and the asymptotic momentum operator $\mathbf{P}_{as}$ are naturally defined by the relation*

$$\exp(-i(H_{as}a_0 - \mathbf{P}_{as}\mathbf{a})) = U_{as}(1, a),$$

where $(1, a)$ is a translation.

It is easy to see that the asymptotic Hamiltonian and the asymptotic momentum operator can be expressed in terms of the operators $a^+(\mathbf{k}, s), a(\mathbf{k}, s)$ by the formulas

$$\mathcal{H}_{as} = \sum_i \sum_{s \in N_i} \int \sqrt{\mathbf{k}^2 + \mu_i^2}\, a^+(\mathbf{k}, s)a(\mathbf{k}, s)d\mathbf{k},$$

$$\mathbf{P}_{as} = \sum_{s \in N} \int \mathbf{k}a^+(\mathbf{k}, s)a(\mathbf{k}, s)d\mathbf{k}.$$

We will now formulate an additional axiom which will allow us to construct a scattering theory in this case (*the strong spectral condition*).

A4. The joint spectrum of the operators $H, \mathbf{P}$ on the subspace C of the space $\mathcal{H}$, containing all the irreducible invariant subspaces of the operators $U(g)$, does not intersect the joint spectrum of these operators on the orthogonal complement of the space C. The spectrum of the energy operator H contains a gap (i.e. there exists an $\epsilon > 0$, such that the single point of the spectrum of the operator H on the ray $(-\infty, \epsilon)$ is the point 0, corresponding to the physical vacuum).

It is clear that the space C equals the direct sum of the subspace A and the single-particle subspace B, therefore, the joint spectrum of the operators $H, \mathbf{P}$ on the space C is identical to the union of the *vacuum spectrum* (i.e. the spectrum of the operators $H, \mathbf{P}$ on the space A) and the *single-particle spectrum* (i.e. the spectrum of the operators $H, \mathbf{P}$ on the space B). It follows from the statements above that the vacuum spectrum consists of the single point 0 and the single-particle spectrum can be written as the union of several sets U_{μ_i}, specified by the relation $\omega^2 - \mathbf{k}^2 = \mu_i^2, \omega \geq 0$.

The orthogonal complement to the space $C = A + B$, referred to as the multi-particle space and denoted by M, the joint spectrum of $H, \mathbf{P}$ on M will be called the *multi-particle spectrum*.

By using these definitions, we can reformulate axiom A4.

A4$'$. The vacuum spectrum, single-particle spectrum, and the multi-particle spectrum are pairwise non-intersecting.

Thus, the spectrum of the operators $H, \mathbf{P}$ on the space $\mathcal{H}$ is structured as follows. It contains the point 0, corresponding to the vector Φ, and several sets U_{μ_i} with $\mu_i > 0$, constituting the single-particle spectrum, and the multi-particle spectrum, which does not contain the point 0 and does not intersect any of sets U_{μ_i}.

Let us now formulate the following statement. *If the axioms 1–4 are satisfied, then the Møller matrices $S_\pm$ and the scattering matrix S can be defined as in Section 10.1; using the ideas of Section 10.1, one can prove the existence of the Møller matrices and the scattering matrix. The Møller matrices and scattering matrix defined in this manner are Lorentz-invariant* (i.e. they satisfy the equations

$$U(g)S_\pm = S_\pm U_{\mathrm{as}}(g), \tag{12.6}$$

$$U_{\mathrm{as}}(g)S = S U_{\mathrm{as}}(g) \tag{12.7}$$

for any transformation $g \in \mathcal{P}$).

Indeed, the definitions and discussions in Section 10.1 can be applied to the case at hand. (The conditions in Section 10.1 are not necessarily filled, since the definition of the single-particle space in Section 10.1 is different from the definition in this section; similarly, the definitions of the space $\mathcal{H}_{\text{as}}$ differ. However, in light of the fact that the single-particle space B defined in this section is contained in the single-particle space in the sense of Section 10.1, the discussions in Chapter 10 do not need to be modified in this case.) The Lorentz invariance of the Møller matrices and the scattering matrix can be proved by the discussions in Section 6, Chapter 6 in Jost (1965); we will sketch another proof in Chapter 13.

Let us note that the results of this section hold if the asymptotically commutative family of operators is replaced by an asymptotically commutative family of operator generalized functions (in the sense of Section 10.5). The domain of definition of these operator generalized functions should be invariant with respect to the operators $U(g)$, where $g \in \mathcal{P}$. Conditions A2 and A3 require corresponding modifications (we ought to require that the vector Φ is cyclic vector for this family and that for every generalized operator function $A(\mathbf{x}, t)$, the corresponding generalized operator functions $A_g(\mathbf{x}, t) = U_g A(\mathbf{x}, t) U_g^{-1}$, where g is an arbitrary element of the group $\mathcal{P}$, should also belong to the family).

12.2   Axiomatics of local quantum field theory

To formulate the axioms of Lorentz-invariant theory, we added extra conditions to the axiomatic scattering theory constructed in Chapter 10. These extra conditions guarantee the Lorentz invariance of the scattering matrix.

Here, we will provide a different set of axiomatics introduced by Haag and Araki.

Let us fix a representation U_g of the Poincaré group $\mathcal{P}$ on the Hilbert space $\mathcal{H}$, and, as in the previous section, let us require the spectral condition (axiom A1).

Let us assume that for every bounded set $\mathcal{O} \subset E^4$, there is a collection $R(\mathcal{O})$ of bounded operators on the space $\mathcal{H}$. We assume that this collection is closed under addition and multiplication of operators, multiplication by a scalar, involution $A \to A^*$, and weak limits (such a collection of operators is called a W^*-algebra).

Let us assume that the algebra $R(\mathcal{O})$ has the following properties:

R1. If $\mathcal{O}_1 \subset \mathcal{O}_2$, then $R(\mathcal{O}_1) \subset R(\mathcal{O}_2)$.

R2. *Covariance with respect to the Poincaré group:*

$$U_g R(\mathcal{O}) U_g^{-1} = R(g\mathcal{O}).$$

R3. *Local commutativity:* If any point of the set $\mathcal{O}_1$ is separated by a space-like interval from any point of the set $\mathcal{O}_2$, then $R(\mathcal{O}_1)$ and $R(\mathcal{O}_2)$ commute with each other.

R4. *Cyclicity:* The physical vacuum Φ is cyclic with respect to the union $\mathcal{R}$ of the algebras $R(\mathcal{O})$.

It is easy to see that the union $\mathcal{R}$ of the algebras $R(\mathcal{O})$ is an asymptotically commutative algebra. Indeed, let $A \in R(\mathcal{O}_1) \subset \mathcal{R}$, $B \in R(\mathcal{O}_2) \subset \mathcal{R}$. Then, by condition R2, we have

$$A(\mathbf{x}, t) = \exp(iHt - i\mathbf{P}\mathbf{x}) A \exp(-iHt + i\mathbf{P}\mathbf{x}) \in R(\mathcal{O}_1 - x),$$

where $x = (\mathbf{x}, t)$. Since the sets $\mathcal{O}_1$ and $\mathcal{O}_2$ are bounded, there exists an a such that for $|\mathbf{x}|^2 - |t|^2 \geq a^2$, any point of the set $\mathcal{O}_1 - x$ is separated from any point of the set $\mathcal{O}_2$ by a space-like interval. Using condition R3, we obtain that for $|\mathbf{x}|^2 - |t|^2 \geq a^2$,

$$[A(\mathbf{x}, t), B] = 0.$$

On the other hand, $\|A(\mathbf{x}, t)\| = \|A\|$, and therefore,

$$\|[A(\mathbf{x}, t), B]\| \leq 2\|A\| \cdot \|B\|.$$

Combining these relations, we see that

$$\|[A(\mathbf{x}, t), B]\| \leq C \frac{1 + |t|^n}{1 + |\mathbf{x}|^n}$$

for any n (to show this, it is enough to note that

$$\frac{1 + |t|^n}{1 + |\mathbf{x}|^n} > c > 0 \text{ for } |\mathbf{x}|^2 - |t|^2 \geq a^2 \Bigg).$$

Furthermore, the algebra $\mathcal{R}$ is clearly invariant with respect to Lorentz transformations.

Therefore, the algebra $\mathcal{R}$ and the representation U_g satisfy all conditions of Section 12.1, except for axiom A4.

If we require that axiom A4 holds, then the considerations provided in Section 12.1 allow us to prove that *there exists a Lorentz-invariant scattering matrix.*

Another important axiomatics is due to Wightman. Again, we require that the Poincaré group is represented in Hilbert space $\mathcal{H}$ and the spectral condition A1 is satisfied.

We assume that an operator generalized function $\phi(x)$ is defined on the space $\mathcal{H}$ (here, $x \in E^4$). More precisely, we assume that for any function $f \in \mathcal{S}(E^4)$, there is an operator $\phi(f)$; all the operators $\phi(f)$ are defined on the same linear set D and transform D into itself; the operator $\phi(f)$ must depend linearly on f and for any vectors $\xi, \eta \in D$, the functional $\langle \phi(f)\xi, \eta \rangle$ must continuously depend on f in the topology of the space $\mathcal{S}$ (i.e. it must define a generalized function from the space $\mathcal{S}'(E^4)$).

As before, we will write $\phi(f) = \int \phi(x)f(x)dx$; the operator generalized function $\phi(x)$ will be referred to as a quantum field.

Let us assume that the quantum field $\phi(x)$ is Hermitian (i.e. the operators $\phi(f)$ and $\phi(\bar{f})$ are Hermitian conjugate).

We say that the representation $U(g)$ of the Poincaré group $\mathcal{P}$ and the operator generalized function $\phi(x)$ (the quantum field) satisfy the Wightman axioms,[1] if the following conditions are satisfied:

W1. *Transformation law of quantum fields*: The operators U_g keep the set D fixed: $U_g D = D$. The quantum field $\phi(x)$ transforms according to the law

$$U(g)\phi(x)U^{-1}(g) = \phi(gx). \tag{12.8}$$

More precisely, equation (12.8) implies that

$$U(g)\phi(f)U^{-1}(g) = \phi(f_g),$$

where

$$f_g(x) = f(g^{-1}x).$$

---

[1]Here, we have formulated the Wightman axioms in the case of a scalar field; for the general case, see, for example, Bogolyubov *et al.* (1969).

W2. *Locality (local commutativity)*: If the points $x, y \in E^4$ are separated by a space-like interval, then $\phi(x)$ commutes with $\phi(y)$.

More precisely, we require that the operators $\phi(f)$ and $\phi(g)$ commute in the case when the support of the functions f and g are separated by a space-like interval.

W3. *Cyclicity*: The vector Φ (the ground state of the energy operator) is a cyclic vector of the family of operators $\phi(f)$.

The Wightman functions $w_m(x_1, \ldots, x_m)$ will be defined by the relation

$$w_m(x_1, \ldots, x_m) = \langle \phi(x_1) \cdots \phi(x_m) \Phi, \Phi \rangle .$$

By the kernel theorem (see Appendix A.7), the function w_n can be viewed as a generalized function from $S'(E^{4m})$.

Using the considerations of Section 8.2, one can prove the reconstruction theorem, which states that knowing the Wightman functions w_m, we can construct a Hilbert space $\mathcal{H}$, a Lorentz group representation, and a quantum field $\phi(x)$. It is easy to work out the conditions on the functions $w_m(x_1, \ldots, x_m)$ which guarantee that these functions are Wightman functions for a quantum field, satisfying the Wightman axioms (these conditions are analogous to conditions 1–6 from Section 8.2).[2]

Let us now show that by adding axiom A4 from Section 12.1 to the Wightman axioms, *we can provide a definition of Møller matrices (hence of the scattering matrix) and prove their existence and Lorentz invariance*. To do this, it is sufficient to consider a family of operator generalized functions $\phi(gx)$, where g runs over a Poincaré group $\mathcal{P}$, and note that this family is asymptotically commutative in the sense of Section 10.5 and it satisfies conditions A1–A4 from Section 12.1. (Axiom A3 follows immediately from axioms W1, axiom A2 from axiom W3, condition D5 from Section 10.5 for $\delta < 1$ follows from local commutativity.)

---

[2]For more details, see Streater and Wightman (2016), Jost (1965), and Bogolyubov *et al.* (1969).

Thus, we can see that all the statements which can be obtained from the axioms in Section 12.1 can be derived from Wightman's axioms. However, there are theorems which can be proved in Wightman's axiomatics that have not been proven (and, apparently, cannot be proven) in the axiomatic formulation in Section 12.1. (This is not surprising; Wightman axiomatized local quantum field theory; the axioms of Section 12.1 are based on weaker assumption of asymptotic commutativity).

The most important statement of this type are dispersion relations obtained first in Bogolyubov *et al.* (1956) and derived from Wightman axioms in Hepp and Epstein (1971).

Let us also mention the CPT theorem, stating that the Wightman axioms imply CPT invariance (i.e. the existence of additional symmetries).

Let us formulate Borchers' theorem:

If two quantum fields $\phi_1(x), \phi_2(y)$ on a Hilbert space satisfy the Wightman axioms and are mutually local (i.e. $[\phi_1(x), \phi_2(y)] = 0$ for any points x, y separated by a space-like interval), then they define the same scattering matrix.

The CPT theorem can be used in the proof of this theorem. However, this theorem has another proof: it follows from the results of Section 10.1 (from the theorem that two asymptotically commutative algebras define the same scattering matrix).

Finally, let us analyze the connection between the Wightman axioms and the Haag–Araki axioms. Let us assume that the operators $\phi(f) = \int f(x)\phi(x)dx$, entering in the Wightman axioms, are not only Hermitian but also essentially self-adjoint for any real smooth function f with compact support. For every such function f, one can construct a family $\mathcal{T}_f$ of bounded operators, arising as functions of the operator $\phi(f)$ (i.e. having the form $\alpha(\phi(f))$, where α is a bounded function). Let us use the symbol $R(\mathcal{O})$ for the smallest W^*-algebra, containing all families $\mathcal{T}_f$, where f is a smooth real function with support in the set $\mathcal{O}$. It is easy to check that the algebras $R(\mathcal{O})$ satisfy the Haag–Araki axioms.

12.3 The problem of constructing a non-trivial example

It is natural to try to construct a Lorentz-invariant quantum theory by means of the quantization of Lorentz-invariant classical theory. Let us recall that in Section 8.3, we considered the question of quantizing a classical theory corresponding to the Hamiltonian

$$\mathcal{H}(\pi, \phi) = \frac{1}{2} \int \pi^2(\mathbf{x}) d\mathbf{x} + V(\phi).$$

It is easy to see that such a classical system is Lorentz invariant if

$$\mathcal{H}(\pi, \phi) = \frac{1}{2} \int \pi^2(\mathbf{x}) d\mathbf{x} - \frac{1}{2} \int \phi(\mathbf{x}) \Delta\phi(\mathbf{x}) d\mathbf{x} + \int U(\phi(\mathbf{x})) d\mathbf{x}. \quad (12.9)$$

Indeed, the Hamiltonian equations for the system, described by the functional Hamiltonian (12.9), have the form

$$\frac{\partial \phi(\mathbf{x}, t)}{\partial t} = \pi(\mathbf{x}, t),$$

$$\frac{\partial \pi(\mathbf{x}, t)}{\partial t} = \Delta\phi(\mathbf{x}, t) - F(\phi(\mathbf{x}, t)),$$

where

$$F(\phi) = \frac{\partial U(\phi)}{\partial \phi}.$$

Thus, the function $\phi(x) = \phi(\mathbf{x}, t)$ satisfies the equation

$$\Box\phi(x) + F(\phi(x)) = 0,$$

which is clearly Lorentz invariant (the equation does not change form when replacing $\phi(x)$ by $\phi(\Lambda x)$, where Λ is a Lorentz transformation).

We can also prove the Lorentz invariance of classical systems, described by the Hamiltonian (12.9), by considering the action functional of such a system. This functional can be presented in the form

$$S = \int \left(\left(\frac{\partial \phi}{\partial t} \right)^2 - (\nabla \phi)^2 \right) dx - \int U(\phi(x)) dx,$$

which is invariant under Lorentz transformations.

Let us now consider the simplest system with the Hamiltonian of the form (12.9), a system with

$$U(\phi(x)) = \frac{1}{2}m^2\phi^2(\mathbf{x}).\tag{12.10}$$

After quantizing this system, we obtain a quantum Hamiltonian

$$H = \frac{1}{2}\int \hat{\pi}^2(\mathbf{x})d\mathbf{x} - \frac{1}{2}\int \hat{\phi}(\mathbf{x})\Delta\hat{\phi}(\mathbf{x})d\mathbf{x} + \frac{1}{2}m^2\int \hat{\phi}^2(\mathbf{x})d\mathbf{x},\tag{12.11}$$

where $\hat{\pi}(\mathbf{x}), \hat{\phi}(\mathbf{x})$ are symbols satisfying the relations

$$\hat{\pi}^+(\mathbf{x}) = \hat{\pi}(\mathbf{x}), \ \hat{\phi}^+(\mathbf{x}) = \phi(\mathbf{x}),$$

$$[\hat{\pi}(\mathbf{x}), \hat{\pi}(\mathbf{y})] = [\hat{\phi}(\mathbf{x}), \hat{\phi}(\mathbf{y})] = 0,$$

$$[\hat{\pi}(\mathbf{x}), \hat{\phi}(\mathbf{y})] = \frac{1}{i}\delta(\mathbf{x} - \mathbf{y}).$$

The Hamiltonian (12.11) can be written in the form (8.26), where $\nu(\mathbf{x}) = -\Delta\delta(\mathbf{x}) + m^2\delta(\mathbf{x})$. An operator realization of Hamiltonians of the form (8.26) (free Hamiltonians) is described in Section 8.3. Using the results of Section 8.3, we can prove that the operator realization of the Hamiltonian (12.11) can be constructed on the Fock space $\mathcal{H} = F(L^2(E^3))$; as the energy operators and the momentum operators, we can select the operators

$$\hat{H} = \int \omega(\mathbf{k})a^+(\mathbf{k})a(\mathbf{k})d\mathbf{k},$$

$$\hat{\mathbf{P}} = \int \mathbf{k}a^+(\mathbf{k})a(\mathbf{k})d\mathbf{k},$$

where $\omega(\mathbf{k}) = \sqrt{\mathbf{k}^2 + m^2}$; the operators (the operator generalized functions) $\hat{\phi}(\mathbf{x}, t)$ are defined by the formula

$$\hat{\phi}(\mathbf{x}, t) = (2\pi)^{-3/2}\int (a^+(\mathbf{k})\exp(i\omega(\mathbf{k})t - i\mathbf{k}\mathbf{x})$$

$$+ a(\mathbf{k})\exp(-i\omega(\mathbf{k})t + i\mathbf{k}\mathbf{x}))\frac{d\mathbf{k}}{\sqrt{2\omega(\mathbf{k})}}.\tag{12.12}$$

The ground state Φ coincides with the Fock vacuum θ.

The operators $\hat{\phi}(x) = \hat{\phi}(\mathbf{x}, t)$ satisfy the Lorentz-invariant equation

$$\Box\hat{\phi}(x) + m^2\hat{\phi}(x) = 0.$$

It is therefore natural to expect that the quantum system is Lorentz invariant. Actually, the representation of the translation group on the space $F(L^2(E^3))$, given rise by the operators H and $\mathbf{P}$, can be extended to a unitary representation of Poincaré group in such a way that

$$U(g)\hat{\phi}(x)U^{-1}(g) = \hat{\phi}(gx)$$

(here, g is a Lorentz transformation and $U(g)$ is the corresponding unitary operator). The necessary Lorentz group representation can be obtained by considering on the space $L^2(E^3)$ the representation of the type $(m, 1)$, described in Appendix A.9. (We should note that the representation on the space $L^2(E^3)$ specifies a representation on the space $F(L^2(E^3))$; we have already used this fact in Section 12.1).

It is easy to check that the operator generalized function $\hat{\phi}(\mathbf{x}, t)$ defined by the formula (12.12) and the representation of the Poincaré group satisfy all the Wightman axioms.

The given quantum system is called a free scalar field with mass m. Corresponding particles are non-interacting bosons with spin 0. The scattering matrix is trivial as it should be for non-interacting particles.

To obtain a non-trivial Lorentz-invariant scattering matrix, we can try to quantize a classical theory with the Hamiltonian of the form (12.9), where the function $U(\phi)$ is not quadratic. Unfortunately, this approach leads to problems related to the fact that in every step one encounters singular expressions which cannot be easily given a precise meaning. For example, the expressions of the form $\hat{\phi}^n(x)$, where $n \geq 2$, and similarly more general expressions $\nu(\hat{\phi}(x))$, where $\nu(\phi)$ is a nonlinear function, are not well defined; therefore, the equation

$$\Box\hat{\phi}(x) + F(\hat{\phi}(x)) = 0,$$

for the field operators $\hat{\phi}(x)$ does not have precise meaning.

Let us suppose that the Hamiltonian (12.9) can be presented in the form $\mathcal{H} = \mathcal{H}_0 + V$, where $\mathcal{H}_0$ is a Hamiltonian of the form (12.10), $V = \int \nu(\phi(\mathbf{x}))d\mathbf{x}$. After quantization, the resulting formal Hamiltonian can be written in the form

$$H = H_0 + V = \frac{1}{2} \int \hat{\pi}^2(\mathbf{x})d\mathbf{x} - \frac{1}{2} \int \hat{\phi}(\mathbf{x})\Delta\hat{\phi}(\mathbf{x})d\mathbf{x}$$

$$+ \frac{1}{2}m^2 \int \hat{\phi}^2(\mathbf{x})d\mathbf{x} + \int \nu(\hat{\phi}(\mathbf{x}))d\mathbf{x}. \qquad (12.13)$$

If we try to express Hamiltonian in terms of the symbols $a^+(\mathbf{k}), a(\mathbf{k})$, as in Section 8.3, and bring the resulting expression to normal form by means of CCR, then except for the standard infinite constants, which we have previously thrown away, we will obtain other infinite summands. Therefore, one often considers the formal Hamiltonian

$$H = H_0 + N \int \nu(\hat{\phi}(\mathbf{x}))d\mathbf{x}, \qquad (12.14)$$

where N is the normal product symbol (i.e. we express $\int \nu(\hat{\phi}(\mathbf{x}))d\mathbf{x}$ in terms of $a^+(\mathbf{x}), a(\mathbf{k})$ by the formulas (8.27), and then reorder the symbols in normal order, assuming they commute). Replacing the formal Hamiltonian (12.13) by the formal Hamiltonian (12.14), we can get rid of some infinities, but other infinities remain. In particular, calculating the Green functions of the Hamiltonian (12.14) in the framework of perturbation theory, we encounter diverging integrals.[3]

Unfortunately, even now, a consistent method of overcoming these challenges in the case of arbitrary function $\nu(\phi)$ or even in the case when the function $\nu(\phi)$ is a polynomial does not exist. However, if the function $\nu(\phi)$ is a polynomial of order ≤ 4 (i.e. $\nu(\phi) = a\phi^3 + b\phi^4$), one can overcome these problems in the framework of perturbation theory. There exists a method that allows us to construct a family of Lorentz-invariant scattering matrices that correspond to the family of Hamiltonians of the form $H_0 + N \int \nu(\hat{\phi}(\mathbf{x}))d\mathbf{x}$. This method was introduced by Feynman; it is called covariant renormalization.

---

[3]We should note that in the one-dimensional case (i.e. in the case when in the Hamiltonian (12.14) integration is performed over a single variable) calculations of Green functions of the Hamiltonian (12.14) do not give rise to diverging integrals.

Let us consider for the sake of concreteness the case when $\nu(\phi) = b\phi^4$. We consider the family of Hamiltonians of the form

$$H = H_0(w) + bV_\Lambda$$
$$= \frac{1}{2}\int \hat{\pi}^2(\mathbf{x})d\mathbf{x} + \frac{1}{2}\int w(\mathbf{x} - \mathbf{y})\hat{\phi}(\mathbf{x})\hat{\phi}(\mathbf{y})d\mathbf{x}d\mathbf{y}$$
$$+ b\int f_\Lambda(\mathbf{x}_1 - \mathbf{x}_4, \mathbf{x}_2 - \mathbf{x}_4, \mathbf{x}_3 - \mathbf{x}_4)$$
$$+ \hat{\phi}(\mathbf{x}_1)\hat{\phi}(\mathbf{x}_2)\hat{\phi}(\mathbf{x}_3)\hat{\phi}(\mathbf{x}_4)d^4\mathbf{x}.$$

Here, $w(\mathbf{x})$ is an arbitrary function, $f_\Lambda(\mathbf{x}_1, \mathbf{x}_2, \mathbf{x}_3) = \Lambda^3 f(\Lambda\mathbf{x}_1, \Lambda\mathbf{x}_2, \Lambda\mathbf{x}_3)$, where the function $f \in \mathcal{S}$ and satisfies the condition $\int f(\mathbf{x}_1, \mathbf{x}_2, \mathbf{x}_3) d^3\mathbf{x} = 1$. It is easy to see that $\lim_{\Lambda\to\infty} f_\Lambda(\mathbf{x}_1, \mathbf{x}_2, \mathbf{x}_3) = \delta(\mathbf{x}_1)\delta(\mathbf{x}_2)\delta(\mathbf{x}_3)$ and therefore in the limit as $\Lambda \to \infty$, the interaction V_Λ becomes $V = \int \phi^4(\mathbf{x})d\mathbf{x}$; the replacement of V by V_Λ is called ultraviolet cutoff. (Defining V_Λ, we can use the normal product in place of the regular product; the class of Hamiltonians under consideration does not change in this case.) Let us select the function $w(\mathbf{x})$ from the condition that energy of the particles, defined by the Hamiltonian $H_0(w) + bV_\Lambda$, equals m for $\mathbf{k} = 0$; we will denote the resulting Hamiltonian $H(m, b, \Lambda)$ (in the framework of perturbation theory, the function $w(\mathbf{x})$ is easy to find). The transition to the Hamiltonian $H(m, b, \Lambda)$ is called mass renormalization (the number m corresponds to the mass of the particles defined by the Hamiltonian). Let us denote the scattering matrix corresponding to the Hamiltonian $H(m, b, \Lambda)$ by $S(m, b, \Lambda)$. In order to obtain a Lorentz-invariant theory, we must take the limit $\Lambda \to \infty$ (remove the ultraviolet cutoff). However, taking the limit $\Lambda \to \infty$, we obtain divergent expressions in the perturbation series.

In order to obtain a non-trivial Lorentz-invariant scattering matrix in the framework of perturbation theory, we should assume that the quantity b (bare charge) changes in the process of removing the ultraviolet cutoff. In the perturbation theory, one can prove the following statement: there exists a function $b(m, \Lambda, g)$ such that the scattering matrix $S(m, b(m, \Lambda, g), \Lambda)$ has a finite, non-trivial, Lorentz-invariant limit $S(m, g)$ as $\Lambda \to \infty$. (In order to find the function $b(m, \Lambda, g)$, we suppose that in the process of removing the cutoff we leave a physical quantity g fixed and express the bare charge in terms of the cutoff parameter and this quantity. We can take as g any matrix element of the scattering matrix; for some particular choice, it is called physical charge or dressed charge.) The result (which we have described in a non-standard form) is the theory of renormalization. The Green functions $G_n(x_1, \ldots, x_n | m, \Lambda, g)$ of the Hamiltonian $H(m, b(m, \Lambda, g), \Lambda)$ do not have a finite limit as $\Lambda \to \infty$, however, we can find a function $Z(m, \Lambda, g)$ such that the functions $Z^{-n/2}(m, \Lambda, g) G_n(x_1, \ldots, x_n | m, \Lambda, g)$ have a finite Lorentz-invariant limit as $\Lambda \to \infty$.

The statements above suggest the following way of constructing objects satisfying the Wightman axioms and giving rise to non-trivial scattering matrices. First, we must find the functions $b(m, \Lambda, g)$ and $Z(m, \Lambda, g)$, such that a non-trivial finite limit

$$\lim_{\Lambda \to \infty} Z^{-n/2}(m, \Lambda, g) w_n(x_1, \ldots, x_n | m, \Lambda, g) = w_n(x_1, \ldots, x_n | m, g)$$

exists. Here, $w_n(x_1, \ldots, x_n | m, g)$ is a Wightman function of the Hamiltonian $H(m, b(m, \Lambda, g), \Lambda)$. One can hope that the functions $w(x_1, \ldots, x_n | m, g)$ defined this way satisfy the conditions necessary for constructing objects obeying the Wightman axioms and that the corresponding scattering matrix is non-trivial. Unfortunately, at present, this hypothesis is unproven. Moreover, in four-dimensional space–time, an example of objects satisfying the Wightman functions and giving rise to a non-trivial scattering matrix is not known.

Clearly, if we can construct a non-trivial scattering matrix in Wightman's axioms, then we will obtain a non-trivial example of a scattering matrix in the axioms of Section 12.1. However, the other direction does not hold — a non-trivial example in the axioms of Section 12.1 can be easier to construct.

The situation in two-dimensional and three-dimensional space–time is much better. A new branch of mathematical physics — constructive field theory — was born from the attempts to find non-trivial examples and in two and three dimensions, it was successful. Two-dimensional conformal field theories were thoroughly analyzed in numerous papers starting with the seminal paper by Belavin *et al.* (1984). Another way to construct two-dimensional theories comes from integrable models.

In all dimensions, the considerations based on supersymmetry and superstring theory led to much deeper understanding of quantum field theories.

We are moving forward!

Chapter 13

Methods of Quantum Field Theory in Statistical Physics

13.1 Quantum statistical mechanics

The equilibrium state in (classical or quantum) statistical mechanics can be defined as a state maximizing the entropy under some given conditions.

In quantum mechanics, we represent a state by the density matrix K; the entropy of this state is defined by the formula $S = -\operatorname{Tr} K \log K$. Denoting the Hamiltonian by $\hat{H}$, we can say that the equilibrium state maximizes the entropy for a given mean energy $E = \operatorname{Tr} \hat{H} K$. (More precisely, we can say that this is an equilibrium state of the canonical ensemble or a Gibbs state.) Let us check that this state has the form

$$\frac{e^{-\beta \hat{H}}}{Z},\qquad(13.1)$$

where β can be interpreted as the inverse temperature $\beta = \frac{1}{T}$ and $Z = \operatorname{Tr} e^{-\beta \hat{H}}$ is called the partition function or the statistical sum. If the spectrum of $\hat{H}$ is discrete, then $Z = \sum e^{-\beta E_i}$ where E_i are the energy levels. In the case when the ground state is non-degenerate, we see that the equilibrium state tends to the ground state and its entropy tends to zero when $T \to 0$.

Note that (13.1) makes sense only if Z is finite (the operator $e^{-\beta \hat{H}}$ belongs to the trace class).

To prove (13.1), we use the method of Lagrange multipliers, reducing our problem to finding the stationary points of the functional

$$L = -\operatorname{Tr} K \log K - \beta \operatorname{Tr}(\hat{H}K - E) - \zeta(\operatorname{Tr} K - 1).$$

Calculating the variation of L, we obtain

$$\delta L = -\operatorname{Tr} \log K \delta K - \operatorname{Tr} \delta K - \beta \operatorname{Tr} \hat{H} \delta K - \zeta \operatorname{Tr} \delta K,$$

hence at the stationary point,

$$-\log K - 1 - \beta \hat{H} - \zeta = 0.$$

This is equivalent to (13.1) with $Z = e^{1+\zeta}$.

Note that in the calculation of variation, we have used the formula

$$\delta \operatorname{Tr} \phi(K) = \operatorname{Tr} \phi'(K)\delta K, \tag{13.2}$$

where ϕ' stands for the derivative of the function ϕ. It is easy to check this formula in the case when $\phi(K) = K^n$; to treat the general case, we can represent ϕ as a limit of polynomials.

The partition function Z does not have any physical meaning, but there are many physical quantities that can be conveniently expressed in terms of Z. For example, differentiating the definition of Z with respect to β, we obtain

$$E = \bar{H} = -\frac{\partial \log Z}{\partial \beta}.$$

The expression for the entropy in equilibrium state is

$$S = \beta E + \log Z.$$

Introducing the notion of free energy by the formula $F = E - TS$, we obtain

$$F = -T \log Z.$$

If the Hamiltonian depends on a parameter λ and $\hat{H}(\lambda) = \hat{H} + \lambda A + \cdots$ then using (13.2), we can write

$$Z(\lambda) = Z + (-\beta)\lambda \operatorname{Tr} A e^{-\beta \hat{H}} + \cdots .$$

Here, $\cdots$ stands for higher order terms with respect to λ. Note that

$$\bar{A} = \frac{\operatorname{Tr} A e^{-\beta \hat{H}}}{Z}$$

is the mean value (expectation value) of the observable A in the equilibrium state. We will also use the notation $\langle A \rangle$ for the expectation value. The expectation values $\langle A_1 \cdots A_n \rangle$ or, more generally, $\langle A_1(t_1) \cdots A_n(t_n) \rangle$ are called correlation functions (here, A_i are observables and $A_i(t)$ are the corresponding Heisenberg operators). If we want to emphasize that the correlation functions are calculated for the equilibrium state with the temperature $T = \frac{1}{\beta}$, we will use the notation $\langle A_1(t_1) \cdots A_n(t_n) \rangle_\beta$.

If A and B are two observables, then

$$\langle A(t) B \rangle_\beta = \langle B A(t + i\beta) \rangle_\beta. \tag{13.3}$$

This equation, called the Kubo–Martin–Schwinger (KMS) condition, can be proved by simple formal manipulations if the Hilbert space $\mathcal{H}$ is finite-dimensional. In the infinite-dimensional case, one should check that the correlation function $\langle B A(t) \rangle_\beta$ can be extended analytically to the strip $0 \leq \Im t \leq \beta$ (here, $\Im$ stands for imaginary part).

It is easy to see that

$$\bar{A} = -T \frac{\partial \log Z}{\partial \lambda} = \frac{\partial F}{\partial \lambda} \tag{13.4}$$

(the derivatives are calculated at the point $\lambda = 0$).

If the Hamiltonian depends linearly on a set of parameters $\lambda_1, \ldots, \lambda_k$, we can calculate the correlation functions by differentiating the free energy F. For example, if $\hat{H} = \hat{H}_0 + \lambda_1 A_1 + \cdots + \lambda_k A_k$, we have

$$\frac{\partial^2 F}{\partial \lambda_i \partial \lambda_j} = \langle A_i A_j \rangle - \langle A_i \rangle \langle A_j \rangle. \tag{13.5}$$

(The derivatives are calculated at the point $\lambda_i = 0$.) The RHS of (13.5) is called a truncated correlation function. Higher truncated correlation functions can be defined as higher derivatives of F. (See Section 10.1 for the definition of truncated correlation functions in terms of correlation functions.)

If we have several integrals of motion, we can consider the equilibrium state fixing the expectation values of all these integrals and looking for the maximal value of entropy under these conditions. For example, we can fix the expectation values of energy and of the number of particles $\hat{N}$. Then we can repeat the above consideration by adding one more Lagrange multiplier. We obtain the following formula for the equilibrium state:

$$\frac{e^{-\beta(\hat{H}-\mu\hat{N})}}{Z}. \tag{13.6}$$

Note that we have denoted the new Lagrange multiplier as $-\beta\mu$ to agree with standard notation (μ can be identified with chemical potential).

13.1.1 *Examples*

Let us consider the multidimensional harmonic oscillator as an example. Introducing the creation and annihilation operators obeying CCR, we can represent the Hamiltonian in the form $\hat{H} = \sum_i \omega_i \hat{a}_i^+ \hat{a}_i + C$, where $C = \sum \omega_i/2$ is the energy of the ground state. We shift the energy scale assuming that the ground state has zero energy: $C = 0$. Then the energy levels are $\sum n_i \omega_i$ where n_i is a non-negative integer. It is easy to calculate the partition function

$$Z = \Pi \frac{1}{1 - e^{-\beta\omega_i}},$$

the free energy

$$F = \frac{1}{\beta} \sum \log(1 - e^{-\beta\omega_i}),$$

and the mean value of energy

$$\bar{H} = \sum \omega_i \bar{n}_i,$$

where $\bar{n}_i = \frac{1}{e^{\beta\omega_i} - 1}$.

The same formulas work for non-interacting identical bosons. In the case of non-interacting identical fermions, we have very similar formulas. The creation and annihilation operators, in this case, obey CAR and the energy levels are given by the formula $\sum \omega_i n_i$ where

$n_i = 0, 1$. The formulas for the partition function, free energy, and mean energy are given by

$$Z = \Pi(1 + e^{-\beta\omega_i}),$$

$$F = -\frac{1}{\beta} \sum \log(1 + e^{-\beta\omega_i}),$$

$$\bar{H} = \sum \omega_i \bar{n}_i,$$

where $\bar{n}_i = \frac{1}{e^{\beta\omega_i}+1}$.

13.2 Equilibrium states of translation-invariant Hamiltonians

In momentum representation, one can write a translation-invariant Hamiltonian in the form

$$H = \sum_{m,n} \int H_{m,n}(k_1, \ldots, k_m \,|\, l_1, \ldots, l_n)$$

$$\times a^+(k_1) \cdots a^+(k_m)a(l_1) \cdots a(l_n)d^m k d^n l, \qquad (13.7)$$

where $H_{m,n}(k_1, \ldots, k_m \,|\, l_1, \ldots, l_n)$ contains a factor $\delta(k_1 + \cdots + k_m - l_1 - \cdots - l_n)$. (We assume that the arguments belong to $\mathbb{R}^D$.)

Even in the simplest case, when $\hat{H} = \int \omega(p)a^+(p)a(p)dp$, the formula for the equilibrium state makes no sense (the integral for the partition function diverges). Moreover, as we have seen, the formula (13.7) in general does not define an operator on Fock space. However, the Hamiltonian with volume cutoff $\hat{H}_\Omega$ specifies an operator in Fock space under some mild conditions and we can define an equilibrium state and correlation functions. In particular, we can consider the correlation functions $^\beta w_n^\Omega(k_1, \epsilon_1, t_1, \ldots, k_n, \epsilon_n, t_n) = \langle a_{k_1}^{\epsilon_1}(t_1) \cdots a_{k_n}^{\epsilon_n}(t_n) \rangle_\beta$ generalizing Wightman functions. (Wightman functions are correlation functions for $T = 0$.)

We can define correlation functions in infinite volume $^\beta w_n = \lim {}^\beta w_n^\Omega$ by taking the limit $\Omega \to \infty$ (removing the volume cutoff). As in Section 8.2, the limit is understood in the sense of generalized functions. Under some mild conditions, one can prove that the correlation functions in infinite volume obey KMS condition (13.3) (one should verify that these functions can be analytically extended to the strip $0 \leq \Im t \leq \beta$).

Note that the limit at hand does not always exist, or, better, there exist many limits, i.e. many equilibrium states in infinite volume. The limiting procedure we have used can be slightly modified in many ways: we can use various volume cutoffs or add to the Hamiltonian $\hat{H}^{\Omega}$ terms that tend to zero as Ω tends to ∞. If the limit exists, we say that it gives correlation functions in the equilibrium state (more precisely, in one of equilibrium states.) We can repeat with minimal changes the considerations of Sections 8.1 and 8.2. Again, as in Section 8.2, we can construct for every β, a Hilbert space $\mathcal{H} = \mathcal{H}(\beta)$, a dense subset D, a vector Φ, operators $\hat{H}, \hat{\mathbf{P}}$, and operator generalized functions $a^{\epsilon}(k, t)$ in such a way that

$$^{\beta}w_n(k_1, \epsilon_1, t_1, \ldots, k_n, \epsilon_n, t_n) = \langle a^{\epsilon_1}(k_1, t_1) \cdots a^{\epsilon_n}(k_n, t_n)\Phi, \Phi \rangle$$

(this is the Reconstruction theorem). The only difference is that the spectral condition should be replaced by the KMS condition.

The operator generalized functions $a^{\epsilon}(k, t)$ obey the Heisenberg equations. This allows us to say that the objects we have constructed specify an operator realization of the equilibrium state. (The definition of an operator realization is the same as in Section 8.1, but the condition that Φ is a ground state is replaced by the KMS condition.)

13.3 Algebraic approach to quantum theory

Quantum mechanics can be formulated in terms of an algebra of observables. The starting point of this formulation is a unital associative algebra $\mathcal{A}$ over $\mathbb{C}$ (the algebra of observables). One assumes that this algebra is equipped with antilinear involution $A \to A^*$. One says that a linear functional ω on $\mathcal{A}$ specifies a state if $\omega(A^*A) \geq 0$ (i.e. if the functional is positive). The space of states will be denoted by $\mathcal{C}$. If $\omega(1) = 1$, we say that the state is normalized. The probability distribution $\rho(\lambda)$ of a real observable $A = A^*$ in a normalized state ω should obey the relation $\omega(A^n) = \int \lambda^n \rho(\lambda)d\lambda$. (In general, this formula does not specify the probability distribution uniquely. However, if for every continuous function f of a real argument, we can define an element $f(A)$ of the algebra $\mathcal{A}$, then the probability distribution is specified uniquely by the formula $\omega(f(A)) = \int f(\lambda)\rho(\lambda)d\lambda$. In particular, $f(A)$ is well defined for every continuous function f if $\mathcal{A}$ is a C^*-algebra.)

In the standard exposition of quantum mechanics, the algebra of observables consists of operators acting on a (pre-)Hilbert space. Every vector x having a unit norm specifies a state by the formula $\omega(A) = \langle Ax, x \rangle$. (More generally, a density matrix K defines a state $\omega(A) = \operatorname{Tr} AK$.) This situation is in some sense universal: for every state ω on $\mathcal{A}$, one can construct a (pre-)Hilbert space $\mathcal{H}$ and a representation of $\mathcal{A}$ by operators on this space in such a way that the state ω corresponds to a vector in this space.

To construct $\mathcal{H}$, one defines inner product on $\mathcal{A}$ by the formula $\langle A, B \rangle = \omega(A^*B)$. The space $\mathcal{H}$ can be obtained from $\mathcal{A}$ by means of factorization with respect to zero vectors of this inner product. The inner product on $\mathcal{A}$ descends to $\mathcal{H}$ providing it with the structure of a pre-Hilbert space. The state ω is represented by a vector Φ of $\mathcal{H}$ that corresponds to the unit element of $\mathcal{A}$. An operator of multiplication from the left by an element $C \in \mathcal{A}$ descends to an operator $\hat{C}$ acting on $\mathcal{H}$; this construction gives a representation of $\mathcal{A}$ in the algebra of bounded operators on $\mathcal{H}$. The algebra of operators of the form $\hat{C}$ where $C \in \mathcal{A}$ will be denoted by $\hat{\mathcal{A}}(\omega)$. (In our definition, $\mathcal{H}$ is a pre-Hilbert space; taking its completion, we obtain a representation of $\mathcal{A}$ by operators in Hilbert space $\bar{\mathcal{H}}$.) The above construction is called Gelfand–Naimark–Segal (GNS) construction.

Although every state of the algebra $\mathcal{A}$ can be represented by a vector in Hilbert space, in general, it is impossible to identify Hilbert spaces corresponding to different states.

Time evolution in the algebraic formulation is specified by a one-parameter group $\alpha(t)$ of automorphisms of the algebra $\mathcal{A}$ preserving the involution. This group acts in the obvious way on the space of states. If ω is a stationary state (a state invariant with respect to time evolution), then the group $\alpha(t)$ descends to a group $U(t)$ of unitary transformations of the corresponding space $\mathcal{H}$. The generator H of $U(t)$ plays the role of the Hamiltonian; it can be considered as a self-adjoint operator in $\bar{\mathcal{H}}$. The vector Φ representing ω obeys $H\Phi = 0$.

We say that the stationary state ω is a ground state if the spectrum of H is non-negative.

Every element B of the algebra $\mathcal{A}$ specifies two operators acting on linear functionals on $\mathcal{A}$:

$$(B\omega)(A) = \omega(BA), \tag{13.8}$$

$$(\tilde{B}\omega)(A) = \omega(AB^*). \tag{13.9}$$

Note that the first of these operators is denoted by the same symbol as the element. The operator B commutes with the operator $\tilde{B}'$. These operators do not preserve positivity, but the operator $\tilde{B}B$ does. This means that the operator sending ω into $\tilde{B}B(\omega)$ acts on the space of states. If a state ω corresponds to a vector Φ in a representation of $\mathcal{A}$ (i.e. $\omega(A) = \langle A\Phi, \Phi\rangle$), then the state corresponding to the vector $B\Phi$ is equal to $\tilde{B}B\omega$.

13.3.1 *Quantum field theory and statistical physics in $\mathbb{R}^d$*

Symmetries of quantum theory in the algebraic formulation are automorphisms of the algebra $\mathcal{A}$ commuting with the involution and the evolution automorphisms $\alpha(t)$. We are especially interested in the case when among the symmetries are operators $\alpha(\mathbf{x}, t)$, where $\mathbf{x} \in \mathbb{R}^d$ and $t \in \mathbb{R}$, that are automorphisms of $\mathcal{A}$ obeying $\alpha(\mathbf{x}, t)\alpha(\mathbf{x}', t') = \alpha(\mathbf{x} + \mathbf{x}', t + t')$. We will use the notation $A(\mathbf{x}, t)$ for $\alpha(\mathbf{x}, t)A$, where $A \in \mathcal{A}$.

These automorphisms can be interpreted as space–time translations. One should expect that starting with a formal translation-invariant Hamiltonian, one can construct an algebra $\mathcal{A}$ and space–time translations $\alpha(\mathbf{x}, t)$ in such a way that the corresponding equilibrium states can be identified with the equilibrium states of the preceding section (this statement can be justified in the framework of perturbation theory).

We say that the algebra $\mathcal{A}$ and space–time translations specify a quantum theory in $\mathbb{R}^d$. We will define particles as elementary excitations of the ground state; the theory of these particles and their collisions is a quantum field theory through an algebraic approach.

The action of a translation group on $\mathcal{A}$ induces an action of this group on the space of states. In relativistic quantum field theory, one should have an action of the Poincaré group on the algebra of observables (hence on states).

Let us now consider a state ω that is invariant with respect to the translation group. We will define (quasi-)particles as "elementary excitations" of ω. To consider collisions of (quasi-)particles, we should require that ω satisfies a cluster property in some sense.

The weakest form of cluster property is the following condition:

$$\omega(A(\mathbf{x}, t)B) = \omega(A)\omega(B) + \rho(\mathbf{x}, t), \tag{13.10}$$

where $A, B \in \mathcal{A}$ and ρ is small in some sense for $\mathbf{x} \to \infty$. For example, we can impose the condition that $\int |\rho(\mathbf{x}, t)|d\mathbf{x} < c(t)$, where $c(t)$ has at most polynomial growth. Note that (13.10) implies asymptotic commutativity in some sense: $\omega([A(\mathbf{x}, t), B])$ is small for $\mathbf{x} \to \infty$.

To formulate a more general cluster property, we introduce the notion of correlation functions in the state ω :

$$w_n(\mathbf{x}_1, t_1, \ldots, \mathbf{x}_n, t_n) = \omega(A_1(\mathbf{x}_1, t_1) \ldots A_n(\mathbf{x}_n, t_n)),$$

where $A_i \in \mathcal{A}$. These functions generalize Wightman functions of relativistic quantum field theory. We consider the corresponding truncated correlation functions $w_n^T(\mathbf{x}_1, t_1, \ldots, \mathbf{x}_n, t_n)$.

We have assumed that the state ω is translation-invariant; it follows that both correlation functions and truncated correlation functions depend on the differences $\mathbf{x}_i - \mathbf{x}_j, t_i - t_j$. We say that the state ω has the cluster property if the truncated correlation functions are small for $\mathbf{x}_i - \mathbf{x}_j \to \infty$. A strong version of the cluster property is the assumption that the truncated correlation functions tend to zero faster than any power of $\min \|\mathbf{x}_i - \mathbf{x}_j\|$. Then its Fourier transform with respect to variables $\mathbf{x}_i$ has the form $\nu_n(\mathbf{p}_2, \ldots, \mathbf{p}_n, t_1, \ldots, t_n)\delta(\mathbf{p}_1 + \cdots + \mathbf{p}_n)$, where the function ν_n is smooth. We will later need a weaker form of cluster property that can be formulated as a requirement that the function ν_n is three times continuously differentiable with respect to $\mathbf{p}_2, \ldots, \mathbf{p}_n$.

Instead of cluster property, one can impose a condition of asymptotic commutativity of the algebra $\hat{\mathcal{A}}(\omega)$. In other words, one should require that the commutator $[\hat{A}(\mathbf{x}, t), \hat{B}]$, where $A, B \in \mathcal{A}$ is small for $\mathbf{x} \to \infty$. For example, we can assume that for every n the norm of this commutator is bounded from above by $C_n(t)(1+\|x\|)^{-n}$, where the function $C_n(t)$ has at most polynomial growth. (This property will be called strong asymptotic commutativity.) Note that in relativistic quantum field theory, in the Haag–Araki or the Wightman formulation, strong asymptotic commutativity can be derived from locality.

Let us show how one can define one-particle excitations of the state ω and the scattering of (quasi-)particles.

The action of the translation group on $\mathcal{A}$ generates a unitary representation of this group on the pre-Hilbert space $\mathcal{H}$ constructed from ω. Generators of this representation $\mathbf{P}$ and $-H$ are identified with the momentum operator and the Hamiltonian. The vector in the space $\mathcal{H}$ that corresponds to ω will be denoted by Φ. If Φ is a ground state, we say that it is the physical vacuum. If ω obeys the KMS condition $\omega(A(t)B) = \omega(BA(t + i\beta))$, we say that ω is an equilibrium state with the temperature $T = \frac{1}{\beta}$.

13.3.2 *Particles and quasiparticles*

We say that a state σ is an excitation of ω if it coincides with ω at infinity. More precisely, we should require that $\sigma(A(\mathbf{x}, t)) \to \omega(A)$ as $\mathbf{x} \to \infty$ for every $A \in \mathcal{A}$. Note that the state corresponding to any vector $A\Phi$, where $A \in \mathcal{A}$ is an excitation of ω; this follows from the cluster property.

One can define a *one-particle state* (a one-particle excitation of the state ω or elementary excitation of ω) as a generalized $\mathcal{H}$-valued function $\Phi(\mathbf{p})$ obeying $\mathbf{P}\Phi(\mathbf{p}) = \mathbf{p}\Phi(\mathbf{p}), H\Phi(\mathbf{p}) = \varepsilon(\mathbf{p})\Phi(\mathbf{p})$. (More precisely, for some class of test functions $f(\mathbf{p})$, we should have a linear map $f \to \Phi(f)$ of this class into $\mathcal{H}$ obeying $\mathbf{P}\Phi(f) = \Phi(\mathbf{p}f), H\Phi(f) = \Phi(\varepsilon(\mathbf{p})f)$, where $\varepsilon(\mathbf{p})$ is a real-valued function called dispersion law. For definiteness, we can assume that test functions belong to the Schwartz space $\mathcal{S}(\mathbb{R}^d)$.) *Let us fix an element $B \in \mathcal{A}$ such that $\hat{B}\Phi = \Phi(\phi)$.* (We assume that ϕ is smooth and does not vanish anywhere.) Recall that we consider $\mathcal{H}$ as a pre-Hilbert space obtained by the GNS construction, therefore, every element of $\mathcal{H}$ can be represented in the form $\hat{B}\Phi$.

Note that for every function $g(\mathbf{x}, t) \in \mathcal{S}(\mathbb{R}^{d+1})$, we have

$$\hat{B}(g)\Phi = \Phi(g_f), \tag{13.11}$$

where $\hat{B}(g) = \int g(\mathbf{x}, t)\hat{B}(\mathbf{x}, t)d\mathbf{x}dt$, $g_f(\mathbf{p}) = \hat{g}(\mathbf{p}, -\varepsilon(\mathbf{p}))f(\mathbf{p})$, and $\hat{g}$ stands for the Fourier transform of g.

It follows that we can always assume that $\hat{B} = \int \alpha(\mathbf{x}, t) \hat{A}(\mathbf{x}, t)d\mathbf{x}dt$, where $\alpha(\mathbf{x}, t) \in \mathcal{S}(\mathbb{R}^{d+1})$, $A \in \mathcal{A}$. We will say that an operator $\hat{B}$ satisfying this assumption and transforming Φ into one-particle state is a good operator. For a good operator, $\hat{B}(\mathbf{x}, t) = \int \alpha(\mathbf{x}' - \mathbf{x}, t' - t)\hat{A}(\mathbf{x}', t')d\mathbf{x}'dt'$, hence this expression is a smooth function of $\mathbf{x}$ and t.

We assume that $\Phi(f)$ is normalized (i.e. $\langle \Phi(f), \Phi(f') \rangle = \langle f, f' \rangle$).

Note that it is possible that there exist several types of (quasi-)particles $\Phi_r(f)$ with different dispersion laws. We say that the space spanned by $\Phi_r(f)$ is one-particle space and denote it $\mathcal{H}_1$. If the theory is invariant with respect to spatial rotations, then the infinitesimal rotations play the role of components of angular momentum. If $d = 3$, a particle with spin s can be described as a collection of $2s + 1$ functions $\Phi_r(\mathbf{p})$ obeying $\mathbf{P}\Phi_r(\mathbf{p}) = \mathbf{p}\Phi_r(\mathbf{p})$, $H\Phi_r(\mathbf{p}) = \varepsilon(\mathbf{p})\Phi_r(\mathbf{p})$. The group of spatial rotations acts in the space spanned by these functions; this action is a tensor product of the standard action of rotations of the argument and irreducible $(2s + 1)$-dimensional representation. (Here, s is half-integer, the representation is two-valued if s is not an integer.) In relativistic theory, an irreducible subrepresentation of the representation of the Poincaré group in $\mathcal{H}$ specifies a particle with spin.

13.3.3 *Scattering*

Let us assume that we have several types of (quasi-)particles defined as generalized functions $\Phi_k(\mathbf{p})$ obeying $\mathbf{P}\Phi_k = \mathbf{p}\Phi_k(\mathbf{p})$, $H\Phi_k(\mathbf{p}) = \varepsilon_k(\mathbf{p})\Phi_k(\mathbf{p})$, where the functions $\varepsilon_k(\mathbf{p})$ are smooth and strictly convex. Take some good operators $B_k \in \mathcal{A}$, obeying $\hat{B}_k\Phi = \Phi(\phi_k)$. Define $\hat{B}_k(f, t)$, where f is a function of $\mathbf{p}$ as $\int \tilde{f}(\mathbf{x}, t)\hat{B}_k(\mathbf{x}, t)d\mathbf{x}$ and, as earlier, $\tilde{f}(\mathbf{x}, t)$ is a Fourier transform of $f(\mathbf{p})e^{-i\varepsilon_k(\mathbf{p})t}$ with respect to $\mathbf{p}$.

Let us consider the vectors

$$\Psi(k_1, f_1, \ldots, k_n, f_n \mid t_1, \ldots, t_n) = \hat{B}_{k_1}(f_1, t_1) \cdots \hat{B}_{k_n}(f_n, t_n)\Phi.$$
$$(13.12)$$

We assume that $f_1, \ldots, f_n$ have compact support. Let us introduce the notation $\mathbf{v}_i(\mathbf{p}) = \nabla \varepsilon_{k_i}(\mathbf{p})$. The set of all vectors $\mathbf{v}_i(\mathbf{p})$ with $f_i(\mathbf{p}) \neq 0$ will be denoted U_i. *We assume that the sets $\overline{U_i}$ (the closures of U_i) do not overlap.* This assumption will be called *NO* condition in what follows.

Then assuming the strong cluster property or asymptotic commutativity (precise formulation will be given below), *one can prove that the vector (13.12) has a limit as $t_i \to \infty$ or $t_i \to -\infty$*. The set spanned by these limits will be denoted $\mathcal{D}_+$ or $\mathcal{D}_-$.

Note that the assumption that the sets $\overline{U_i}$ do not intersect (*NO* condition) can be omitted if the space–time dimension is ≥ 4. In these dimensions, we can drop the *NO* condition defining the sets $\mathcal{D}_\pm$.

The existence of the limit of the vectors (13.12) allows us to define Møller matrices. We introduce the space $\mathcal{H}_{as}$ as a Fock representation for the operators $a_k^+(f), a_k(f)$ obeying canonical commutation relations (CCR).

We define Møller matrices S_- and S_+ as operators defined on $\mathcal{H}_{as}$ and taking values in $\bar{\mathcal{H}}$ by the formula

$$\Psi(k_1, f_1, \ldots, k_n, f_n \mid \pm\infty) = S_\pm(a_{k_1}^+(\bar{f}_1\bar{\phi}_{k_1}) \cdots a_{k_n}^+(\bar{f}_n\bar{\phi}_{k_n})\theta).$$
$$(13.13)$$

This formula specifies $S_\pm$ on a dense subset of the Hilbert space $\mathcal{H}_{as}$. These operators are isometric, hence they can be extended to $\mathcal{H}_{as}$ by continuity.

One can say that the vector

$$e^{-iHt}\Psi(k_1, f_1, \ldots, k_n, f_n \mid \pm\infty)$$
$$= \Psi(k_1, f_1 e^{-i\varepsilon_{k_1}t}, \ldots, k_n, f_n e^{-i\varepsilon_{k_n}t} \mid \pm\infty) \qquad (13.14)$$

describes the evolution of a state corresponding to a collection of n particles with wave functions $f_1\phi_1 e^{-i\varepsilon_{k_1}t}, \ldots, f_n\phi_n e^{-i\varepsilon_{k_n}t}$ as $t \to \pm\infty$.

Note that we use the notation $a_k(f) = \int f(\mathbf{p}) a_k(\mathbf{p}) d\mathbf{p}$, $a_k^+(f) = (a_k(f))^* = \int \bar{f}(\mathbf{p}) a_k^+(\mathbf{p}) d\mathbf{p}$, hence $a_k^+(\bar{f}) = \int f(\mathbf{p}) a^+(\mathbf{p}) d\mathbf{p}$.

The existence of the limit in (13.12) can be derived from the strong cluster property if we impose an additional condition $\hat{B}_k^* \Phi = 0$ or the condition $\hat{B}_k^* \Phi = \Phi_k(\nu_k)$.

Let us sketch the proof that the limit in (13.12) exists, assuming for simplicity that the times t_i are equal: $t_i = t$. Let us denote the LHS of (13.12) as $\Psi(t)$. It is sufficient to prove that the norm of the derivative of this vector with respect to t is a summable function: $\int \|\dot{\Psi}(t)\| dt < \infty$.

Note that $\|\dot{\Psi}(t)\|^2$ can be expressed in terms of correlation functions, hence in terms of truncated correlation functions. If we assume the strong cluster property, then $\|\dot{\Psi}(t)\|$ tends to zero faster than any power of $|t|$, this implies the existence of the limit and the fact that $\|\Psi(\pm\infty) - \Psi(t)\|$ also tends to zero faster than any power.

If we are interested only in the existence of the limit, it is sufficient to assume a weaker version of cluster property (three continuous derivatives of the function ν_n). This is sufficient to prove that $\|\dot{\Psi}(t)\| < C|t|^{-\frac{3}{2}}$. Every factor in an arbitrary term of the expansion for $\|\dot{\Psi}(t)\|^2$ has the form

$$I_{k,l}(t) = \langle (\dot{\hat{B}}_{i_1}(f_{i_1}, t))^* \cdots (\dot{\hat{B}}_{i_k}(f_{i_k}, t))^*$$

$$\times \dot{\hat{B}}_{j_1}(f_{j_1}, t) \cdots \dot{\hat{B}}_{j_l}(f_{j_l}, t) \rangle^T,$$

where the dots above the operators denote differentiation with respect to time, which can enter in one of the first k operators and in one of the last l operators. We can suppose without loss of generality that $k \geq 1, l \geq 1$ (otherwise $I_{k,l} = 0$; this follows from the remark that one-particle states are orthogonal to Φ.). If each factor in the given term has $k = l = 1$, then the term is equal to zero because one of the factors either has the form

$$\langle (\dot{\hat{B}}_i(f_i, t))^* \dot{\hat{B}}_j(f_j, t) \rangle^T = \langle (\dot{\hat{B}}_i(f_i, t))^* \dot{\hat{B}}_j(f_j, t) \Phi, \Phi \rangle,$$

or the form

$$\langle (\dot{B}_i(f_i, t))^* \dot{B}_j(f_j, t) \rangle^T = \langle (\dot{B}_i(f_i, t))^* \dot{B}_j(f_j, t) \Phi, \Phi \rangle,$$

and both expressions are zero.

In all other cases, the factor tends to zero as $t \to \infty$ faster than any power of t. This follows from Lemma 10.2 of Section 10.1.

If the space–time dimension is ≥ 4, we can prove the existence of the limit of vectors (13.12) without the NO condition; the proof generalizes the considerations used in Section 10.1 for $d = 3$.

The definition of $S_\pm$ that we gave specifies these operators as multi-valued maps (for example, we can use different good operators in the construction and it is not clear whether we get the same answer). However, we can check that this map is isometric and every multi-valued isometric map is really single-valued (see Section 10.1). In particular, this means that the definition does not depend on the choice of good operators.

To prove that the map is isometric, we express the inner product of two vectors of the form $\Psi(t)$ in terms of truncated correlation functions. Only two-point truncated correlation functions survive in the limit $t \to \pm\infty$. This allows us to say that the map is isometric (see Section 10.1 for a more detailed proof in a slightly different situation).

Let us define in- and out-operators by the formulas

$$a_{\rm in}(f)S_- = S_- a(f), \quad a_{\rm in}^+(f)S_- = S_- a^+(f),$$

$$a_{\rm out}(f)S_+ = S_+ a(f), \quad a_{\rm out}^+(f)S_+ = S_+ a^+(f).$$

(For simplicity of notations, we consider the case when we have only one type of particles. If we have several types of particles, in- and out-operators as well as the operators a^+, a are labeled by a pair (k, f), where f is a test function and k characterizes the type of particle.) These operators are defined on the image of S_- and S_+ correspondingly. One can check that

$$a_{\rm in}^+(\bar f \bar\phi) = \lim_{t \to -\infty} \hat B(f,t), \quad a_{\rm out}^+(\bar f \bar\phi) = \lim_{t \to \infty} \hat B(f,t). \tag{13.15}$$

The limit is understood as a strong limit. It exists on the set of vectors of the form (13.13) (with NO assumption for $d < 3$). The proof follows immediately from the fact that taking the limit $t_i \to \infty$

in the vectors (13.12), we can first take the limit for $i > 1$ and then the limit $t_1 \to \infty$.

The formula (13.15) can be written in the following way:

$$a_{\text{in}}^+(\bar{f}\bar{\phi})\Psi(f_1,\ldots,f_n\,|-\infty) = \Psi(f,f_1,\ldots,f_n\,|-\infty),$$
$$a_{\text{out}}^+(\bar{f}\bar{\phi})\Psi(f_1,\ldots,f_n\,|\infty) = \Psi(f,f_1,\ldots,f_n\,|\infty).$$

Similarly,

$$a_{\text{in}}(f)\Psi(\phi^{-1}f,f_1,\ldots,f_n\,|-\infty) = \Psi(f_1,\ldots,f_n\,|-\infty), \qquad (13.16)$$
$$a_{\text{out}}(f)\Psi(\phi^{-1}f,f_1,\ldots,f_n\,|\infty) = \Psi(f_1,\ldots,f_n\,|\infty). \qquad (13.17)$$

If the operators S_+ and S_- are unitary, we say that the theory has a particle interpretation. In this case (and also in the more general case when the image of S_- coincides with the image of S_+), we can define the scattering matrix

$$S = S_+^* S_-.$$

The scattering matrix is a unitary operator in $\mathcal{H}_{as}$. Its matrix elements in the basis $|p_1,\ldots,p_n\rangle = \frac{1}{n!}a^+(p_1)\cdots a^+(p_n)\theta$ (scattering amplitudes) can be expressed in terms of in- and out-operators.

$$S_{mn}(\mathbf{p}_1,\ldots,\mathbf{p}_m\,|\,\mathbf{q}_1,\ldots,\mathbf{q}_n)$$
$$= \langle a_{\text{in}}^+(\mathbf{q}_1)\cdots a_{\text{in}}^+(\mathbf{q}_n)\Phi, a_{\text{out}}^+(\mathbf{p}_1)\cdots a_{\text{out}}^+(\mathbf{p}_m)\Phi\rangle. \qquad (13.18)$$

Effective cross-sections can be expressed in terms of the squares of scattering amplitudes.

Note that only when ω is a ground state can one hope that the particle interpretations exists. In other cases, instead of a scattering matrix and cross-sections, one should consider inclusive scattering matrix and inclusive cross-sections (see the following).

Formula (13.18) is proved only for the case when all values of momenta $\mathbf{p}_i, \mathbf{q}_j$ are distinct. More precisely, this formula should be understood in the sense of generalized functions and as test functions we should take collections of functions $f_i(\mathbf{p}_i), g_j(\mathbf{q}_j)$ with non-overlapping $\overline{U(f_i)}, \overline{U(g_j)}$. Let us write this in more detail:

$$S_{mn}(f_1, \ldots, f_m \,|\, g_1, \ldots, g_n)$$

$$= \int d^m\mathbf{p}\, d^n\mathbf{q} \prod f_i(\mathbf{p}_i) \prod g_j(\mathbf{q}_j) S_{mn}(\mathbf{p}_1, \ldots, \mathbf{p}_m \,|\, \mathbf{q}_1, \ldots, \mathbf{q}_n)$$

$$= \langle a_{\mathrm{in}}^+(\bar{g}_1) \cdots a_{\mathrm{in}}^+(\bar{g}_n)\Phi, a_{\mathrm{out}}^+(\bar{f}_1) \cdots a_{\mathrm{out}}^+(\bar{f}_m)\Phi \rangle.$$

Using (13.15), we obtain

$$S_{mn}(f_1, \ldots, f_m \,|\, g_1, \ldots, g_n)$$

$$= \lim_{t\to\infty,\tau\to-\infty} \langle \hat{B}(f_m\phi^{-1}, t)^* \cdots \hat{B}(f_1\phi^{-1}, t)^* \hat{B}(g_1\phi^{-1}, \tau)$$

$$\cdots \hat{B}(g_n\phi^{-1}, \tau))\Phi, \Phi \rangle$$

$$= \lim_{t\to\infty,\tau\to-\infty} \omega(B(f_m\phi^{-1}, t)^* \cdots B(f_1\phi^{-1}, t)^* B(g_1\phi^{-1}, \tau)$$

$$\cdots B(g_n\phi^{-1}, \tau)),$$

where $B(f, t)^* = \int d\mathbf{x}\, B^*(\mathbf{x}, t)\overline{\tilde{f}(\mathbf{x}, t)}$.

Note that in the same way, we can obtain a more general formula

$$S_{mn}(f_1, \ldots, f_m \,|\, g_1, \ldots, g_n)$$

$$= \lim_{t\to\infty,\tau\to-\infty} \langle \hat{B}_{m+1}(f_1\phi_{m+1}^{-1}, \tau)$$

$$\cdots \hat{B}_{m+n}(g_n\phi_{m+n}^{-1}, \tau))\Phi, \hat{B}_1(f_1\phi_1^{-1}, t) \cdots B_m(f_m\phi_m^{-1}, t)\Phi \rangle$$

$$= \lim_{t\to\infty,\tau\to-\infty} \omega(B_m(f_m\phi_m^{-1}, t)^* \cdots B_1(f_1\phi_1^{-1}, t)^* B_{m+1}(g_1\phi_{m+1}^{-1}, \tau)$$

$$\cdots B_{m+n}(g_n\phi_{m+n}^{-1}, \tau)), \tag{13.19}$$

where B_i are different good operators and $B_i\Phi = \Phi(\phi_i)$.

13.3.4 *Asymptotic behavior of $\langle \hat{Q}(\mathbf{x}, t)\Psi, \Psi' \rangle$*

Our next goal is to calculate the asymptotic behavior as $t \to \infty$ of the expression $\langle \hat{Q}(\mathbf{x}, t)\Psi, \Psi' \rangle$ where $Q \in \mathcal{A}$.

We assume that the *NO* condition is satisfied and that vectors Ψ, Ψ' belong to $\mathcal{D}_+$. It is sufficient to consider the case

$$\Psi = \lim_{t\to\infty} \Psi(t),\, \Psi(t) = \Psi(f_1,\ldots,f_n\,|\,t),$$

$$\Psi' = \lim_{t\to\infty} \Psi'(t),\, \Psi(t) = \Psi(f_1',\ldots,f_m'\,|\,t).$$

(To simplify the notations, we consider the case of one type of particles.) We assume that the strong cluster property is satisfied. Then the differences between Ψ and $\Psi(t)$, and between Ψ' and $\Psi'(t)$ are negligible for $t \to \infty$ (we neglect terms tending to zero faster than any power of t, more precisely terms that are less than $C_n t^{-n}$ where C_n does not depend on $\mathbf{x}$). We see that it is sufficient to study the asymptotic behavior of

$$\langle \hat{Q}(\mathbf{x},t)\Psi(t), \Psi'(t)\rangle$$
$$= \langle \hat{Q}(\mathbf{x},t)B(f_1,t)\cdots B(f_n,t)\Phi, B(f_1')\cdots B(f_1',t)\cdots B(f_m',t)\Phi\rangle$$
$$= \omega(B^*(f_m',t)\cdots B^*(f_1',t)\hat{Q}(\mathbf{x},t)B(f_1,t)\cdots B_n(f_n,t)).$$

We decompose this expression in terms of truncated correlation functions. It is clear that every non-negligible truncated function should contain equal number of B's and B^*'s. Using this remark and (13.17), we obtain

$$\langle \hat{Q}(\mathbf{x},t)\Psi, \Psi'\rangle = \langle \Psi, \Psi'\rangle M + \int d\mathbf{p}d\mathbf{p}'\langle a_{\text{out}}(\mathbf{p})\Psi, a_{\text{out}}(\mathbf{p}')\Psi'\rangle N(\mathbf{p},\mathbf{p}')$$

$$+ \int d\mathbf{p}\langle a_{\text{out}}(\mathbf{p})\Psi, \Psi'\rangle T_1(\mathbf{p})$$

$$+ \int d\mathbf{p}'\langle \Psi, a_{\text{out}}(\mathbf{p}')\Psi'\rangle T_2(\mathbf{p}') + R, \tag{13.20}$$

where R is negligible,

$$M = \omega(Q) = \langle \hat{Q}\Phi, \Phi\rangle,\, N(\mathbf{p},\mathbf{p}') = \langle \hat{Q}(\mathbf{x},t)\Phi(\mathbf{p}), \Phi(\mathbf{p}')\rangle,$$

$$T_1(\mathbf{p}) = \langle \hat{Q}(\mathbf{x},t)\Phi(\mathbf{p}), \Phi\rangle,\, T_2(\mathbf{p}') = \langle \hat{Q}(\mathbf{x},t)\Phi, \Phi(\mathbf{p}')\rangle.$$

Similar formulas can be written for $t \to -\infty$.

13.3.5 *Scattering theory from asymptotic commutativity*

In this section, we derive the existence of limit in (13.12) from asymptotic commutativity of the algebra $\hat{\mathcal{A}}(\omega)$. This approach is useful in relativistic theory where asymptotic commutativity follows from locality.

Let us assume that the operators $\hat{B}_k$ and $\hat{B}_l$ asymptotically commute:

$$\|[\hat{B}_k(\mathbf{x},t),\hat{B}_l(\mathbf{x}',t)]\| < \frac{C}{1+|\mathbf{x}-\mathbf{x}'|^a},$$

where $a > 1$. We also suppose the condition of asymptotic commutativity is satisfied for the spatial and time derivatives of these operators.

Instead of these conditions, we can assume that $\hat{B}_k = \int g_k(\mathbf{x},t)\hat{A}_k(\mathbf{x},t)d\mathbf{x}dt$ and

$$\|[\hat{A}_k(\mathbf{x},t),\hat{A}_l(\mathbf{x}',t')]\| < \frac{C(t-t')}{1+|\mathbf{x}-\mathbf{x}'|^a},$$

where g_k are functions from Schwartz space, $C(t)$ is a function of at most polynomial growth and $a > 1$.

The statement we need follows from the following fact:

$$\int \|[\hat{B}_{k_i}(f_i,t),\dot{\hat{B}}_{k_j}(f_j,t)]\|dt < \infty. \tag{13.21}$$

To check this, we note that $\dot{\Psi}(t)$ is a sum of n terms; every term contains a product of several operators $\hat{B}$ and one operator $\dot{\hat{B}}$. The estimate for the norm of $\dot{\Psi}(t)$ follows from (13.21) and (13.11) and the remark that the norms of operators $\hat{B}_{k_i}(f_i,t)$ are bounded. (We change the order of operators in the summand under consideration in such a way that $\dot{\hat{B}}$ is from the right. Then we note that $\dot{\hat{B}}(f,t)\Phi = 0$ as follows from (13.11).)

Equation (13.21) remains to be proved. First of all, we note that one can prove the estimate

$$|\tilde{f}_k(\mathbf{x},t)| \le C|t|^{-\frac{D}{2}} \tag{13.22}$$

(the proof for $d = 3$ given in Section 10.2 works for any d).

Let us consider the integral

$$\int d\mathbf{x}\,d\mathbf{x}'\,\tilde{f}_{k_i}(\mathbf{x},t)\tilde{f}_{k_j}(\mathbf{x}',t)[\hat{B}_{k_i}(\mathbf{x},t),\dot{\hat{B}}_{k_j}(\mathbf{x}',t)]. \tag{13.23}$$

First of all, we consider the integral (13.23) over the domain $\Gamma(t)$ defined in the following way. Take non-overlapping compact sets $G \subset \mathbb{R}^d, G' \subset \mathbb{R}^d$. We say that the point $(\mathbf{x},\mathbf{x}')$ belongs to $\Gamma(t)$ if there exist points $\mathbf{v} \in G, \mathbf{v}' \in G'$ such that $|\mathbf{v} - \frac{\mathbf{x}}{t}| < C|t|^{-\rho}, |\mathbf{v}' - \frac{\mathbf{x}'}{t}| < C|t|^{-\rho}$ where $\rho = \frac{1}{2} - \epsilon$. (We will apply our estimate in the case when G is the set of points of the form $\nabla\varepsilon_{k_i}(\mathbf{p})$ where $\mathbf{p}$ belongs to the support of f_{k_i} and G' is defined in similar way using the function f_{k_j}.) For large t, the volume of this domain is less than $t^{2d(1-\rho)}$ and the norm of the integrand is less than $t^{-d} \times t^{-a}$. (We have used the fact that the distance between x and x' grows linearly with t in the integration domain, hence the norm of the commutator in (13.23) is less than t^{-a}.) This allows us to say that the norm of the integral does not exceed $t^{-a+\epsilon}$.

To estimate the integral (13.21) over the complement to $\Gamma(t)$, we consider the integral

$$\int \exp(-i\varepsilon(\mathbf{p})t + i\mathbf{px})\mu\left(t^\rho\left(\mathbf{v}(\mathbf{p}) - \frac{\mathbf{x}}{t}\right)\right)f(\mathbf{p})d\mathbf{p}, \tag{13.24}$$

where $\mu(\mathbf{x})$ is a smooth function equal to zero for $|\mathbf{x}| \le \nu_1$ and equal to one for $|\mathbf{x}| \ge \nu_2$. One can prove that for every r, the absolute value of this integral does not exceed $C|t|^{-(1-2\rho)r}$ and for $|x| > bt$, it does not exceed $C|x|^{-r}|t|^{-(1-2\rho)r}$.

The integral (10.24) is equal to $\tilde{f}(\mathbf{x},t)$ if $t^\rho(\mathbf{v}(\mathbf{p}) - \frac{\mathbf{x}}{t}) \ge \nu_2$ for $\mathbf{p} \in \text{supp}\,f$. This means that the estimate of this integral can be applied to at least one of the factors $\tilde{f}_{k_i}(\mathbf{x},t), \tilde{f}_{k_j}(\mathbf{x}',t)$ in the integrand of the integral (13.23) on the complement to the domain $\Gamma(t)$. Using this fact and the inequality

$$\int d\mathbf{x}\,|\tilde{f}_k(\mathbf{x},t)| \le C|t|^{\frac{d}{2}},$$

we obtain that the norm of the integral (13.23) over the complement to $\Gamma(t)$ tends to zero faster than any power of $|t|$ as $t \to \pm\infty$. Combining this estimate with the estimate for the integral over $\Gamma(t)$

and taking $\epsilon < a - 1$, we obtain the estimate for the norm of (13.23): for large $|t|$, it is less than $|t|^{-b}$ where $b > 1$. The same arguments work if the dot (time derivative) in (13.23) is removed.

To prove (13.21), we note that calculating $\dot{\hat{B}}(f, t)$ we should take into account that f depends on t. It is easy to check that

$$\dot{\hat{B}}(f, t) = \int d\mathbf{x} \tilde{g}(\mathbf{x}, t) \hat{B}(\mathbf{x}, t) + \int d\mathbf{x} \tilde{f}(\mathbf{x}, t) \dot{\hat{B}}(\mathbf{x}, t),$$

where $g(\mathbf{p}) = -i\varepsilon(\mathbf{p}) f(\mathbf{p})$. Using this expression, we obtain that the commutator in (13.21) can be written as a sum of (13.23) and similar expression with removed dots. Both summands do not exceed $|t|^{-b}$ with $b > 1$ for large $|t|$. This implies (13.21).

Note that the proof based on asymptotic commutativity also allows us to estimate the speed of convergence to the limit. In the case of strong asymptotic commutativity, the difference between $\Psi(t)$ and $\Psi(\pm\infty)$ tends to zero faster than any power of t.

13.3.6 *Green functions and scattering: LSZ*

Let us start with scattering of elementary excitations of ground state (of particles). In this case, the scattering matrix can be expressed in terms of on-shell values of Green functions. The Green function in translation-invariant stationary state ω is defined by the formula

$$G_n = \omega(T(A_1(\mathbf{x}_1, t_1) \cdots A_r(\mathbf{x}_r, t_r))),$$

where $A_i \in \mathcal{A}$ and T stands for time ordering. More precisely, this is a definition of Green function in $(\mathbf{x}, t)$-representation, taking Fourier transform with respect to $\mathbf{x}$, we obtain Green functions in $(\mathbf{p}, t)$-representation; taking in these functions inverse Fourier transform with respect to t, we obtain Green functions in in $(\mathbf{p}, \epsilon)$-representation. Due to translation-invariance of ω, we obtain that in $(\mathbf{x}, t)$-representation, the Green function depends on differences $\mathbf{x}_i - \mathbf{x}_j$, in $(\mathbf{p}, t)$-representation, it contains a factor $\delta(\mathbf{p}_1 + \cdots + \mathbf{p}_r)$. Similarly, in $(\mathbf{p}, \epsilon)$, we have the same factor and the factor $\delta(\epsilon_1 + \cdots + \epsilon_r)$. We omit both factors talking about poles of Green functions.

For $n = 2$ in $(\mathbf{p}, \epsilon)$-representation, G_2 has the form

$$G(\mathbf{p}_1, \epsilon_1 \mid A, A')\delta(\mathbf{p}_1 + \mathbf{p}_2)\delta(\epsilon_1 + \epsilon_2).$$

Poles of function $G(\mathbf{p}, \epsilon \mid A, A')$ correspond to particles, the dependence of the position of the pole on $\mathbf{p}$ specifies the dispersion law $\varepsilon(\mathbf{p})$ (we consider poles with respect to the variable ϵ for fixed $\mathbf{p}$). This follows from Källén–Lehmann representation (Section 8.1), but this can be obtained also from the following considerations.

We will prove that the scattering amplitudes can be obtained as on-shell values of Green functions (LSZ formula). To simplify notations, we consider the case when we have only one single-particle state $\Phi(\mathbf{p})$ with dispersion law $\varepsilon(\mathbf{p})$. We assume that the elements $A_i \in \mathcal{A}$ are chosen in such a way that the projection of $\hat{A}_i\Phi$ on the one-particle space has the form $\Phi(\phi_i) = \int \phi_i(\mathbf{p})\Phi_i(\mathbf{p})d\mathbf{p}$ where $\phi_i(\mathbf{p})$ is a non-vanishing function. We introduce the notation $\Lambda_i(\mathbf{p}) = \phi_i(\mathbf{p})^{-1}$.

Let us consider Green function

$$G_{mn} = \omega(T(A_1^*(\mathbf{x}_1, t_1) \cdots A_m^*(\mathbf{x}_m, t_m)A_{m+1}(\mathbf{x}_{m+1}, t_{m+1})$$

$$\cdots A_{m+n}(\mathbf{x}_{m+n}, t_{m+n}))$$

in $(\mathbf{p}, \epsilon)$-representation. It is convenient to change slightly the definition of $(\mathbf{p}, t)$- and $(\mathbf{p}, \epsilon)$-representation changing the signs of variables $\mathbf{p}_i$ and ϵ_i for $1 \leq i \leq m$ (for variables corresponding to the operators A_i^*). This convention agrees with the conventions in Section 9.2. Multiplying the Green function in $(\mathbf{p}, \epsilon)$-representation by

$$\prod_{1 \leq i \leq m} \overline{\Lambda_i(\mathbf{p}_i)}(\epsilon_i + \varepsilon(\mathbf{p}_i)) \prod_{m < j \leq m+n} \Lambda_j(\mathbf{p}_j)(\epsilon_j - \varepsilon(\mathbf{p}_j)),$$

and taking the limit $\epsilon_i \to -\varepsilon_i(\mathbf{p}_i)$ for $1 \leq i \leq m$ and the limit $\epsilon_j \to \varepsilon_j(\mathbf{p}_j)$ for $m < j \leq m+n$, we obtain on-shell Green function denoted by σ_{mn}. We prove that it coincides with scattering amplitudes:

$$\sigma_{m,n}(\mathbf{p}_1, \ldots, \mathbf{p}_{m+n}) = S_{mn}(\mathbf{p}_1, \ldots, \mathbf{p}_m \mid \mathbf{p}_{m+1}, \ldots, \mathbf{p}_{m+n}). \quad (13.25)$$

First of all, we note that the on-shell Green function can be expressed in terms of the asymptotic behavior of the Green function in $(\mathbf{p}, t)$-representation: if for $t \to \pm\infty$ and fixed $\mathbf{p}$, this behavior is described by linear combination of exponent $e^{i\lambda t}$, then the location of poles is determined by the exponent indicators λ and the coefficients in front of exponents determine the residues. Using this statement, one can show that

$$\lim_{t\to\infty,\tau\to-\infty}\omega(B_m(f_m\phi_m^{-1},t)^*\cdots B_1(f_1\phi_1^{-1},t)^*B_{m+1}(g_1\phi_{m+1}^{-1},\tau)$$

$$\cdots B_{m+n}(g_n\phi_{m+n}^{-1},\tau)),\tag{13.26}$$

where B_i are good operators can be expressed in terms of on-shell Green function as

$$\int d^{m+n}\mathbf{p}f_1(\mathbf{p}_1)\ldots f_m(\mathbf{p}_m)g_1(-\mathbf{p}_{m+1})$$

$$\cdots g_n(-\mathbf{p}_{m+n})\sigma_{m,n}(\mathbf{p}_1,\ldots,\mathbf{p}_{m+n}).$$

Using (13.19), we obtain (13.25) in the case when A_i are good operators. We will show that the general case can be reduced to the case when operators A_i are good.

Let us suppose that one-particle spectrum does not overlap with multi-particle spectrum. (We represent the space $\bar{\mathcal{H}}$ as a direct sum of one-dimensional space $\mathcal{H}_0$ spanned by Φ, one-particle space $\mathcal{H}_1$ and the space $\mathcal{M}$ called multi-particle space. Our condition means that the joint spectrum of $\mathbf{P}$ and H in $\mathcal{M}$ (multi-particle spectrum) does not overlap with the joint spectrum of $\mathbf{P}$ and H in $\mathcal{H}_0+\mathcal{H}_1$. If the theory has particle interpretation, then our condition means that $\varepsilon(\mathbf{p}_1+\mathbf{p}_2)<\varepsilon(\mathbf{p}_1)+\varepsilon(\mathbf{p}_2)$.[1] Then we can construct an operator B transforming Φ into one-particle state (a good operator) using the formula $B=\int\alpha(\mathbf{x},t)A(\mathbf{x},t)d\mathbf{x}dt$ where $A\in\mathcal{A}$ and the projection of $A\Phi$ onto one-particle state does not vanish. Namely, we should assume that the support of $\hat{\alpha}(\mathbf{p},\omega)$ (of the Fourier transform of α) does not intersect the multi-particle spectrum and does not contain 0. Moreover, the operator we constructed obeys $\hat{B}^*\Phi=0$.

---

[1]The physical meaning of this condition: the energy conservation law forbids the decay of a particle. This condition is not always satisfied, however, stability of a particle is always guaranteed by some conservation laws. Our considerations can be applied in this more general situation.

Now, we can apply this construction to the operators A_i and verify that on-shell Green functions corresponding to good operators B_i coincide with Green functions corresponding to the operators A_i. (We note that the correlation functions of operators B_i can be expressed in terms of correlation functions of operators A_i. As we mentioned already, on-shell Green function can be expressed in terms of asymptotic behavior as $t \to \pm\infty$ of the Green function in $(\mathbf{p}, t)$-representation. It is easy to find the relation between this behavior for the Green functions of operators A_i and the behavior for the Green functions of operators B_i.)

For relativistic theories with a mass gap, one can derive asymptotic commutativity and cluster property from Haag–Araki or Wightman axioms. This proves the existence of scattering matrix. One can prove that scattering matrix is Lorentz-invariant; let us sketch the proof of this fact.

The operator $\hat{B}(f, t)$ used in our construction can be written as an integral of $\tilde{f}(\mathbf{x}, t) B(\mathbf{x}, t)$, where $\tilde{f}(\mathbf{x}, t)$ is a positive frequency solution of the Klein–Gordon equation over the hyperplane $t = constant$. We will define the operator $\hat{B}(f, \rho)$ integrating the same integrand over another hyperplane ρ. One can replace the operators $\hat{B}(f, t)$ by the operators $\hat{B}(f, \rho)$ in (13.12) and prove that the expression we obtained has a limit if the hyperplanes ρ_i tend to infinity in time direction (for example, if they have form $\alpha t + \mathbf{ax} = constant$ and the constant tends to infinity). This means that we can use $\hat{B}(f, \rho)$ in the definition of Møller matrices and scattering matrix. The same arguments that were used to prove that the limit does not depend on the choice of good operators can be applied to verify that the new construction gives the same Møller matrices. This implies Lorentz invariance because the Lorentz group acts naturally on operators $\hat{B}(f, \rho)$.

13.3.7 *Generalized Green functions; the inclusive scattering matrix*

Let us define generalized Green functions (GGreen functions) in the state ω by the following formula where $B_i \in \mathcal{A}$:

$$G_n = \omega(MN),$$

where

$$N = T(B_1(\mathbf{x}_1, t_1) \cdots B_n(\mathbf{x}_n, t_n))$$

stands for chronological product (times decreasing) and

$$M = T^{\mathrm{opp}}(B_1^*(\mathbf{x}_1', t_1') \cdots B_n^*(\mathbf{x}_n', t_n'))$$

stands for antichronological product (times increasing).

One can give another definition of GGreen functions introducing the operator

$$Q = T(B_1(\mathbf{x}_1, t_1) \cdots B_n(\mathbf{x}_n, t_n)\tilde{B}_1(\mathbf{x}_1', t_1') \cdots \tilde{B}_n(\mathbf{x}_n', t_n')),$$

where the operators $B_i, \tilde{B}_i$ act on the space of linear functionals on $\mathcal{A}$. (Recall (13.8) and (13.9) that operators B and $\tilde{B}$ act on linear functionals defined on $\mathcal{A}$; they transform $\omega(A)$ into $\omega(BA)$ and in $\omega(AB^*)$ correspondingly.) It is easy to check that

$$G_n = (Q\omega)(1).$$

Let us define inclusive S-matrix as on-shell GGreen function. We will show that in the case when the theory has particle interpretation, inclusive cross-section can be expressed in terms of inclusive S-matrix.

Recall that the inclusive cross-section of the process $(M, N) \to (Q_1, \ldots, Q_m)$ is defined as a sum (more precisely, a sum of integrals) of effective cross-sections of the processes $(M, N) \to (Q_1, \ldots, Q_m, R_1, \ldots, R_n)$ over all possible $R_1, \ldots, R_n$. If the theory does not have particle interpretation, this formal definition of inclusive cross-section does not work, but still the inclusive cross-section can be defined in terms of probability of the process $(M, N \to (Q_1, \ldots, Q_n+ \text{something else}))$ and expressed in terms of inclusive S-matrix.

Let us consider the expectation value

$$\nu(a_{\mathrm{out},k_1}^+(\mathbf{p}_1)a_{\mathrm{out},k_1}(\mathbf{p}_1) \cdots a_{\mathrm{out},k_m}^+(\mathbf{p}_m)a_{\mathrm{out},k_m}(\mathbf{p}_m)), \qquad (13.27)$$

where ν is an arbitrary state. This quantity is the probability density in momentum space for finding m outgoing particles of the types $k_1, \ldots, k_n$ with momenta $\mathbf{p}_1, \ldots, \mathbf{p}_m$ plus other unspecified outgoing particles. It gives inclusive cross-section if $\nu = \nu(t)$ describes the evolution of a state represented as a collection of incoming particles.

It will be convenient to consider more general expectation value

$$\nu(a^+_{\text{out},k_1}(f_1)a_{\text{out},k_1}(g_1)\cdots a^+_{\text{out},k_m}(f_m)a_{\text{out},k_m}(g_m)), \qquad (13.28)$$

where f_i, g_i are test functions. This expression can be understood as a generalized function

$$\nu(a^+_{\text{out},k_1}(\mathbf{p}_1)a_{\text{out},k_1}(\mathbf{q}_1)\cdots a^+_{\text{out},k_m}(\mathbf{p}_m)a_{\text{out},k_m}(\mathbf{q}_m));$$

we get (13.27) taking $\mathbf{p}_i = \mathbf{q}_i$.

As earlier, we assume that we have several types of (quasi-)particles $\Phi_k(\mathbf{p})$ obeying $\mathbf{P}\Phi_k = \mathbf{p}\Phi_k(\mathbf{p}), H\Phi_k(\mathbf{p}) = \varepsilon_k(\mathbf{p})\Phi_k(\mathbf{p})$, where the functions $\varepsilon_k(\mathbf{p})$ are smooth and strictly convex. Good operators $B_k \in \mathcal{A}$ obey $\hat{B}_k\Phi = \Phi(\phi_k)$. The operator $\hat{B}_k(f,t)$, where f is a function of $\mathbf{p}$, defined as $\int \tilde{f}(\mathbf{x},t)\hat{B}_k(\mathbf{x},t)d\mathbf{x}$ (as always $\tilde{f}(\mathbf{x},t)$ is a Fourier transform of $f(\mathbf{p})e^{-i\varepsilon_k(\mathbf{p})t}$ with respect to $\mathbf{p}$).

Now we can calculate (13.28). First of all, we take as ν the state corresponding to the vector (13.14). We are representing this vector in terms of good operators B_i using the formula

$$\Psi(k_1, f_1, \ldots, k_n, f_n \,|\, \infty) = \lim_{t\to-\infty} \Psi(k_1, f_1, \ldots, k_n, f_n \,|\, t)$$

$$= \lim_{t\to-\infty} \hat{B}_{k_1}(f_1, t)\cdots\hat{B}_{k_n}(f_n, t)\Phi.$$

The corresponding state ν considered as linear functional on $\mathcal{A}$ can be expressed in terms of the state ω corresponding to Φ. (We should use the remark that the state corresponding to the vector $A\Phi$ can be written as $\tilde{A}A\omega$.) Expressing the out-operators by the formula (13.15), we obtain the expression of (13.28) in terms of GGreen functions on-shell.

13.4 *L*-functionals

In the approach of Section 13.2, for every equilibrium state, we should construct a Hilbert space depending on the temperature, CCR or CAR are represented in this space, the equilibrium state is described by a vector in the space. In this section, we will describe a formalism of *L*-functionals (positive functionals on Weyl algebra) that can be used to describe the states corresponding to vectors and density matrices in all representations of CCR. This formalism can be used to calculate physical quantities for equilibrium state in the framework of an analog of Feynman diagram technique.

Note that there exists an obvious generalization of all these results to the case of fermions. In this case, L-functionals are positive functionals on the algebra with generators satisfying canonical anticommutation relations (CAR) (Clifford algebra), α and α^* in the definition of L-functional are anticommuting variables.

Let us consider a representation of CCR in Hilbert space $\mathcal{H}$. Here, we understand CCR as relations

$$[a_k, a_l^+] = \hbar\delta_{kl}, [a_k, a_l] = [a_k^+, a_l^+] = 0,$$

where k, l run over a discrete set M. To a density matrix K (or more generally, to any trace class operator in $\mathcal{H}$), we can assign a functional $L_K(\alpha^*, \alpha)$ defined by the formula

$$L_K(\alpha^*, \alpha) = \operatorname{Tr} e^{-\alpha a^+} e^{\alpha^* a} K. \tag{13.29}$$

Here, αa^+ stands for $\sum \alpha_k a_k^+$ and $\alpha^* a$ for $\sum \alpha_k^* a_k$, where k runs over M. This formula makes sense if $\alpha\alpha^* = \sum |\alpha_k|^2 < \infty$. (This follows from the remark that $e^{-\alpha a^+ + \alpha^* a} = e^{\hbar\alpha\alpha^*/2} e^{-\alpha a^+} e^{\alpha^* a}$ is a unitary operator in $\mathcal{H}$). We can apply (13.29) also in the case when K is an arbitrary operator of trace class (not necessarily a density matrix).

One can say that L_K is a generating functional of correlation functions.

One can also consider a more general case when CCR are written in the form

$$[a(k), a^+(k')] = \hbar\delta(k, k'), [a(k), a(k')] = [a(k)^+, a(k')^+] = 0,$$

k, k' run over a measure space M. We are using the exponential form of CCR; in this form, a representation of CCR is specified as a collection of unitary operators $e^{-\alpha a^+ + \alpha^* a}$ obeying appropriate commutation relations. Here, $\alpha(k)$ is a complex function on the measure space M, the expressions of the form $\alpha^* a, \alpha a^+$ can be written as integrals $\int \alpha^*(k) a(k) dk, \int \alpha(k) a^+(k) dk$ over M. In the space of CCR $a(k), a^+(k')$ become Hermitian conjugate generalized operator functions. We always assume that α is square integrable, then the expression (13.29) is well defined.

Knowing L_K, we can calculate $\langle A \rangle_L = \operatorname{Tr} AK$ where A is an element of the unital associative algebra $\mathcal{A}$ generated by a_k, a_l^+ (Weyl algebra). Hence, we can consider L_K as a positive linear functional $\langle A \rangle_L$ on the algebra $\mathcal{A}$. (We consider $\mathcal{A}$ as an algebra with involution $^+$, positivity means that $\langle A^+ A \rangle_L \geq 0$. Recall that positive linear functional ω on unital algebra with involution is called a normalized state if $\omega(1) = 1$.)

A functional $L(\alpha^*, \alpha)$ satisfying the condition $L(-\alpha^*, -\alpha) = L^*(\alpha, \alpha^*)$ and positivity condition is called physical L-functional. We say that such a functional is normalized if $L(0,0) = 1$. Normalized physical L-functionals are in one-to-one correspondence with states on $\mathcal{A}$. We will consider also the space $\mathcal{L}$ of all functionals $L(\alpha^*, \alpha)$ (of all L-functionals). Every trace class operator K in a representation space of CCR specifies an element of $\mathcal{L}$ by the formula (13.29).

We can define an antilinear involution $L \to \tilde{L}$ on the space $\mathcal{L}$ of functionals $L(\alpha^*, \alpha)$ by the formula

$$(\tilde{L})(\alpha^*, \alpha) = L^*(-\alpha, -\alpha^*). \tag{13.30}$$

Physical L-functionals are invariant with respect to this involution. It is easy to check that

$$\tilde{L}_K = L_{K^+}.$$

Every normalized vector $\Phi \in \mathcal{H}$ specifies a density matrix and hence a physical L-functional. Conversely, every physical L-functional corresponds to a vector in some representation of CCR, given by GNS construction (see the preceding section). One can characterize this representation by the requirement that it contains a cyclic vector Φ obeying $L(A) = \langle A\Phi, \Phi \rangle$. However, an L-functional can be obtained from many density matrices in many representations of CCR.

An action of Weyl algebra $\mathcal{A}$ on $\mathcal{L}$ can be specified by operators

$$b^+(k) = \hbar c_1^+(k) - c_2(k), b(k) = c_1(k)$$

obeying CCR. Here, $c_i^+(k)$ are multiplication operators by α_k^*, α_k and $c_i(k)$ are derivatives with respect to α_k^*, α_k. This definition is prompted by relations

$$L_{a(k)K} = b(k)L_K, \quad L_{a^+(k)K} = b^+(k)L_K. \tag{13.31}$$

Applying the involution $L \to \tilde{L}$, we obtain another representation of $\mathcal{A}$ on $\mathcal{L}$. It is specified by the operators

$$\tilde{b}^+(k) = -\hbar c_2^+(k) + c_1(k), \tilde{b}(k) = -c_2(k),$$

obeying CCR and satisfying

$$L_{Ka^+(k)} = \tilde{b}(k)L_K, \quad L_{Ka(k)} = \tilde{b}^+(k)L_K. \tag{13.32}$$

More generally, for any $A \in \mathcal{A}$, we have an operator acting in $\mathcal{L}$ denoted by the same symbol and obeying $A(L_K) = L_{AK}$ (the left multiplication by A in $\mathcal{A}$ specifies the action of A on linear functionals). We can define also an operator $\tilde{A}$ using the formula $\tilde{A}(L) = A(\tilde{L})$. If $L = L_K$, then $\tilde{A}L = L_{KA^+}$. It is easy to check that $\widetilde{A+B} = \tilde{A} + \tilde{B}, \widetilde{AB} = \tilde{A}\tilde{B}, A\tilde{B} = \tilde{B}A$. Operators $\tilde{A}$ specify another action of Weyl algebra on $\mathcal{L}$ that commutes with the original action; one can say that the direct sum of two Weyl algebras acts on $\mathcal{L}$.

Note that the space of physical L-functionals is not invariant with respect to the operators A and $\tilde{A}$, however, it is invariant with respect to the operators $\tilde{A}A$ where $A \in \mathcal{A}$. It is easy to check that

$$\tilde{A}AL_K = L_{AKA^+}.$$

Let us consider a Hamiltonian H in a space of representation of CCR. We will write H in the form

$$H = \sum_{m,n} \sum_{k_i,l_j} H_{m,n}(k_1, \ldots, k_m \,|\, l_1, \ldots, l_n)a_{k_1}^+ \cdots a_{k_m}^+ a_{l_1} \cdots a_{l_n}.$$

$$\tag{13.33}$$

There are two operators in $\mathcal{L}$ corresponding to H:

$$H = \sum_{m,n} \sum_{k_i,l_j} H_{m,n}(k_1, \ldots, k_m \,|\, l_1, \ldots, l_n)b_{k_1}^+ \cdots b_{k_m}^+ b_{l_1} \cdots b_{l_n}$$

$$\tag{13.34}$$

(we denote it by the same symbol) and

$$\tilde{H} = \sum_{m,n} \sum_{k_i, l_j} H_{m,n}(k_1, \ldots, k_m \,|\, l_1, \ldots, l_n) \tilde{b}^+_{k_1} \cdots \tilde{b}^+_{k_m} \tilde{b}_{l_1} \cdots \tilde{b}_{l_n}.$$

$$(13.35)$$

The equation of motion for the L-functional $L(\alpha^*, \alpha)$ has the form

$$i\hbar \frac{dL}{dt} = \hat{H}L = HL - \tilde{H}L. \qquad (13.36)$$

(We introduced the notation $\hat{H} = H - \tilde{H}$.) It corresponds to the equation of motion for density matrices; this follows from the formula

$$\hat{H}L_K = L_{HK - KH^+}.$$

Note that often the equations of motion for L-functionals make sense even in the situation when the equations of motion in the Fock space are ill defined. This is related to the fact that vectors and density matrices from all representations of CCR are described by L-functionals. This means that by applying the formalism of L-functionals, we can avoid the problems related to the existence of inequivalent representations of CCR. It is well known, in particular, that these problems arise for translation-invariant Hamiltonians; in perturbation theory, these problems appear as divergences related to infinite volume. Therefore, in the standard formalism, it is necessary to consider at first a Hamiltonian in finite volume Ω (to make volume cutoff or, in another terminology, infrared cutoff) and to take the limit $\Omega \to \infty$ in physical quantities (see Sections 8.2 and 13.2 for more details). In the formalism of L-functionals, we can work directly in infinite volume.

In general, the ground state of formal Hamiltonian (13.7) does not belong to Fock space, but the corresponding L-functional is well defined. (Note that we always assume that ultraviolet divergences are absent.) Similarly, the equilibrium states for different temperatures belong to different Hilbert spaces, but all of them are represented by well-defined L-functionals.

In what follows, we consider (13.33) as a formal expression; we assume that it is formally Hermitian. There is no necessity to assume that (13.33) specifies a self-adjoint operator in one of the representations of CCR.

Let us make some remarks about adiabatic evolution in the formalism of L-functionals. Let us take a Hamiltonian that depends on time t, but is changing adiabatically (very slowly). More formally, we can assume that the Hamiltonian depends on a parameter g and $g = h(at)$ where $a \to 0$. Let us take a family of L-functionals $L(g)$, where $L(g)$ is a stationary state of the Hamiltonian $H(g)$, i.e.

$$\hat{H}(g)L(g) = 0.$$

It is obvious that $L(h(at))$ obeys the equation of motion up to terms tending to zero as $a \to 0$. (Note that a similar statement is wrong in the standard Hilbert space formulation of quantum mechanics because the vector corresponding to a state is defined up to phase factor. We can only say that for a smooth family of eigenvectors $\Psi(g)$, there exists a phase factor $C(g)$ such that $C(h(at))\Psi(h(at))$ obeys the equation of motion up to terms tending to zero as $a \to 0$.)

13.4.1 *Translation-invariant Hamiltonians in the formalism of L-functionals; one-particle states*

In what follows, we consider translation-invariant Hamiltonians.

We say that an L-functional σ is an excitation of a translation-invariant L-functional ω if it coincides with ω at infinity. More precisely, we should require that $\sigma(T_\mathbf{x}\alpha^*, T_\mathbf{x}\alpha)) \to \omega(\alpha^*, \alpha)$ as $\mathbf{x} \to \infty$. Here, $T_\mathbf{x}$ stands for spatial translation.

We assume that the L-functional ω has cluster property. The weakest form of cluster property is the requirement that $\langle AT_\mathbf{x}B\rangle_\omega - \langle A\rangle_\omega\langle B\rangle_\omega$ tends to zero as $\mathbf{x} \to \infty$ for $A, B \in \mathcal{A}$.

If Φ is a vector corresponding to ω in GNS representation space $\mathcal{H}$, then the state corresponding to any vector $A\Phi$ where $A \in \mathcal{A}$ is an excitation of ω; this follows from cluster property.

We can define elementary excitations (called particles if ω is a ground state and quasiparticles if ω is a general translation-invariant stationary state) in the same way as in Section 13.3.2.

13.4.2 *Quadratic Hamiltonians*

Let us consider as an example the simplest translation-invariant Hamiltonian

$$H_0 = \int \omega(k)a^+(k)a(k)dk, \tag{13.37}$$

where k runs over $\mathbb{R}^d$. It can be approximated by a Hamiltonian of the form $\sum \omega_k a_k^+ a_k$ having a finite number of degrees of freedom. The equilibrium state of the latter Hamiltonian can be represented by density matrix $\Omega(T)$ in the Fock space; it is easy to check that

$$a_k\Omega(T) = e^{-\hbar\frac{\omega_k}{T}}\Omega(T)a_k, \quad a_k^+\Omega(T) = e^{\frac{\hbar\omega_k}{T}}\Omega(T)a_k^+. \tag{13.38}$$

Applying (13.31) and (13.32), we obtain equations for the corresponding L-functional; taking the limit, we obtain for the L-functional corresponding to the equilibrium state in infinite volume

$$c_1(k)L_T = e^{-\frac{\hbar\omega(k)}{T}}\left(-\hbar c_2^+(k) + c_1(k)\right)L_T, \tag{13.39}$$

hence

$$c_1(k)L_T = -n(k)c_2^+(k)L_T, \quad c_2(k)L_T = -n(k)c_1^+(k)L_T, \tag{13.40}$$

where

$$n(k) = \frac{\hbar}{e^{\frac{\hbar\omega(k)}{T}} - 1}. \tag{13.41}$$

We obtain

$$L_T = e^{-\int \alpha^*(k)n(k)\alpha(k)dk}. \tag{13.42}$$

If we are interested in equilibrium state for given density, we should replace $\omega(k)$ with $\omega(k) - \mu$ in (13.41) (here, μ stands for chemical potential).

The Hamiltonian $\hat{H}$ governing the evolution of L-functionals can be written in the form

$$\hat{H} = \int \omega(k)b^+(k)b(k)dk - \int \omega(k)\tilde{b}^+(k)\tilde{b}(k)dk$$

$$= \int \hbar(\omega(k)c_1^+(k)c_1^+(k)dk - \omega(k)c_2^+(k)c_2(k))dk.$$

It follows that the vector generalized functions $\Phi_1(k) = c_1^+(k)\Phi$, $\Phi_2(k) = c_2^+(k)\Phi$ in the space of GNS representation corresponding to (13.42) are one-(quasi)particle excitations. (Note that in this statement, $n(k)$ in the formula (13.42) is an arbitrary function; then the corresponding state is not an equilibrium, but it is still a stationary translation-invariant state.)

Very similar considerations allow us to prove that the L-functional corresponding to an equilibrium state of general quadratic Hamiltonian $H = a^+ M a^+ + a^+ N a + a R a$ is Gaussian (has the form $e^{\alpha^* A \alpha^* + \alpha^* B \alpha + \alpha C \alpha}$).

13.4.3   *Perturbation theory*

Let us assume now that the Hamiltonian H is represented as a sum of quadratic Hamiltonian H_0 and interaction Hamiltonian $H_{\text{int}} = gV$. Then in the formalism of L-functionals, one can introduce in the standard way the evolution operator $\hat{U}(t, t_0)$, the evolution operator in the interaction picture $\hat{S}(t, t_0)$ and the operator $\hat{S}_a$, the analog of adiabatic S-matrix. If $\hat{H}_a$ governs the evolution of L-functional for the Hamiltonian $H_0 + h(at)H_{\text{int}}$, then we denote by $\hat{U}_a(t, t_0)$ the operator transforming the L-functional at the moment t_0 into the L-functional at the moment t (the evolution operator). The operator $\hat{S}_a(t, t_0)$ is defined by the formula

$$\hat{S}_a(t, t_0) = e^{\frac{i}{\hbar}\hat{H}_0 t}\hat{U}_a(t, t_0)e^{-\frac{i}{\hbar}t\hat{H}_0 t_0}.$$

(We assume that $h(t)$ is a smooth function equal to 1 in the neighborhood of 0 and to 0 in the neighborhood of infinity. It is increasing for negative t and decreasing for positive t.) The operator $\hat{S}_a$ is defined as $\hat{S}_a(\infty, -\infty)$.

The perturbation theory for operators $\hat{S}_a(t, t_0)$ can be constructed in the standard way: we apply the formula

$$\hat{S}_a(t, t_0) = T \exp\left(-\frac{i}{\hbar}h(at)\hat{H}_{\text{int}}(t)\right).$$

(We use the notation $A(t) = e^{\frac{i}{\hbar}\hat{H}_0 t}A e^{-\frac{i}{\hbar}\hat{H}_0 t}$.)

If we are interested only in the action of these operators on functionals represented by polynomials of α, α^* with smooth coefficients tending to zero at infinity (smooth functionals in the terminology of Tyupkin (1973)), the diagram techniques can be described as follows. The vertices come from $-\frac{i}{\hbar}h(at)\hat{H}_{\text{int}}$. To find the propagator, we calculate the T-product of two operators of the form $b_i^{(+)}(t)$ and express it in normal form with respect to the operators c_i^+, c_i (i.e. the operators c_i are from the right). The propagator (that can be considered as 4×4 matrix) is equal to the numerical part of this expression. In other words, the propagator is given by the formula

$$\langle T(b(k_1,t_1,\sigma_1)b(k_2,t_2,\sigma_2))\rangle_{L=1}.$$

Here, $b(k,0,\sigma)$ is one of the operators $b^+, b, \tilde{b}^+, \tilde{b}$.

Let us define the adiabatic generalized Green functions (GGreen functions) by the formula

$$G_n^a(k_1,t_1,\sigma_1,\ldots,k_n,t_n,\sigma_n)$$
$$= \langle T(b(k_1,t_1,\sigma_1)\cdots b(k_n,t_n,\sigma_n)\hat{S}_a(\infty,-\infty)))\rangle_{L=1} \quad (13.43)$$

As usual, the perturbative expansion for these functions can be constructed by the same rules as for adiabatic S-matrix, but the diagrams have n external vertices.

Note that $\hat{S}_a(\infty,-\infty)1 \to 1$ and $\hat{S}_a(0,-\infty)1 \to \mathbf{L}$ as $a \to 0$. Here, $\mathbf{L}$ denotes the L-functional corresponding to the ground state of the Hamiltonian H. (This follows immediately from similar statement for the adiabatic evolution operators $\hat{U}_a(\infty,-\infty)$ and $\hat{U}_a(0,-\infty)$ and from the fact that the L-functional $L = 1$ corresponds to the ground state of H_0.) We obtain that the adiabatic GGreen function $G_n^a(k_1,t_1,\sigma_1,\ldots,k_n,t_n,\sigma_n)$ tends to the GGreen function

$$G_n(k_1,t_1,\sigma_1,\ldots,k_n,t_n,\sigma_n) = \langle T(\mathbf{b}(k_1,t_1,\sigma_1)\cdots \mathbf{b}(k_n,t_n,\sigma_n))\rangle_{\mathbf{L}}$$
$$(13.44)$$

as $a \to 0$.

(Here, we use the notation

$$\mathbf{b}(k,t,\sigma) = \hat{S}^{-1}(t,0)b(k,t,\sigma)\hat{S}(t,0) = \hat{U}^{-1}(t,0)b(k,0,\sigma)\hat{U}(t,0)$$

for the analog of Heisenberg operators.) This means that we can construct the perturbation expansion for the GGreen function

$G_n(k_1, t_1, \sigma_1, \ldots, k_n, t_n, \sigma_n)$ taking the limit $a \to 0$ in the diagrams for $G_n^a(k_1, t_1, \sigma_1, \ldots, k_n, t_n, \sigma_n)$. The only modification of diagrams is in internal vertices: now, these vertices are governed by $-\frac{i}{\hbar}\hat{H}_{\mathrm{int}}$.

Similar procedure can be applied for the calculation of the action of operators $\hat{S}_a(t, t_0)$ on the space of functionals represented as a product of a smooth functional and Gaussian functional $\Lambda = e^\lambda$, where λ is a quadratic expression in terms of α^*, α. We assume that Λ is translation invariant and stationary with respect to the evolution corresponding to the Hamiltonian H_0, i.e. $\hat{H}_0\Lambda = 0$. (Here, H_0 is a translation-invariant quadratic Hamiltonian not necessarily of the form (13.37).) It is easy to check imposing some non-degeneracy conditions that there exist such operators $\hat{c}_i^+(k), \hat{c}_i(k), i = 1, 2$ obeying CCR that Λ can be characterized as a functional satisfying the conditions $\hat{c}_i(k)\Lambda = 0, i = 1, 2, \Lambda(0, 0) = 0$. Then the perturbative expression for the action of $\hat{S}_a(t, t_0)$ on the space under consideration can be obtained by means of the diagram technique with propagators described in the same way as for $\Lambda = 1$, the only difference is that instead of normal form with respect to the operators $c_i^+(k), c_i(k)$, we should consider normal form with respect to the operators $\hat{c}_i^+(k), \hat{c}_i(k), i = 1, 2$. Equivalently, we can define the propagator by the formula

$$\langle T(b(k_1, t_1, \sigma_1)b(k_2, t_2, \sigma_2))\rangle_\Lambda.$$

Again, we can define adiabatic GGreen functions

$$G_n^a(k_1, t_1, \sigma_1, \ldots, k_n, t_n, \sigma_n)_\Lambda$$
$$= \langle T(b(k_1, t_1, \sigma_1) \cdots b(k_n, t_n, \sigma_n)\hat{S}_a(\infty, -\infty)))\rangle_\Lambda \qquad (13.45)$$

corresponding to Λ and introduce the diagram technique for their calculation. The GGreen functions corresponding to Λ can be defined either as limits of adiabatic GGreen functions as $a \to 0$ or by the formula

$$G_n(k_1, t_1, \sigma_1, \ldots, k_n, t_n, \sigma_n)_{\boldsymbol{\Lambda}} = \langle T(\mathbf{b}(k_1, t_1, \sigma_1) \cdots \mathbf{b}(k_n, t_n, \sigma_n)))\rangle_{\boldsymbol{\Lambda}},$$
$$(13.46)$$

where $\boldsymbol{\Lambda} = \lim_{a \to 0} \hat{U}_a(0, -\infty)\Lambda$ denotes the stationary state of the Hamiltonian H that we obtain from the stationary state Λ of the Hamiltonian H_0 adiabatically switching the interaction on. The diagrams representing the functions (13.46) have n external vertices, the internal vertices come from $-\frac{i}{\hbar}\hat{H}_{\text{int}}$, the propagator is equal to

$$\langle T(b(k_1, t_1, \sigma_1)b(k_2, t_2, \sigma_2))\rangle_\Lambda.$$

In particular, we can take H_0 of the form (13.37) and

$$\Lambda = e^{-\int \alpha^*(k)n(k)\alpha(k)dk}. \tag{13.47}$$

(All translation-invariant Gaussian functionals that are stationary with respect to the Hamiltonian (13.37) have this form.) Then we obtain the following formulas for the propagator:

$$\langle T(b^+(t)b(\tau))\rangle_\Lambda = \theta(t - \tau)e^{i\hbar\omega(k)(t-\tau)}n(k)$$
$$+ \theta(\tau - t)e^{i\hbar\omega(k)(t-\tau)}(n(k) + \hbar)r,$$
$$\langle T(b^+(t)\tilde{b}^+(\tau))\rangle_\Lambda = e^{i\hbar\omega(k)(t-\tau)}(n(k) + \hbar),$$
$$\langle T(\tilde{b}^+(t)\tilde{b}(\tau))\rangle_\Lambda = \langle T(b^+(t)b(\tau))\rangle_\Lambda,$$
$$\langle T(b(t)\tilde{b}(\tau))\rangle_\Lambda = e^{i\hbar\omega(k)(t-\tau)}n(k).$$

All other entries vanish.

It follows from the above formulas that the diagrams we constructed coincide with the diagrams of Keldysh and TFD formalisms (see Chu and Umezawa (1994) for review of both formalisms).

We have noted already that the L-functional corresponding to an equilibrium state of quadratic Hamiltonian is Gaussian. Assuming that the equilibrium state of the Hamiltonian $H_0 + gV$ can be obtained from the equilibrium state of H_0 by means of adiabatic evolution, we can say that the diagram technique we have described allows us to calculate the GGreen functions in the equilibrium state.

13.4.4 *GGreen functions*

We have constructed the diagram technique for GGreen functions. As in the standard technique, we can express all diagrams in terms of connected diagrams; moreover, connected diagrams can be expressed in terms of 1 PI diagrams. (One says that a diagram is one particle irreducible (1 PI) if it remains connected when we remove one of the edges. Calculating a 1 PI diagram, we do not take into account the contributions of external edges.)

The contribution of a disconnected diagram is equal to the product of the contributions of its components (up to some factor taking into account the symmetry group of the diagram). The two-point GGreen function

$$G_2(k_1, t_1, \sigma_1, k_2, t_2, \sigma_2)_\Lambda = \langle T(\mathbf{b}(k_1, t_1, \sigma_1)\mathbf{b}(k_2, t_2, \sigma_2)))\rangle_\Lambda \quad (13.48)$$

obeys the Dyson equation

$$\begin{aligned}
G_2(&k_1, t_1, \sigma_1, k_2, t_2, \sigma_2)_\Lambda \\
&= G_2(k_1, t_1, \sigma_1, k_2, t_2, \sigma_2)_\Lambda \\
&\quad + \int dk_2' dt_2' d\sigma_2' dk_2'' dt_2'' d\sigma_2'' G_2(k_1, t_1, \sigma_1, k_2', t_2', \sigma_2')_\Lambda \\
&\quad \times M(k_2', t_2', \sigma_2', k_2'', t_2'', \sigma_2'')_\Lambda \\
&\quad \times G_2(k_2'', t_2'', \sigma_2'', k_2, t_2, \sigma_2)_\Lambda
\end{aligned} \quad (13.49)$$

connecting it with the propagator and self-energy operator (mass operator) M (the integration over discrete parameters is understood as summation). The generalized mass operator M is defined as a sum of 1 PI diagrams. The Green functions and mass operator can be regarded as kernels of integral operators, hence the Dyson equation can be represented in operator form

$$G_2^\Lambda = G_2^\Lambda + G_2^\Lambda M^\Lambda G_2^\Lambda$$

or

$$(G_2^\Lambda)^{-1} = (G_2^\Lambda)^{-1} + M^\Lambda. \quad (13.50)$$

It is useful to write this equation in (k, ϵ)-representation. (Here, ϵ stands for the energy variable.) Due to translational invariance in this representation, the operators entering Dyson equation are operators of multiplications by a matrix function of (k, ϵ). We identify the operators with these matrix functions. The (quasi-)particles are related to the poles of the matrix function $G_2^\Lambda(k, \epsilon)$. Recall that we assume that the Hamiltonian H is represented as a sum of quadratic Hamiltonian $H_0 = \int \omega(k)a^+(k)a(k)dk$ and interaction Hamiltonian $H_{\text{int}} = gV$. The poles of the GGreen function for $g = 0$ (of the propagator) are located at the points $\pm\omega(k)$ (we set $\hbar = 1$); the dependence of these poles on g can be found in the framework of the perturbation theory; the location of these poles will be denoted $\pm\omega(k\,|\,g)$. (Note that we cannot apply the perturbation theory directly, but we can use it to find zeros of the RHS of (13.50). The function $\omega(k\,|\,g)$ can be regarded as energy of (quasi-)particle. Only in the ground state, one can hope that this function is real (thermal quasiparticles are in general unstable).

13.4.5 *Adiabatic S-matrix*

The Dyson equation can also be written for adiabatic GGreen functions; they can be used to describe the asymptotic behavior of these functions for $a \to 0$. As we have noted, the adiabatic GGreen functions tend to GGreen functions as $a \to 0$, but they do not converge uniformly. However, the adiabatic self-energy operator converges uniformly; this allows us to analyze the asymptotic behavior of GGreen functions. (The same is true for conventional Green functions). The reason for the uniform convergence is the fact that matrix function M in (k,t)-representation tends to zero as $t \to \infty$. Conventional adiabatic Green functions were analyzed by Likhachev *et al.* (1970), the same method was applied by Tyupkin (1973) to obtain the approximation for adiabatic GGreen functions. These results were used to obtain the renormalized scattering matrix from adiabatic scattering matrix. Note that all these considerations are based on the assumption that the functions $\omega(k\,|\,g)$ are real, therefore rigorously they can be applied only to the scattering of particles (of elementary excitations of the ground state). Nevertheless, they make sense as approximate formulas if the quasiparticles are almost stable (we should assume that the collision time is much less than the lifetime of quasiparticles and choose a in such a way that $\frac{1}{a} \ll$ the lifetime of quasiparticles, but $\frac{1}{a} \gg$ the collision time).

The following statements were derived by Tyupkin (1973) from the results of Schwarz (1967) and Likhachev *et al.* (1970) in the framework of perturbation theory.

The scattering matrix in the formalism of L-functionals can be defined as an operator on the space of smooth L-functionals by the formula

$$\hat{S} = \lim_{a \to 0} V_a \hat{S}_a V_a, \tag{13.51}$$

where $\hat{S}_a = \hat{S}_a(\infty, -\infty)$ stands for the adiabatic S-matrix,

$$V_a = \exp i \int r_a(k)(c_1^+(k)c_1(k) - c_2^+(k)c_2(k))dk$$

and the function r_a is chosen from the requirement that $\hat{S}$ acts trivially on one-particle states. (One can give an explicit expression for r_a in terms of one-particle energies $\omega(k \,|\, g)$; namely $r_a(k) = \frac{1}{a}\int_{-\infty}^{0}(\omega(k \,|\, h(\tau)) - \omega(k))d\tau$.)

One can prove the existence of the limit in (13.51) *in the framework of perturbation theory.* (One should impose the condition $\omega(k_1 + k_2) < \omega(k_1) + \omega(k_2)$. This condition means that one-particle spectrum does not overlap with multi-particle spectrum.)

The conventional renormalized S-matrix was related by Likhachev *et al.* (1970) to the adiabatic S-matrices. (To obtain the S-matrix, one should multiply the adiabatic S-matrix in finite volume by factors similar to V_a, take the limit when the volume tends to infinity, and then take the limit $a \to 0$.) Using the methods of Likhachev *et al.* (1970), one can relate $\hat{S}$ to the conventional renormalized S-matrix S; we obtain $\hat{S}L_K = L_{SKS^{-1}}$. Using this formula, one can express inclusive cross-section in terms of $\hat{S}$ (see Section 13.4.6).

13.4.6   *Scattering of (quasi-)particles; inclusive cross-section*

This subsection is independent of the rest of this section (we use only the definition of *L*-functional).

Let us start with the situation of quantum field theory when the standard scattering matrix S is well defined as an operator acting in the Fock space of asymptotic states. (Strictly speaking, we should denote the operators acting in this space as $a_{\text{in}}(k), a_{\text{in}}^+(k)$, but we use shorter notations $a(k), a^+(k)$.) We assume that the scattering matrix as well as Møller matrices S_-, S_+ are unitary. Considering the Fock space as a representation of CCR, we assign an L-functional L_K to a density matrix K in the Fock space using the formula (13.47). We define the scattering matrix in the space of L-functionals by the formula

$$\hat{S}L_K = L_{SKS^*}. \tag{13.52}$$

If the density matrix K corresponds to a vector Ψ, we can represent the RHS of (13.52) as

$$\mathrm{Tr}\, e^{-\alpha a^+} e^{\alpha^* a} SKS^*$$

$$= \mathrm{Tr}\, e^{\alpha^* a} SKS^* e^{-\alpha a^+} = \langle e^{\alpha^* a} S\Psi, e^{-\alpha^* a} S\Psi\rangle$$

$$= \sum_n \int dp_1 \cdots dp_n \langle e^{\alpha^* a} S\Psi | p_1, \ldots, p_n\rangle \overline{\langle e^{-\alpha^* a} S\Psi | p_1, \ldots, p_n\rangle},$$

where $|p_1, \ldots, p_n\rangle = \sqrt{\frac{1}{n!}} a^+(p_1) \cdots a^+(p_n)\theta$ constitute an orthonormal basis in Fock space. (Here, $\theta = |0\rangle$ stands for Fock vacuum.) The expression $\langle a(k_1) \cdots a(k_m) S\Psi | p_1, \ldots, p_n\rangle = \sqrt{\frac{(m+n)!}{n!}} \langle S\Psi | k_1, \ldots, k_m, p_1, \ldots, p_n\rangle$ can be interpreted as the scattering amplitude of the process $\Psi \to (k_1, \ldots, k_m, p_1, \ldots, p_n)$ (after dividing by numerical factor). We see that the LHS is expressed in terms of scattering amplitudes. In particular, writing it in the form

$$\sum_{m,m'} \int dk_1 \cdots dk_m dk'_1 \cdots dk'_{m'} \frac{(-1)^m}{m!m'!} \alpha(k_1)$$

$$\cdots \alpha(k_m)\alpha^*(k'_1)\alpha'_{m'}(k'_{m'})\hat{\sigma}_{m,m'}(k_1, \ldots, k_m | k'_1, \ldots, k'_{m'} | \Psi),$$

we obtain

$$\hat{\sigma}_{m,m'}(k_1, \ldots, k_m | k'_1, \ldots, k'_{m'} | \Psi)$$

$$= \sum_n \int dp_1 \cdots dp_n \frac{\sqrt{(m+n)!}\sqrt{(m'+n)!}}{n!}$$

$$\times \langle S\Psi | k_1, \ldots, k_m, p_1, \ldots, p_n\rangle \overline{\langle S\Psi | k'_1, \ldots, k'_{m'}, p_1, \ldots, p_n\rangle}.$$

If $m = m'$ and $k_i = k'_i$, this expression is proportional to the inclusive cross-section $\Psi \to k_1, \ldots, k_m$. If the initial state Ψ has definite momentum q, then $\langle S\Psi \,|\, k_1, \ldots, k_m, p_1, \ldots, p_n \rangle$ is a product of delta-function $\delta(\sum k_i + \sum p_j - q)$ coming from momentum conservation and a function that will be denoted $\rho_m(k_1, p_1, \ldots, p_n \,|\, \Psi)$. Now,

$$\hat{\sigma}_{m,m'}(k_1, \ldots, k_m | k_1', \ldots, k_{m'}' | \Psi)$$

$$= \tilde{\sigma}_{m,m'}(k_1, \ldots, k_m | k_1', \ldots, k_{m'}' | \Psi)$$

$$\times \delta(k_1' + \cdots + k_m' - (k_1 + \cdots + k_m)),$$

where

$$\tilde{\sigma}_{m,m'}(k_1, \ldots, k_m | k_1', \ldots, k_{m'}' | \Psi)$$

$$= \sum_n \int dp_1 \cdots dp_n \frac{\sqrt{(m+n)!}\,\sqrt{(m'+n)!}}{n!}$$

$$\times \rho_m(k_1, \ldots, k_m, p_1, \ldots, p_n \,|\, \Psi)\overline{\rho_{m'}(k_1', \ldots, k_n', p_1, \ldots, p_n \,|\, \Psi)}$$

$$\times \delta(k_1 + \cdots + k_m + p_1 + \cdots + p_n - q).$$

The inclusive cross-section is proportional to $\tilde{\sigma}_{m,m}(k_1, \ldots, k_m | k_1, \ldots, k_m | \Psi)$ in this case.

We will call $\hat{S}$ inclusive S-matrix. Let us show that this matrix can be calculated in terms of GGreen functions (more precisely, in terms of on-shell values of these functions). Let us take a density matrix K in the representation space of CCR. We assume that the momentum and energy operators (infinitesimal spatial and time translations) act on this space and K is translation invariant. This allows us to define Heisenberg operators $a^+(k,t), a(k,t)$ where k is the momentum variable and t is the time variable. Then the corresponding GGreen function can be defined as $\mathrm{Tr}\, BAK$. Here, A denotes chronological product (T-product) of Heisenberg operators (the times are decreasing) and B stands for antichronological product T^{opp} of Heisenberg operators (the times are increasing).

If the density matrix K corresponds a vector Φ, the GGreen function can be represented in the form

$$\langle A\Phi, B^*\Phi \rangle,$$

where B^* stands for chronological product of Hermitian conjugate operators.

Let us now consider the case when Φ is the ground state. Then one can obtain an expression of $\hat{S}$ in terms of GGreen functions that is analogous to the LSZ formula. It will be derived from some identities that were used in the proof of LSZ, namely, we can use the identity

$$[\cdots [S, a_{\text{in}}(k_1, \sigma_1)] \cdots a_{\text{in}}(k_n, \sigma_n)]$$

$$= (-1)^n S_- S_+^* \int dt_1 \cdots dt_n L_n \cdots L_1 T(a(k_1, t_1, \sigma_1) \cdots a(k_n, t_n, \sigma_n)),$$

$$(13.53)$$

where S_-, S_+ are Møller matrices, the scattering matrix S is represented in terms of in-operators $a_{\text{in}}(k, 1) = a_{\text{in}}^+(k)$, $a_{\text{in}}(k, -1) = a_{\text{in}}(k)$, operators $a(k, t, \sigma)$ are Heisenberg operators $a^+(k, t), a(k, t)$, the operators $\int dt_i L_i$ in (k, ϵ)-representation can be interpreted as "on-shell operators." (To apply the operator $\int dt_i L_i$ in (k, ϵ)-representation, we should multiply by $i\Lambda(k_i, \sigma_i)(\epsilon_i - \omega(k_i))$ and take the limit $\epsilon_i \to \omega(k_i)$. Here, $\omega(k)$ stands for the location of the pole of the two-point Green function and Λ can be expressed in terms of the residue in this pole. Note that in the transition to (k, ϵ)-representation, we are using direct Fourier transform for $\sigma = -1$ and inverse Fourier transform for $\sigma = 1$.)

The identity (13.53) can be obtained, for example, from (32.17) of Section 9.2 (by means of conjugation with S_-).

It follows from (13.53) that Green function defined as vacuum expectation value of chronological product $T(a(k_1, t_1, \sigma_1) \cdots a(k_n, t_n, \sigma_n))$ is related to scattering amplitude: one should take the Fourier transform with respect to time variables and "go on-shell" in the sense explained above. This gives the LSZ formula (see the preceding section for more general approach). We remark that these considerations also go through in the case when instead of vacuum expectation value $\langle 0|A|0 \rangle$ where $|0\rangle$ obeys $a_{\text{in}}(k)|0\rangle = 0$ (represents physical vacuum), we can take matrix elements $\langle 0|A|p_1, \ldots, p_n \rangle$ where $|p_1, \ldots, p_n\rangle = \frac{1}{\sqrt{n!}} a_{\text{in}}^+(p_1) \cdots a_{\text{in}}^+(p_n)|0\rangle$. This remark allows us to express "on-shell" GGreen functions as sesquilinear combinations of scattering amplitudes; comparing this expression with the formula for $\hat{S} L_K$, we obtain the expression of $\hat{S}$ in terms of GGreen functions "on-shell" (an analog of LSZ formula). Indeed, we can consider

GGreen function as vacuum expectation value of chronological product A multiplied by antichronological product B. Using the fact that $|p_1, \ldots, p_n\rangle$ constitute a generalized orthonormal basis, we can say that

$$\langle 0|BA|0\rangle = \sum_n \int dp_1 \cdots dp_n \langle 0|B|p_1, \ldots, p_n\rangle \langle p_1, \ldots, p_n|A|0\rangle.$$

$$(13.54)$$

This representation allows us to express on-shell GGreen functions in terms of scattering amplitudes. (The antichronological product is related to the chronological one by Hermitian conjugation.)

Let us write explicit expressions obtained this way. We represent S in normal form

$$S = \sum_{r,s} \frac{1}{r!s!} \int dp_1 \cdots dp_r dq_1 \cdots q_s \sigma_{r,s}(p_1, \ldots, p_r \,|\, q_1, \ldots, q_s) a_{\mathrm{in}}^+(p_1)$$

$$\cdots a_{\mathrm{in}}^+(p_r) a_{\mathrm{in}}(q_1) \cdots a_{\mathrm{in}}(q_s).$$

We assume that all p_i's are distinct and all q_j's are distinct, then the coefficient functions in normal form coincide with scattering amplitudes $\langle S a_{\mathrm{in}}^+(q_1) \cdots a_{\mathrm{in}}^+(q_s)\theta \,|\, a_{\mathrm{in}}^+(p_1) \cdots a_{\mathrm{in}}^+(p_r)\theta\rangle$. Let us take $\sigma_i = 1$ for $i \leq m$, $\sigma_i = -1$ for $i > m$ in (13.53). Introducing the notation $q_i = k_{i-m}$, we obtain that

$$\langle p_1, \ldots, p_n|LHS|0\rangle = \frac{1}{\sqrt{n!}} \sigma_{m+n,l}(p_1, \ldots, p_n, k_1, \ldots, k_m \,|\, q_1, \ldots, q_l).$$

Here, LHS stands for the LHS of (13.53). Now, we can apply (13.53) and (13.54) to identify on-shell GGreen functions with matrix entries of the inclusive scattering matrix $\hat{S}$.

Our considerations used LSZ relations that are based on the conjecture that the theory has particle interpretation. Moreover, the very definition of $\hat{S}$ that we have applied requires the existence of conventional S-matrix. However, using (13.51) as the definition of the inclusive scattering matrix $\hat{S}$, one can prove the relation between $\hat{S}$ and on-shell GGreen functions analyzing diagram techniques for these objects. This proof can also be applied in the case when the theory does not have particle interpretation. Moreover, the same ideas can be applied to quasiparticles considered as elementary excitations of translation-invariant stationary state. These excitations are related to the poles of the two-point GGreen function; we can define the inclusive scattering matrix as on-shell GGreen function or generalizing (13.51). Of course, this definition makes sense only if the quasiparticles are (almost) stable.

In the preceding section, we have analyzed the inclusive scattering matrix in the framework of algebraic approach to quantum theory (see also Schwarz (2019c)).

Appendix

A.1   Hilbert spaces

Let us consider a set $\mathcal{H}$ equipped with two operations: addition of two elements and multiplication of elements by a complex number. We say that $\mathcal{H}$ is a *linear space* if the following conditions are satisfied: (1) $x+y = y+x$; (2) $(x+y)+z = x+(y+z)$; (3) $\lambda(x+y) = \lambda x+\lambda y$; (4) $\lambda(\mu x) = (\lambda\mu)x, (\lambda+\mu)x = \lambda x + \mu x$; (5) $0 \cdot x = 0, 1 \cdot x = x$ (where $x, y, z \in \mathcal{H}$; λ, μ are complex numbers). The elements of the linear space are called vectors. We say the space is spanned by a subset S if every element of the space can be represented as a linear combination of the elements of the subset S. The dimension of a linear space is the minimal number of vectors spanning the space. If this number is infinite, then we say that the space is infinite dimensional. The symbol 0 denotes the vector satisfying the conditions $x + 0 = x$ for all x in $\mathcal{H}$.

A function on $\mathcal{H}$ taking values in the set of complex numbers is called a *linear functional* if $f(\lambda x + \mu y) = \lambda f(x) + \mu f(y)$.

We say that a linear space is *pre-Hilbert* if it is equipped with the inner product $\langle x, y\rangle$, satisfying the following axioms: (1) $\langle x, y\rangle = \overline{\langle y, x\rangle}$; (2) $\langle \lambda x, y\rangle = \lambda\langle x, y\rangle$; (3) $\langle x+y, z\rangle = \langle x, z\rangle + \langle y, z\rangle$; (4) $\langle x, x\rangle \geq 0$ and $\langle x, x\rangle = 0$ only if $x = 0$ (here, $x, y, z \in \mathcal{H}$ and λ is a complex number).

We say that a bijective map is isomorphic if it preserves all operations defined in a pre-Hilbert space.

The number $\|x\| = \sqrt{\langle x, x\rangle}$ is called the norm of vector x. The vector x is normalized if $\|x\| = 1$. Two vectors are orthogonal if $\langle x, y\rangle = 0$.

One says that the sequence $x_n \in \mathcal{H}$ convergences to the element $x \in \mathcal{H}$ (denoted by $x = \lim x_n$), if $\lim_{n \to \infty} \|x - x_n\| = 0$. We say that the sequence $x_n \in \mathcal{H}$ weakly converges to the vector $x \in \mathcal{H}$ (denoted $x = \text{wlim}\, x_n$) if for all elements $y \in \mathcal{H}$ the sequence $\langle x_n, y \rangle$ tends to $\langle x, y \rangle$.

A sequence is called a Cauchy sequence if $\lim_{m,n \to \infty} \|x_m - x_n\| = 0$.

The set $M \subset \mathcal{H}$ is closed in $\mathcal{H}$ if every point $x \in \mathcal{H}$ that can be represented as a limit of a sequence $x_n \in M$ is itself an element of M. A set M is dense in $\mathcal{H}$ if every vector $x \in \mathcal{H}$ can be represented as a limit of elements of M.

A pre-Hilbert space $\mathcal{H}$ is called Hilbert space if every Cauchy sequence converges. Every pre-Hilbert space can be embedded in a Hilbert space. In other words, for every pre-Hilbert space $\mathcal{H}$, one can construct a Hilbert space $\tilde{\mathcal{H}}$, called the completion of $\mathcal{H}$, and an isomorphic map α of the space $\mathcal{H}$ onto a dense subset of the space $\tilde{\mathcal{H}}$ (isomorphic embedding of $\mathcal{H}$ into $\tilde{\mathcal{H}}$).

The completion $\tilde{\mathcal{H}}$ is unique in the following sense: if $\mathcal{H}_1$ and $\mathcal{H}_2$ are two completions and α_1 and α_2 are isomorphic embeddings of $\mathcal{H}$ into $\mathcal{H}_1$ and $\mathcal{H}_2$, then there exists an isomorphic map α of $\mathcal{H}_1$ onto $\mathcal{H}_2$ satisfying $\alpha \alpha_1 = \alpha_2$.

A pre-Hilbert space is called separable if it contains a countable dense subset. We will consider only separable spaces.

A.2   Systems of vectors in a pre-Hilbert vector space

A subset $M \subset \mathcal{H}$ is called a linear subspace if every linear combination $\lambda x + \mu y$ with $x, y \in M$ is also in M.

A system of vectors $\xi_a \in \mathcal{H}$ is called orthonormal if $\langle \xi_\alpha, \xi_\beta \rangle = \delta_{\alpha\beta}$ (we consider systems that are either finite or countable). A orthonormal system is called an orthonormal basis if linear combinations of elements of this system are dense in $\mathcal{H}$.

The quantities $\langle x, \xi_a \rangle$ are called the Fourier coefficients of x in the orthonormal basis ξ_a. For every vector $x \in \mathcal{H}$, the series $\sum_a \langle x, \xi_a \rangle \xi_a$ (called the Fourier series of x in the orthonormal basis ξ_a) converges to the vector x. The scalar product $\langle x, y \rangle$ can be expressed in terms of the Fourier coefficients as follows:

$$\langle x, y \rangle = \sum_\alpha \langle x, \xi_\alpha \rangle \, \overline{\langle y, \xi_\alpha \rangle}.$$

In particular, $\langle x, x \rangle = \sum_\alpha |\langle x, \xi_\alpha \rangle|^2$.

If $\mathcal{H}$ is a Hilbert space, then the sequence c_α is a sequence of Fourier coefficients of vector x if and only if $\sum_\alpha |c_\alpha|^2 < \infty$.

A.3 Examples of function spaces

1. The space l_n^2, also denoted as $\mathbb{C}^n$, is the n-dimensional space of rows of n complex numbers equipped with the inner product

$$\langle x, y \rangle = \sum_{i=1}^{n} x_i \bar{y}_i$$

   (here, $x = (x_1, \ldots, x_n) \in \mathbb{C}^n, y = (y_1, \ldots, y_n) \in \mathbb{C}^n$).
2. The space l^2 consists of sequences of complex numbers $x = (x_1, \ldots, x_i, \ldots)$, satisfying the condition $\sum_{i=1}^{\infty} |x_i|^2 < \infty$. Addition and multiplication by a complex number are defined coordinate-wise and the scalar product is defined as $\langle x, y \rangle = \sum_{i=1}^{\infty} x_i \bar{y}_i$.

   The spaces l_n^2 and l^2 are Hilbert spaces.

   If $\mathcal{H}$ is a Hilbert space and ξ_α is an orthonormal basis for $\mathcal{H}$, then we can construct an isomorphic map to the space l_n^2 (if $\mathcal{H}$ is finite dimensional) and l^2 (if $\mathcal{H}$ is infinite dimensional). The isomorphism transforms a vector x to its sequence of Fourier coefficients in the basis ξ_α, obtained as $\langle x, \xi_\alpha \rangle$.
3. The space $C^2(E^r)$ consists of functions on an r-dimensional space E^r that are continuous and square integrable (i.e. the functions satisfy the condition $\int |f(\xi_1, \ldots, \xi_r)|^2 d\xi_1, \ldots, d\xi_r < \infty$). Multiplication and addition of functions are defined in the standard way. The scalar product of two functions f and g in $C^2(E^r)$ is given by

$$\langle f, g \rangle = \int f(\xi_1, \ldots, \xi_r) \overline{g(\xi_1, \ldots, \xi_r)} d\xi_1, \ldots, \xi_r$$

$$= \int f(\xi) \overline{g(\xi)} d\xi.$$

The space $C^2(E^r)$ is a pre-Hilbert space. Its completion is $L^2(E^r)$.

4. The space $C^2(E^r \times S)$, where S denotes a finite set, consists of functions $f(\xi, s)$ that are continuous with respect to the variable ξ and are square integrable (i.e. satisfying the condition $\sum_s \int |f(\xi, s)|^2 d\xi < \infty$). Here, $\xi = (\xi_1, \ldots, \xi_r) \in E^r$; $s \in S, d\xi = d\xi_1 \ldots d\xi_r$. The scalar product is defined by

$$\langle f, g \rangle = \sum_s \int f(\xi, s)\overline{g(\xi, s)} d\xi.$$

The function $f(\xi, s)$ can be considered as a column of k functions $f_s(\xi) = f(\xi, s)$ depending on the variable $\xi \in E^r$ (k is the number of elements in S).

The completion of $C^2(E^r \times S)$ is denoted by the symbol $L^2(E^r \times S)$.

Let us fix a family $\mathcal{B}$ of subsets of M that contains with every two subsets A, B the subsets $A \cup B, A \cap B$, and $A \setminus B$ (we call this family a ring of subsets). Let us also assume that the set M is a countable union of subsets belonging to the family $\mathcal{B}$. A countably additive measure μ on a family $\mathcal{B}$ is specified by an assignment of non-negative numbers $\mu(A)$ to every set $A \in \mathcal{B}$.

We require that for any countable sequence of sets $A, A_1, \ldots,$ $A_n, \ldots$ from $\mathcal{B}$ we have $\mu(A) = \sum_{i=1}^{\infty} \mu(A_i)$ if

(1) $A = \cup_{i=1}^{\infty} A_i$,
(2) the sets $A_1, \ldots, A_n, \ldots$ are pairwise disjoint. (In other words, we assume that the measure is countably additive.)

The set M, equipped with the family $\mathcal{B}$ and the countably additive measure μ, specifies a *measure space*.

A set $R \subset M$ (not necessarily belonging to $\mathcal{B}$) is called a set of measure zero if for every $\epsilon > 0$, there exists a set $A \in \mathcal{B}$, such that $R \subset A$ and $\mu(A) < \epsilon$. One says that a sequence of functions $f_n(x)$ converges to a function $f(x)$ almost everywhere on a set M if the set of points $x \in M$, such that the sequence $f_n(x)$ does not converge to $f(x)$, is a set of measure zero. In general, if a relation is satisfied everywhere except a measure zero set, then we say that it is satisfied almost everywhere.

For functions defined on a measure space, one can define a notion of Lebesgue integral. A function f defined on a set $A \in \mathcal{B}$ is called a simple function, if the set of values of the function is finite $\{y_1, \ldots, y_n\}$ and the sets A_i consisting of points where the function f takes the value y_i belong to the family $\mathcal{B}$. The Lebesgue integral of a simple function f is defined by the formula

$$\int_A f(x)d\mu = \sum_{i=1}^{r} y_i \mu(A_i).$$

A function $f(x)$ that can be represented as a limit of a sequence of simple functions converging almost everywhere is called measurable. In what follows, we consider only measurable functions. Let us say that two measurable functions are equivalent if their difference is equal to zero almost everywhere. We do not distinguish equivalent measurable functions.

If a measurable function f is bounded, then it can be represented as a limit of a sequence of simple functions $f_n(x)$ that are bounded from above by a constant, $|f_n(x)| \leq C$. The Lebesgue integral of the function f on a set $A \in \mathcal{B}$ can be defined by

$$\int_A f(x)d\mu = \lim_{n\to\infty} \int_A f_n(x)d\mu.$$

One can prove that this limit always exists and does not depend on the choice of the sequence $f_n(x)$.

If the function f is unbounded and non-negative, then one can define its Lebesgue integral by the relation

$$\int_A f(x)d\mu = \lim_{n\to\infty} \int_A f_n(x)d\mu,$$

where $f_n(x) = f(x)$ for $f(x) \leq n$ and $f_n(x) = 0$ for $f(x) \geq n$.

If the set A does not belong to the family $\mathcal{B}$, but can be represented as $A = \bigcup_{i=1}^{\infty} A_i$, where $A_i \in \mathcal{B}$, then the Lebesgue integral of a non-negative function f over A is defined by

$$\int_A f(x)d\mu = \lim_{n\to\infty} \int_{A_1 \cup \cdots \cup A_n} f(x)d\mu.$$

(For any non-negative measurable function, the Lebesgue integral exists, but is not necessarily finite.)

A measurable function is called Lebesgue summable on the set $A \subset M$, if the Lebesgue integral of the function $|f(x)|$ is finite. For a summable real function, the Lebesgue integral is defined by the relation

$$\int_A f(x)d\mu = \int_A |f(x)|d\mu - \int_A (|f(x)| - f(x))d\mu.$$

If the function $f(x)$ is complex, then the Lebesgue integral is defined by $\int_A \mathrm{Re}(f(x))d\mu + i \int_A \mathrm{Im}(f(x))d\mu$, where $\mathrm{Re}(f(x)), \mathrm{Im}(f(x))$ are real and imaginary parts of $f(x)$.

We note some important properties of the Lebesgue integral:

1. Let $f_n(x)$ be a sequence of functions converging almost everywhere to the function $f(x)$ and suppose there exists a function $g(x)$ dominating the functions $f_n(x)$ (i.e. $|f_n(x)| \leq g(x)$). Then $\lim_{n \to \infty} \int_A f_n(x)d\mu = \int_A f(x)d\mu$.

   This theorem allows us to interchange the limit and Lebesgue integration.

2. Let M_1 and M_2 be measure spaces. Then in the space $M_1 \times M_2$ consisting of the pairs (x_1, x_2), where $x_1 \in M_1$ and $x_2 \in M_2$, one can construct a countably additive measure μ defined by $\mu(A_1 \times A_2) = \mu_1(A_1)\mu_2(A_2)$ (here, μ_i denotes the measure on M_i; the measure μ is called the product measure of μ_1 and μ_2).

   If the function $f(x_1, x_2)$ is a function on the product $M_1 \times M_2$ that is summable with respect to the measure μ, then for almost all x_2 the function $f(x_1, x_2)$ is a summable function of $x_1 \in M_1$. The double integral of the function f (i.e. the integral over the measure μ) can be expressed as the repeated integral

$$\int_{A_1 \times A_2} f(x_1, x_2)d\mu = \int_{A_2} d\mu_2 \int_{A_1} f(x, x_2)d\mu_1$$

   (Fubini's theorem).

Using the Lebesgue integral, one can introduce a notion of measure for some subsets that do not belong to the family $\mathcal{B}$. Namely, for every subset $A \subset M$, one can construct a function $\chi_A(x)$ that is equal to 1 for $x \in A$ and 0 for $x \notin A$ and set $\mu_L(A) = \int_M \chi_A(x)d\mu$. This measure μ_L is called the Lebesgue measure. It is defined for the set A if the function χ_A is measurable (then the set A is called measurable).

For every measure set M, we can construct a space $L^2(M)$ of square-integrable functions[1] (i.e. measurable functions such that the Lebesgue integral of the function $|f^2(x)|$ over the set M is finite). The scalar product on the space $L^2(M)$ can be defined by means of the Lebesgue integral[2]

$$\langle f, g \rangle = \int f(x)\overline{g(x)}d\mu.$$

One can prove that the space $L^2(M)$ is a Hilbert space.

For the Euclidean space E^r, we take the standard volume as a measure. Then, for functions that are integrable in the standard sense, the conventional integral (Riemann integral) coincides with the Lebesgue integral. The space $L^2(E^r)$ is the completion of the space $C^2(E^r)$.

If S is a finite set, then a measure of its subset A is by definition the number of points in A. The space $L^2(S)$ is isomorphic to the space l_k^2 where k is the number of points in S. In the set $E^r \times S$, we define the measure as the product of measure in E^r and S. The space $L^2(E^r \times S)$ constructed with this measure is the completion of the space $C^2(E^r \times S)$.

A.4 Operations with Hilbert spaces

The direct sum $\mathcal{H}_1 + \mathcal{H}_2$ of Hilbert spaces $\mathcal{H}_1$ and $\mathcal{H}_2$ is a Hilbert space, whose elements consist of pairs (h_1, h_2) where $h_1 \in \mathcal{H}_1, h_2 \in \mathcal{H}_2$, with addition, multiplication by a complex number, and the

---

[1]Two equivalent measurable functions specify the same element of the space of $L^2(M)$.

[2]Integrating over the whole measure space M, we use the notation $\int \phi(x)d\mu$ instead of the notation $\int_M \phi(x)d\mu$.

scalar product defined by

$$(h_1, h_2) + (h_1', h_2') = (h_1 + h_1', h_2 + h_2'),$$

$$\lambda(h_1, h_2) = (\lambda h_1, \lambda h_2),$$

$$\langle (h_1, h_2), (h_1', h_2') \rangle = \langle h_1, h_1' \rangle + \langle h_2, h_2' \rangle.$$

Let us assume that subspaces $\mathcal{H}_1$ and $\mathcal{H}_2$ of the Hilbert space $\mathcal{H}$ satisfy the following conditions: (1) every two vectors h_1 and h_2 are orthogonal and (2) every $h \in \mathcal{H}$ can be represented as a sum $h = h_1 + h_2$ of two vectors $h_1 \in \mathcal{H}_1$ and $h_2 \in \mathcal{H}_2$. Then one says that the Hilbert space $\mathcal{H}$ is represented as a direct sum of subspaces $\mathcal{H}_1$ and $\mathcal{H}_2$. In this case, one can construct a natural isomorphism of the direct sum $\mathcal{H}_1 + \mathcal{H}_2$ and the space $\mathcal{H}$, transforming the vector $(h_1, h_2) \in \mathcal{H}_1 + \mathcal{H}_2$ into the vector $h_1 + h_2 \in \mathcal{H}$.

The direct sum $\sum_i \mathcal{H}_i$ of a countable sequence of Hilbert spaces $\mathcal{H}_1, \ldots, \mathcal{H}_i, \ldots$ is defined as the space of sequences $(h_1, \ldots, h_i, \ldots)$, satisfying the condition $\sum_i \|h_i\|^2 < \infty$ (here, $h_i \in \mathcal{H}_i$). Linear combination and scalar product are defined by

$$\lambda(h_1, \ldots, h_i, \ldots) + \lambda'(h_1', \ldots, h_i', \ldots)$$

$$= (\lambda h_1 + \lambda' h_1', \ldots, \lambda h_i + \lambda' h_i', \ldots),$$

$$\langle (h_1, \ldots, h_i, \ldots), (h_1', \ldots, h_i', \ldots) \rangle$$

$$= \sum_i \langle h_i, h_i' \rangle.$$

The Hilbert space $\mathcal{H}$ is said to be a tensor product of Hilbert spaces $\mathcal{H}_1$ and $\mathcal{H}_2$, if there exists a bilinear map $\alpha(h_1, h_2)$ mapping $\mathcal{H}_1 \times \mathcal{H}_2$ to $\mathcal{H}$ and having the following properties:

(1) $\langle \alpha(h_1, h_2), \alpha(h_1', h_2') \rangle = \langle h_1, h_1' \rangle \cdot \langle h_2, h_2' \rangle$;

(2) the set of linear combinations of vectors of the form $\alpha(h_1, h_2)$ is dense in $\mathcal{H}$. We say that a map $\alpha(h_1, h_2)$ from $\mathcal{H}_1 \times \mathcal{H}_2$ into $\mathcal{H}$ is bilinear if it is linear with respect to h_1 for a fixed h_2 and is linear with respect to h_2 for fixed h_1.

The tensor product is defined by the above properties up to a natural isomorphism. (If $\mathcal{H}$ and $\mathcal{H}'$ are two tensor products of $\mathcal{H}_1$ and $\mathcal{H}_2$

and α_1 and α_2 are corresponding bilinear maps, then there exists an isomorphism λ of Hilbert spaces $\mathcal{H}$ and $\mathcal{H}'$, such that $\lambda\alpha = \alpha'$.)

The tensor product of spaces $\mathcal{H}_1$ and $\mathcal{H}_2$ will be denoted by the symbol $\mathcal{H}_1 \otimes \mathcal{H}_2$ and the vector $\alpha(h_1, h_2)$ by the symbol $h_1 \otimes h_2$.

It is easy to check that the tensor product of the spaces l_m^2 and l_n^2 is isomorphic to the space l_{mn}^2. The space l_{mn}^2 can be realized as a space of $m \times n$ matrices. Then the bilinear map α assigns to the vectors $(x_1, \ldots, x_m) \in l_m^2$ and $(y_1, \ldots, y_n) \in l_n^2$ the matrix A with the elements $A_{ij} = x_i y_j$.

The tensor product of n copies of space $\mathcal{H}$ is called the nth tensor power of the space and is denoted by $\otimes \mathcal{H}^n$. The formal definition is via taking repeated tensor products

$$\otimes \mathcal{H}^n = (\ldots((\mathcal{H} \otimes \mathcal{H}) \otimes \mathcal{H}) \cdots \otimes \mathcal{H}).$$

Every permutation π of the indices $(1, \ldots, n)$ naturally corresponds to an isomorphic map ρ_π of the space $\otimes \mathcal{H}^n$ on itself.

The subspace $\mathcal{H}_s^n$ of the space $\otimes \mathcal{H}^n$ consisting of vectors that satisfy the condition $\rho_\pi x = x$ for all permutations π is called an nth symmetric power of $\mathcal{H}$. The nth antisymmetric power of $\mathcal{H}_a^n$ of the space $\mathcal{H}$ is defined as the subspace consisting of vectors obeying $\rho_\pi x = (-1)^{\gamma(\pi)} x$ for all π (where $\gamma(\pi)$ stands for the parity of the permutation π).

Note that

$$L^2(M_1) \otimes L^2(M_2) = L^2(M_1 \times M_2)$$

(the measure in the space $M_1 \times M_2$ is defined as the product of measures in M_1 and M_2). The bilinear map α in the definition of the tensor product sends the functions $f_1 \in L^2(M_1)$, $f_2 \in L^2(M_2)$ to the function $f(x_1, x_2) = f_1(x_1) f_2(x_2) \in L^2(M_1 \times M_2)$. The symmetric (antisymmetric) power of the space $L^2(M)$ can be realized as the space of symmetric (antisymmetric) square integrable functions of the variables $x_1, \ldots, x_n \in M$.

A.5 Operators on Hilbert spaces

We consider linear operators acting on a Hilbert space $\mathcal{H}_1$ and taking values in a Hilbert space $\mathcal{H}_2$ as maps from dense subspace $D \subset$

$\mathcal{H}_1$ into $\mathcal{H}_2$, satisfying the condition of linearity (i.e. the operator A sends every vector $x \in D$ to the vector $Ax \in \mathcal{H}_2$ and satisfies $A(\lambda_1 x_1 + \lambda_2 x_2) = \lambda_1 A x_1 + \lambda_2 A x_2$ where $x_1, x_2 \in D$). The domain of definition for the operator A is denoted by D_A. The set of vectors in $\mathcal{H}_2$ that can be written as Ax for $x \in \mathcal{H}_1$ is called the range of A or the image of A.

We say that the operator A is bounded if there exists a number K, such that $\|Ax\| \leq K\|x\|$ for all $x \in D_A$. The operator norm of A is defined as $\sup_x \frac{\|Ax\|}{\|x\|}$. If the operator A is bounded, then it can be continuously extended to the whole space $\mathcal{H}_1$ (i.e. there exists a unique bounded operator on $\mathcal{H}_1$ that coincides with A on D_A).

The multiplication operator by the number λ will be denoted by the same symbol. In particular, the identity operator, considered as a multiplication operator by 1, will be denoted by the symbol 1.

A linear combination $C = \lambda_1 A_1 + \lambda_2 A_2$ of operators A_1 and A_2 is defined by $Cx = \lambda_1 A_1 x + \lambda_2 A_2 x$. Operator multiplication $F = A_1 A_2$ of operators A_1 and A_2 is defined as composition $Fx = A_1(A_2 x)$.

The linear combination of operators A_1 and A_2 acting from $\mathcal{H}_1$ into $\mathcal{H}_2$ is defined if the intersection of their domains of definition D_{A_1} and D_{A_2} is everywhere dense in $\mathcal{H}_1$. Similarly, the product of operators A_1 and A_2 is defined if A_2 maps $\mathcal{H}_0$ into $\mathcal{H}_1$, A_1 maps $\mathcal{H}_1$ into $\mathcal{H}_2$ and the set of $x \in D_{A_2}$ for which $A_2 x \in D_{A_1}$ is dense in $\mathcal{H}_0$.

If the range of an operator A, transforming $\mathcal{H}_1$ into $\mathcal{H}_2$, is dense in $\mathcal{H}_2$ and the operator is injective (only $0 \in \mathcal{H}_1$ goes to $0 \in \mathcal{H}_2$), then we can define the inverse operator A^{-1}, for which $A^{-1}y = x$, if $y = Ax$.

If the operator A transforms $\mathcal{H}_1$ into $\mathcal{H}_2$ and the operator B transforms $\mathcal{H}_2$ into $\mathcal{H}_1$, then the two are called conjugate if $\langle Ax, y \rangle = \langle x, By \rangle$ for all $x \in D_A, y \in D_B$.

If two operators B_1 and B_2 are both conjugate with the operator A and both are defined on the vector $y \in \mathcal{H}_2$, then $B_1 y = B_2 y$ (this follows from the density of D_A in $\mathcal{H}_1$). We will denote by A^* the conjugate operator to A that is defined for all $y \in \mathcal{H}_2$ for which there exists $z \in \mathcal{H}_1$ that satisfies $\langle Ax, y \rangle = \langle x, z \rangle$ (in other words, A^* has the maximum domain of definition of all operators conjugate to A).

Note that we demand that an operator is defined on a dense set, therefore, the operator A^* does not always exist.

If the operator A is bounded, then the operator A^* always exists and is bounded, with $\|A^*\| = \|A\|$.

The operator U transforming $\mathcal{H}_1$ into $\mathcal{H}_2$ is called an isometry if it preserves inner products: $\langle Ux, Uy \rangle = \langle x, y \rangle$. For an operator to be an isometry, it is necessary and sufficient for it to satisfy $U^*U = 1$. An isometry is called a unitary operator if it is surjective (the range coincides with $\mathcal{H}_2$). The condition for unitarity can be written in the form $UU^* = U^*U = 1$. A unitary operator defines an isomorphism between $\mathcal{H}_1$ and $\mathcal{H}_2$, while an isometry defines an isomorphism between $\mathcal{H}_1$ and a subspace of $\mathcal{H}_2$.

If $\mathcal{H}_1 = \mathcal{H}_2 = \mathcal{H}$, then operators on $\mathcal{H}$ and their associated conjugate operators are defined on the same space $\mathcal{H}$. In this case, we define the following notions. An operator A is called Hermitian if it is conjugate to itself (i.e. $\langle Ax, y \rangle = \langle x, Ay \rangle$ for all $x, y \in D_A$). An operator A is called self-adjoint if $A^* = A$.

If A is Hermitian, B is self-adjoint, and $D_A \subset D_B$ and $Ax = Bx$ for all $x \in D_A$, then the operator B is called a self-adjoint extension of A. If a Hermitian operator has only one self-adjoint extension, then it is called essentially self-adjoint.

Self-adjoint operators are important in quantum mechanics because they correspond to physical quantities. Usually, in physics books, only Hermitian operators are discussed; the notion of the self-adjoint operator is identified with the notion of the Hermitian operator. This does not cause many issues, since most Hermitian operators encountered in physics turn out to be essentially self-adjoint. Note, however, that if a Hermitian operator is not essentially self-adjoint, then a choice of self-adjoint extension in physics problems leads to different results (quite often this choice corresponds to a choice of boundary conditions.)

An operator A is called bounded from below if there exists a constant K such that $\langle Ax, x \rangle \geq K \langle x, x \rangle$ for any vector $x \in D_A$. If $K = 0$, then we call A positive semidefinite, and if for $x \neq 0$ the expression $\langle Ax, x \rangle$ is positive, then we call A positive definite.

Hermitian operators that are bounded from below always have self-adjoint extensions. One of these extensions is called the Friedrichs extension A_μ; it can be defined for the operator A by the following extremal property: if B is a self-adjoint extension of A and λ is a number such that $B + \lambda$ is positive definite, then A_μ satisfies $\langle (A_\mu + \lambda)^{-1}x, x \rangle \leq \langle (B + \lambda)^{-1}x, x \rangle$. Friedrichs extension A_μ is bounded from below if A is bounded from below (in fact, if $\langle Ax, x \rangle \geq K \langle x, x \rangle$ then $\langle A_\mu x, x \rangle \geq K \langle x, x \rangle$).

The concrete operators we consider in this book are often given as formal expressions that combine simpler operators (such as multiplication and differentiation in the space $L^2(E^n)$ or annihilation and creation operators on Fock space). For full rigor, one should specify the domain of all operators considered. Instead, however, for simplicity, we agree that operators in the space $L^2(E^n)$ are defined on the space $\mathcal{S}(E^n)$ of rapidly decaying smooth functions (more detailed definition of $\mathcal{S}(E^n)$ is given in A.6) and operators in the Fock space $F(L^2(E^n))$ are defined on the set of Fock states specified by a sequence of functions $f_k(x_1, \ldots, x_k) \in \mathcal{S}(E^{nk})$ (we assume that only a finite number of these functions are not equal to 0).

Saying that a formal operator expression specifies a self-adjoint operator, we have in mind that this expression specifies a Hermitian operator that is either essentially self-adjoint or is bounded from below. In the latter case, the self-adjoint operator defined by the expression at hand is the Friedrichs extension of the corresponding Hermitian operator.

When the formal operator expression does not define a Hermitian operator or specifies a Hermitian operator that does not have a self-adjoint extension, then we say that this formal expression does not specify a self-adjoint operator.

We say that the operator P is a projection operator if $P = P^*$ and $P^2 = P$. Projections on a Hilbert space are in one-to-one correspondence with the subspaces of the Hilbert space (every projection corresponds to its range).

Let us consider the example of the operator $\hat{a}(x)$ of multiplication by a measurable function $a(x)$ in the space $L^2(M)$ where M is a measure space (the operator $\hat{a}(x)$ transforms $f(x)$ into the function

$a(x)f(x)$ and is defined on the functions $f(x) \in L^2(M)$ such that the function $a(x)f(x)$ is square integrable). It is easy to check that $(\hat{a}(x))^* = \overline{\hat{a}(x)}$. It follows that the operator $\hat{a}(x)$ is (1) self-adjoint if the function $a(x)$ is real, (2) unitary if $|a(x)| = 1$, and (3) is a projection if the function $a(x)$ takes only the values 0 and 1.

Note that the operator $\hat{a}(x)$ can be considered on a smaller domain; the adjoint operator does not depend on the choice of the domain. It follows from this remark that for real $a(x)$ the operator $\hat{a}(x)$ is essentially self-adjoint on every domain.

The operator of multiplication is universal in some sense. Namely, one can prove the following important theorem.

For every self-adjoint or unitary operator A in the space $\mathcal{H}$, one can find a measure space M and an isomorphism α of the space $\mathcal{H}$ onto the space $L^2(M)$ that transforms the operator A into a multiplication operator by a function $a(x)$ (i.e. $\alpha A \alpha^{-1} = \hat{a}(x)$). The function $a(x)$ is real if A is self-adjoint and $|a(x)| = 1$ if A is a unitary operator.

The theorem allows us to define a function of a self-adjoint or a unitary operator. Namely, if $\phi(\lambda)$ is a function of a real variable, then we can define the operator $\phi(A)$, where A is a self-adjoint operator, as the operator of multiplication by $\phi(a(x))$ using the isomorphism α described above (more precisely, if $\alpha A \alpha^{-1} = \hat{a}(x)$, then $\alpha \phi(A) \alpha^{-1} = \hat{\phi}(a(x))$). (We assume that the function $\phi(a(x))$ is measurable; this is always the case if the function $\phi(\lambda)$ is Borel measurable, i.e. it is measurable with respect to the family $\mathcal{B}$ generated by open sets, their countable intersections, and unions.)

Note the following obvious relations:

1. If $h(\lambda) = \gamma f(\lambda) + \mu g(\lambda)$; $r(\lambda) = f(\lambda)g(\lambda)$; $s(\lambda) = f(g(\lambda))$, then $h(A) = \gamma f(A) + \mu g(A)$; $r(A) = f(A)g(A)$; $s(A) = f(g(A))$.
2. If $B = U^{-1}AU$, where U is a unitary operator on the space $\mathcal{H}$, then $\phi(B) = U^{-1}\phi(A)U$.
3. If $\phi(\lambda)$ is a bounded function and $|\phi(\lambda)| \le M$, then $\phi(A)$ is a bounded operator with $\|\phi(A)\| \le M$.
4. $[\phi(A)]^* = \overline{\phi(A)}$; if the function $\phi(\lambda)$ is real, then the operator $\phi(A)$ is self-adjoint, and if $|\phi(\lambda)| = 1$, then $\phi(A)$ is a unitary operator.

In particular, if A is self-adjoint, then the operator $\exp(itA)$ is unitary.

If ϕ is a measurable function on the circle $|z| = 1$, and the operator A is unitary, then the operator $\phi(A)$ can be defined precisely in the same way and has similar properties.

If A is a self-adjoint operator, it is convenient to consider projections $E_\mu = e_\mu(A)$, where $e_\mu(\lambda)$ is a function equal to 1 for $\lambda \leq \mu$ and 0 for $\lambda > \mu$. Projections E_μ are called spectral projections of the operator A. It follows from the relation $1 = \int de_\mu$ that

$$A = \int_{-\infty}^{\infty} \mu \, dE_\mu$$

(this formula is called the spectral decomposition of the operator A; for finite a and b, the integral $\int_a^b \mu \, dE_\mu$ is defined as the limit of integral sums

$$\sum \mu_k (E_{\mu_{k+1}} - E_{\mu_k}),$$

where $\mu_0 = a < \mu_1 < \cdots < \mu_n = b$; $\Delta\mu_k = \mu_{k+1} - \mu_k \to 0$. The integral $\int_{-\infty}^{\infty} \mu \, dE_\mu$ is defined as the strong limit $\int_a^b \mu \, dE_\mu$ where $a \to \infty, b \to -\infty$.

We say that a real number λ does not belong to the spectrum of a self-adjoint operator A, if for some $\epsilon > 0$ we have $E_{\lambda+\epsilon} = E_{\lambda-\epsilon}$. It is easily checked that the spectrum of the operator $\hat{a}(x)$ of multiplication by the function $a(x)$ in the space $L^2(M)$ coincides with the range of $a(x)$. Recall that we do not distinguish equivalent measurable functions, therefore, the range of measurable functions should be defined in such a way that the set does not change when we replace a function by an equivalent function. We say that the number λ belongs to the range of the function $a(x)$ if for every function $b(x)$ differing from $a(x)$ only on a set of measure 0 and every $\epsilon > 0$, there exists an $x \in M$ such that $\lambda - \epsilon < b(x) < \lambda + \epsilon$.

Using this statement and the previously defined isomorphism transforming a self-adjoint operator into an operator of multiplication by a function, it is easy to prove the following statements:

1. If the number β doesn't belong to the spectrum of a self-adjoint operator H, then

$$\int \exp(-i\beta a)\exp(iHa)da = 0.$$

More precisely, for any $x, y \in \mathcal{H}$

$$\int \exp(-i\beta a)\langle\exp(iHa)x, y\rangle\, da = 0.$$

2. Let us suppose that for every smooth finite function $\chi(\omega)$ with support on (a, b) we have

$$\int \tilde{\chi}(t)\exp(iHt)dt = 0,$$

where

$$\tilde{\chi}(t) = \int \exp(-i\omega t)\chi(\omega)d\omega.$$

Then the spectrum of the operator H lies outside the interval (a, b).

Two operators A and B, defined on the whole space $\mathcal{H}$, are called commuting if $AB = BA$. The same definition can be used in the case where operators A and B are defined on the same set D and transform it into itself. However, for self-adjoint operators, one should use another definition of commuting operators. One says that two self-adjoint operators A and B in the Hilbert space $\mathcal{H}$ commute if $\phi(A)\psi(B) = \psi(B)\phi(A)$ for all bounded functions $\phi(\lambda), \psi(\lambda)$ (if this is the case, we say that $AB = BA$). Note that the operators $\phi_1(A)$ and $\phi_2(A)$ commute (here, $\phi_1(\lambda)$ and $\phi_2(\lambda)$ are arbitrary real functions and A is a self-adjoint operator).

One can prove the following theorem.

For any family of self-adjoint commuting operators $A_1, \ldots, A_n, \ldots$ in a Hilbert space $\mathcal{H}$, one can find a measure space M and an isomorphism α between the space $\mathcal{H}$ and the space $L^2(M)$ transforming each of these operators into the operator of multiplication by a function (i.e. $\alpha A_i \alpha^{-1} = \hat{a}_i(x)$).

As in the case of one operator, this theorem allows us to define functions of families of operators. Namely, if $A_1, \ldots, A_n$ are commuting self-adjoint operators and $f(\lambda_1, \ldots, \lambda_n)$ is a measurable function of n real variables, then $f(A_1, \ldots, A_n)$ is defined as the operator corresponding to the operator of multiplication by the function $f(a_1(x), \ldots, a_n(x))$ when we apply the isomorphism α.

If Δ is an open set in n-dimensional space, we define $e_\Delta(x_1, \ldots, x_n)$ to be the function that is equal to 1 if $(x_1, \ldots, x_n) \in \Delta$ and 0 otherwise.

The projection operators $E_\Delta = e_\Delta(A_1, \ldots, A_n)$ are called the spectral projections of the family of commuting self-adjoint operators $A_1, \ldots, A_n$.

We say that the point $(\lambda_1, \ldots, \lambda_n)$ of an n-dimensional space belongs to the spectrum (more precisely, the joint spectrum) of the family of commuting self-adjoint operators $A_1, \ldots, A_n$, if for every neighborhood U of the point $(\lambda_1, \ldots, \lambda_n)$, the spectral projection $E_U = e_U(A_1, \ldots, A_n)$ does not equal 0.

If the operators $A_1, \ldots, A_n$ can be realized in the space $L^2(M)$ as operators of multiplication by the functions $a_1(x), \ldots, a_n(x)$, then the joint spectrum of these operators consists of all points of the form $(a_1(x), \ldots, a_n(x))$ (more precisely, the point $(\lambda_1, \ldots, \lambda_n)$ belongs to the spectrum if and only if for all functions $b_i(x)$ equivalent to the functions $a_i(x)$ and any neighborhood U of the point $(\lambda_1, \ldots, \lambda_n)$ there exists a point $x \in M$ such that $(b_1(x), \ldots, b_n(x)) \in U$).

Let us discuss the question of the convergence of a sequence of operators. We define three types of convergence, assuming for brevity that all operators are bounded.

A sequence of operators A_n converges to the operator A uniformly (or in norm), if $\|A - A_n\| \to 0$.

A sequence of operators A_n converges to the operator A strongly if $\lim A_n x = Ax$ for any vector x.

A sequence of operators A_n converges to the operator A weakly if $\lim \langle A_n x, y \rangle = \langle Ax, y \rangle$ for any vectors x and y.

Convergence in norm will be denoted by the symbol $A = \mathrm{nlim}\, A_n$ or $A_n \implies A$, strong convergence will be denoted by the symbol $A = \mathrm{slim}\, A_n$ or $A_n \to A$, and weak convergence will be denoted by $A = \mathrm{wlim}\, A_n$ or $A_n \rightharpoonup A$.

It is clear that convergence in norm implies strong convergence and strong convergence implies weak convergence.

Note the following statements.

(a) if for a dense set of vectors x the limit $\operatorname{slim} A_n x$ exists and the sequence $\|A_n\|$ is bounded, then the sequence of operators A_n is strongly convergent;

(b) if the limit $\lim \langle A_n x, y \rangle$ exists for all x from a dense subset X and for all y from a dense subset Y and the sequence $\|A_n\|$ is bounded, then the sequence A_n is weakly convergent;

(c) if $A_n \to A, B_n \to B$ and the sequence $\|A_n\|$ is bounded, then $A_n B_n \to AB$ (the analog of this statement for weak convergence does not hold); if $A_n \to A$ and the sequence $\|A_n^{-1}\|$ is bounded, then $A_n^{-1} \to A^{-1}$;

(d) if A_n is a sequence of unitary operators and $A_n \implies A$, then the operator A is unitary; if $A_n \to A$, then the operator A is isometric; if $A_n \rightharpoonup A$, then we can only say that $\|A\| \le 1$;

(e) if $A_n \rightharpoonup A$, then $A_n^* \rightharpoonup A^*$ (for strong convergence, the analogous statement does not hold; however, if A_n is a sequence of unitary operators that strongly converges to the unitary operator A, then $A_n^* = A_n^{-1} \to A^* = A^{-1}$).

Very often, it is useful to consider a family of operators transforming a fixed linear subspace D that is dense in the Hilbert subspace into itself (in other words, the operators are defined on D and their range is contained within D); the family of operators having this property will be denoted $\mathcal{N}_D$. The family $\mathcal{N}_D$ is closed with respect to linear combination and multiplication of operators. If for an operator $A \in \mathcal{N}_D$, there exists an operator $B \in \mathcal{N}_D$ satisfying the condition $\langle Ax, y \rangle = \langle x, By \rangle$ for all $x \in D, y \in D$, then we say that the operator B is D-conjugate to the operator A and we denote it by A^+.

The strong limit of a sequence $A = \operatorname{slim} A_n$ of operators $A_n \in \mathcal{N}_D$ is defined as the operator satisfying the condition $Ax = \lim A_n x$ for all $x \in D$; the weak limit $A = \operatorname{wlim} A_n$ is defined by the condition $\langle Ax, y \rangle = \lim \langle A_n x, y \rangle$ for any $x, y \in D$.

A.6 Locally convex linear spaces

A seminorm on a linear space R is a non-negative function $p(x)$ that satisfies the conditions $p(\lambda x) = |\lambda| p(x)$, $p(x + y) \le p(x) + p(y)$ for all $x, y \in R$ and for any real number λ. A seminorm that vanishes only for $x = 0$ is called a norm.

A locally convex topology on a linear space R is specified by a system A of seminorms $p_\alpha(x)$ that satisfies the following conditions: (1) for every two seminorms $p_\alpha, p_\beta \in A$, one can find a seminorm p_γ such that $p_\gamma \ge p_\alpha, p_\gamma \ge p_\beta$; (2) for every point $x \in R$, one can find a seminorm $p_\alpha \in A$ that does not vanish at the point x. A linear space equipped with a locally convex topology is called a locally convex topological linear space or simply a locally convex space, and the system A is called the defining system of seminorms. If the defining system consists of a single seminorm (which is automatically a norm), then the space is called a normed space.

The function $f(x)$ on a locally convex space R is continuous in the topology of this space if for every $\epsilon > 0$ and every point $x_0 \in R$, there exists a seminorm p_α in the defining system of seminorms for R and a $\delta > 0$ such that for any point x satisfying the condition $p_\alpha(x - x_0) < \delta$, we have $|f(x) - f(x_0)| < \epsilon$. As an example of a locally convex space, we can consider the space $\mathcal{S}(E^r)$ of smooth functions of r variables, having all derivatives tending to zero faster than any polynomial.

The topology on the space $\mathcal{S}(E^r)$ is specified by the system of seminorms

$$\|\phi\|_{\alpha,\beta} = \sup_{x \in E^r} |x^{(\alpha)} D^{(\beta)} \phi(x)|$$

(here, $\alpha = (\alpha_1, \ldots, \alpha_r)$, $\beta = (\beta_1, \ldots, \beta_r)$ are arbitrary sets of non-negative integers, $x^{(\alpha)} = x_1^{\alpha_1} \ldots x_r^{\alpha_r}$, $D^{(\beta)} = \frac{\partial^{\beta_1 + \cdots + \beta_r}}{\partial x_1^{\beta_1} \ldots \partial x_r^{\beta_r}}$).

We say that two systems of seminorms on a linear space are equivalent (i.e. they specify the same topology) if every function that is continuous in one system is necessarily continuous in the other. In other words, one can consider different defining systems of seminorms on the same locally convex space; among these systems, there exists a maximal system consisting of all continuous seminorms. Instead of using the precise term "a seminorm that is continuous with respect to the locally convex topology in R", we will use the shorter term "a seminorm in R".

A sequence ξ_n of elements of the locally convex space R converges to the element ξ if for every seminorm p in R the sequence $p(\xi_n - \xi)$ tends to zero. A sequence ξ_n is called Cauchy if for every seminorm p in R we have $\lim_{m\to\infty,n\to\infty} p(\xi_m - \xi_n) = 0$. In a complete locally convex space, every Cauchy sequence is convergent.

A.7 Generalized functions (distributions)

Physicists define generalized functions $f(x)$ as symbols that only make sense under the sign of an integral $\int f(x)\phi(x)dx$, where $\phi(x)$ is a "good" function. The mathematical formalization of this definition is given as follows.

Let M be a measure space and R any linear subspace of the space of measurable functions on M (the elements of R are called test functions). A generalized function on M (more precisely, a generalized function with numerical values) is defined as a linear functional $f(\phi)$ on the space R. A generalized function on the space M is denoted by $f(x)$ and we define $\int f(x)\phi(x)dx = f(\phi)$. Note that $\int f(x)\phi(x)dx$ should be understood as another way of writing the number $f(\phi)$.

It is often convenient to consider the space of test functions R endowed with a topology, and generalized functions as continuous linear functionals on R. We will consider the topology on the space of test functions only in one situation — when defining generalized functions of moderate growth.

Clearly, the definition of generalized functions depends on the choice of the space of test functions R. The set of generalized functions, corresponding to a given space of test functions R, we denote by R'. The set R' is naturally equipped with the structure of a linear space. If $R \subset L^2(M)$, then any function $g \in L^2(M)$ defines a linear functional on R by the formula $g(\phi) = \int_M g(x)\phi(x)d\mu$. Moreover, this integral also makes sense for other functions g. In other words, under certain conditions, a conventional function generates a generalized function, which will be denoted by the same symbol.[3]

Generalized functions of n variables $x_1, \ldots, x_n \in M$ will be defined as functionals $f(\phi_1, \ldots, \phi_n)$, depending on n elements $\phi_1, \ldots, \phi_n \in R$ and linear in every argument (in other words, generalized functions of n variables are defined as linear functionals on the tensor product of n copies of the space R). We will denote a generalized function of n variables by the symbol $f(x_1, \ldots, x_n)$ and will write that $f(\phi_1, \ldots, \phi_n) = \int f(x_1, \ldots, x_n)\phi(x_1) \ldots \phi(x_n)dx_1 dx_2 \ldots dx_n$. The set of generalized functions of n variables will be denoted by R'_n.

Let us consider some examples of generalized functions.

1. Let $C(E^r)$ be the space of continuous functions on the Euclidean space E^r. By the symbols $\delta(\mathbf{x}, \mathbf{a})$ and $\delta(\mathbf{x} - \mathbf{a})$, we will denote the generalized function (the functional on the space $C(E^r)$) that maps a function $\phi(\mathbf{x}) \in C(E^r)$ to $\phi(\mathbf{a})$, the function's value at the point $\mathbf{a}$ (the point $\mathbf{a} \in E^r$ plays the role of a parameter). In other words, the function $\delta(\mathbf{x} - \mathbf{a})$ (δ-function) satisfies the relation $\int \delta(\mathbf{x} - \mathbf{a})\phi(\mathbf{x})d\mathbf{x} = \phi(\mathbf{a})$. It is clear that for any subspace $R \subset C(E^r)$, a δ-function is a generalized function from the set R'.

2. Let $C_n(E^r)$ be the space of n-times continuously differentiable functions on the space E^r. The generalized function $\delta^{(\alpha)}(\mathbf{x} - \mathbf{a}) \in (C_n(E^r))'$ is defined by the relation

$$\int \delta^{(\alpha)}(\mathbf{x} - \mathbf{a})\phi(\mathbf{x})d\mathbf{x} = (-1)^{|\alpha|}(D^{(\alpha)}(\phi))(\mathbf{a})$$

(here, $\alpha = (\alpha_1, \ldots, \alpha_k)$ is a collection of non-negative integers satisfying $|\alpha| = \alpha_1 + \cdots + \alpha_k \leq n$, $D^{(\alpha)}(\phi) = \frac{\partial^{|\alpha|}}{\partial x_1^{\alpha_1} \ldots \partial x_n^{\alpha_n}}\phi$).

3. Let us assume that a space R is contained in the space $L^2(M)$, then we can define a generalized function of two variables

---

[3]Let us emphasize that while $\int f(x)\phi(x)dx$ is understood as another way of writing the number $f(\phi)$ and is defined if ϕ is a test function and f is a generalized function, in this section we use the notation $\int f(x)\phi(x)d\mu$ for the Lebesgue integral of the function $f(x)\phi(x)$. Clearly, the Lebesgue integral is defined when the function $f(x)\phi(x)$ is summable. For integrals in the Euclidean space E^r in both cases, we use the same notation $\int f(\mathbf{x})\phi(\mathbf{x})d\mathbf{x}$.

$\delta(x, y) \in R_2'$ by means of the relation

$$\int \delta(x, y)\phi(x)\psi(y)dxdy = \int \phi(x)\psi(x)d\mu.$$

4. Every operator A in the space $L^2(M)$ specifies a generalized function of two variables $A(x, y) \in (D_A)_2'$ by the formula

$$\int A(x, y)\phi(x)\psi(y)dxdy = \langle A\psi, \bar{\phi} \rangle,$$

and the function $A(x, y)$ is called the kernel of the operator A.

The operator A can be represented as an integral operator with the kernel $A(x, y)$. In other words, we have

$$(A\psi)(x) = \int A(x, y)\psi(y)dy. \tag{A.1}$$

The equation (A.1) should be understood as an abbreviation of the following relation:

$$\int (A\psi)(x)\phi(x)d\mu = \int A(x, y)\psi(y)\phi(x)dxdy.$$

5. Considering numerical generalized functions on the space E^r, we restrict ourselves to generalized functions of moderate growth, defined as continuous functionals on the space $\mathcal{S}(E^r)$ of smooth functions having derivatives that tend to zero at infinity faster than any power of the norm of x. The topology in this space is described in A.6.

The linear space of generalized functions of moderate growth is denoted by $\mathcal{S}'(E^r)$.

Every numerical function $a(x)$ that grows at infinity no faster than a polynomial (i.e. for large enough x, we have $|a(x)| \leq C\|x\|^n$) and is summable in any bounded domain specifies a generalized function $a(\phi)$, belonging to the space $\mathcal{S}'$, by the formula

$$a(\phi) = \int a(x)\phi(x)dx$$

(the integral is understood as a Lebesgue integral).

Let us define some operations on generalized functions.

A generalized function $f \in \mathcal{S}'$ is called the limit of a sequence $f_n \in \mathcal{S}'$ of generalized functions, if for every function $\phi \in \mathcal{S}$, the limit of the sequence $f_n(\phi)$ is equal to $f(\phi)$ (in other words, $\lim \int f_n(x)\phi(x)dx = \int f(x)\phi(x)dx$).

The derivative $f_i = \frac{\partial f}{\partial x_i}$ of a generalized function $f \in \mathcal{S}'$ is defined by the formula

$$f_i(\phi) = f\left(-\frac{\partial \phi}{\partial x_i}\right)$$

or, in a different notation, by the formula

$$\int \frac{\partial f(x)}{\partial x_i}\phi(x)dx = -\int f(x)\frac{\partial \phi}{\partial x_i}dx.$$

The Fourier transform of a generalized function f is defined by the relation

$$\tilde{f}(\phi) = f(\tilde{\phi}), \quad \text{where } \tilde{\phi}(x) = (2\pi)^{-\frac{r}{2}} \int \exp(i\langle \mathbf{k}, x\rangle)\phi(\mathbf{k})d\mathbf{k}.$$

Note that the derivative of a function $f \in \mathcal{S}'$ and the Fourier transform of this function always exist and belong to the space $\mathcal{S}'$ (this is obvious because together with the function $\phi \in \mathcal{S}$, the functions $\frac{\partial \phi}{\partial x_i}$ and $\tilde{\phi}$ also belong to $\mathcal{S}$).

The following important statement is called the kernel theorem.

Let us consider the functional $f(\phi, \psi)$ of $\phi \in \mathcal{S}(E^r), \psi \in \mathcal{S}(E^n)$, belonging to the space $\mathcal{S}'(E^r)$ for fixed ψ and $\mathcal{S}'(E^n)$ for fixed ϕ. Then, there exists one and only one functional $\hat{f} \in \mathcal{S}'(E^{r+n})$, obeying

$$\hat{f}(\phi \otimes \psi) = f(\phi, \psi)$$

(the symbol $\phi \otimes \psi$ denotes the function

$$\phi(x_1, \ldots, x_r)\psi(y_1, \ldots, y_n) \in \mathcal{S}(E^{r+n})).$$

Vector generalized functions and operator generalized functions are defined analogously to numerical generalized functions.

Namely, to define a vector generalized function, one should construct for every function ϕ, from the space of test functions R, a vector $f(\phi)$, from a linear space F, that linearly depends on ϕ.

(In other words, a vector generalized function is an operator defined on the space R and taking values in the linear space F.)

To define an operator generalized function, we should construct for every function ϕ, from the space of test functions R, an operator $A(\phi)$ acting linearly on elements in the space $\mathcal{H}_1$ and transforming it into the space $\mathcal{H}_2$ (we assume that all operators $A(\phi)$ have the same domain D). One can say that an operator generalized function is an operator defined on the space of test functions R, taking values on the linear space of operators with domain $D \subset \mathcal{H}_1$ and taking values in $\mathcal{H}_2$.

For vector and operator generalized functions, we will use the same notation as for numerical generalized functions

$$f(\phi) = \int f(x)\phi(x)dx.$$

To describe some examples of vector generalized functions, we use the following general constructions.

Let us suppose that a linear space L is embedded in a linear space K and on the measure space M we have specified a function $f(x)$ with values in K. Let us also assume that for every test function $\phi \in R$ there exists an integral $\int f(x)\phi(x)d\mu$ and this integral[4] belongs to the space L. Then, it is clear that the correspondence $f(\phi) = \int f(x)\phi(x)d\mu$ can be considered as a generalized function with values in L, generated by the conventional function $f(x)$ with values in K. We will use the same notation $f(x)$ for this generalized function.

Let us consider for example the case where $x \in E^r, R = L^2(E^r) = L, K = \mathcal{S}'(E^r)$. Consider the functions $\delta_a(x) = \delta(x-a)$ and $\phi_a(x) = (2\pi)^{-r/2} \exp(i\langle a, x\rangle)$ from $K = \mathcal{S}'(E^r)$ that depend on the parameter $a \in E^r$ (i.e. we consider the functions on E^r with values in $\mathcal{S}'(E^r)$). If $f(a) \in R = L^2(E^r)$, then $\int f(a)\delta_a(x)da = f(x) \in L = L^2(E^r)$ and $\int f(a)\phi_a(x)da = \tilde{f}(x) \in L = L^2(E^r)$. In other words, we have constructed vector generalized functions $\delta_a(x), \phi_a(x)$ with values in $L^2(E^r)$.

---

[4]The integral in this formula can be understood in any sense; it is important only that it is linear with respect to ϕ.

Let us suppose that D is a linear subspace in $\mathcal{H}_1$ and the space $\mathcal{H}_2$ is embedded in the space $\tilde{\mathcal{H}}_2$. Further, let us assume that to every point x of a measure space M, we have constructed an operator $A(x)$ with domain D and taking values in $\tilde{\mathcal{H}}_2$. Finally, if we assume that for every function ϕ, the operator $A(\phi) = \int A(x)\phi(x)d\mu$ transforms D into $\mathcal{H}_2$, then the conventional operator function $A(x)$ generates a generalized operator function $A(\phi)$ with values in the set of operators transforming D into $\mathcal{H}_2$.

In particular, if $M = E^r$, $R = L^2(E^r)$, $D = \mathcal{H}_1 = F_n$, $\mathcal{H}_2 = F_{n+1}$, where F_n is the space of symmetric (antisymmetric) square-integrable functions $f(\mathbf{x}_1, \ldots, \mathbf{x}_n)$ depending on variables $\mathbf{x}_1, \ldots,$ $\mathbf{x}_n \in E^r$, and $\tilde{\mathcal{H}}_2 = \mathcal{S}'(E^{r(n+1)})$ is the space of generalized functions of moderate growth, depending on the variables $\mathbf{x}_1, \ldots, \mathbf{x}_{n+1} \in E^r$, then the operator function $a_n^+(\mathbf{x})$ on E^n, where $a_n^+(\mathbf{x})$ is an operator transforming the function $f(\mathbf{x}_1, \ldots, \mathbf{x}_n) \in F_n$ into the function $\sqrt{n+1}P[f(\mathbf{x}_1, \ldots, \mathbf{x}_n)\delta(\mathbf{x}_{n+1} - \mathbf{x})] \in \tilde{\mathcal{H}}_2$, specifies an operator generalized function, acting from F_n to F_{n+1}. This is clear because the operator $\int \phi(\mathbf{x})a^+(\mathbf{x})d\mathbf{x}$ transforms F_n into F_{n+1} (in the above formula, P denotes the operator of symmetrization (antisymmetrization)). This operator generalized function is considered in Section 3.2.

If the space of test functions R is dense in $L^2(M)$, then the vector generalized function $f(x)$ with values in $\mathcal{H}$ is called normalized (or, more precisely, δ-normalized) if the following condition is satisfied:

$$\langle f(x), f(y) \rangle = \delta(x, y)$$

(to make this equation precise, we should integrate it, multiplying by test functions $\phi(x), \psi(y)$ and writing it in the form

$$\left\langle \int f(x)\phi(x)dx, \int f(y)\psi(y)dy \right\rangle = \int \delta(x, y)\phi(x)\overline{\psi(y)}dxdy$$

$$= \int \phi(x)\overline{\psi(x)}d\mu$$

or, in the shorter form $\langle f(\phi), f(\psi) \rangle = \langle \phi, \psi \rangle$).

This means that for a generalized function f that is δ-normalized, we can construct an isometric map of the space R into $\mathcal{H}$, sending the function $\phi \in R$ into the vector $f(\phi) \in \mathcal{H}$. This map can be extended continuously to an isometric map of the space $L^2(M)$ into $\mathcal{H}$, with the extended map denoted by the same symbol f. If linear combinations of vectors $f(\phi)$ are dense in $\mathcal{H}$, then the isometric map of $L^2(M)$ into $\mathcal{H}$ is an isomorphism (i.e. it is a unitary map onto $\mathcal{H}$).

A δ-normalized generalized vector function f having this property is called a generalized basis. In this case, the statements given above imply that every vector $a \in \mathcal{H}$ can be expanded with respect to the generalized basis (i.e. in other words a has a unique representation in the form

$$a = f(\phi) = \int \phi(x) f(x) dx,$$

where $\phi \in L^2(M)$). If $a = f(\phi)$, then the function $\phi(x)$ can be written in the form

$$\phi(x) = \langle a, f(x) \rangle.$$

The generalized vector functions δ_a and ϕ_a defined above can be considered as generalized bases.

If A is an operator on the space $\mathcal{H}$, then the generalized function

$$\langle x|A|y \rangle = \langle Af(y), f(x) \rangle$$

is called the matrix of the operator A in the generalized basis $f(x)$.

Note the following useful statement: let us consider a sequence A_n of operators, assuming that the norms $\|A_n\|$ form a bounded sequence and the matrices $\langle x|A_n|y \rangle$ converge to the matrix $\langle x|A|y \rangle$, then the operators A_n weakly converge to the operator A $\Big[$saying that the matrices $\langle x|A_n|y \rangle$ converge to $\langle x|A|y \rangle$, we have in mind that for all functions $\phi, \psi \in R$, where R is a dense subspace of $L^2(M)$, the following relation holds:

$$\lim \int \langle x|A_n|y \rangle \, \phi(x)\psi(y) dx dy = \int \langle x|A|y \rangle \, \phi(x)\psi(y) dx dy \Big].$$

Let us fix the space of test functions R and a dense linear subspace D in the Hilbert space $\mathcal{H}$. We will consider only the generalized operator functions that transform D into itself (i.e. in other words, we assume that the domain of the operator $A(\phi)$, where ϕ is a test function, coincides with D and all vectors $A(\phi)a$, where $a \in D$, are contained in D).

The set of all operator functions of this kind will be denoted by $\mathcal{M}_D$.

We call generalized operator functions $A \in \mathcal{M}_D, B \in \mathcal{M}_D$ conjugate if for any $a \in D, b \in D$

$$\langle A(x)a, b \rangle = \langle a, B(x)b \rangle$$

(more precisely, if $\langle A(\phi)a, b \rangle = \langle a, B(\overline{\phi})b \rangle$). In this case, we will use the symbol $B = A^+$.

The product of generalized operator functions $A \in \mathcal{M}_D, B \in \mathcal{M}_D$ is defined as the generalized operator function

$$C(x, y) = A(x)B(y).$$

To be more precise, the generalized operator function C maps a test function $\alpha(x, y)$ of the form

$$\alpha(x, y) = \sum_v \phi_v(x)\psi_v(y), \tag{A.2}$$

where $\phi_v, \psi_v \in R$, into the operator

$$C(\alpha) = \int (\alpha(x, y)A(x)B(y)dxdy = \sum_v A(\phi_v)B(\psi_v).$$

In some cases, the generalized operator function $C(x, y)$ can be defined on a broader class of test functions than functions in the form (A.2). As an example, consider the following statement (the operator analog for the kernel theorem).

Let us suppose that the space of test functions is the space $\mathcal{S}(E^r)$. Let us fix in the set $\mathcal{M}_D$ a subset $\mathcal{R}_D$, consisting of operator generalized functions $A \in \mathcal{M}_D$, satisfying the following conditions: (a) the functional $\langle A(\phi)a, b \rangle$ for any $a, b \in D$ is continuous in the topology of the space $\mathcal{S}$ (in other words, the numerical generalized functions $\langle A(x)a, b \rangle$ belong to the space $\mathcal{S}'$); (b) there exists an operator generalized function $A^+ \in \mathcal{M}_D$ conjugate to A.

If $A_1, \ldots, A_p \in \mathcal{R}_D$, then to every function $\psi \in \mathcal{S}(E^{pr})$ we can assign an operator

$$C(\psi) = \int \psi(x_1, \ldots, x_p) A_1(x_1) \ldots A_p(x_n) d^p x, \qquad (A.3)$$

defined on the set D. Namely, we should take

$$C(\psi) = \sum_{k=1}^{s} A_1(\phi_1^{(k)}) \ldots A_p(\phi_p^{(k)})$$

in the case when

$$\psi = \sum_{k=1}^{s} \phi_1^{(k)}(x_1) \ldots \phi_p^{(k)}(x_p), \qquad (A.4)$$

and in general, we should represent ψ as limit of a sequence ψ_n of functions of the form (A.4), converging to ψ in the topology of $\mathcal{S}$ and take

$$C(\psi) = \operatorname{slim} C(\psi_n). \qquad (A.5)$$

The existence of this limit follows from the following estimates:

$$\begin{aligned}
\|C(\psi_m)a - C(\psi_n)a\|^2 &= \langle C(\psi_m)a - C(\psi_n)a, C(\psi_m)A - C(\psi_n)a \rangle \\
&= \big\langle (C^+(\psi_m)C(\psi_m) + C^+(\psi_n)C(\psi_n) \\
&\quad - C^+(\psi_m)C(\psi_n) - C^+(\psi_n)C(\psi_m))a, a \big\rangle \\
&= \int \rho_{m,n}(x_1, \ldots, x_p | y_1, \ldots, y_p) \big\langle A_p^+(x_p) \\
&\quad \ldots A_1^+(x_1) A_1(y_1) \ldots A_p(y_p) a, a \big\rangle \, d^p x \, d^p y,
\end{aligned}$$
$$(A.6)$$

where

$$\begin{aligned}
\rho_{m,n}(x_1, \ldots, x_p | y_1, \ldots, y_p) &= \overline{\psi}_m(x_1, \ldots, x_p)\psi_m(y_1, \ldots, y_p) \\
&\quad + \overline{\psi}_n(x_1, \ldots, x_p)\psi_n(y_1, \ldots, y_p) \\
&\quad - \overline{\psi}_n(x_1, \ldots, x_p)\psi(y_1, \ldots, y_p) \\
&\quad - \overline{\psi}_m(x_1, \ldots, x_p)\psi_n(y_1, \ldots, y_p).
\end{aligned}$$

It is easy to see that for $m \to \infty, n \to \infty$, the function $\rho_{m,n}$ tends to zero in the topology of the space $\mathcal{S}(E^{2pr})$.

Using the kernel theorem, one can conclude from this fact that the expression A.6 tends to zero for $m \to \infty, n \to \infty$. This means that the sequence $C(\psi_n)a$ is a Cauchy sequence, hence it is convergent for every $a \in D$. This proves that the sequence of operators $C(\psi_n)$ is strongly convergent; hence, we can define the operator $C(\psi)$ by means of the formula (A.5).

Let us consider now the set D' consisting of linear combinations of vectors of the form $C(\psi)a$, where $a \in D$, and $C(\psi)$ is an operator of the form (A.3), constructed by means of the operator analog of the kernel theorem. Let us show that the expression of the form (A.3) specifies an operator with domain D' by the formula

$$C(\psi)(C'(\psi')a) = (C(\psi)C'(\psi'))a,$$

where

$$C'(\psi') = \int \psi'(x_1, \ldots, x_q) A_1'(x_1) \ldots A_q'(x_q) d^q x,$$

$$C(\psi)C'(\psi') = \int \psi(x_1, \ldots, x_p) \psi'(x_1', \ldots, x_q') A_1(x_1)$$

$$\ldots A_p(x_p) A_1'(x_1') \ldots A_q'(x_q') d^p x d^q x'.$$

To check that this definition is correct, we should verify that the vector $C(\psi)d$ does not depend on the representation of the vector d in the form $d = C'(\psi')a$. We should also check that the constructed operator is linear. The proof is based on the following remark. Assume that the vector ξ_n has the form

$$\xi_n = \sum_{i=1}^{s(n)} C_n^{(i)}(\phi_n^{(i)}) a_i,$$

$\xi_n \to 0$, and the sequence

$$\eta_n = \sum_{i=1}^{s(n)} (C(\psi_n) C_n^{(i)}(\phi_n^{(i)})) a_i,$$

is convergent. Then $\eta_n \to 0$. $\Big[$ Here, $a_i \in D$, $\phi_n^{(i)} \in \mathcal{S}$, $\psi_n \in \mathcal{S}$, $\psi_n \to \psi$ in the topology of $\mathcal{S}$, $A_k^{(i)} \in \mathcal{R}_D$,

$$C_n^{(i)}(\phi_n^{(i)}) = \int \phi_n^{(i)}(y_1, \ldots, y_{p(i)}) A_1^{(i)}(y_1) \ldots A_{p(i)}^{(i)}(y_{p(i)}) d^{p(i)} y. \Big]$$

This is true because for every vector $\zeta \in D$

$$|\langle \eta_n, \zeta \rangle| = |\langle \xi_n, C^+(\psi_n)\zeta \rangle| \le \|\xi_n\| \|C^+(\psi_n)\zeta\| \to 0.$$

A.8 Eigenvectors and generalized eigenvectors

We call a vector $\phi \in \mathcal{H}$ an eigenvector of the operator A, acting on the space $\mathcal{H}$, if $A\phi = \lambda\phi$ and $\phi \ne 0$. The number λ is called an eigenvalue, corresponding to the eigenvector ϕ. The set $\mathcal{H}_\lambda$ of eigenvectors for a given eigenvalue λ is in fact a linear subspace; the dimension of this space is called the multiplicity of the eigenvalue λ. An eigenvalue with multiplicity one is called non-degenerate or simple.

One says that the operator A has discrete spectrum if linear combinations of its eigenvectors are dense in the space $\mathcal{H}$.

The eigenvalues of a self-adjoint operator are real; the eigenvectors corresponding to distinct eigenvalues of a self-adjoint operator are orthogonal.

If a self-adjoint operator has discrete spectrum, then the spectrum of this operator is the closure of its set of eigenvalues. The spectral projection E_μ is equal to a sum of projections on the linear subspace $\mathcal{H}_\lambda$, where $\lambda \le \mu$.

Suppose that M is a measure space, R is a space of test functions defined on M, and $f \in R'$ is a generalized vector function of the variable $x \in M$ with values in $\mathcal{H}$. Then the function f is called a generalized eigenfunction of an operator A, acting on the space $\mathcal{H}$, if

$$Af(x) = a(x)f(x).$$

More precisely, the above equation means that

$$A \int f(x)\phi(x)dx = \int f(x)a(x)\phi(x)dx,$$

i.e. $Af(\phi) = f(a\phi)$. The last equation must be satisfied in the cases when it makes sense (i.e. when $\phi \in R$ and $a\phi \in R$).

The generalized vector function $\phi_a = (2\pi)^{-r/2} \exp(iax)$, defined in (A.6), turns out to be a generalized eigenfunction for the operators $\frac{1}{i}\frac{\partial}{\partial x_1}, \ldots, \frac{1}{i}\frac{\partial}{\partial x_r}$ (and more broadly for differential operators with constant coefficients):

$$\frac{1}{i}\frac{\partial}{\partial x_k}\phi_a = a_k\phi_a.$$

Note the following important statement.

For any commuting self-adjoint operators $A_1, \ldots, A_n$, acting on the Hilbert space $\mathcal{H}$, we can find a generalized basis of functions that are eigenfunctions for each of the operators (more precisely, we can find a generalized basis f that is δ-normalized ($\langle f(x), f(y) \rangle = \delta(x, y)$) and satisfies the equation $A_i f(x) = a_i(x)f(x)$ for every $i = 1, \ldots, n$).

The proof of this statement is easily obtained if we recall that there exists an isomorphism α of the space $\mathcal{H}$ and the space $L^2(M)$, by which the operators A_i are transformed into operators of multiplication by the function $a_i(x)$. Indeed, the generalized vector function f can be represented by the formula

$$f(\phi) = \int f(x)\phi(x)dx = \alpha^{-1}\phi,$$

where $\phi \in L^2(M)$.

A.9 Group representations

Suppose that for every group element g of a group G, there is a corresponding unitary operator T_g on the Hilbert space $\mathcal{H}$, such that the identity operator corresponds to the unit of the group and the multiplication of operators corresponds to the multiplication operation in the group:

$$T_e = 1, \quad T_{gh} = T_g T_h. \tag{A.7}$$

We call this correspondence a unitary representation of the group G.

A subspace $\mathcal{H}_1 \subset \mathcal{H}$ is called invariant if every operator T_g transforms $\mathcal{H}_1$ into itself. A unitary representation $\mathcal{H}$ is called irreducible if the space $\mathcal{H}$ has no non-trivial (differing from the zero subspace and the full space $\mathcal{H}$) invariant subspace.

Two unitary representations T_g and T'_g on the spaces $\mathcal{H}$ and $\mathcal{H}'$ are equivalent if there exists an isomorphism α between the spaces $\mathcal{H}$ and $\mathcal{H}'$ satisfying

$$T'_g \alpha = \alpha T_g.$$

For any self-adjoint operator A, we can construct a unitary representation of the group of real numbers E^1 by assigning to every number $t \in E^1$ the operator $T_t = \exp(iAt)$ (we use addition as the group operation on E^1 so that equation (A.7) takes the form

$$T_0 = 1, \ T_{t+\tau} = T_t T_\tau).$$

Conversely, given a unitary representation T_t of the group E^1 such that for any two vectors $a, b \in \mathcal{H}$, the function $\langle T_t a, b \rangle$ is continuous or at least measurable, then we can find a self-adjoint operator A such that $T_t = \exp(iAt)$; this operator can be obtained by the formula $A = \frac{1}{i} \lim_{t \to 0} \frac{T_t - 1}{t}$ (Stone's theorem).

A family of operators satisfying the conditions of Stone's theorem is called a one-parameter group of unitary operators and the operator A is called the generator (or an infinitesimal generator) of the group.

A Poincaré group $\mathcal{P}$ is the group of affine transformations of Minkowski space

$$x' = \Lambda x + a, \tag{A.8}$$

preserving the space–time interval and the sign of x_0.

(Minkowski space is to be understood as a four-dimensional space with a scalar product $(x, y) = x_0 y_0 - \langle \mathbf{x}, \mathbf{y} \rangle = x_0 y_0 - x_1 y_1 - x_2 y_2 - x_3 y_3$. The space–time interval is defined as $\sqrt{(x - y, x - y)}$. We write vectors from Minkowski space as $x = (x_0, \mathbf{x})$, where $x_0 \in E^1$, $\mathbf{x} \in \mathbf{E}^3$.)

The transformation (A.8), viewed as an element of the group $\mathcal{P}$, will be denoted by (Λ, a).

Let T_g be a unitary representation of the group $\mathcal{P}$ such that for any vectors $a, b \in \mathcal{H}$ the function $\langle T_g a, b \rangle$ is continuous in g (we will only consider unitary representations satisfying this continuity condition).

The group $\mathcal{P}$ contains a subgroup of translations $(1, a)$, consisting of transformations $x' = x + a$. This subgroup is commutative. By Stone's theorem, the unitary operator $T_{(1,a)}$, corresponding to the translations $(1, a)$, can be represented in the form

$$T_{(1,a)} = \exp[-i(a_0 H - a_1 P_1 - a_2 P_2 - a_3 P_3)]$$

$$= \exp[-i(a_0 H - \mathbf{aP})],$$

where H, P_1, P_2, P_3 are commuting self-adjoint operators.

The operator H is called the energy operator and the vector operator $\mathbf{P} = (P_1, P_2, P_3)$ is the momentum operator.

It is easy to check that the joint spectrum of the operators $H, \mathbf{P}$ is invariant with respect to homogenous Lorentz transformations.

One of the simplest representations of the group $\mathcal{P}$ can be written in the following way.

Let us consider the set U_m in four-dimensional space specified by the relation $p_0^2 - \mathbf{p}^2 = m^2$, $p_0 \geq 0$. To every function $f \in L^2(E^3)$, we will assign a function $\check{f}$ on the space U_m defined by the formula $\check{f}(p_0, \mathbf{p}) = f(\mathbf{p})\sqrt{p_0}$. This assignment can be viewed as an isomorphism between the space $L^2(E^3)$ and some space of functions on U_m. This space of functions on U_m is denoted by $L^2(U_m)$; the inner product on $L^2(U_m)$ is defined by the formula

$$\langle \check{f}, \check{g} \rangle = \int \check{f}(p_0, \mathbf{p})\overline{\check{g}(p_0, \mathbf{p})}\frac{d\mathbf{p}}{p_0}.$$

On the space $L^2(U_m)$, we define a unitary representation of the group $\mathcal{P}$ by the formula

$$(U(\Lambda, a)\check{f})(p) = \exp(i(a, p))\check{f}(\Lambda^{-1}p) \tag{A.9}$$

(to check that the formula (A.9) defines a unitary representation, observe that the expression $p_0^{-1}d\mathbf{p}$ defines a Lorentz-invariant measure on U_m). The resulting representation is continuous; the spectrum of the operators $H, \mathbf{P}$ on the space $L^2(U_m)$ coincides with the set U_m.

By the isomorphism between $L^2(E^3)$ and $L^2(U_m)$, demonstrated above, there exists equivalent unitary representation $V(\Lambda, a)$

on $L^2(E^3)$:

$$(V(\Lambda, a)\check{f}) = U(\Lambda, a)\check{f}.$$

Let us describe (up to equivalence) all the irreducible representations of the group $\mathcal{P}$ in which the energy operator H is non-negative and not equal to zero.

For every non-negative number m, there exists an infinite series of unitary irreducible representations, where the joint spectrum of the operators $H, \mathbf{P}$ in these representations is the set U_m. The representation of this series will be denoted by the symbols (m, σ), where σ is a non-negative integer for positive m and any integer for $m = 0$.

A representation of the type (m, σ) can be realized via unitary operators on the space $L^2(E^3 \times S)$, where S is a set consisting of $2\sigma + 1$ elements for positive m and of one element for $m = 0$. Here, the energy and momentum operators on $L^2(E^3 \times S)$ are given by the formulas

$$Hf(\mathbf{p}, s) = \sqrt{\mathbf{p}^2 + m^2} f(\mathbf{p}, s), \tag{A.10}$$

$$\mathbf{P}f(\mathbf{p}, s) = \mathbf{p}f(\mathbf{p}, s). \tag{A.11}$$

For example, the representation described above is a representation of type $(m, 1)$.

Let us note that in addition to the single-valued unitary representations of $\mathcal{P}$ described here, there exist two-valued unitary representations. Irreducible two-valued representations, where the energy operator is non-negative and is not identically zero, can be described in the same way as single-valued (with the difference that σ must be a half-integer).

Bibliography

Araki, H. and Haag, R. (1967). Collision cross sections in terms of local observables, *Communications in Mathematical Physics* **4**, 2, pp. 77–91.

Belavin, A. A., Polyakov, A. M., and Zamolodchikov, A. B. (1984). Infinite conformal symmetry in two-dimensional quantum field theory, *Nuclear Physics B* **241**, 2, pp. 333–380.

Berezin, F. (2012). *The Method of Second Quantization*, Vol. 24 (Elsevier).

Bethe, H. A. (1947). The electromagnetic shift of energy levels, *Physical Review* **72**, 4, p. 339.

Bogolyubov, N., Logunov, A., and Todorov, I. (1969). *Fundamentals of the Axiomatic Approach in Quantum Field Theory*, Nauka: Moscow [In Russian].

Bogolyubov, N., Medvedev, A., and Polivanov, M. (1956). *Problems of the Theory of Dispersion Relations* (Princeton University).

Bogolyubov, N. and Parasyuk, O. (1955). A theory of multiplication of causative singular functions, *Doklady Akademii Nauk SSSR* **100**, pp. 25–28.

Chu, H. and Umezawa, H. (1994). A unified formalism of thermal quantum field theory, *International Journal of Modern Physics A* **9**, 14, pp. 2363–2409.

Dyson, F. J. (1949). The radiation theories of Tomonaga, Schwinger, and Feynman, *Physical Review* **75**, 3, p. 486.

Faddeev, L. (1975). Hadrons from leptons, Tech. rep., VA Steklov Mathematics Institute.

Faddeev, L. D. (1963). Mathematical questions in the quantum theory of scattering for a system of three particles, *Trudy Matematicheskogo Instituta imeni VA Steklova* **69**, pp. 3–122.

Fateev, V. A. and Shvarts, A. (1973). On axiomatic scattering theory, *Theoretical and Mathematical Physics* **14**, 2, pp. 112–124.

Fedoryuk, M. V. (1971). The stationary phase method and pseudodifferential operators, *Russian Mathematical Surveys* **26**, 1, p. 65.

Feynman, R. P. (2005). Space-time approach to non-relativistic quantum mechanics, in *Feynman's Thesis — A New Approach To Quantum Theory* (World Scientific: Singapore), pp. 71–109.

Haag, R. (1958). Quantum field theories with composite particles and asymptotic conditions, *Physical Review* **112**, 2, p. 669.

Haag, R. and Kastler, D. (1964). An algebraic approach to quantum field theory, *Journal of Mathematical Physics* **5**, 7, pp. 848–861.

Hepp, K. (1965). On the connection between the LSZ and Wightman quantum field theory, *Communications in Mathematical Physics* **1**, 2, pp. 95–111.

Hepp, K. (1966). Proof of the Bogoliubov–Parasiuk theorem on renormalization, *Communications in Mathematical Physics* **2**, 1, pp. 301–326.

Hepp, K. and Epstein, H. (1971). *Analytic Properties of Scattering Amplitudes in Local Quantum Field Theory* [Russian translation].

Hunziker, W. and Sigal, I. M. (2000). The quantum n-body problem, *Journal of Mathematical Physics* **41**, 6, pp. 3448–3510.

Jaffe, A. (2000). Constructive quantum field theory, *Mathematical Physics*, pp. 111–127.

Jost, R. (1965). *The General Theory of Quantized Fields*, in Lectures in Applied Mathematics, Vol. IV (American Mathematical Society).

Kato, T. (2013). *Perturbation Theory for Linear Operators*, Vol. 132 (Springer Science & Business Media).

Kuroda, S. T. (1959). On the existence and the unitary property of the scattering operator, *Il Nuovo Cimento (1955–1965)* **12**, 5, pp. 431–454.

Lehmann, H., Symanzik, K., and Zimmermann, W. (1955). Zur vertexfunktion in quantisierten feldtheorien, *Il Nuovo Cimento (1955–1965)* **2**, 3, pp. 425–432.

Likhachev, V., Tyupkin, Y. S., and Shvarts, A. (1970). The adiabatic S matrix and quasiparticles, *Theoretical and Mathematical Physics* **2**, 1, pp. 1–20.

Likhachev, V., Tyupkin, Y. S., and Shvarts, A. (1972). Adiabatic theorem in quantum field theory, *Theoretical and Mathematical Physics* **10**, 1, pp. 42–55.

Osterwalder, K. and Schrader, R. (1973). Axioms for Euclidean Green's functions, *Communications in Mathematical Physics* **31**, 2, pp. 83–112.

Polyakov, A. (1974). Spectrum of particles in quantum field theory, *JETP Letters* **20**, pp. 430–433.

Robertson, W. *et al.* (1980). *Topological Vector Spaces*, Vol. 53 (Cambridge University Press Archive).

Ruelle, D. (1962). On asymptotic condition in quantum field theory, *Helvetica Physica Acta* **35**, 3, p. 147.

Schwarz, A. (2019a). Geometric approach to quantum theory, arXiv: 1906.04939.

Schwarz, A. (2019b). Inclusive scattering matrix and scattering of quasiparticles, arXiv:1904.04050.

Schwarz, A. (2019c). Scattering in algebraic approach to QFT inclusive scattering matrix, arXiv:1910.00000.

Schwarz, A. S. (1967). A new formulation of the quantum theory, *Doklady Akademii Nauk* **173**, 4, pp. 793–796.

Schwarz, A. S. and Tyupkin, Y. S. (1987). Measurement theory and the Schroedinger equation, in *Quantum Field Theory and Quantum Statistics: Essays in Honour of the Sixtieth Birthday of ES Fradkin. V. 1.*

Schwinger, J. (1958). *Selected Papers on Quantum Electrodynamics* (Courier Corporation).

Sigal, I. M. and Soffer, A. (1987). The n-particle scattering problem: Asymptotic completeness for short-range systems, *Annals of Mathematics*, **126**, 1, pp. 35–108.

Soffer, A. (2006). Soliton dynamics and scattering, *International Congress of Mathematicians*, Vol. 3, pp. 459–471.

Streater, R. F. and Wightman, A. S. (2016). *PCT, Spin and Statistics, and All That* (Princeton University Press).

Takhtadzhyan, L. A. and Faddeev, L. D. (1974). Essentially nonlinear one-dimensional model of classical field theory, *Theoretical and Mathematical Physics* **21**, 2, pp. 1046–1057.

Tao, T. (2009). Why are solitons stable? *Bulletin of the American Mathematical Society* **46**, 1, pp. 1–33.

Tiupkin, I. S., Fateev, V. A., and Shvarts, A. (1975). Classical limit of scattering matrix in quantum field theory, *Akademii Nauk SSSR Doklady*, Vol. 221, pp. 70–73.

Tomonaga, S.-i. (1946). On a relativistically invariant formulation of the quantum theory of wave fields. *Progress of Theoretical Physics* **1**, 2, pp. 27–42.

Tyupkin, Y. S. (1973). On the adiabatic definition of the S matrix in the formalism of L functionals, *Theoretical and Mathematical Physics* **16**, 2, pp. 751–756.

Tyupkin, Y. S., Fateev, V., and Shvarts, A. (1975). On the existence of heavy particles in gauge field theories, *Pis'ma v Zhurnal Ehksperimental'noj i Teoreticheskoj Fiziki* **21**, 1, pp. 91–93.

Tyupkin, Y. S. and Shvarts, A. (1972). On the adiabatic change of a stationary state, *Theoretical and Mathematical Physics* **10**, 2, pp. 172–175.

Velo, G. and Wightman, A. S. (2012). *Constructive Quantum Field Theory II*, Vol. 234 (Springer Science & Business Media).

Wightman, A. S. (1956). Quantum field theory in terms of vacuum expectation values, *Physical Review* **101**, 2, p. 860.

Index